# PREFACE

## The Diseased Plant

Each host-pathogen interaction is a grim struggle for survival between the two concerned organisms. Over time, however, surviving plants and their pathogens in their natural niches reach a condition of equilibrium in which both confronting organisms are assured of survival. This is so because elimination of the host endangers existence of the pathogen as well so that pathogens have been kept in check by the necessity of host conservation. Observations with brome stem leaf mottle and holcus transitory mottle viruses (both are phleum mottle virus strains) indicated that the least damaging parasitic relationship between the two best benefits the host as well as the pathogen for survival and dispersal. However, such situations normally occur where man has not interfered or where subsistence agriculture prevails. Subsistence agriculture employs genetically heterogeneous plants under conditions of low fertility and low moisture. Diseases rarely occur in epidemic proportions in these stands because these plant populations are in a state of ecological balance.

Modern agriculture, on the other hand, is technology based and causes perturbations in the naturally equilibrated ecological situations. Intensive monoculture of high yielding varieties over extensive areas coupled with improved tillage, water management, and extensive use of fertilizers have created micro- and macroenvironments extremely favorable to rampaging pathogens. This large-scale culture of genetically uniform crops in highly unnatural conditions inevitably invites its own mass destruction by parasites. This has already resulted in the devastation of various crops by many catastrophic and threatening plant diseases and in many plant disease epidemics in developed and developing countries. Several of these diseases have since been tamed, but others are still continuing their destructive dance. The losses can be immense, and the areas engulfed could stretch over thousands of acres. All this and much else is brought out in all starkness in the first chapter.

The existence of two independent but interacting entities, the host and the pathogen, results in the coming into existence of a new entity, the diseased plant or aegricorpus. Many of its aspects, from physiognomy to yield, are different from the healthy plant. Its physiology is different, and this, in turn, excites the development of various "patho" aspects of the disease. This is covered in Chapter 2. A group of proteins, collectively called pathogenesis-related proteins, produced in plants by various stresses, including the stress of virus infection, are now an important and developing field of research (Chapter 3.)

## The Virus Pathogen

The symptoms can appear only if the virus is able to move out of the initially inoculated infected cell to adjacent or distant tissues. That is, the virus must have the genetic potentiality to do so in order to maneuver its own movement away from the infection court. This was a difficult premise to believe in, but the experimental evidence generated within the last decade clearly establishes it to be so (Chapter 4).

Plant viruses, to be successful pathogens, must continuously adapt to changing ecological, physiological, biological, and environmental conditions. They must continuously evolve to do so or, in other words, must continuously form variants which are more suitable to new conditions. Each new condition has its own virus variant which is best suited to it. This happens because plant viruses show great variability (Chapter 5). Simultaneously, all viruses show a specific host range which is not subject to easy change. This is so because host range is also based on viral genome (Chapter 6). Studies involving pseudorecombinants have established it clearly and recombinant investigations on cauliflower mosaic virus have confirmed this.

Plant viruses, like all organisms, also must move out to new ecological niches and to new green pastures where competition is less. They must therefore mediate their own movement away from a focal point. In fact, plant viruses actively control and mediate their own transmission through various types of vectors (Chapters 7 and 8), but particularly the insect vectors. Transmission of viruses by vectors in many cases is therefore not a random process. It is a specific event, a specific phenomenon, in which viral genome and/or virus capsid is directly involved. Participation of the viral capsid has recently been clearly established in virus transmission by aphids (Chapter 7) and, by extension, can be expected to be true of many other virus-specific vector systems, although evidence in this regard is still inadequate.

All the above characters positively help viruses to be efficient invaders and successful pathogens.

Epidemiology is the study of populations of pathogens in populations of host and of the resulting disease under the influence of environment and human interference. It is a very complex discipline dealing with populations of different individuals and their varying interaction over a whole range of environmental conditions. Plant virus epidemiology is still more complex because vector populations are actively and intimately involved in the spread of plant virus diseases. Epidemiological studies are amenable to mathematical analysis and modeling, since epidemiology deals with populations, and each population in turn is influenced by a host of factors. This leads to a multivariate situation and generates complex data sets. It is a baffling and daunting prospect, but the development of a conceptual model can help to simplify matters. This is done in Chapter 9.

**Disease Control**

The basic justification for the existence of plant pathology is plant disease control: the successful devising of plant disease control measures, which cut down disease losses to within acceptable limits, is the best practical contribution of plant pathology to human welfare, and in the case of plant virus diseases, certain milestones are already there. Disease management envisages integration of various methods of disease control for managing a plant disease (Chapter 10). Knowledge of pathogen ecology helps devise an appropriate disease management strategy and better optimization of the various control measures. It also helps devise various disease forecasting systems, which, in the ultimate analysis, is the best way to manage a disease (Chapter 10).

Recent research has updated several of the old well-tried control mechanisms as well as produced new plant disease control strategies which seem at present to be the future hope. Tissue culture (Chapter 11) and disease resistance (Chapter 12) belong to the former category, while DNA recombinant technology belongs to the latter category. Recombinant technology is still the emerging field and its potential is vast (Chapter 13).

# Plant Viruses

## Volume II
## Pathology

Editor

**C. L. Mandahar, Ph.D.**
Professor
Department of Botany
Panjab University
Chandigarh, India

CRC Press
Boca Raton   Ann Arbor   Boston

**Library of Congress Cataloging-in-Publication Data**

(Revised for volume 2)

Plant viruses.

   Includes bibliographies and index.
   Contents: v. 1. Structure and replication --
v. 2. Pathology.
   1. Plant viruses.  I. Mandahar, C. L.
QR351.P584 1989    576′.6483    88-4333
ISBN 0-8493-6947-9 (v. 1)
ISBN 0-8493-6948-7 (v. 2)

   Direct all inquiries to CRC Press, Inc., 2000 Corporate Blvd., N.W., Boca Raton, Florida, 33431.

© 1990 by CRC Press, Inc.

International Standard Book Number 0-8493-6947-9 (Vol. I)
International Standard Book Number 0-8493-6948-7 (Vol. II)

Library of Congress Card Number 88-4333
Printed in the United States

# THE EDITOR

**Dr. C. L. Mandahar, M.Sc., Ph.D.,** is a Professor and Chairman of the Botany Department, Panjab University, Chandigarh, India. He graduated from Panjab University where he also obtained his Ph.D.

Dr. Mandahar has published more than 100 research papers, review articles, and chapters in books. He is the author of three other books: one of which is a textbook on plant viruses which is now in its second edition. Dr. Mandahar has been the editor of the Plant Virus Section of the *Indian Journal of Virology*, is on the editorial board of some other journals, has delivered many lectures by invitation, and is a member of several societies.

He has been a recipient of research grants from the Indian Council of Agricultural Research, the Department of Science and Technology, and the University Grants Commission.

# CONTRIBUTORS

**George N. Agrios, Ph.D.**
Professor and Chairman
Department of Plant Pathology
University of Florida
Gainesville, Florida

**R. S. S. Fraser, Ph.D., D. Sc.**
Head of Station
Institute of Horticultural Research
Littlehampton, West Sussex, England

**I. D. Garg, Ph.D.**
Scientist
Division of Plant Pathology
Central Potato Research Institute
Shimla, India

**Stewart M. Gray, Ph.D.**
Research Plant Pathologist,
 ARS, USDA and
Assistant Professor
Department of Plant Pathology
Cornell University
Ithaca, New York

**Lisa Haley**
Plant Science Technology
Monsanto Agricultural Company
St. Louis, Missouri

**Kerry F. Harris, Ph.D.**
Professor
Department of Entomology
College of Agriculture and Life Sciences
Texas A & M University
College Station, Texas

**Cynthia Hemenway, Ph.D.**
Plant Science Technology
Monsanto Company
St. Louis, Missouri

**Wojciech Kaniewski, Ph.D.**
Plant Science Technology
Monsanto Company
St. Louis, Missouri

**S. M. Paul Khurana, Ph.D.**
Head
Division of Plant Pathology
Central Potato Research Institute
Shimla, India

**E. Clifford Lawson**
Plant Science Technology
Monsanto Company
St. Louis, Missouri

**C. L. Mandahar, Ph.D.**
Professor and Chairman
Department of Botany
Panjab University
Chandigarh, India

**F. Mohamed**
Department of Plant Pathology
University of Missouri
Columbia, Missouri

**Keith M. O'Connell**
Plant Science Technology
Monsanto Company
St. Louis, Missouri

**Patricia R. Sanders**
Plant Science Technology
Monsanto Company
St. Louis, Missouri

**James E. Schoelz, Ph.D.**
Assistant Professor
Department of Plant Pathology
College of Agriculture
University of Missouri
Columbia, Missouri

**O. P. Sehgal, Ph.D.**
Professor
Department of Plant Pathology
College of Agriculture
University of Missouri
Columbia, Missouri

**Richard Stace-Smith, Ph.D.**
Research Scientist
Research Station
Agriculture Canada
Vancouver, British Columbia, Canada

**P. E. Thomas, Ph.D.**
Research Plant Pathologist
Agricultural Research Service
United States Department of Agriculture
Prosser, Washington

**Nilgun E. Tumer, Ph.D.**
Plant Science Technology
Monsanto Company
St. Louis, Missouri

**K. Roger Wood, D.Phil.**
Senior Lecturer
School of Biological Sciences
University of Birmingham
Birmingham, England

# ACKNOWLEDGMENTS

The existence of this book is due to the direct or indirect collective participation of scores of people. They helped me in their own respective ways to complete it within the stipulated time. It is impossible to mention them all by name but some names do stand out because of their intimate and continued involvement.

I am thankful to all the contributors who responded to my invitation. This book could not have been produced without their participation. I express my gratitude to T. W. Carroll, R. W. Fulton, Kerry F. Harris, R. C. Sinha, and R. Stace-Smith for helping me to finalize this series of two volumes. Drs. S. M. Paul Khurana, N. Rishi, and I. D. Garg were helpful in several ways. My thanks to them. Three people, whose consideration saw the manuscripts of the two volumes through and to whom many thanks are due, are stenographer D. N. Sharma, photographer Gurcharan Singh, and artist M. C. Mankoo.

Prof. Karl Maramorosch has always taken keen personal interest in the progress of this series. This has been a perennial source of encouragement and I am beholden to him for this. Prof. O. P. Sehgal acted as an indispensable outpost, reconnoitering the ground, contacting various contributors based in the U.S. and Canada, and even making decisions where necessary. His active involvement with this volume was a major factor in finalizing it as per schedule. I gratefully acknowledge all this.

Many plant virologists and publishers have permitted me and other contributors to make use of their copyrighted material in various chapters of this volume. We are all obliged to them for this. The text would have been incomplete without this illustrative material. Material of the following authors has been used. G. N. Agrios, J. F. Antoniw, J. G. Atabekov, J. F. Bol, M. K. Brakke, B. J. C. Cornelissen, J. W. Davies, A. M. M. de Laat, R. S. S. Fraser, B. Fritig, A. J. Gibbs, R. I. Hamilton, L. Hirth, P. Kaesberg, M. Legrand, M. Matsuoka, A. F. Murant, Y. Ohashi, T. Ohno, R. Rezelman, K. Richards, A. W. Robards, I. M. Roberts, D. J. Robinson, W. Rochow, H. L. Sanger, K. Tomenius, A. van Kammen, L. C. Van Loon, M. H. V. Van Regenmortel, H. E. Waterworth, M. Weintraub, C. Wetter, and R. F. White. Many thanks to them all and also to the following publishers and the sources from which copyrighted material has been taken:

1. Academic Press, London — *Physiological and Molecular Plant Pathology*.
2. Academic Press, New York — *Advances in Virus Research, Journal of Ultrastructure Research, Virology*.
3. American Association for Advancement of Science, Washington, D.C. — *Science*.
4. American Phytopathological Society — *Phytopathology*.
5. American Society for Microbiology — *Molecular and Cellular Biology*.
6. American Society for Plant Physiologists — *Plant Physiology*.
7. Elsevier/North-Holland Biomedical Press, Amsterdam — *Handbook of Plant Virus Infections and Comparative Diagnosis*, Kurstak, E., Ed., 1981.
8. Elsevier Science Publishers, B. V., Amsterdam — *Gene*.
9. European Molecular Biology Organization and IRL Press, Oxford — *EMBO Journal*.
10. John Wiley & Sons, U.K. — *Plant Resistance to Viruses*, Evered, D. and Harnnet, S., Eds., 1987.
11. Macmillan Publishers, London — *Nature*.
12. Pergamon Press, New York — *Phytochemistry*.
13. S. Karger, Basel — *Intervirology*.
14. Society for General Microbiology and Cambridge University Press, England — *Journal of General Virology*.
15. Springer-Verlag, Berlin — *Molecular and General Genetics; Intercellular Communication in Plants: Studies on Plasmodesmata*, Gunning, B. E. S. and Robards, A. W., Eds., 1976.

Lastly, my thanks to my friends Profs. S. S. Kumar and G. S. Rawla who were always solicitous to my needs, and to my wife Aruna and children Vinod, Atul, and Abha who together provided me the necessary environment.

# TABLE OF CONTENTS

Chapter 1

## ECONOMIC CONSIDERATIONS

### G. N. Agrios

## TABLE OF CONTENTS

## I. INTRODUCTION

Plant viruses are of considerable interest to the science of biology, and their study has contributed significantly to the elucidation of several mysteries of traditional and molecular biology. In addition, plant viruses are of great practical significance because they infect plants,cause plant diseases, and result in economic loss. Some kinds of plants, when infected with certain viruses, develop no obvious symptoms of disease.[1] Generally, however, most infections of plants with viruses, whether systemic or partially systemic, lead to the appearance of certain symptoms on the plant. The symptoms, whether visible or otherwise measurable, lead to the development of disease of a greater or lesser severity. Diseased plants generally grow more slowly, attain a smaller size, and produce less fruit than healthy plants.[2] Such reductions in plant growth and yield are almost always accompanied by economic loss to the producer. Numerous cases of economic loss caused by virus diseases of plants have been documented.[1-11]

Since virus infections of plants often cause economic loss, measures are taken to produce plants free of viruses or to protect them from becoming infected with viruses. These measures may include the development of plant varieties resistant to the virus;[12] indexing of plants;[13-15] and use of only virus-free seed[16] or virus-free nursery stock;[17] employment of certain cultural practices,[18] such as crop rotations, that reduce plant infection by some viruses; or attempts to avoid or control potential vectors of the virus.[3,8,19] All of these measures, of course, cause economic loss by the fact that they are costly to develop and to apply; they limit the selection of plant varieties that may be available for planting, and some, such as vector control, may add toxic chemicals to our environment. Some of the resulting economic loss may be borne by the government (really, money paid by the taxpayers), by agricultural industries (which pass it on to the consumers), or by the grower, who will either suffer from the reduced income or will also pass it on to the consumer as higher prices.

Of course, knowledge about plant viruses, the diseases they cause, and the measures that may be taken to avoid or reduce losses from virus diseases, is developed over many years. Such knowledge is developed by thousands of people working at universities, experiment stations, industries, etc. The cost of supporting all these people, their laboratories, greenhouses, and fields, and the cost of purchasing the sophisticated equipment and supplies necessary for virus work must also be added to the economic loss caused by plant viruses. These costs are generally borne by the public as taxes paid to the government, or as higher prices paid to the industries carrying out the research.

Estimating the economic losses caused by plant viruses has been very difficult. One could probably survey all pertinent universities, agricultural experiment stations, and industries and obtain an indication of the budgets allocated to the study of plant viruses, the diseases they cause, and their control. But such study has yet to be made and, therefore, the costs of these efforts remain unknown. Also, there is no information on the cost of the cultural or chemical measures taken by the growers in order to avoid, reduce, or control yield losses caused by plant viruses. Finally, there is generally incomplete and often contradictory information on the size of yield losses caused by viruses. More particularly, there is incomplete information on the size of the eventual economic losses to individual growers, to a region, or to a nation that may be caused by specific virus diseases on specific crop plants. In this chapter an effort will be made to analyze the types of losses caused by plant viruses, to estimate the costs of these losses, to examine the relationship of the losses to the mode of virus spread in nature, then describe some specific losses caused by viruses on some crops, and to estimate the economic costs of management of plant virus diseases.

## II. TYPES OF LOSSES

Plant viruses may cause economic loss by reducing yield quantity and/or quality, reducing seed germination, reducing seedling and young plant growth, by interfering with bud take, rooting of cuttings, or germination of parts of vegetatively propagated plants, and by necessitating the application of certain cultural practices and control measures.

### A. Effect on Yield Quantity

It is generally quite difficult to measure the amount of crop yield lost to virus infections in a particular field or area. The main difficulties stem from (1) the lack of comparable infected and uninfected crops of the same variety in adjacent fields; (2) the fact that virus infections are systemic and decrease plant yield much more severely if they infect young, growing plants than plants that are fully developed; (3) the fact that virus infections often cause symptoms that can be confused with symptoms caused by other pathogens; (4) the fact that if a relatively small percentage of plants become infected young and remain stunted, the surrounding plants may grow larger and compensate for a considerable portion of the yield loss of the infected plant; and (5) the fact that even symptomless or near-symptomless infections of crop plants result in significant losses of yield.

Table 1 shows the crop yields lost to viruses in California (reported in 1963),[20] in the entire U.S. (1965),[21] and in North Carolina (1988).[22] A quick comparison of the percentages of crops lost to viruses in California and the U.S. shows that, although in some crops viruses cause less than 10% and in others more than 70% of the losses due to diseases, in most crops, virus-induced losses account for about 30 to 50% of the total losses caused by diseases. It must be pointed out here that in both the California and the U.S. surveys, only diseases caused by fungi, bacteria, and viruses were considered. On the other hand, in the 1987 North Carolina survey, in which losses caused by nematodes, abiotic agents, and complexes were considered, viruses caused only 12.5% of all the losses. In the North Carolina survey, viruses caused 20.1% of the total losses caused by fungi, bacteria, and viruses combined.

The quantity of yield lost to virus infection varies drastically with prevalence of the virus in the crop, the crop cultivar, the strain of the virus, and the age of the plants at the time they become infected. Any one of these factors may vary in each location and so the losses also vary.

Individual plants are often infected concurrently with two or more viruses, or with one or two viruses and, in addition, with one or two fungi, bacteria, nematodes, etc. Such doubly or triply infected plants often are affected much more severely, and yield much less than the total yield reduction suggested by adding the yield losses caused by each of the pathogens.[23]

In all host/virus combinations, plants suffer greater losses in yield when they become infected while still young than when they have reached their full size. For example, soilborne wheat mosaic virus and wheat streak mosaic virus can each cause up to 100% yield loss in fields of susceptible wheat cultivars if all the plants become infected with the virus while young.[24] For sugar beets, it has been estimated that there is loss of 3% of yield for every

**Table 1**
**CROP YIELDS LOST TO VIRUS INFECTIONS IN THE U.S.**

| Crop | California (1963) | | | | North Carolina (1987) | | | | U.S. (1965) | |
|---|---|---|---|---|---|---|---|---|---|---|
| | % Crop lost to viruses | % Crop lost to all diseases | 1963 Value lost crop (in thousands) | Equivalent crop acres lost to viruses | % Crop lost to viruses | % Crop lost to all diseases | 1987 Value of lost crop (in thousands) | Equivalent crop acres lost to viruses | % Crop lost to viruses | % Crop lost to all diseases |
| **Field crops** | | | | | | | | | | |
| Clover | — | — | — | — | 30 | 50 | 23,000 | 144,000 | 5 | 24 |
| Barley | 5 | 14.6 | 3,763 | 71,000 | — | — | — | — | 5.8 | 14 |
| Oats | 5 | 8.5 | 155 | 5,050 | 2.5 | 2.5 | 127 | 600 | 4.6 | 21 |
| Wheat | 3 | 9.8 | 460 | 9,500 | 1.0 | 2.6 | 454 | 4,400 | 1.6 | 14 |
| Soybean | — | — | — | — | 1.0 | 15.8 | 2,313 | 14,040 | 0.8 | 14 |
| Tobacco (B) | — | — | — | — | 2.5 | 9.3 | 714 | 190 | — | — |
| Tobacco Flue | — | — | — | — | 1.6 | 11.2 | 12,636 | 3,385 | 1.4 | 11 |
| Sugar beets | 6.2 | 10.5 | 4,620 | 18,000 | — | — | — | — | 9 | 16 |
| **Vegetables** | | | | | | | | | | |
| Beans | 2.0 | 5 | 247 | 1,530 | — | — | — | — | 4 | 20 |
| Carrots | 1 | 2 | 202 | 244 | — | — | — | — | | |
| Cucurbits | 8.5 | 21.2 | 4,962 | 7,300 | — | — | — | — | 5 | 12 |
| Lettuce | 12.3 | 25.4 | 10,370 | 14,800 | 25 | 45 | 3,683 | 1,202 | 3.5 | 16 |
| Peas (green) | 2 | 8 | 51 | 252 | — | — | — | — | 6 | 23 |
| Peppers | 9 | 11.5 | 757 | 1,000 | 5 | 41 | 728 | 433 | 2.5 | 14 |
| Potatoes | 3 | 12.5 | 1,685 | 3,000 | 1 | 18 | 200 | 164 | 6.2 | 19 |
| Sweet potatoes | 6 | 25 | 361 | 582 | 10 | 34 | 9,100 | 3,500 | 3.5 | 18 |
| Tomatoes | 1 | 11 | 1,101 | 1,600 | 1.3 | 34 | 157 | 52 | 6 | 21 |
| **Fruits** | | | | | | | | | | |
| Apples | 4 | 9 | 535 | 535 | — | — | — | — | 0.2 | 8 |
| Cherries | 11 | 19 | 726 | 1,530 | — | — | — | — | 10.6 | 24 |
| Citrus | — | — | — | — | — | — | — | — | 16.4 | 25 |
| Grapefruit | 2.5 | 14.7 | 163 | 353 | — | — | — | — | — | — |
| Lemon | 8.7 | 9.4 | 4,100 | 4,295 | — | — | — | — | — | — |
| Orange | 7.1 | 9.3 | 8,310 | 12,930 | — | — | — | — | — | — |
| Grapes | 6.2 | 26 | 10,172 | 29,000 | — | — | — | — | 13.6 | 27 |
| Nectarines | 11 | 18.4 | 587 | 1,072 | — | — | — | — | — | — |
| Peaches | 4 | 13.1 | 2,238 | 4,542 | — | — | — | — | 12.3 | 38 |
| Plums | 4 | 14 | 2,047 | 5,677 | — | — | — | — | 1.2 | 14 |
| Raspberries | 1 | 2.6 | 29 | 23 | — | — | — | — | 12.4 | 38 |
| Strawberries | 2 | 13.6 | 1,010 | 230 | — | — | — | — | 5 | 26 |

week the plant shows symptoms.[25] Similarly, tobacco plants infected with tobacco mosaic virus (TMV) at transplanting, 1 month later, or 2 months later, produced yields that were 33, 25, and 4% lower than yields of uninoculated plants.[26] Pepper plants inoculated with cucumber mosaic virus on 6/22, 7/20, 7/30, and 8/10 produced green pepper yields that were, respectively, 76, 58, 36, and 19% lower than the yields of uninoculated plants.[27] It usually happens, however, that in any given field or area, only a few of the young plants are, or become, naturally infected with a virus early in the season; as a result, the proportion of plants producing the poorest yields, because of very early infection by the virus, is relatively small. As the season advances, increasingly more plants become infected with virus, but with every passing week before infection the virus-caused yield loss per plant decreases; that is, although the number of plants that become infected with the virus increases with time, the percentage of yield loss caused on sequentially infected individual plants decreases with time. The significance, therefore, of early spread of viruses in crops on yield reductions in these crops is quite obvious.

The incidence and prevalence of viruses on different crops varies considerably with the crop, the virus, the location, and the year. Vegetatively propagated crops, such as potatoes, strawberries, many ornamentals, and many fruit trees, are often thoroughly infected with latent viruses at the time of propagation, but such latent viruses usually cause relatively small reductions in yield.[5,6,28] Other crops, e.g., barley, may carry seed-borne viruses, such as barley stripe mosaic virus, in up to 50% or more of the seed, with little or no spread of the virus in the field. Still other crops, e.g., lettuce, may carry a virus such as lettuce mosaic virus in only a small percentage of the seed, but, in the field, the virus is rapidly spread to other lettuce plants by aphids so that up to 80 to 100% of the plants may be infected by harvest time.[16] In many plant/virus combinations, the crop seed is free of virus and the first plants become infected with virus brought into the field by vectors, usually insects, that carry the virus from infected weeds or other crops. In most such cases, the virus spreads slowly at first, but as the season advances virus spread to new plants increases rapidly until most or all plants in the field become infected.[7,29] Examples of such virus spread include cucumber mosaic virus on pepper,[30] sugar beet yellows and sugar beet mosaic viruses on sugar beets,[29] watermelon mosaic virus on squash,[31,32] and many others.

The effect of virus incidence on yield may be influenced dramatically by the host variety and by the virus strain. For example, a 100% incidence of potato spindle tuber viroid (PSTV) caused 20.3% yield loss in 'Katadin' potatoes in Florida, but 68.5% loss in 'Triumph' potatoes in Louisiana.[33] Similarly, mild PSTV strains caused 17 to 24% yield loss, while severe strains caused 64% loss.[34]

## B. Effect on Yield Quality

Virus infections not only reduce the amount of the crop produced, but, quite frequently, the produced crop (fruit, leaf, flower, tuber, etc.) is of inferior quality in terms of size, shape, color, texture, and, in some cases, taste.[30,35] Such produce is either left unharvested in the field or it must be sorted and culled upon storage or marketing, or it is downgraded at much reduced prices. In some cases, entire carloads of produce such as lettuce were rejected in the markets of New York and San Francisco because of poor quality caused by lettuce mosaic virus.[36,37]

## C. Effect on Germination and Growth of Propagative Organs

Virus-infected plants often have fewer fruit and seeds than healthy plants. This occurs not only because virus-infected plants are less vigorous, but also because the haploid megasporophytes and microsporophytes of infected plants are adversely affected by the virus and result in poor pollination and dramatically reduced fruit set. For example, pollination of sweet cherry trees with pollen from prune dwarf virus-infected trees reduces fruit set by 25 to 75%.[38]

Virus-infected seeds sometimes do not germinate, or germinate poorly and produce week plants.[39] Plants produced from virus-infected seeds or other propagative parts nearly always remain stunted, develop severe symptoms and/or abnormal growth, and produce markedly reduced yields. Such plants further produce virus-infected seeds or other propagative organs leading to a progressive "degeneration" of the crop that eventually produces little and generally low-quality yield. This type of degeneration is particularly obvious in crops that are propagated vegetatively and, because of it, are often virus infected, for example, potatoes, strawberries, sugarcane, chrysanthemum, and others.[5,6,40]

Because most fruit trees and some ornamentals are propagated by grafting the desired horticultural variety (scion) on a hardy rootstock, the graft union between scion and rootstock often becomes the battleground for a hypersensitive reaction to the virus when the latter is introduced through one or the other of the two components of the plant. For example, in tristeza disease of citrus, infected citrus trees on sour orange rootstock decline and die when infected with citrus tristeza virus. This occurs as a result of a hypersensitive-like reaction that develops near the graft union of the sour orange rootstocks in response to the citrus tristeza virus which is usually introduced into orange leaves by aphids and spreads throughout the tree and into the rootstock.[41] On the other hand, in graft union necrosis and decline disease of pome fruits caused by tomato ringspot virus, the virus is spread by nematode vectors that introduce it into young roots of tolerant rootstocks. When the virus reaches the scion, a hypersensitive-like response occurs near the graft union, the trees decline, and eventually may die.[42] However, in the stem-pitting disease of stone fruits, which is also caused by tomato ringspot virus (TomRSV), the seedling rootstocks themselves are sensitive to the virus, and when nematodes transmit the virus to rootstocks, the latter develop stem pitting and decline, eventually resulting in decline of the entire tree.[43] Cherry leaf roll virus in California causes the blackline disease of English walnut when propagated on 'Northern California Black' and 'Paradox' walnut root stocks.[43a,43b] The virus is spread horizontally from plant to plant by infected pollen and is a serious threat to the walnut industry in California.

### D. Differences of Losses in Annual and Perennial Crops

Depending on the particular plant/virus combination, losses from viruses can be severe in both annual and perennial crops. Viruses of annual crops usually come into the crop via the seed or via vectors such as aphids. A few viruses, e.g., tobacco mosaic, come into crops by contact. Within annual crops, viruses are further spread rather rapidly by aphids or other vectors. By the end of the season, a variable number (up to 100%) of crop plants may have been infected with the virus for varying periods of time, each of them thereby suffering proportional reductions in yield. The loss caused by a virus on an annual crop may be small or great; however, the loss ceases when the crop is harvested.

In perennial crops, viruses are usually brought into the crop with the propagative material.[5,6,9,11,44] Therefore, in such crops, the virus is present within the plants throughout their lives, that is, both during their developmental period and during their productive years. Depending on the host cultivar/virus strain combination, plants may show few adverse effects, or they may remain stunted and may produce few blossoms and few fruits of poor quality. In later years, host growth and yield may improve or they may continue to be poor; in some cases, infected hosts continue to decline and produce much reduced yields, or they die prematurely.[29,42,43]

In a few perennial crops, the virus is brought into the crop by vectors such as aphids, other insects, or nematodes. Such viruses cause some of the most destructive diseases of perennial crops and include citrus tristeza,[41] plum pox (sharka) of stone fruits,[35] fanleaf of grapevines,[45] stem pitting of stone fruit and graft union necrosis and decline of pome fruits,[46] swollen shoot of cocoa,[47] and a few others. Perennial plants, of course, are exposed to infection by vector-borne or pollen-transmitted viruses for many continuous years; therefore, usually all of them become infected with virus long before the end of the normal life span of the plants. Losses by individual trees and by entire fields affected by the above diseases are usually great. In addition, because of necrotic reactions of varying degrees occurring at the graft union of such trees following infection with the virus, many infected trees decline and die. Replacement of poorly yielding or dying perennial crops is costly, not only because new nursery trees must be purchased and planted, but more importantly because revenue is lost for several years while the expenses of growing and protecting the nonbearing young trees continue.

## III. RELATIONSHIP OF LOSSES TO MODE OF VIRUS TRANSMISSION IN NATURE

Regardless of how viruses are transmitted, some of the viruses in each group are capable of causing severe economic losses. Unquestionably, however, most viruses causing frequent and severe losses on their host crops are transmitted by insect vectors that multiply rapidly and can move efficiently from plant to plant and from field to field.

### A. Losses Caused by Viruses That Lack Vectors
*1. Vegetatively Transmitted Viruses*
Such viruses are widespread among fruit trees and grapevines; small fruits, such as strawberries and raspberries; ornamentals such as rose, chrysanthemum, and carnation; and in other vegetatively propagated crops such as potatoes. Many of these viruses cause relatively mild, indistinguishable symptoms and therefore are difficult or more often impossible to avoid during propagation. As a result, often all the plants produced carry the virus. Yields of infected plants may be reduced by 10 to 80% in one to a few years. However, severely affected plants are avoided in the next propagation cycle, and so the cycle begins again with plants infected with mostly mild strains. Since many of these plants are perennials, the results are often continually declining yields.

*2. Seed- and Pollen-Transmitted Viruses*
A few seed-transmitted viruses, e.g., barley stripe mosaic virus, are so readily transmitted through the seed (up to 90%) that, although they lack vectors, they are widespread in some areas and cause losses ranging from 17 to 31%.[48] Some seed-transmitted viruses are also spread from plant to plant via infected pollen, for example, prune dwarf and *Prunus* necrotic ringspot viruses.[49,50] Such viruses infect their hosts either through the seed producing the rootstock or, when the scion begins to produce blossoms, through transmission of the virus via pollen to the blossoms of healthy trees. The combination of transmission of these viruses by seed and pollen with the perennial nature of the hosts of many such viruses usually results in thorough infection of all plants in a field within a few years, and subsequent severe yield loss for many years afterwards. Many seed-transmitted viruses are also transmitted by insect or nematode vectors. In the latter cases, seed transmission serves mainly to introduce the virus in a few plants in the field. The viruses are then transmitted by the vectors and become much more widespread in the field, infecting more plants each season, and consequently causing much greater losses.

*3. Contact-Transmitted Viruses*
Tobacco mosaic virus (TMV), and its very close relative tomato mosaic virus, are the only viruses which, although transmitted strictly by contact, cause severe losses (10 to 35%) on their hosts (tobacco, tomato, and a few other plants). Potato virus X (PVX) and cucumber mosaic virus (CMV) are also readily transmitted by contact. However, PVX is also trans-

mitted vegetatively through infected potato seed pieces, while CMV is transmitted readily by numerous species of aphids. These means of spread seem to be much more common and important for these viruses than is contact transmission.

## B. Losses Caused by Viruses That Have Vectors

### 1. Viruses Transmitted by Airborne Vectors: Insects and Mites

Unquestionably the most common, most frequent, and most severe economic losses are caused by viruses transmitted by insects. Losses by mite-transmitted viruses are much less common. Of the insect-transmitted viruses, the most losses are caused by aphid-transmitted viruses and to a much lesser extent by viruses transmitted by leafhoppers, beetles, mealy bugs, whiteflies, and thrips.[1,2] Mite-transmitted viruses cause common and severe losses in cereals.[24]

Numerous economically very important diseases are caused by aphid-transmitted viruses such as CMV, bean yellow mosaic, beet yellows, maize dwarf mosaic, potato leaf roll (PLRV), potato virus Y (PVY), lettuce mosaic, citrus tristeza, plum pox, and many others. Among the most important leafhopper-transmitted viruses are those causing sugar beet curly top, rice tungro, and rice dwarf. Other extremely economically important viruses that are transmitted by insects include cocoa swollen shoot, transmitted by mealybugs, and tomato spotted wilt, transmitted by thrips. The losses caused by these and many other diseases may vary from moderate to severe. In some cases they cause rapid and total destruction of the crop (e.g., CMV, lettuce mosaic, beet yellows); in others, e.g., citrus tristeza, they cause a rapid or slow decline and possible death of the trees. The importance of these viruses stems, of course, from the ability of the vectors to spread them very rapidly from a few plants in or around the field which may be infected early in the season, to most or all susceptible plants in the field within a few weeks.

### 2. Viruses Transmitted by Soilborne Vectors: Nematodes and Fungi

In spite of the slow movement of these vectors through the soil, they transmit several economically important viruses. Nematodes, for example, transmit tomato ringspot virus (TomRSV), which causes stem pitting and graft union diseases in stone fruits and pome fruits; fanleaf virus, which causes severe losses in grapevines; arabis mosaic, which severely damages raspberries, and others. In these diseases, the perennial nature of the host allows the slow-moving nematodes to leave infected plants and reach adjacent healthy hosts and to slowly spread the virus from infected to healthy plants. Among the fungus-transmitted viruses, soilborne wheat mosaic has been devastating in areas of Kansas. Both nematode- and fungus-transmitted viruses cause severe losses where they occur, but they are usually quite localized and lack the potential to cause sudden major epidemics. The areas of their occurrence and importance, however, may increase over the years because they tend to become permanently entrenched in the fields into which they spread.

## IV. EXAMPLES OF LOSSES CAUSED BY VIRUSES IN CROPS

All crops, whether annual or perennial, and regardless of where in the world they are grown, are affected by several viruses. Depending on the crop and its particular variety, the virus and its particular strain, the availability of vectors early in the season, the environmental conditions, and the cultural practices, the crop may suffer minor to severe losses every year. In the following pages only a few of the virus diseases that have caused or are still causing constant or spectacular losses will be mentioned. It should be noted here that countless other viruses affect these and other crops and any of them may cause minor or severe losses on the crop it affects in a particular location in any given year.

## A. Losses Caused by Some Viruses in Cereals

Several of the viruses that infect cereals cause considerable losses on these crops. Barley yellow dwarf virus (BYDV) is the most common and most widespread virus in small grains. It causes severe losses in barley and oats, somewhat less severe losses in wheat, and it also reduces foliage yield of forage grasses.[24,51,51a] In the U.K., 60 to 70% of cereal plants are usually infected with BYDV, causing an average annual loss of 5 to 10% in grain yield worth approximately $120 million to 200 million.[52] The same virus has been estimated to cause annual grain losses worth approximately $30 million in Australia and more than half the $90 million losses caused by viruses in cereals in the U.S.[53] Early infection of cereals (oats, barley, and even wheat) with severe strains of BYDV may result in crop failure. Regional or national losses from BYDV in the U.K., New Zealand, and Canada have been estimated at up to 10%.[54-56]

Another important virus of cereals, wheat streak mosaic virus, is prevalent in the central Great Plains of the U.S., where it destroys a significant percentage of the wheat crop annually. In Kansas alone it often reduces wheat yields by approximately 1 million tons per year, worth more than $100 million annually.[57] In 1988, wheat streak mosaic virus caused losses estimated at $140 million. Other common cereal viruses include barely stripe mosaic virus, which may infect up to 90% of the plants in affected fields and often causes yield losses up to 30%;[48] and soilborne wheat mosaic virus, which in the 1950s caused severe wheat yield losses of up to 30% in Kansas, up to 50% in Missouri,[58] and almost complete failure of susceptible varieties of wheat in Illinois.[59]

Of the other cereals, maize (corn) is frequently affected by, among others, maize dwarf mosaic, maize chlorotic dwarf, maize mosaic, maize streak, and maize rayado fino viruses. Losses varying from slight to 90% have been observed in fields infected with one or more of these viruses.[7] Maize rayado fino virus is widespread and is becoming increasingly important in maize in the tropical and subtropical areas of the Americas.[60] Yield losses of 40 to 50% have been recorded in some locally adapted maize cultivars in Central America, but losses up to 100% may occur in the same areas in some introduced or newly developed maize genotypes.[60]

Sugarcane suffered heavy, 30 to 40% losses from sugarcane mosaic virus before resistant varieties became available. Today, Fiji disease causes losses on sugarcane in many islands from Australia to the Philippines.[61] It also occurs in some south Asian countries and threatens sugarcane areas in the Americas. It causes severe losses because infected planting material produces very stunted plants of no value, and plants of sensitive cultivars when infected in one crop may become no more than stunted stools in the subsequent ratoon crop. Disease incidence and losses in an area of Australia built up to 100% in the 1970s as a result of the introduction of a cultivar whose resistance broke down under high inoculum pressure.[61]

Rice is affected by several viruses also. In some parts of Japan, 30 to 60% of rice areas contain plants infected with viruses such as rice dwarf.[62] Rice tungro is present throughout the south and southeast Asia and is the most important virus disease in the region.[63] The disease is usually endemic, but occasionally becomes epidemic and destroys large areas of rice, as happened in the mid-1960s in the Central Plain of Thailand,[63a] where yields of about 150,000 acres were reduced by about 50%, and in 1971 in the Philippines when an estimated rice yield of 450,000 tons were lost due to tungro infection. Rice hoja blanca causes similarly severe losses in North and South America and in Japan.

## B. Losses Caused by Some Viruses in Field Crops (Sugar Beets, Tobacco, Legumes)
### 1. Sugar Beets

In the first quarter of this century, yield of sugar per acre in California sugar beets dropped from about 2.0 tons per acre in 1910 to a little more than 1 ton in 1925 to 1930.[64] In areas that were devastated by curly top, beet production was abandoned. In 1934, varieties resistant to curly top were introduced, and along with improvements in other areas, by 1950 sugar yields increased to 3.0 tons per acre.[64a] The California State Department of Agriculture adopted an insecticide spray control on weed stands outside the cultivated areas at an annual cost of about $1 million. New, more virulent virus strains, however, keep appearing.[64,65]

Beet yellows virus (BYV), beet western yellows virus (BWYV), and beet mosaic virus (BMV) were also present in California. The first two were probably the more destructive, and of those, BWYV was much more widespread. They caused severe losses (up to 25%) starting about 1950 through the late 1960s.[29,65] The yellows viruses were also severe elsewhere.[29] For example, in the 1950s, potential sugar yield in England was reduced by 50% if sugar beets became infected with BYV in June and July. Similar losses were reported from The Netherlands, France, Germany, and Sweden.[29] BMV was also reported to cause losses ranging from trace to 50%.[66] In the late 1960s, two sugar beet hybrids with moderate resistance to virus yellows were introduced in California. At the same time, a beet-free period between harvesting and sowing of the new crop was established. In subsequent years, beet sugar production increased by about 1 ton per acre compared to the yields of the previous 15 years.[65] In the mid-1970s, the virus yellows complex was reported to reduce sugar beet yields by about 25% in the U.K. annually, amounting to a loss of about £14 million per year.[67]

## 2. Tobacco

TMV, tobacco etch, tobacco vein mottle, and tobacco ringspot viruses often infect 10 to 40% and sometimes 80% or more of the tobacco plants in a field,[68-70] causing losses ranging from 12 to 36%.[26,70,71] In 1987, 75% of all tobacco plants in North Carolina were infected with TMV and other mosaic-causing viruses, resulting in a loss to the state of $12.5 million for that year.[22] In the early 1980s, severe strains of PVY affected 73 tobacco fields in a single valley in Chile. The resulting yield losses ranged from 13 to 58% in different fields, with an average yield loss for all fields of 34%.[72]

## 3. Legumes

Legumes in general (beans, cowpeas, peanuts, peas, and soybeans) are infected by numerous viruses. These viruses are more or less host specific. Viruses like bean yellow mosaic, cowpea aphid-borne mosaic, peanut mottle, peanut stunt, pea mosaic, pea seedborne mosaic, soybean mosaic, and others are often present in most or all fields of the appropriate host.[73-75] When present, they infect from trace to 50 or 80% of the plants, and cause losses that vary in the different fields from slight to very severe, depending on how young the plants were when they became infected with the virus.[74,76,77] For example, bud necrosis disease of groundnuts (peanuts) is one of the most damaging peanut diseases in India.[78] A 50 to 100% incidence of the virus in peanuts occurs 2 to 3 weeks after the peak of thrips flights, and yields can be reduced up to 90%. Similarly, cowpea severe mosaic virus incidence in cowpeas ranged from 16 to 100% in commercial cowpea fields in Costa Rica and other tropical regions, causing yield reductions up to 50 to 90%.[79]

## C. Losses Caused by Viruses in Potatoes and Other Tuberous Plants
### 1. Potatoes

Several viruses infect potatoes wherever potatoes are grown. The viruses frequently infect most or all plants in a field and, singly or in combination, cause tuber yield losses that vary from slight (5 to 10%) to severe (up to 80% loss or more) depending on the virus, virus strain, variety, and environmental conditions.[6,80,81] Most of the losses are caused by secondary infections when infected seed tubers are planted. Among the most common and important potato viruses are PVY, PLRV, PVX, and PSTV. For example, fields planted with a high percentage of tubers infected with PLRV produced only 35% as much as fields planted with healthy tubers.[82] Even when a grower started with virus-free seed, but then replanted the tubers in successive years, incidence of PVX and PLRV were 39, 72, 78, and 88% in each of the successive years while yield reductions in the same years were 29, 62, 76, and 83%.[83] Alone, PVX and several other potato viruses cause generally mild symptoms and losses of 8 to 20%, but when they infect certain potato varieties, or infect in combinations of more than one virus, yields are often reduced quite drastically. It has been estimated that PVX reduces crop yield by 0.3% for each 1% incidence of diseased plants.[84] PSTV can reduce yields up to 65%,[33,34,85,86] although some mild PSTV strains reduce yield only by 17 to 24%.[34] Similarly, the yield of some varieties was reduced less than that of others, e.g., PSTV-infected 'Triumph' plants in Louisiana produced 68.5% less, while PSTV-infected 'Katadin' plants in Florida produced 20.3% less than corresponding uninfected plants.[33]

## 2. Cassava

Cassava is the most important food crop in Africa, where more than 50 million tons of fresh cassava tubers are produced each year. Cassava mosaic virus is the most spectacular disease of cassava. It occurs every year and is widespread, infecting almost 100% of the plants. The disease is devastating. It reduces edible root weight by 40 to 85% compared to roots of uninfected plants.[87-89]

## D. Losses Caused by Some Viruses in Vegetables

Each of the many important vegetables is affected by several viruses that cause losses of varying severity. Some viruses, such as cucumber mosaic (CMV) and watermelon mosaic in cucurbits, TMV in tomato, turnip mosaic in cabbage, lettuce mosaic in lettuce, and bean yellow mosaic in beans, can be expected to appear and cause severe losses in susceptible crops every year. Occasional outbreaks of just about any virus, however, and occasional severe losses, can be expected to be caused by any virus on one of its hosts in some field or valley. Lettuce big vein, possibly caused by a viroid, has been reported to affect from 8 to 70% of the plants in many fields in California and to cause losses of more than $30 million to California growers.[36] Another virus, lettuce mosaic virus, used to infect up to 90% of the plants in the fields of California and Arizona, and growers abandoned entire fields because of lettuce mosaic. Many times, carloads of mosaic-infected lettuce were rejected in the markets of big cities.[36a,37] Celery mosaic also caused severe yield losses (50 to 70%) and losses in quality in transit in California in the 1930s.[90a] Establishment of a celery-free period increased yields twofold to threefold.[91] In Hawaii in the 1960s, tomato spotted wilt virus (TSWV) caused severe losses in lettuce in the summer and even more severe losses in tomatoes, causing cessation of tomato culture. Subsequently, however, TSWV has increased in incidence and severity and presently it causes losses of up to 50 to 90% in lettuce fields during all seasons of the year.[92]

## 1. Tomatoes

Tomatoes are frequently affected by several viruses, and often 100% of the plants become infected. Depending on the growth stage of the plant at the time of infection, yield losses, sometimes compounded by lowered quality in the remaining fruit, may vary from 18 to 50% when caused by TMV,[93] from 18 to 66% by CMV,[94] from 25 to 83% loss by tomato yellows,[95] to almost total loss by tobacco etch virus.[96] Tomato yellow leaf curl virus affects tomato production in many Middle Eastern countries. It causes losses ranging from 50 to 75% in many regions, making tomato production unprofitable during the autumn.[97]

## 2. Peppers

Worldwide, peppers often suffer severe yield reductions from one or more viruses, of which the most common are CMV, PVY, and tobacco etch.[27,30,98,99-101] Pepper yield losses from viruses have been estimated to reach 15 to 70% in various fields in different years and to depend primarily on how early in the season the viruses are brought into the crop by the insect vectors.[27,30]

### 3. Cucurbits

Wherever they are grown, cucurbits are annually attacked by one or more of the viruses causing cucumber mosaic, watermelon mosaic, zucchini yellow mosaic, squash mosaic, and some other diseases. Virus incidence is zero or very low early in the season, but increases more or less rapidly, usually reaching 70 to 100% by the end of the season.[31] In susceptible cucurbits and other crops, in which epidemics are caused primarily by stylet-borne, aphid-transmitted viruses like those causing cucumber mosaic, watermelon mosaic, and zucchini yellow mosaic, generally all plants become infected by a virus within 2 to 3 weeks after 5% of the plants are found infected.[102] In the early 1980s in southern California, 85 to 90% of the melon and watermelon plants became infected with watermelon mosaic virus-2 each year, while some viruses, e.g., watermelon mosaic virus-1, were absent some years, but infected up to 40% of the plants other years.[103,104] Zucchini yellow mosaic virus appeared for the first time in 1982 in 2% of the plants, then in 10% in 1983 and in 40% in 1984. Many plants had multiple infections, and yield losses for various fields ranged from 25 to 100%, the overall yields of melons in these areas decreasing by 40 to 50%.[103,104] In an outbreak of watermelon mosaic virus in central New York, yield losses of winter squash in some large commercial fields were estimated at 50 to 70%.[32]

Numerous reports exist of virus incidences of 40 to 100% being common in many crops, e.g., onion yellow dwarf virus in onion[105] and in leek,[106] various pea viruses in pea,[107] bean mosaic viruses in bean,[108] and others. When virus incidence is that high, yield losses are almost always quite severe.

### E. Losses Caused by Some Viruses in Fruit Trees

All types of fruit trees, citrus, pome fruits, stone fruits, and the various tropical fruits such as cocoa, coconuts, papayas, etc., often suffer continuous and unspectacular losses from viruses, because many or most trees are frequently infected with mild or latent viruses. Such losses are usually estimated at about 8 to 20% of the potential yield of the trees.[9,44,109-111] Within each type of fruit tree, however, outbreaks of certain viruses occur which become widespread and severe. Such virus outbreaks cause continuous, spectacular, and catastrophic losses to the particular type of tree affected and to the economy of the state or country involved.

### 1. Citrus

The most catastrophic virus disease of citrus is tristeza.[111] Tristeza killed more than 10 million (60%) of the trees on sour orange in Argentina between 1930 and 1945, and within 12 years it spread to all citrus-producing areas of Brazil and killed more than 6 million trees.[41,112] More than 5 million trees have been killed by tristeza in Spain in the last 30 years. The disease has killed more than 3 million sweet orange on sour orange trees in California, and presently tristeza kills or makes unproductive an estimated 250,000 to 350,000 sweet orange on sour orange trees in Florida every year.[113] Totally, between 40 million and 50 million trees were killed or became unproductive in the last 50 to 60 years. The total economic loss is difficult to estimate, but considering that many of the affected trees were of bearing age, the cost of annual yield lost until it could be replaced by equivalent size trees has been estimated at a minimum of $10 to $30 per tree. This would give a global loss of $500 million to $1.5 billion caused entirely by the citrus tristeza virus.

### 2. Pome Fruits

In pome fruits, viral diseases like apple mosaic and pear vein yellows reduce yield by 12 to 20% or more and, because such losses are not spectacular, the viruses are already quite prevalent in orchards.[9,114] Other viral diseases, such as apple scar skin, cause severe yield loss and blemishes in the fruit of the affected trees, but, because this results in almost total economic loss, affected trees are removed. In the early 1960s, apple stem-pitting virus became important in North America because it destroyed hundreds of thousands of apple trees that had been propagated on the winter-hardy, but stem-pitting virus-sensitive rootstock Virginia Crab. In the 1970s and 1980s, a similar situation developed when apple varieties sensitive to TomRSV were propagated on clonal rootstocks, such as MM 106, which are tolerant to the nematode-transmitted TomRSV.[46] Hundreds of thousands of affected trees developed stem pitting and a brown line at the graft union, declined, and had to be removed.

### 3. Stone Fruits

Stone fruits (peaches, plums, nectarines, cherries, and apricots), wherever they are grown, suffer severe losses from one or another of the many stone fruit viruses every year. Some viruses may be somewhat localized, e.g., peach mosaic virus is now of economic importance only in western Colorado and California, while *Prunus* stem-pitting disease, caused by TomRSV, is prevalent in the mid-Atlantic states, and plum pox virus, the cause of plum pox or sharka disease, is so far confined to Europe and the Near East. On the other hand, viruses like *Prunus* necrotic ringspot (PNRSV) and prune dwarf are distributed throughout the world.[11,43,115,116] The importance of these stone fruit viruses stems from the fact that they are spread from tree to tree by vectors (peach mosaic by mites, TomRSV by nematodes, plum pox by aphids) or by pollen (PNRSV and prune dwarf virus). For example, peach mosaic virus, which is vectored by an eriophyid mite, spread from 7 trees to 30,500 trees within 4 years in Colorado.[11a] In the first 5 years after an infected tree-removal program was begun in California, 204,000 mosaic-infected peach trees were destroyed. Even today, peach mosaic virus limits the culture of many popular but susceptible freestone peach cultivars in the southwestern U.S.[11a]

PNRSV, in its many strains, infects all *Prunus* species and, alone or together with prune dwarf virus, causes severe losses worldwide on susceptible varieties of all its hosts. At least 80% of the sweet cherry trees in England and from 78 to 92% of sweet cherries in parts of New Zealand were found infected with one or more of the stone fruit viruses, primarily PNRSV. It was estimated that the potential crop was reduced by at least 30%.[115,116]

PNRSV is spread by infected pollen and by infected seeds. Pollination with infected pollen results in considerably reduced (by 25 to 75%) fruit set.[38] Presence of PNRSV in the rootstock or scionwood causes reduced bud take and poor graft unions.[49,117,118] The virus begins to spread when the trees are fully grown and blossom profusely. By the time the orchard is 12 to 15 years old, particularly in sour cherry and peach orchards, 90 to 100% of the trees are infected with PNRSV.[116,119,120] Fruit yields by infected trees are reduced by 30 to 50% and infected trees of susceptible varieties may exhibit dieback and decline.[49,115,118,119,121-123] In the late 1980s, about 28% of young, not yet blossoming peach trees in some areas of California were infected with *Prunus* ringspot virus or prune dwarf virus or both.[124] In some orchards, 90% of the trees were infected. In order to eliminate or reduce the incidence of viruses in nursery stock, legislation was enacted in California, effective January 1, 1988, for a tax of 0.25% on all stone, pome, and nut trees and grapevines sold. Part of that revenue will be used to pay for nursery indexing and certification programs aimed to reduce virus incidence in nursery stock.

*Prunus* stem-pitting disease, caused by the nematode-transmitted TomRSV, became prevalent in peach orchards in the mid-Atlantic states in the late 1960s and in the 1970s.[43] In some orchards 12 to 25% of the trees were affected, and in several 4- to 13-year-old orchards over 75% of the trees showed stem-pitting symptoms.[43,125] Affected trees become commercially worthless. Younger trees die in 2 to 3 years, while older trees may survive and produce fruit of poor quality for some years before they too decline and die.[43]

Plum pox or sharka disease affects plums, peaches, nectarines, and apricots and, so far, it is confined to Europe, Turkey, and Syria. Observed for the first time in Bulgaria in 1932, plum pox was found in Yugoslavia in 1936, and by 1970 it had spread northward through central Europe to Germany, France, England, Czechoslovakia, and Poland and southward to Greece. The economies of whole countries in southeastern Europe are affected. By 1970, more than 20% of the 75 million plum trees in Yugoslavia were infected, as were similar percentages of plum trees in Czechoslovakia and of plum and peach trees in Greece. In 1982, sharka was found for the first time in northern Italy,[126] and now it is present in Spain and Portugal and in the Near East.[127] In severely infected areas, 70 to 100% of the trees were infected.[35] Affected plum trees produce fruit that exhibit spots and rings which later become necrotic and sunken (poxes) as the fruit ripens. Much of the affected fruit drops prematurely, 20 to 30 d before the normal maturity date, and fruit remaining on the trees after the drop are commercially useless because they lack flavor and are low in sugar. Peach and apricot fruit of susceptible varieties do not usually drop, but they develop rings and spots, become disfigured, and are of poor quality and usually unmarketable. Yield losses and overall economic losses have been catastrophic to plum and peach growers in large geographical areas of the affected countries. In these areas, all infected trees had to be removed by the growers and replaced with resistant varieties or with other crops.[35]

*4. Tropical Fruits*

Cocoa swollen shoot virus reduces yields and debilitates trees that die or linger on. Cocoa swollen shoot disease is very serious in Ghana, Nigeria, and Togo. It is less serious in the Ivory Coast and Sierra Leone. In eastern Ghana, 118,000 tons of cocoa were harvested in 1936/37, but after swollen shoot virus became established and spread, only 38,700 tons of cocoa were harvested in 1955/56.[47] In 1947, 50 million trees were infected in Ghana. By 1961, 104 million swollen shoot-infected cocoa trees had been removed, equal to 69.3% of all cocoa trees in Ghana. Approximately 187 million swollen shoot-infected cocoa trees had been eradicated between 1946 and 1985, and in 1985 about 40 million additional trees were infected with the virus and were awaiting removal.[128]

One of the most catastrophic virus or viroid diseases of fruit trees is the cadang-cadang disease of coconut palms in the Philippines. In the 1950s, cadang-cadang caused annual copra losses worth about $15 million to $20 million. The disease destroyed at least 12 million coconut trees between 1926 to 1971.[129] In 1978 and 1980, respectively, approximately 391,000 and 209,000 new cases of disease were detected. It is estimated that approximately 30 million coconut palm trees had been killed by cadang-cadang by 1982.[130]

**F. Losses Caused by Some Viruses in Small Fruits and Grapevines**

Each of several viruses of strawberries, e.g., strawberry mottle and strawberry mild yellow edge, causes losses which in most strawberry varieties range from 17 to 30%.[131,132] When these viruses occur together or with some other virus, which is common, losses are even greater. Also, losses tend to increase each year after the time of infection.[5,40,133]

Grape fanleaf virus, which apparently occurs wherever grapes are grown, makes grapevines less and less fruitful and usually reduces by 50% or more the weight of grapes produced by susceptible varieties. It further reduces the quality and marketability of the available grapes.[45] Another grape virus with worldwide distribution, grape leaf roll virus, reduces vine growth, number and size of grape clusters, color of berries, and sugar content of grapes. Leaf roll alone reduces yield by 20% and causes an annual loss of about 5% of the total grape crop in California and considerably more in parts of Europe and elsewhere.[134,135]

**G. Losses Caused by Some Viruses in Forage Crops**

Clovers, alfalfa, forage grasses, etc., are often infected by viruses, usually 30 to 80% of the plants being infected with one or more viruses. Quite generally, therefore, forage crops suffer great losses from viruses in the form of reduced plant growth, shortened longevity (from 10 to 20 years to 3 to 6 years), forage yield reduction, increased susceptibility to other diseases, and reduction in winter hardiness.[136-139] Forage yield reductions in the field vary with the forage crop, the location, and the kind and incidence of the virus involved. However, virus incidence of 30 to 60% in forage grasses or legumes is common and forage yield reductions of 23 to 55% have been measured.[140-142] In North Carolina in 1987, one or more viruses were found to infect approximately 80% of all white clover plants, reducing yields by 30%, at an average annual statewide loss of $23 million.[22]

### H. Losses Caused by Some Viruses in Ornamentals

Numerous viruses infect and cause losses in the many ornamentals grown commercially. Some of the most common losses are caused by CMV on gladiolus and chrysanthemum, carnation ringspot on carnations, geranium leaf curl on geranium,[143] rose mosaic and related viruses on rose.[144] Several other viruses cause frequent and severe (10 to 30%) losses in bulb crops, in orchids, etc.[20,21,145]

## V. ECONOMIC COSTS OF PLANT VIRUS MANAGEMENT

In addition to the direct yield losses caused by viruses, which are usually borne as financial losses by the producers and as reduced produce and higher prices by consumers, plant viruses also cause losses as a result of the costs they incur to growers and/or to the public at large for their management. To these should probably be added the costs of research carried out at research centers or universities studying plant viruses. It must be noted, however, that funds expended for research and management of plant viruses are not really lost, but rather they are used by and support the livelihood of the people carrying out these functions. In the U.S., where there are approximately 300 plant virologists, a rough estimate of plant virology research costs, including salaries of scientists and support personnel, equipment, supplies, laboratories, fields, greenhouses, etc., would be about $36 million per year. This amount is, of course, paid for by the public in the form of federal and state taxes.

In addition to the research costs, countries, states, and certain industries carry out several functions aiming at reducing potential losses from plant virus diseases. For example, the federal government and several states have quarantine personnel and facilities in order to detect and eliminate imported viruses before they are released and spread in the state or the country.

Some states have voluntary or compulsory indexing programs for certain viruses of specific crop plants. For example, California has an indexing program for lettuce seed, and only seed lots containing no virus-carrying seeds per 30,000 seeds tested are certified for sale.[16] There are also indexing schemes for detecting and certifying as "virus-free" or "virus-tested" budwood, seed, or nursery stock of certain stone fruits, pome fruits, small fruits and grapes, of certain ornamentals, and of potatoes.[5,11,15,17,80,145a] As with lettuce seed, indexing for virus in these crops used to be, and in some crops still is, carried out by bioassay, that is, by inoculating susceptible indicator plants with sap or grafts from the plants being tested. In recent years, serological methods, particularly ELISA, have been replacing bioassays. More recently, use of DNA probes, corresponding to portions of the viral nucleic acid, are used to detect virus in plant sap.[146] The cost for virus indexing is often borne by the particular seed industries. In several states and countries, however, indexing costs are borne by the government (taxpayers). With some crops, e.g., strawberries, virus-free plants are produced via micropropagation. This involves culturing in the laboratory tissue explants obtained from a few plants kept virus-free in insect-free screenhouses and tested for virus at frequent intervals. The costs of all these procedures are passed on to consumers as higher prices.

Much of plant virus management depends on finding and/or developing varieties of crops resistant to viruses. For example, most cucumbers are now resistant to CMV, some tomatoes and tobaccos are resistant to TMV, many sugar beets are resistant to curly top virus, and so on.[12] Management by resistant varieties, however, involves prolonged, costly searches for resistance genes followed by time and efforts required to incorporate the resistance genes into the horticulturally desirable, but susceptible varieties. The latter then must be multiplied, tested for resistance to other pathogens, and finally distributed to growers. Even after a resistant variety is established, the search for varieties with different resistance genes must go on to replace resistant varieties that subsequently become susceptible to new virus strains.[147] The costs involved in this type of virus management include the costs of searching for and incorporating new genes for resistance into susceptible varieties, the higher costs of resistant seed, and also the losses suffered in the absence of resistance, or following breakdown of resistance, because of unavailability of resistant seed. The search for plants with resistance (repellance) to virus vectors is still in its infancy, but that too involves similar costs.

Some plant viruses can be managed successfully by certain cultural practices, such as keeping new crops separated sufficiently from older, infected crops in space or time. For example, young peach orchards can be protected from *Prunus* ringspot virus if they are planted more than 500 yards from older infected orchards. Even losses caused by some aphid-transmitted viruses, e.g., potato leaf roll, can be reduced drastically by cultivating the seed-producing crop in isolated areas free of the virus. Also, celery in California and Florida is kept free of celery mosaic virus, and wheat in Alberta, Canada is kept free of wheat streak mosaic virus by providing a celery-free and wheat-free period, respectively, at a time in the year at which no inoculum is available elsewhere to be carried by the vector to the new crop when that appears.[57,65,148] Separating crops in space or time involves hidden costs of unused land or time that could have otherwise been used to produce the particular crops.

Several other cultural practices, such as use of border, barrier, or buffer crops, crop rotations, reflective surfaces (mulches), sticky traps, oil sprays, etc., have been used to reduce crop losses caused by viruses spread by different vectors. Although the effectiveness of these methods has been variable, the costs involved in applying each are always substantial.[149-151]

Management of viral diseases of plants through chemical control of the virus vectors is attempted primarily when the virus is transmitted by the vector in a persistent manner, or when the virus spreads mainly from sources within the field. Control of viruses is obtained in some cases by spraying the areas in which the vector overwinters rather than the crop fields themselves.[64,64a] Soilborne viruses transmitted by nematodes or fungi can be controlled to some extent by soil fumigation with nematicides.[19] In all chemical control measures, the costs of pesticides and of their application are considerable and therefore must be weighed against their benefits, which are not always apparent.

In some crops, e.g., tomato, citrus, and papaya, losses caused by some viruses, e.g., TMV, citrus tristeza, and papaya ringspot, respectively, can be reduced significantly by inoculating the plants with mild strains of the respective viruses before the plants become infected with severe strains of the same virus (cross-protection).[152,153] Although the cost of the procedure is relatively small, even mild strains cause some losses and there are also the dangers of mutations, double infections, and infections of other, more susceptible, hosts. The disadvantages and costs of cross-protection can apparently be eliminated by incorporating the virus gene that codes for the virus coat protein into the genome of susceptible plants via genetic engineering techniques. Such plants produce virus coat protein and show marked cross-protection from severe virus strains without the plants containing any virus at all.[154] New discoveries in the molecular plant pathology area may further improve control of plant viruses, for example, via introduction and perpetuation in the plant of such competing nucleic acids as modified satellite RNAs or anti-sense RNAs. Although the costs of developing such methods are high, the eventual potential gains are likely to be enormous.

## VI. CONCLUSIONS

Plant viruses may and usually do cause significant losses in crop yield and quality wherever crops are grown. Crop losses may vary from year to year, from one area to another, and from one crop to another. Crop losses generally result in economic loss to individual growers who suffer the crop loss and to consumers. Annual crop losses from viruses may range from almost unnoticeable to complete destruction of the crop. Crop and economic losses from plant viruses are sometimes so great that growers are forced to abandon the culture of that crop. Crop losses from viruses can be avoided by growing virus-free seed or nursery stock, growing resistant varieties, following certain cultural practices, and attempting to avoid or reduce the vectors of viruses. Losses from viruses have been reduced drastically in several crops in many developed countries, but continue to be great in developing countries. New techniques of virus detection and identification, and new knowledge on epidemiological behavior of plant viruses have helped greatly in developing several, more or less effective, virus-control strategies and thereby in reducing crop and economic losses. Additional research is needed to further develop these strategies and to discover new ones. Discovery and incorporation in susceptible crops of plant genes for resistance to virus infection, or to virus vectors, is probably the most effective and most promising virus-control strategy; developments in molecular plant pathology may further increase the effectiveness and importance of this approach. In the meantime, however, more research and greater efforts are needed to learn the identity and to study the epidemiology of all plant viruses throughout the world, since viruses spread, naturally or by man, and cause diseases and losses in crops in areas where these viruses did not exist before.

## REFERENCES

1. **Matthews, R. E. F.,** *Plant Virology,* 2nd ed., Academic Press, New York, 1981.
2. **Walkey, D. G. A.,** *Applied Plant Virology,* John Wiley & Sons, New York, 1985.
3. **Anon.,** *Proc. Workshop on Epidemiology of Plant Virus Diseases, Orlando, FL,* Clemson University, Clemson, S.C., 1986.
4. **Bos, L.,** Crop losses caused by virus, *Crop Prot.,* 1, 263, 1982.
5. **Converse, R. H., Ed.,** *Virus Diseases of Small Fruits,* Agric. Handbook No. 631, Agricultural Research Service, U.S. Department of Agriculture, Washington, D.C., 1987, 277.
6. **de Bokx, J. A., Ed.,** *Viruses of Potatoes and Seed-Potato Production,* Pudoc, Wageningen, The Netherlands, 1981.
7. **Gordon, D. T., Knoke, J. K., and Scott, G. E., Eds.,** Virus and Viruslike Diseases of Maize in the United States, South. Coop. Ser. Bull. No. 247, Ohio Agriculture Research and Development Center, Wooster, OH, 1981.
8. **Plumb, R. T. and Thresh, J. M., Eds.,** *Plant Virus Epidemiology: The Spread and Control of Insect-Borne Viruses,* Blackwell Scientific, Oxford, 1983.
9. **Posnette, A. F., Eds.,** Virus Diseases of Apples and Pears, Tech. Commun. No. 30, Commonwealth Agriculture Bureau, Farnham Royal, Bucks, England, 1963.
10. **Sherf, A. F. and McNabb, A. A.,** *Vegetable Diseases and Their Control,* 2nd ed., John Wiley & Sons, New York, 1986, 728.
11. U.S. Department of Agriculture, *Virus Diseases and Noninfectious Disorders of Stone Fruits in North America,* Agric. Handbook No. 437, U.S. Department of Agriculture, Washington, D.C., 1976.
11a. **Pine, T. S.,** Peach mosaic, in *Virus Diseases and Noninfectious Disorders of Stone Fruits in North America,* Agric. Handbook No. 437, U.S. Department of Agriculture, Washington, D.C., 1976, 61.
12. **Fraser, R. S. S.,** Genes for resistance to plant viruses, *CRC Crit. Rev. Plant Sci.,* 3, 257, 1986.
13. U.S. Department of Agriculture, *Indexing Procedures for 15 Virus Diseases of Citrus Trees,* Agric. Handbook No. 333, U.S. Department of Agriculture, Washington, D.C., 1968.

14. **Converse, R. H., Adams, A. N., Barbara, D. J., Clark, M. F., Casper, R., Hepp, R. F., Martin, R. R., Morris, T. J., Spiegel, S., and Yoshikawa, N.,** Laboratory detection of viruses and mycoplasmalike organisms in strawberry, *Plant Dis.,* 72, 744, 1988.
15. **Raju, B. C. and Olson, C. J.,** Indexing systems for producing clean stock for disease control in commercial floriculture, *Plant Dis.,* 69, 189, 1985.
16. **Grogan, R. G.,** Control of lettuce mosaic with virus-free seed, *Plant Dis.,* 64, 446, 1980.
17. **Hollings, M.,** Disease control through virus-free stock, *Annu. Rev. Phytopathol.,* 3, 367, 1965.
18. **Thresh, J. M.,** Cropping practices and virus spread, *Annu. Rev. Phytopathol.,* 20, 193, 1982.
19. **Lamberti, F.,** Combatting nematode vectors of plant viruses, *Plant Dis.,* 65, 113, 1981.
20. **Anon.,** *Estimates of Crop Losses and Disease-Control Costs in California, 1963,* University of California Experiment Station, Davis, CA, 1963, 102.
21. **Anon.,** Losses in Agriculture, Agric. Handbook No. 291, U.S. Department of Agriculture, Washington, D.C., 1965, 120.
22. **Main, C. E. and Gurtz, S. K.,** *1987 Estimates of Crop Losses in North Carolina Due to Plant Diseases & Nematodes,* Special Publ. No. 7, Department of Plant Pathology, North Carolina State University, Raleigh, NC, 1988, 209.
23. **Ross, J. P.,** Effect of single and double infection of soybean mosaic and bean pod mottle viruses on soybean yield and seed characters, *Plant Dis. Rep.,* 59, 806, 1968.
24. **Slykhuis, J. T.,** Virus and virus-like diseases of cereal crops, *Annu. Rev. Phytopathol.,* 14, 189, 1976.
25. **Hull, R.,** The health of the sugar beet crop in Great Britain, *J. R. Agric. Soc. Engl.,* 122, 101, 1961.
26. **McMurtrey, J. E., Jr.,** Effect of mosaic diseases on yield and quality of tobacco, *J. Agric. Res. (Washington, D.C.),* 38, 257, 1929.
27. **Agrios, G. N., Walker, M. E., and Ferro, D. N.,** Effect of cucumber mosaic virus inoculation at successive weekly intervals on growth and yield of pepper (*Capsicum annuum*) plants, *Plant Dis.,* 69, 52, 1985.
28. **Mink, G. I. and Shay, J. R.,** Latent viruses in apple, *Purdue Univ. Agric. Exp. Stn. Res. Bull.,* 756, 1962.
29. **Bennett, C. W.,** Sugar beet yellows disease in the United States, *U.S. Dep. Agric. Tech. Bull.,* 1218, 63, 1960.
30. **Agrios, G. N., Walker, M. E., Ferro, D. N., and Corredor, D.,** Virus incidence and spread in pepper field plots treated with reflective mulch and oil, *Phytopathology,* 73 (Abstr.), 361, 1983.
31. **Komm, D. A. and Agrios, G. N.,** Incidence and epidemiology of viruses affecting cucurbit crops in Massachusetts, *Plant Dis. Rep.,* 62, 746, 1978.
32. **Provvidenti, R. and Schroeder, W. T.,** Epiphytotic of watermelon mosaic among cucurbitaceae in central New York in 1969, *Plant Dis. Rep.,* 54, 744, 1970.
33. **Le Clerg, E. L., Lombard, P. M., Eddins, A. H., Cook, H. T., and Campbell, J. C.,** Effect of different amounts of spindle tuber and leaf roll on yield of Irish potatoes, *Am. Potato J.,* 21, 60, 1944.
34. **Sing, R. P., Finnie, R. E., and Bagnall, R. H.,** Losses due to the potato spindle tuber virus, *Am. Potato J.,* 48, 262, 1971.
35. **Sutic D. and Pine, T. S.,** Sarka (plum pox) disease, *Plant Dis. Rep.,* 52, 253, 1968.
36. **Kontaxis, D. G.,** Big vein disease of lettuce in Imperial Valley, *Calif. Agric.,* 32, 16, 1978.
36a. **Patterson, C. L., Grogan, R. G., and Campbell, R. N.,** Economically important diseases of lettuce, *Plant Dis.,* 70, 982, 1986.
37. **Shields, I. J., Foster, R. E., and Keener, P. D.,** Control lettuce mosaic in Arizona, *Ariz. Ext. Serv. Circ.,* 258, 1957.
38. **Way, R. D. and Gilmer, R. M.,** Fruit set reductions in cherry trees pollinated by trees with sour cherry yellows, *Phytopathology,* 53(3), 352, 1963.
39. **Sinclair, J. B., Ed.,** *Compendium of Soybean Diseases,* American Phytopathological Society, St. Paul, MN, 1982.
40. **Maas, J. E., Ed.,** *Compendium of Strawberry Diseases,* American Phytopathological Society, St. Paul, MN, 1984.
41. **Bennett, C. W. and Costa, A. S.,** Tristeza disease of citrus, *J. Agric. Res. (Washington, D.C.),* 78, 207, 1949.
42. **Tuttle, M. A. and Gotlieb, A. R.,** Apple union necrosis: histopathology and distribution of tomato ringspot virus in affected trees, Vt. Agric. Exp. Stn. RR 36, University of Vermont, Burlington, 1984.
43. **Mircetich, S. M. and Fogle, H. W.,** Peach stem pitting, in *Virus Diseases and Noninfectious Disorders of Stone Fruits in North America,* Agric. Handbook No. 437, U.S. Department of Agriculture, Washington, D.C., 1976, 77.
43a. **Mircetich, S. M. and Rowhani, A.,** The relationship of cherry leaf roll virus and blackline disease of English walnut trees, *Phytopathology,* 74, 423, 1984.
43b. **Mircetich, S. M., Refsguard, J., and Matheron, M. E.,** Blackline of English walnut trees traced to graft transmitted virus, *Calif. Agric.,* 34, 8, 1980.

44. U.S. Department of Agriculture, *Virus Diseases and Other Disorders with Viruslike Symptoms of Stone Fruits in North America,* Agric. Handbook No. 10, U.S. Department of Agriculture, Washington, D.C., 1951.

45. **Vuittenez, A.,** Fanleaf of grapevine, in *Virus Diseases of Small Fruits and Grapevines,* Frazier, N. W., Ed., University of California, Agriculture Science, Berkeley, CA, 1970, 217.

46. **Rosenberger, D. A., Harrison, M. B., and Gonsalves, D.,** Incidence of apple union necrosis, tomato ringspot virus, and *Xiphinema* vector species in Hudson Valley orchards, *Plant Dis.,* 67, 356, 1983.

47. **Owasu, G. K.,** The cocoa swollen shoot disease problem in Ghana, in *Plant Virus Epidemiology: The Spread and Control of Insect-Borne Viruses,* Plumb, R. T. and Thresh, J. M., Eds., Blackwell Scientific, Oxford, 1983, 73.

48. **Eslick, R. F.,** Yield reductions in Glacier barley associated with a virus infection, *Plant Dis. Rep.,* 37, 290, 1953.

49. **Nyland, G., Gilmer, R. M., and Moore, J. D.,** "Prunus" ring spot group, in *Virus Diseases and Noninfectious Disorders of Stone Fruits in North America,* Agric. Handbook No. 437, U.S. Department of Agriculture, Washington, D.C., 1976, 104.

50. **Gilmer, R. M., Nyland, G., and Moore, J. D.,** Prune dwarf, in *Virus Diseases and Noninfectious Disorders of Stone Fruits in North America,* Agric. Handbook No. 437, U.S. Department of Agriculture, Washington, D.C., 1976, 179.

51. **Bruehl, G. W.,** *Barley Yellow Dwarf, A Virus Disease of Cereals,* Monogr. No 1, American Phytopathological Society, St. Paul, MN, 1961, 52.

51a. **Yount, D. J., Martin, J. M., Carroll, T. W., and Zaske, S. K.,** Effects of barley yellow dwarf virus on growth and yield of small grains in Montana, *Plant Dis.,* 69, 487, 1985.

52. **Carr, A. J. H.,** Infection by viruses and subsequent host damage, in *Plant Diseases: Infection, Damage and Loss,* Wood, R. K. S. and Jellis, G. J., Eds., Blackwell Scientific, Oxford, 1984, 199.

53. **Ralph, W.,** Bresatic is developing a suite of DNA probes to detect plant viruses, *Genetic Engineering News,* January 1988, 12.

54. **Plumb, R. T.,** Barley yellow dwarf virus — a global problem, in *Plant Virus Epidemiology: The Spread and Control of Insect-Borne Viruses,* Plumb, R. T. and Thresh, J. M., Eds., Blackwell Scientific, Oxford, 1983, 185.

55. **Gill, C. C.,** Assessment of losses on spring wheat naturally infected with barley yellow dwarf virus, *Plant Dis.,* 64, 197, 1980.

56. **Smith, H. C.,** Control of barley yellow dwarf in cereals, *N.Z. J. Agric. Res.,* 6, 229, 1963.

57. **Atkinson, T. G. and Grant, M. N.,** An evaluation of streak mosaic losses in winter wheat, *Phytopathology,* 57, 188, 1967.

58. **Sill, W. H., Jr. and Talens, L. T.,** The 1957 soil-borne wheat mosaic epiphytotic in Kansas, *Plant Dis. Rep.,* 42, 513, 1958.

59. **Koehler, B., Bever, W. M., and Bonnett, O. T.,** Soil-borne wheat mosaic, *Ill. Agric. Exp. Stn. Bull.,* 556, 567, 1952.

60. **Gamez, R.,** The ecology of maize rayado fino virus in the American tropics, in *Plant Virus Epidemiology: The Spread and Control of Insect-Borne Viruses,* Plumb, R. T. and Thresh, J. M., Eds., Blackwell Scientific, Oxford, 1983, 267.

61. **Egan, B. T. and Hall, P.,** Monitoring the Fiji disease epidemic in sugarcane at Bundaberg, Australia, in *Plant Virus Epidemiology: The Spread and Control of Insect-Borne Viruses,* Plumb, R. T. and Thresh, J. M., Eds., Blackwell Scientific, Oxford, 1983, 287.

62. **Kiritani, K.,** Changes in cropping practices and the incidence of hopper-borne diseases of rice in Japan, in *Plant Virus Epidemiology: The Spread and Control of Insect-Borne Viruses,* Plumb, R. T., and Thresh, J. M., Eds., Blackwell Scientific, Oxford, 1983, 239.

63. **Ling, K. C., Tiongco, E. R., and Flores, Z. M.,** Epidemiological studies of rice tungro, in *Plant Virus Epidemiology: The Spread and Control of Insect-Borne Viruses,* Plumb, R. T. and Thresh, J. M., Eds., Blackwell Scientific, Oxford, 1983, 249.

63a. **King, T. H.,** Occurrence and distribution of diseases and pests of rice and their control in Thailand, *FAO Plant Prot. Bull.,* 1970, 390.

64. **Duffus, J. E.,** Epidemiology and control of curly top diseases of sugarbeet and other crops, *Plant Virus Epidemiology: The Spread and Control of Insect-Borne Viruses,* Plumb, R. T. and Thresh, J. M., Eds., Blackwell Scientific, Oxford, 1983, 297.

64a. **Bennett, C. W.,** *The Curly Top Disease of Sugarbeet and Other Plants,* Monogr. No. 7, American Phytopathological Society, St. Paul, MN, 1971, 81.

65. **Duffus, J. E.,** Epidemiology and control of aphid-borne virus diseases in California, in *Plant Virus Epidemiology: The Spread and Control of Insect-Borne Viruses,* Plumb, R. T. and Thresh, J. M., Eds., Blackwell Scientific, Oxford, 1983, 221.

66. **Severin, H. H. P. and Drake, R. M.,** Sugar-beet mosaic, *Hilgardia,* 18, 483, 1948.

67. **Heathcote, G. D.**, Effect of virus yellows on yield of some monogerm cultivars of sugar beet, *Ann. Appl. Biol.*, 88, 145, 1978.
68. **Gooding, G. V., Jr. and Todd, F. A.**, Virus diseases of flue-cured tobacco in North Carolina, *Plant Dis. Rep.*, 50, 308, 1966.
69. **Gooding, G. V., Jr. and Todd, F. A.**, Virus diseases of burley tobacco in North Carolina, *Plant Dis. Rep.*, 51, 409, 1967.
70. **Gooding, G. V., Jr., Main, C. E., and Nelson, L. A.**, Estimating losses caused by tobacco vein mottling virus in burley tobacco, *Plant Dis.*, 65, 889, 1981.
71. **Johnson, C. S., Main, C. E., and Gooding, G. V., Jr.**, Crop loss assessment for flue-cured tobacco cultivars infected with tobacco mosaic virus, *Plant Dis.*, 67, 881, 1983.
72. **Latorre, B. A., Andrade, O., Penaloza, E., and Escaffi, O.**, A severe outbreak of potato virus Y in Chilean tobacco, *Plant Dis.*, 66, 893, 1982.
73. **Collins, M. H., Murphy, J. F., Witcher, W., and Barnett, O. W.**, Survey of cowpeas in South Carolina for six viruses, *Plant Dis.*, 68, 561, 1984.
74. **Paguio, O. R. and Kuhn, C. W.**, Incidence and source of inoculum of peanut mottle virus and its effect on peanut, *Phytopathology*, 64, 60, 1974.
75. **Pitre, H. N., Patel, V. C., and Keeling, B. L.**, Distribution of bean pod mottle disease on soybeans in Mississippi, *Plant Dis. Rep.*, 63, 419, 1979.
76. **Ross, J. P.**, Response of early- and late-planted soybeans to natural infection by bean pod mottle virus, *Plant Dis.*, 70, 222, 1986.
77. **Suteri, B. D. and Srivastava, B.**, Effect of soybean yellow mosaic on growth and yield of soybean, *(Glycine max)*, *Plant Dis. Rep.*, 63, 151, 1979.
78. **Reddy, D. V. R., Amin, P. W., McDonald, D., and Ghanekar, A. M.**, Epidemiology and control of groundnut bud necrosis and other diseases of legume crops in India caused by tomato spotted wilt virus, in *Plant Virus Epidemiology: The Spread and Control of Insect-Borne Viruses*, Plumb, R. T. and Thresh, J. M., Eds., Blackwell Scientific, Oxford, 1983, 93.
79. **Gamez, R. and Moreno, R. A.**, Epidemiology of beetle-borne viruses of grain legumes in Central America, in *Plant Virus Epidemiology: The Spread and Control of Insect-Borne Viruses*, Plumb, R. T. and Thresh, J. M., Eds., Blackwell Scientific, Oxford, 1983, 103.
80. **Hooker, W. J., Ed.**, *Compendium of Potato Diseases*, American Phytopathological Society, St. Paul, MN, 1981.
81. **Rich, A. E.**, *Potato Diseases*, Academic Press, New York, 1983.
82. **Harper, F. R., Nelson, G. A., and Pittman, U. J.**, Relationship between leaf roll symptoms and yield in netted gem potato, *Phytopathology*, 65, 1242, 1975.
83. **Kolbe, W.**, Ergebnisse eines Kartoffel — Nachbauversuches, *Der Kartoffelbau*, 32, 352, 1981.
84. **Bonde, R.**, Potato X virus causes large losses; better seed is the answer, *Maine Farm Res.*, 2, 10, 1953.
85. **Hunter, J. E. and Rich, A. E.**, The effect of potato spindle tuber virus on growth and yield of Saco potatoes, *Am. Potato J.*, 41, 113, 1964.
86. **Tien, P.**, Viroids and viroid diseases in China, in *Subviral Pathogens of Plants and Animals: Viroids and Prions*, Maramorosch, K. and McKelvey, J. J., Jr., Eds., Academic Press, Orlando, FL, 1985, 123.
87. **Bock, K. R.**, Epidemiology of cassava mosaic disease in Kenya, in *Plant Virus Epidemiology: The Spread and Control of Insect-Borne Viruses*, Plumb, R. T. and Thresh, J. M., Eds., Blackwell Scientific, Oxford, 1983, 337.
88. **Fauquet, C. and Fargette, D.**, A summary of the epidemiology of African cassava mosaic virus, in *Proc. Workshop on Epidemiology of Plant Virus Diseases, Orlando, FL*, Clemson University, Clemson, S.C., 1986, VII, 1.
89. **Seif, A. A.**, Effect of cassava mosaic virus on yield of cassava, *Plant Dis.*, 66, 661, 1982.
90. **Coakley, S. M., Campbell, R. N., and Kimble, K. A.**, Lettuce Mosaic Virus, *Calif. Agric.*, September, 1973.
90a. **Severin, H. H. P. and Freitag, J. H.**, Western celery mosaic, *Hilgardia*, 11, 495, 1938.
91. **Milbrath, D. G.**, Control of western celery mosaic, *Calif. Dep. Agric. Bull.*, 37, 1948.
92. **Cho, J. J., Mitchell, W. C., Mau, R. F. L., and Sakimura, K.**, Epidemiology of tomato spotted wilt virus disease on crisphead lettuce in Hawaii, *Plant Dis.*, 71, 505, 1987.
93. **Broadbent, L.**, Epidemiology and control of tomato mosaic virus, *Annu. Rev. Phytopathol.*, 14, 189, 1976.
94. **Alexander, L. J.**, Effect of tobacco mosaic disease on the yield of unstaked tomatoes, *Phytopathology*, 42, 463, 1952.
95. **Zitter, T. A. and Everett, P. H.**, Effect of an aphid-transmitted yellowing virus on yield and quality of staked tomatoes, *Plant Dis.*, 66, 456, 1982.
96. **Zitter, T. A. and Tsai, J. H.**, Viruses infecting tomato in southern Florida, *Plant Dis.*, 65, 787, 1981.
97. **Makkouk, K. M. and Laterrot, H.**, Epidemiology and control of tomato yellow leaf curl virus, in *Plant Virus Epidemiology: The Spread and Control of Insect-Borne Viruses*, Plumb, R. T. and Thresh, J. M., Eds., Blackwell Scientific, Oxford, 1983, 315.

98. **Benner, C. P., Kuhn, L. W., Demski, J. W., Dobson, J. W., Colditz, P., and Nutter, F. W., Jr.,** Identification and incidence of pepper viruses in northwestern Georgia, *Plant Dis.*, 69, 999, 1985.

99. **Conti, M. and Masenga, V.,** Identification and prevalence of pepper viruses in northwest Italy, *Phytopathol. Z.*, 90, 212, 1977.

100. **Lana, A. F. and Peterson, J. F.,** Identification and prevalence of pepper viruses in southern Quebec, *Phytoprotection*, 61, 13, 1980.

101. **Zitter, T. A. and Ozaki, H. Y.,** Aphid-borne vegetable viruses controlled with oil sprays, *Proc. Fla. State Hortic. Soc.*, 91, 287, 1978.

102. **Lecoq, H., Clauzel, J. M., and Pitrat, M.,** Comparative epidemiology of three cucurbit viruses (CMV, WMV2, and ZYMV) in susceptible and partially resistant melon cultivars in France, in *Proc. Workshop on Epidemiology of Plant Virus Diseases*, Orlando, FL, 1986, I, 16.

103. **Nameth, S. T., Dodds, J. A., Paulus, A. O., and Kishaba, A.,** Zucchini yellow mosaic virus associated with severe diseases of melon and watermelon in Southern California desert valleys, *Plant Dis.*, 69, 785, 1985.

104. **Nameth, S. T., Dodds, J. A., Paulus, A. O., and Laemmlen, F. F.,** Cucurbit viruses of California: an ever-changing problem, *Plant Dis.*, 70, 8, 1986.

105. **Fischer, H. U. and Lockhart, B. E. L.,** High incidence of onion yellow dwarf in areas of commercial onion production in Morocco, *Plant Dis. Rep.*, 58, 252, 1974.

106. **Bos, L.,** A serious outbreak of onion yellow dwarf virus in leek, *Gewasbescherming*, 3, 81, 1972.

107. **Hagedorn, D. J.,** *Virus Diseases of Pea, Pisum sativum*, Monogr. No. 9, American Phytopathological Society, St. Paul, MN, 1974.

108. **Nelson, R.,** *Investigations in the Mosaic Disease of Bean*, Tech. Bull. No. 118, Agricultural Experiment Station, East Lansing, MI, 1932.

109. **Klotz, L. J., Calavan, E. C., and Weathers, L. G.,** Viruses and viruslike diseases of citrus, *Calif. Agric. Exp. Stn. Circ.*, 559, 1, 1972.

110. **Rich, A. E.,** Influence of dapple apple and stem pitting viruses on tree growth and fruit yields over a fourteen-year period, *Plant Dis. Rep.*, 51, 293, 1967.

111. **Whiteside, J. O., Garnsey, S. M., and Timmer, L. W.,** *Compendium of Citrus Diseases*, APS Press, American Phytopathological Society, St. Paul, MN, 1988.

112. **Bar-Joseph, M., Roistacher, C. N., and Garnsey, S. M.,** The epidemiology and control of citrus tristeza disease, in *Plant Virus Epidemiology: The Spread and Control of Insect-Borne Viruses*, Plumb, R. T. and Thresh, J. M., Eds., Blackwell Scientific, Oxford, 1983, 61.

113. **Brlansky, R. H., Pelosi, R. R., Garnsey, S. M., Youtsey, C. O., Lee, R. F., Yokomi, R. K., and Sonoda, R. M.,** Tristeza quick decline epidemic in South Florida, *Proc. Fla. State Hortic. Soc.*, 99, 66, 1986.

114. **Fridlund, P. R.,** Incidence, systemic nature, and spread of the pear vein yellows virus in the Yakima Valley, Washington, *Plant Dis. Rep.*, 57, 483, 1973.

115. **Posnette, A. F., Cropley, R., and Swait, A. A. J.,** The incidence of virus diseases in English sweet cherry orchards and their effect on yield, *Ann. Appl. Biol.*, 61, 351, 1968.

116. **Wood, G. A. and Fry, P. R.,** Virus infection in Central Otago sweet cherry trees, *N.Z. J. Agric. Res.*, 15, 172, 1972.

117. **Agrios, G. N. and Buchholtz, W. F.,** Virus effect on union and growth of peach scions on *Prunus besseyi* and *P. tomentosa* understocks, *Iowa State J. Sci.*, 41, 385, 1967.

118. **Cochran, L. C., Hutchins, L. M., Milbrath, J. A., Stout, G. L., and Zeller, S. M.,** Ring spot, in *Virus Diseases and Other Disorders with Viruslike Symptoms of Stone Fruits in North America*, Agric. Handbook No. 10, U.S. Department of Agriculture, Washington, D.C., 1951, 71.

119. **Davidson, T. R. and George, J. A.,** Effects of necrotic ring spot and sour cherry yellows on the growth and yield of young sour cherry trees, *Can. J. Plant Sci.*, 45, 525, 1965.

120. **Demski, J. W. and Boyle, J. S.,** Spread of necrotic ringspot virus in a sour cherry orchard, *Plant Dis. Rep.*, 52(12), 972, 1968.

121. **Klos, E. J. and Parker, K. G.,** Yields of sour cherry affected with ringspot and yellows viruses, *Phytopathology*, 50, 412, 1960.

122. **Parker, K. G., Brase, K. D., Schmid, G., Barksdale, T. H., and Allen, W. R.,** Influence of ring spot virus on growth and yield of sour cherry, *Plant Dis. Rep.*, 43, 380, 1959.

123. **Wells, J. M., Kirkpatrick, H. C., and Parish, C. L.,** Symptomatology and incidence of prunus necrotic ringspot virus in peach orchards in Georgia, *Plant Dis.*, 70, 444, 1986.

124. **Uyemoto, J. K., Luhn, C. F., Asai, W., Beede, R., Beutel, J. A., and Fenton, R.,** Incidence of ilarviruses in young peach trees in California, *Plant Dis.*, 73, 217, 1989.

125. **Mink, G. I. and Howell, W. E.,** Occurrence and distribution of stem pitting of sweet cherry trees in Washington, *Plant Dis.*, 64, 551, 1980.

126. **Conti, M., Roggero, P., Casetta, A., and Lenzi, R.,** Epidemiology and vectors of plum pox (sharka) in northwest Italy, in *Proc. Workshop on Epidemiology of Plant Virus Diseases*, Orlando, FL, Clemson University, Clemson, S.C., 1986, VI, 4.

127. **Dunez, J.,** Preliminary observations on virus and virus-like diseases of stone-fruit trees in the Mediterranean and Near East countries, *FAO Plant Prot. Bull.,* 34, 43, 1986.
128. **Owusu, G. K. and Thresh, J. M.,** The cacao swollen shoot virus eradication campaign in Ghana, in *Proc. Workshop on Epidemiology of Plant Virus Diseases, Orlando, FL,* Clemson University, Clemson, S.C., 1986, III, 11.
129. **Price, W. C.,** Cadang-cadang of coconut — a review, *Plant Sci.,* 3, 1, 1971.
130. **Zelazny, B. and Pacumbaba, E.,** Incidence of cadang-cadang of coconut palm in the Philippines, *Plant Dis.,* 66, 547, 1982.
131. **Mellor, F. C. and Frazier, N. W.,** Strawberry mottle, in *Virus Diseases of Small Fruits and Grapevines,* Frazier, N. W., Ed., University of California Division of Agricultural Science, Berkeley, CA, 1970, 4.
132. **Mellor, F. C. and Frazier, N. W.,** Strawberry mild yellow-edge, in *Virus Diseases of Small Fruits and Grapevines,* Frazier, N. W., Ed., University of California Division of Agricultural Science, Berkeley, CA, 1970, 14.
133. **Frazier, N. W., Ed.,** *Virus Diseases of Small Fruits and Grapevines,* Frazier, N. W., Ed., University of California Division of Agricultural Science, Berkeley, CA, 1970.
134. **Goheen, A. C.,** Grape leafroll, in *Virus Diseases of Small Fruits and Grapevines,* Frazier, N. W., Ed., University of California Division of Agricultural Science, Berkeley, CA, 1970, 209.
135. **Pearson, R. C. and Goheen, A. C., Eds.,** *Compendium of Grape Diseases,* APS Press, American Phytopathological Society, St. Paul, MN, 1988.
136. **Barnett, O. W. and Diachun, S.,** Virus diseases of clovers, in *Clover Science and Technology,* Agronomy Monogr. No. 25, Agronomy, Crop Science, Soil Science Societies, Madison, WI, 19, 235, 1986.
137. **Goth, R. W. and Wilcoxson, R. D.,** Effects of bean yellow mosaic on survival and flower formation in red clover, *Crop Sci.,* 2, 426, 1962.
138. **Pratt, M. J.,** Reduced winter survival and yield of clover infected with clover yellow mosaic virus, *Can. J. Plant Sci.,* 47, 289, 1976.
139. **Watson, R. D. and Guthrie, J. W.,** Virus-fungus interrelationships in a root rot complex in red clover, *Plant Dis. Rep.,* 48, 723, 1964.
140. **Campbell, L. C. and Moyer, J. W.,** Yield responses of six white clover clones to virus infection under field conditions, *Plant Dis.,* 68, 1033, 1984.
141. **Crill, P., Hagedorn, D. J., and Hanson, E. W.,** Incidence and effect of alfalfa mosaic virus on alfalfa, *Phytopathology,* 60, 1432, 1970.
142. **Gibson, P. B., Barnett, O. W., Burrows, P. M., and King, F. D.,** Filtered air enclosures exclude vectors and enable measurement of effects of viruses on white clover in the field, *Plant Dis.,* 66, 142, 1982.
143. **Jones, L. K.,** Leaf curl and mosaic of geranium, *Wash. Agric. Exp. Stn. Bull.,* 390, 1940.
144. **Wong, S.-M., Horst, R. K., Kawamoto, S. O., and Weaber, K. F.,** Occurrence of apple mosaic virus and prunus necrotic ringspot virus in roses in Sonnenberg Rose Garden, New York, USA, in *Proc. Workshop on Epidemiology of Plant Virus Diseases, Orlando, FL,* Clemson University, Clemson, S.C., 1986, II, 19.
145. **Lawson, R. H.,** Controlling virus diseases of major international flower and bulb crops, *Plant Dis.,* 65, 780, 1981.
145a. **Fridlund, P. R.,** The IR-2 program for obtaining virus-free fruit trees, *Plant Dis.,* 64, 831, 1980.
146. **Anon.,** Virus diseases: a dilemma for plant breeders, a symposium, *HortScience,* 20, 833, 1985.
147. **Paguio, O. R., Kuhn, C. W., and Boerma, H. R.,** Resistance-breaking variants of cowpea chlorotic mottle virus in soybean, *Plant Dis.,* 72, 768, 1988.
148. **Zitter, T. A.,** Epidemiology of aphid-borne viruses, in *Aphids as Virus Vectors,* Harris, K. F. and Maramorosch, K., Eds., Academic Press, New York, 1977, 385.
149. **Harpaz, I.,** Non-pesticidal control of vector-borne diseases, in *Pathogens, Vectors and Plant Diseases: Approaches to Control,* Harris, K. F. and Maramorosch, K., Eds., Academic Press, New York, 1982, 1.
150. **Simons, J. M.,** Use of oil sprays and reflective surfaces for control of insect-transmitted plant viruses, in *Pathogens, Vectors and Plant Diseases: Approaches to Control,* Harris, K. F. and Maramorosch, K., Eds., Academic Press, New York, 1982, 71.
151. **Zitter, T. A. and Simmons, J. N.,** Management of viruses by alteration of vector efficiency and by cultural practices, *Annu. Rev. Phytopathol.,* 18, 289, 1980.
152. **Rast, A. T. B.,** M11-16, an artificial mutant of tobacco mosaic virus for seedling inoculation of tomato crops, *Neth. J. Plant Pathol.,* 78, 110, 1972.
153. **Costa, A. S. and Muller, G. W.,** Tristeza control by cross protection: a U.S.-Brazil cooperative success, *Plant Dis.,* 64, 538, 1980.
153a. **Yeh, S. D., Gonsalves, D., Wang, H. L., Namba, R., and Chiu, R. J.,** Control of papaya ringspot virus by cross protection, *Plant Dis.,* 72, 375, 1988.
154. **Powell, A. P., Nelson, R. S., Barun, D., Hoffmann, N., Rogers, S. G., Fraley, R. T., and Beachy, R. N.,** Delay of disease development in transgenic plants that express the tobacco mosaic virus coat protein gene, *Science,* 232, 738, 1986.

Chapter 2

# PATHOPHYSIOLOGICAL ALTERATIONS

## K. Roger Wood

### TABLE OF CONTENTS

# I. INTRODUCTION

Viral infection of a plant can lead to a multitude of symptoms, some immediately apparent and others requiring microscopic examination. Many are accompanied by profound biochemical and physiological changes. The magnitude of these perturbations can vary from the negligible to the fatal, depending principally on the host and the virus, but also on the environmental conditions in which the plant is maintained. Many aspects of these effects, representing a large number of combinations of host and virus, have been investigated and described in detail.[1-5] However, although there is a lot of well-documented information on the effects of virus infection, there is very little on the interaction between the genome of the pathogen or the polypeptides it encodes and the host targets or receptors which initiate the sequence of events which follow. Only recently have appropriate techniques become available to allow such experiments to be done. In the following pages, some of the biochemical and physiological perturbations in virus-infected plants are described, and in Section VIII, an attempt is made to evaluate them in the context of the phenotypic expression of infection. Recent experimental approaches to the determination of the primary events leading to this expression are also described.

# II. NUCLEIC ACIDS AND PROTEINS

## A. Host Nucleic Acids

During a productive viral infection, there is an inevitable superimposition of viral nucleic acid biosynthesis and there can, as a consequence, be dramatic changes to normal host metabolism. However, except in some cases such as plants infected by some tobamoviruses, the amount of pathogen nucleic acid produced is small in relation to that of the host. Even so, during a period of maximum viral RNA biosynthesis, for example, the rate of viral RNA synthesis can exceed that of host species. It may be that, in some cases, the demands of the biosynthesis of viral species compete significantly with the biosynthesis of normal host species for essential precursors. Effects on host nucleic acid biosynthesis, however, are likely to be rather more sophisticated in the vast majority of cases.

## 1. DNA

The relatively few reports on this suggest that virus infection can significantly influence the capacity of a cell for DNA synthesis, provided the cell becomes infected at an appropriate stage in its development. DNA synthesis was markedly reduced in, for example, TobRSV-infected bean root tips[6] and in young tobacco leaves infected with TMV[7] and in BSMV-infected barley.[8] It also appears that in some cases virus infection can lead to modification of host DNA. For example, there is evidence for virus-induced mutation in BSMV-infected maize.[9,10] Although the mechanisms involved remain uncertain, a mobilization of transposons or other regulatory elements has been suggested.

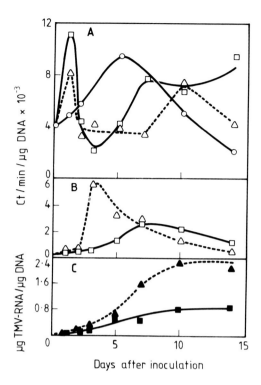

FIGURE 1.   rRNA and TMV RNA synthesis after inoculation of 10-cm long tobacco leaves. (A) $^{32}$P-Incorporation into RNA of healthy (O-O), *vulgare* (△-△), or *flavum* (□-□) TMV-infected leaves. (B) $^{32}$P-Incorporation into *vulgare* (△-△) or flavum (□-□) RNA. (C) Leaf content of *vulgare* (▲-▲) or *flavum* (■-■) RNA. (From Fraser, R. S. S., *J. Gen. Virol.*, 18, 267, 1973. With permission.)

## 2. *RNA*

Studies on the effects of virus infection on host RNA metabolism have concentrated principally on plants infected with RNA viruses; in surprisingly few cases have virus and host RNA species been separated and the net amounts and rates of biosynthesis and degradation of the various host RNA species monitored. One exception is tobacco infected with TMV, which has been studied in detail by Frazer and his colleagues,[2,3,11,12] and less comprehensively by several other groups.[13-15] However, the exceptionally high level of virus RNA biosynthesis does make this a rather atypical situation. The effect of infection by TMV strains causing either relatively mild (*vulgare*) or severe (*flavum*) symptoms on various aspects of host nucleic acid metabolism was investigated. Important observations to emerge from these investigations included the difference in effect on chloroplast and cytoplasmic ribosomal RNA species and the fact that the age of the leaf at the time of infection was of critical importance.

Infection by TMV was observed to have a marked effect on both biosynthesis and degradation of cytoplasmic ribosomal RNA (cyt rRNA). In expanding (10 cm) tobacco leaves, the pattern of biosynthesis of cyt rRNA was as indicated in Figure 1. When inoculated with either the *vulgare* or *flavum* strain of TMV, however, the pattern changed dramatically. There was an initial stimulation in biosynthesis followed by a reduction to a level below that of uninfected tissue within about 2 d after infection. The rate of biosynthesis in infected tissue remained below that of controls until after the period of maximum viral RNA synthesis,

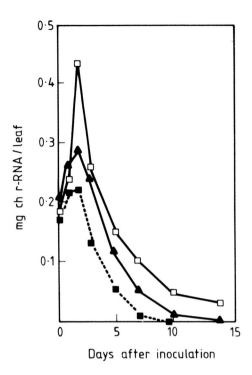

FIGURE 2.   Chloroplast rRNA content of tobacco leaf
(10 cm) after inoculation with *vulgare* (■-■) or *flavum*
(▲-▲) TMV; (□-□), control. (From Fraser, R. S. S.,
*Virology*, 47, 261, 1972. With permission.)

when the rate began to rise again above that in uninfected tissue. When young leaves became
systemically infected with *vulgare*, the rate of cyt rRNA synthesis declined rapidly as soon
as virus replication began, and remained at a low level. Virus infection of older leaves also
had a significant effect on the rate at which cyt rRNA was lost as the tissue aged. Infection
with both strains reduced the rate of degradation, with the *flavum* strain, which produced
the more severe symptoms, having the greater effect. Turnover was also reduced.

The effect on chloroplast ribosomal RNA (ch rRNA) was quite different.[7,11,14,15] Normally,
there is a decrease in the ch rRNA content of tobacco leaf tissue as it ages (Figure 2). The
effect of virus infection was to accelerate this depletion in the ch rRNA content. Infection
by either the *flavum* or *vulgare* strain quickly reduced the rate of ch rRNA biosynthesis,
and the magnitude of the reduction was a reflection of the severity of the symptoms induced
rather than of virus accumulation. The *flavum* strain also appeared to accelerate degradation.
When very small leaves became systemically infected, accumulation of ch rRNA was restricted, but not totally inhibited, by both strains.

A similar picture emerged when biosynthesis of 70S and 80S ribosomes was monitored,
the U1 or *vulgare* strain of TMV inhibiting 70S ribosome accumulation.[15] Additionally,
chloroplasts isolated from TMV-infected tobacco had a reduced capacity for RNA biosynthesis. From the few observations which have been made, it would appear that the effect
on tRNA is similar to that on cyt rRNA. TMV infection reduced the rate of loss from more
mature leaves, *flavum* inducing a greater loss than *vulgare*.[3]

All these observations are not inconsistent with the view that conditions in the infected
leaf are favorable for virus multiplication.[3] The initial stimulation in the biosynthesis of cyt
rRNA provides an adequate supply of ribosomes required for production of a considerable
quantity of capsid protein. During the period of active viral RNA biosynthesis, there is a

reduction in synthesis of cyt rRNA, followed by a return to more normal levels after the period of maximum demand for viral RNA synthesis has passed. Despite the fact that the *vulgare* strain accumulates to a higher level than does the *flavum* strain, the depression in host cyt rRNA synthesis is approximately the same in plants infected with either one, suggesting that other factors in addition to competition for precursors may be important. The reduction in the rate of cyt rRNA degradation presumably maintains the availability of a larger number of 80S ribosomes for biosynthesis of viral proteins. The observation that there is a higher rate of degradation in *vulgare*-infected leaves than in those infected by the *flavum* strain could be a reflection of a lesser demand for RNA precursors in *flavum*-infected tissue. These observations are clearly open to alternative explanations, and there is no information on the mechanisms by which the changes are regulated.

Although none are so comprehensive as those undertaken on TMV-infected tobacco, investigations on hosts which become systemically infected with other viruses have revealed similar trends, particularly with respect to chloroplast ribosomes. BSMV, for example, caused a reduction in ch rRNA content of barley leaves and a significant redirection of RNA synthesis towards virus RNAs.[16,17] In LNYV-infected *Nicotiana glutinosa*, content of 70S ribosomes decreased from about the time symptoms appeared, a trend which was reflected in a decreased rate of biosynthesis of 23S and 16S rRNA and a reduction in leaf growth. Unlike the situation in TMV-infected tobacco, however, the magnitude of the observed decreases did not appear to depend on the severity of the symptoms induced by different strains.[18] There were also decreases in the 80S ribosome concentrations in the yellow areas of TYMV-infected Chinese cabbage[19] and in TSWV-infected 'White Burley' tobacco leaves,[20] while 70S ribosome concentration appeared to be unaffected by virus infection, at least at the times investigated. In CMV-infected 'Xanthi nc' tobacco, however, there was an accelerated reduction in both 70S and 80S ribosome content, particularly in young leaves developing severe mosaic symptoms.[21]

In general, therefore, we see a reduction in the rate of biosynthesis and possibly accelerated degradation of ch rRNA and 70S ribosomes, the magnitude of which is positively correlated with symptom severity. It is not impossible that the consequent reduction in chloroplast protein synthesis is a factor in symptom induction. Cytoplasmic ribosomes are required for virus multiplication and may survive more successfully. When leaves become infected at an early stage of development, then normal development may be retarded.

Virus infection almost invariably leads to a stimulation in ribonuclease activity, although in inoculated leaves it is difficult to establish whether the stimulus is inoculation damage or the stress of virus infection.[22] However, there may also be rises in systemically infected leaves. In TYMV-infected Chinese cabbage, for example, ribonuclease activity was stimulated in areas exhibiting symptoms and supporting virus replication.[23] It may be that the stimulation in activity of one or more ribonucleases is involved in enhanced degradation of ch rRNA, but there is no evidence for it.

## B. Host Proteins

Virus infection, whether producing an inapparent or, at the other extreme, severe necrotic infection, has a profound effect on host protein synthesis. The pattern of protein synthesis in uninfected tissue can of course vary markedly depending on the stage of development. However, at whatever stage it may be when it becomes infected, there is now the superimposition of the requirement for biosynthesis of new proteins encoded by the viral genome. If the virus is one that reaches a relatively modest concentration, then in terms of the total protein biosynthesis of the plant there may need to be only a small redirection to accommodate viral protein synthesis. This may even be true at the cellular level. The concentration of virus proteins other than capsid will be negligible, and even the amount of capsid protein will be small in relation to total protein synthesis. In other cases, however, the situation may be quite different, and TMV again provides one such example. This virus replicates to unusually high levels in infected cells. If the host responds hypersensitively and replication is limited to a relatively few cells, then in terms of the total plant protein the effect is negligible. If, however, the infection is a systemic one then the total amount of viral protein produced can represent a high percentage of the total protein of the plant. In order to achieve this, a major adjustment in the normal priorities of the host is required, at least during the period of active viral replication. Although, again, amounts of virus-encoded, nonstructural proteins required will be negligible, the amounts of capsid protein required are enormous; a decrease in rate of host protein biosynthesis or a stimulation in degradation is required. Fraser and Gerwitz,[24] for example, reported a reduction of 50 to 75% in the incorporation of labeled histidine into host protein in TMV-infected 'Samsun' tobacco.[24] As demands for viral protein synthesis declined, there was a return to normal, indicating that at least in this system, capacity for host protein synthesis was not permanently impaired. This and other observations[25,26] confirmed, however, that at lest a temporary adjustment was required. Total protein synthesis, expressed as the sum of both host and viral contributions and measured by uptake of $^3$H-leucine, was not affected, however. Host protein synthesis may also be reduced in other combinations of virus and host, including LNYV-infected *N. glutinosa*.[27]

Whether these observations reflect successful competition of viral RNA species for 80S ribosomes remains unclear for the moment. At least in TMV-infected tobacco, it seems as though the total host mRNA content, assessed as polyA + RNA, remains relatively constant.[24] There may be, however, substantial qualitative and indeed quantitative changes within this mRNA population. In addition to any redirection of host protein synthesis which may be required to meet the demands of virus replication, there may be very significant qualitative and quantitative changes to the complement of proteins normally present in healthy tissue. They include, for example, ribonucleases and proteases, enzymes involved in phenylpropanoid biosynthesis such as phenylalanine ammonia lyase, oxidative enzymes such as the polyphenoloxidases and peroxidases, and proteins involved in photosynthesis, including ribulose-1,5-bisphosphate carboxylase/oxygenase. In most cases, alterations in the normal levels of these proteins have been assessed by comparing enzymic activity of either crude or semipurified preparations from control and infected tissue. The magnitude of the deviation from normal tends to reflect symptom severity rather than level of virus multiplication; many examples are quoted in the appropriate sections of this chapter. Virus-induced changes in the protein complement of the host have also been assessed by the use of physical separation techniques, particularly one- or two-dimensional gel electrophoresis. Sometimes, in investigations on peroxidase isoenzymes, for example, the two have been combined. Using this sort of approach, both qualitative and quantitative changes have been observed in, for example, barley and wheat infected with WSMV or BSMV,[28] TMV-infected tobacco,[29] and CMV-infected cucumber.[30] Often, however, the difficulties in positively identifying virus-encoded proteins introduced uncertainties into the interpretation of any changes observed. Viroids, which do not encode proteins, can also induce significant changes to the complement of host proteins. In these cases, therefore, interpretation of results is not subject to the same uncertainties, and substantial changes in the host protein profile of tomatoes infected with PSTV, for example, have been reported.[31]

Although both qualitative and quantitative changes in host proteins have been observed on gels, in few instances has the biological significance of these proteins been determined. One of the few exceptions is the group of proteins produced in plants by various stresses, including the stress of virus infection. They are collectively referred to as pathogenesis-related (PR) proteins, or stress proteins, and several are synthesized in comparatively large quantities in plants responding hypersensitively (see Chapter 3). They were first identified by gel electrophoresis of extracts from TMV-infected tobaccos; several have recently been purified[32] and found to have chitinase or glucanase activity.[33,34] An alternative approach to the investigation of perturbations in the regulation of host protein biosynthesis is to isolate mRNAs and clone cDNA copies, an approach which has permitted further characterization of the genes encoding PR proteins and their translation products.

Other features of host metabolism relevant to protein biosynthesis can also be affected by virus infection. In appropriate hosts, there can be, for example, effects on nitrogen fixation. In both CMV-infected pea[35] and in BCMV-infected mung bean,[36] both nitrogenase activity and leghemoglobin concentration were decreased, while nitrate reductase activity and amino acid and nitrate nitrogen concentration were increased. In CMV-infected peas, WCMV-infected clover,[37] and PMV-infected peanuts[38] there also appeared to be fewer nodules than on control plants. The causes of these changes are not known, but have been suggested to be related to decreases in abscisic acid (ABA) concentration.

Finally, amino acid and amide concentrations have been determined in several virus-infected hosts, and in almost every case an increase in their concentration has been observed. Increases in both were observed in, for example, TSWV- and TMV-infected tomato[39] and in LNYV-infected *N. glutinosa*.[18] There was an increase in free amino acids in MDMV-infected corn[40] and in SMV-infected soybean,[41] in proline in RTV-infected rice,[42] and glutamine and asparagine in CMV-infected cowpea.[43] Again, the cause of the enhancement in concentration of these compounds is not clear. It could be the result of a reduction in protein synthesis, for example, or a stimulation of protein degradation. There have also been indications of a virus-induced shift from carbohydrate synthesis (see Section III.C).

## C. Viral Nucleic Acids

The virus genome can be either double- or single-stranded (ss) DNA or RNA, although the vast majority are ss RNA of (+) sense. The ss(+)RNA genomes replicate via partially double-stranded replicative intermediates. Since mRNAs of eukaryotic cells are generally considered to be monocistronic, open-reading frames (ORFs) distal to the 5′-terminus of the ss(+)RNA genomes are normally expressed via the generation in vivo of subgenomic RNAs, which may be produced by selective transcription from (−) strands.[44] The structures and biosynthesis of several viral genomes, both RNA and DNA, of transcripts from DNA genomes and of subgenomic RNAs have been investigated and in many cases their complete nucleotide sequences determined. Details are beyond the scope of this chapter and have been extensively reviewed elsewhere (see also Volume I of this title).

## D. Viral Proteins

With the exception of the capsid, the quantities of other virus-encoded proteins in infected cells are negligible in comparison to those of the host. Their biosynthesis is nevertheless normally essential for virus replication, and considerable effort is being devoted to their identification and to the determination of their function. With few exceptions, which include proteins encoded by TMV,[45] the synthesis of these proteins in vivo has been inferred from the results of in vitro translation of genome-related RNAs or, more recently, by identification of ORFs in the genomic nucleic acid sequences. The reason for this was that during the usual gel electrophoresis separation techniques, with or without radioactive labeling of the proteins, it was virtually impossible to identify virus-encoded proteins in the presence of several hundred host proteins. However, antisera to viral nonstructural proteins, prepared usually from synthetic peptides or from fusion proteins obtained from cloned cDNA sequences, are now becoming available. With such antisera, it is now possible to identify nonstructural proteins produced in vivo in extracts prepared from infected tissue and protoplasts.[46] Using appropriate microscopic techniques it is also becoming possible to determine their location in infected cells.[47,48] Again, details have been reviewed elsewhere. It is, however, relevant to record, if briefly, the functions of some of them.

Although plants seem to possess RNA-dependent RNA polymerase activity,[49,50] it is generally considered that the polymerases involved in viral genome replication are at least in part encoded by the viral genome.[51] It is suggested, for example, that the 183K and 126K proteins encoded by the 5′-proximal cistron of TMV RNA are involved in genomic replication, and replicases or their components encoded by the genomes of CPMV, BMV,[52] and TYMV[53] have been identified. CaMV is considered to replicate via an RNA intermediate, using a replicase encoded by gene V.[54] A replicase is associated with particles of LNYV, which has a ss(−)RNA genome. Proteins may also be required for aphid transmission; the product of gene III of CaMV is considered to be an aphid transmission factor.[55] The RNA genomes of some viruses are translated into polyproteins; their proteolytic processing requires virus-encoded protease activity.[56] Many viruses, including caulimoviruses and potyviruses, form characteristic intracellular inclusion bodies, and virus-encoded proteins may often constitute a significant proportion of their protein complement. P66, encoded by CaMV gene VI, for example, is a major component of CaMV inclusion bodies (and may also have an additional function; see Section VIII.A). There is also evidence to suggest that some viruses, at least, require a genome-encoded protein to permit their movement between cells. Potentiation of movement of one virus by the presence of a second provided circumstantial evidence.[57] There is, however, more direct evidence to suggest that the 30K protein encoded by TMV RNA, produced only transiently during replication, is necessary for cell-to-cell movement[58,59] (see also Chapter 4).

## III. PHOTOSYNTHESIS

When the interaction between virus and host results in the formation of chlorotic or yellow leaf tissue, then there is almost inevitably an associated effect on many aspects of chloroplast ultrastructure and function, the magnitude of which usually correlates with the degree of symptom severity. Photosynthesis is affected in various ways, and over many years there have been several descriptive studies on changes to one or more features of the complex photosynthetic process in several virus/host combinations. Many of these are described in this section.

### A. Rate of Photosynthesis

Perhaps the most commonly observed perturbation is a decrease in net photosynthetic rate of affected tissue when compared on the basis of, for example, leaf area, to that of control tissue. Decreases have been observed in many hosts, including tobacco infected with TobRSV,[60] TMV,[61] or TEV,[62,63] tomato infected with TAV,[64] BCTV,[65] or ToYMV,[66] wheat and barley infected with BYDV,[67-69] BYV-infected beet,[70] MDMV-infected maize,[71] and peanut infected with PGMV.[72] Commonly, effects are maximum at the time of maximum symptom severity. In PGMV-infected peanut, for example, the decrease reached a maximum in the mid period of infection, but then activity recovered as the affected leaf recovered and symptoms declined. It may be that at least partial recovery could account for the surprisingly unchanged rate of photosynthesis observed in a few cases.[73,74]

The biochemical and physiological changes induced by virus infection which in turn result in these observed alterations in photosynthetic rate are less easy to define. There may be important changes in the ability of plants to absorb $CO_2$, in the light energy photosystem, and in the concentration and activity of the enzymes involved in $CO_2$ fixation and subsequent interconversion. Any or all of these may also be related to the alterations which occur in the development and structure of chloroplasts in affected tissue.

Ultrastructural aberrations in virus-infected leaf tissue can range from the drastic to the insignificant, depending not only on host and virus, but also, critically, on the developmental stage at which the leaf becomes infected. Systemic infection of young developing leaves often leads to very dramatic deficiencies in both form and function. In leaves which are approaching full development at the time of infection, however, ultrastructural changes may, in the extreme, be difficult to detect. These changes have been fully documented elsewhere[2,4] and will not be considered further here. In addition to changes in morphology, there have also been several reports of changes in chloroplast number. There were, for example, decreases in chloroplast number per unit area in TEV-infected tobacco,[75] SqMV-infected squash,[74] and in MDMV-infected corn.[71] This could make a significant contribution to change in photosynthetic rate, but is unlikely to be either a complete or universal explanation.

The few experiments in which $CO_2$ diffusion has been studied do not allow any generalizations to be made. Although an increased resistance to diffusion was observed in BYV-infected beet,[76] possibly as a result of reduced stomatal aperture, diffusion did not appear to be affected in TAV-infected tomato.[64]

Since ribulose-1,5-bisphosphate carboxylase/oxygenase is the key enzyme involved in $CO_2$ fixation in $C_3$ plants, changes in either its activity or concentration could have important consequences. Indeed, a reduction in its rate of synthesis has been reported to occur in many hosts, including tobacco infected with TMV,[77] Chinese cabbage infected with TYMV, and beet infected with BYV.[76] A reduced activity of this enzyme has also been recorded.[78] Phosphoenol pyruvate carboxylase (PEP-carboxylase) can also fix $CO_2$, and it might be expected that in certain circumstances a change in its concentration or activity could either enhance or compensate for effects due to changes in ribulose-1,5-bisphosphate carboxylase/oxygenase. PEP-carboxylase activity increased in TMV-infected tobacco,[78] in TYMV-infected Chinese cabbage,[79] and in CMV-infected cowpea,[80] although it apparently did not change in BYV-infected beet[76] and decreased in TSWV-infected tobacco.[81] A stimulation in activity of alanine-aspartate transferase was also reported to occur in TSWV-infected Chinese cabbage.[79]

When $CO_2$ concentration is low and ribulose-1,5-bisphosphate carboxylase/oxygenase is acting as an oxygenase, then the resulting photorespiration might also effectively reduce net photosynthesis. However, in BYV-infected beet, photorespiration was also reduced,[70,76] although it was increased in SCMV-infected sugarcane.[82]

## B. Chlorophyll and Photochemical Reactions

Chlorophylls play a central role in the capture of light energy, resulting ultimately in the production of ATP and NADPH, which in turn are utilized in the carbon reduction reactions. If the chlorophyll content or the integrity of the complexes in which it functions is disturbed, then alterations in photosynthetic capacity are a necessary consequence.

It is perhaps not surprising, therefore, that the observed reduction in photosynthetic rates have usually been shown to be accompanied by decreases in chlorophyll content of affected leaf tissue. This is particularly true of leaves which have become systemically infected when young, develop yellow mosaic, and exhibit chloroplast abnormalities which involve aberrations in the thylakoid systems. Reduction in chlorophyll has been observed in tobacco infected with TMV,[83] CMV,[84] TSWV,[81] or TobRSV,[60] in wheat and barley infected with BYDV[67] or BSMV,[8] in tomato infected with TAV[64] or ToYMV,[66] and in corn infected with MDMV.[71,85] Infection of soybean with SMV, bean with BCMV, and lettuce with LMV also caused significant reduction in chlorophylls a and b.[86] In PGMV-infected peanut, both total chlorophyll concentration[87] and chlorophyll a/b ratio were reduced, with maximum reduction coinciding with maximum symptom severity and maximum reduction in photosynthetic rate.[72,88,89] The authors concluded that the reduction in chlorophyll content, although not great, at least contributed to the reduction in photosynthesis. There was in addition a reduced activity of photosystem II, associated not only with a decrease in chlorophyll a and chlorophyll-protein complexes, but also with a substantially reduced plastoquinone concentration. Quantitative changes in other thylakoid-associated proteins, particularly of photosystem II, were also observed,[89] although their significance is yet to be determined.

In *Tolmiea menziesii* infected with CMV and TBSV, the ratio of chlorophyll a/b was increased,[90] although total chlorophyll content was decreased. It is of interest that in TEV-infected tobacco, where the chloroplast number was reported to decrease, the chlorophyll content of the chloroplasts remained unaffected.[75]

Decrease in chlorophyll concentration is often accompanied by a stimulation in chlorophyllase activity, and since chlorophyllase can catalyze the dephytylation of chlorophyll, it has been suggested that it is an enhanced breakdown of chlorophyll which is a major contributor to its decrease in concentration. In tobacco infected with TMV, for example, stimulation in chlorophyllase was correlated with reduction in chlorophyll in leaves infected with strains causing symptoms of differing severity.[83] However, activity was reported to decrease in tobacco infected with CMV[91] and did not change in infected beet.[92] In TYMV-infected Chinese cabbage, a reduction in biosynthesis was suggested as the cause of reduced chlorophyll content. In the majority of cases, however, the relative contributions of a reduced rate of chlorophyll biosynthesis and an enhanced rate of degradation, and the role, if any, of chlorophyllase in either have not been evaluated. Decrease in chlorophyll content is also often associated with a decrease in carotenoids.[86,93]

Substantially reduced activity of components of the light reactions has also been revealed by changes in Hill reaction activity and/or photophosphorylation. Decreases have been reported to occur, for example, in chloroplasts from BYV-infected beet[94] and from TMV-infected tobacco,[95] but not from TEV-infected tobacco.[96] In TYMV-infected Chinese cabbage, a decrease late in infection was preceded by an earlier stimulation in both Hill activity and in photophosphorylation.[97] Noncyclic photophosphorylation was also decreased in PGMV-infected peanut.

## C. Products of Photosynthesis

The photosynthetic changes outlined in the previous sections may also be accompanied by related metabolic changes in the utilization of the products. There is often, for example, a decrease in the sugar content of leaves, and a concomitant shift towards the production of amino acids and organic acids such as citrate and malate. There was reported to be a preferential synthesis of amino acids and organic acids at the expense of sugars in TYMV-infected Chinese cabbage[79] and in SqMV-infected squash.[74] However, although there can be a reduction in biosynthesis of some soluble carbohydrates,[98] in other situations carbohydrates can accumulate. There were, for example, significant increases in carbohydrate content of PLRV-infected potato, BYDV-infected barley and wheat,[68,69] and in tomato infected with ToYMV[6] or TBSV.[99] It has been suggested that in these cases there may be an impairment of translocation (see also Section V.B).

Accumulation of starch is also a common feature of virus-infected tissue, chloroplasts often containing large numbers of enlarged granules. There is evidence to suggest that both synthesis and degradation could be affected; in TMV-infected tobacco, starch accumulation was higher than normal after a period of darkness, but lower in light conditions. However, amylase activity has been reported to increase in TMV-infected tobacco in the absence of light, but in other virus/host combinations decreases have been observed. It is clear that concentration of carbohydrate, in common with that of many other plant constituents referred to here, is dependent on many factors including rate of biosynthesis, utilization, and transport, and can vary substantially with environmental conditions, symptom severity, and time after infection.

It is apparent therefore that virus infection can have a profound effect on the photosynthetic process and related metabolic pathways. There is usually a decrease in photosynthetic rate, often of a magnitude which cannot be simply explained by reduction in chlorophyll concentration. It is often, but not necessarily, correlated with symptom severity. However, with very few exceptions, comprehensive studies have not been done and it is not yet possible to define the relative importance (or indeed relevance) of many of the perturbations which

**Table 1**
**STIMULATION OF RESPIRATION**

| Before symptom appearance | | | Coincident with symptom appearance | | |
|---|---|---|---|---|---|
| **Host** | **Virus** | **Ref.** | **Host** | **Virus** | **Ref.** |
| | | | **A. Systemic** | | |
| *Hordeum vulgare* | BMV | 100 | *Hordeum vulgare* | BYDV | 67, 102 |
| *Phaseolus vulgaris* cv. The Prince[a] | TMV | 101 | *Cucumis sativus* | CMV | 103 |
| *Vigna sinensis*[a] | TMV | 101 | *Nicotiana tabacum* | PVX,TEV | 62, 73, 104 |
| | | | *Phaseolus vulgaris* cv. Bountiful | AlMV,SBMV | 105 |
| | | | *Zea mays* | MDMV | 106 |
| | | | *Lycopersicon esculentum*[a] | ToYMV | |
| | | | **B. Hypersensitive** | | |
| *Gomphrena globosa* | PVX | 107 | *Gomphrena globosa* | PVX | 110 |
| *Phaseolus vulgaris* | TMV | 108 | *Nicotiana glutinosa* | TMV | 73, 111 |
| *Nicotiana tabacum* cv. Xanthi | TMV | 109 | *Nicotiana sylvestris* | TMV | 112 |
| *Phaseolus vulgaris* cv. The Prince | TNV | 101 | *Phaseolus vulgaris* GN59 | AlMV,SBMV | 105 |
| *Vigna sinensis* | TNV | 101 | *Datura stramonium* | TMV | 110 |
| | | | *Vicia faba* | CMV | 110 |

[a]    Timing uncertain.

have been observed. Still less is it possible to describe how these changes are initiated. Further consideration is given to this question in Section VIII.A.

## IV. RESPIRATION

The effects of virus infection on various aspects of respiration, as with effects on photosynthesis, have been investigated in plants of several species infected with a selection of different viruses, although here, perhaps tobacco has figured even more prominently. In most cases, there appears to be at least a temporary stimulation, but as in other cases the timing of the estimates may well be critically important. Respiration in hosts responding hypersensitively, or permitting systemic invasion, has been assessed; results of some of these experiments are described below, with Table 1 providing a more comprehensive summary.

### A. Respiration Rate
*1. Systemic Hosts*
Systemic infection has usually been found to stimulate, often only marginally, the rate of respiration when measured either as oxygen consumption or carbon dioxide evolution (Table 1). That the relationship between respiration in infected and healthy tissue can depend on timing is illustrated by the work of Takahashi and Hirai.[113] Epidermis was stripped at various times from TMV-inoculated or control tobacco leaves and respiration rate estimated. Respiration, expressed here on a fresh weight basis, was either greater, the same as, or less than that of healthy tissue at various times during the course of the infection. It may be that in some cases apparent stimulation is greater when expressed on a fresh weight rather than a dry weight basis, due to the greater dry weight/fresh weight ratio of infected tissue;[67] this would clearly not be unique to measurements of respiration.

Stimulation of respiration has variously been detected either slightly before[100,101] or, more commonly, coincident with the appearance of visible symptoms. Inevitably, however, the timing of symptom appearance is subject to uncertainty. In a few cases, including tobacco infected with TMV,[114] respiration was reportedly reduced in infected tissue. When results were expressed in a manner which reflected respiration rate per cell, there was also a decrease in tomato infected with the aucuba strain of TMV, a combination which resulted in severe stunting.

Some attempts have been made to correlate changes in respiratory activity, both with extent of virus replication and with symptom severity. Unfortunately, there have been very few investigations in which degree of virus accumulation, symptom severity, and change in respiration rate have been estimated in the same series of experiments. Dwurazna and Weintraub[104] found stimulation of respiration in tobacco infected with several strains of PVX to correlate positively with symptom severity, rather than virus multiplication. Owen,[62,73] using different viruses, found the degree of TMV multiplication in tobacco to be greater than that of either PVX or TEV, although both symptom severity and stimulation of respiratory activity were less. In CMV-infected cucumbers of differing susceptibility, stimulation in respiratory activity was also correlated with symptom severity.[103]

Since TYMV is able to replicate in the absence of photosynthesis, respiration would appear to be able to supply the energy required for virus replication.[115] However, in TMV-infected tobacco protoplasts, replication was reduced when photosynthesis was inhibited.[116] The evidence, although limited, suggests that respiratory changes in plants are a response to virus-induced changes in the host, rather than a response to the energy demands of virus replication.

### 2. Hypersensitive Hosts

Here again, there is usually a virus-induced stimulation of respiratory activity, often of a magnitude greater than that observed in systemic hosts (Table 1). Stimulation has also been reported to occur either in advance of or simultaneously with appearance of visible lesions, although, as with systemic symptoms, precise timing is difficult.

The early enhancement tends to be followed by a decrease as mitochondria in infected tissue degenerate. It may be that the stimulation observed in the early stages of infection can be at least partly accounted for by an increase in the number of mitochondria. However, observations on this are conflicting. Weintraub et al.[117] reported an enhancement in mitochondrial number in the early stages of infection of *N. glutinosa* with TMV, while numbers declined later in infection as respiratory activity decreased. However, although mitochondrial proliferation was observed in *Gomphrena globosa*,[118] others failed to find evidence for it in *N. glutinosa*[119] or tobacco.[120] The significance of changes in mitochondrial number is therefore difficult to assess.

It has also been suggested[117] that stimulation of oxidase activity, particularly that of polyphenoloxidase, which accompanies lesion formation, contributes to the observed stimulation in oxygen uptake. The contribution from this source would be small in the early stages of lesion development, at the time of maximum stimulation in respiratory activity, but greater as symptoms develop and polyphenoloxidase activity increases.

### B. Respiratory Pathways and Enzymes

In virus/host combinations in which the tissue becomes necrotic, there is some evidence that a stimulation of the pentose phosphate pathway contributes to the enhancement of respiration. There was a stimulation of activity of key enzymes of this pathway, glucose-6-phosphate dehydrogenase and 6-phosphogluconate dehydrogenase, in both PVX-[120] and TMV-infected[121] tobacco. The observed decreases in $C_6/C_1$ ratios in these combinations and in SBMV-infected bean provided further confirmatory evidence, although decreases were not found in AlMV-infected bean. There is little evidence for a stimulation of this pathway in the absence of necrosis. In TMV-infected tobacco, for example, neither $C_6/C_1$ ratios nor activities of glucose-6-phosphate dehydrogenase nor 6-phosphogluconate dehydrogenase were affected.[122,123]

# V. VASCULAR TRANSPORT

## A. Water Relations: Uptake, Translocation, and Transpiration

Virus infection frequently results in a reduced water content of leaf tissue. This is true in TobRSV-infected cowpea[124] and in TMV-infected tobacco,[125] for example. In tomato infected with TAV[126] or TMV,[127] however, infected leaves were reported to be more turgid than uninfected control ones.[128] A reduction in water content could be due to a reduction in the ability of the plant to absorb water through the root system, an impairment of translocation, or a stimulation in transpiration. In general, however, studies in this area have been fragmentary and incomplete, so that in the majority of cases it is not possible to identify the site of the origin of the observed changes.

In a few instances, evidence points towards an impairment of root function. There was, for example, a marked change in permeability of root cells of tobasco pepper early in infection with TEV,[129] a change which was associated with degeneration of cortical cells and plastids. As infection progressed, plants became severely wilted, unusual for a viral infection. Necrosis of phloem and cambial cells became apparent,[130] although, since turgor could be restored by cutting stems and submerging the cut ends in water,[131] it was probably not the primary cause of the wilt. There are also reports of increased root cell permeability in cases where wilting does not occur, by far the most common situation. These include peas infected with BYMV or BCMV[132] and squash hypocotyls infected with SqMV.[133] It may be that, in some cases, virus-induced alterations to vascular tissue can affect translocation from roots. There is, however, little information on their relevance to water transport, although there is some, albeit contradictory, evidence to suggest that they may be involved in restriction of carbohydrate translocation.

Finally, there is also evidence that transpiration can be affected in virus-infected leaves; it is usually, but not always, reduced. This might suggest that alterations in transpiration rate, often accompanied by reduction in stomatal aperture, are a consequence of reduced turgor, rather than vice versa. In BYDV-infected barley, the degree of reduction in transpiration appears to be correlated with symptom severity, susceptible plants with pronounced symptoms exhibiting a greater reduction than resistant ones.[134] Transpiration in areas of the leaf with severe symptoms was greater than in areas less severely affected. Transpiration rate was reduced both in BCTV-infected sugar beet and in MDMV-infected maize. In the latter case, reduction was suggested to be due to reduced stomatal aperture caused, in turn, by a reduction in $K^+$ uptake by guard cells.[135] In tomato leaves systemically infected with TAV, where turgor was not reduced by virus infection, there were changes neither in transpiration rate nor stomatal opening.[64]

## B. Carbohydrate Transport

In some instances, including BCMV-infected sunn hemp,[136] SBMV-infected cowpea, and TYMV-infected Chinese cabbage, for example, virus infection is associated with a reduction in the soluble carbohydrate content of leaf tissue. Conversely, in other instances there are significant increases in the levels of carbohydrates, particularly reducing sugars, despite the fact that photosynthetic rate may be reduced. Included in this group are BYV-infected beet,[76,137] BCTV-infected tomato[65,138] and beet, PLRV-infected potato[139] and *Physalis floridana*,[140] and barley infected with BYDV.[69] Not unexpectedly, concentrations change during the course of infection and vary between leaves. In tomato infected with ToYMV,[66] for example, concentrations of starch and sugars were higher in lower leaves of infected plants than in corresponding healthy ones at both early and later stages of infection, while in upper leaves the reverse was true when infection was well advanced.

The cause of the observed enhancement of carbohydrate levels is more difficult to determine, depending as it does on relations between biosynthesis, utilization, and translocation. Although it remains a possibility in some instances, the evidence for impairment of translocation playing a significant role is generally not good. In most cases, ability to translocate carbohydrates is unaffected. BYV did not influence the movement of carbohydrate from infected tissue and although phloem necrosis was observed in potato infected with PLRV, which multiplies in phloem tissue, it apparently occurred after increases in carbohydrate concentration. There was no evidence of phloem damage at any stage which could restrict carbohydrate translocation in barley infected with BYDV.

It is not inconceivable, however, that other viruses which replicate in phloem tissue may cause sufficient damage to interfere with transport. The geminiviruses ToYMV,[141] EMV,[142] and BGMV[143] all cause significant cytological changes in phloem tissue. BCTV also caused phloem degeneration.[144] Although necrosis was not considered to be an important contributory factor to impairment of carbohydrate translocation, it may be that physiological disturbances associated with the degeneration of phloem tissue are sufficiently marked to have an effect. Lastly, it has been suggested that in plant tissue infected with PLRV, BYV, or BCTV, for example, the conversion of sucrose to hexoses, which are less readily translocated, contributes to the enhancement of the leaf carbohydrate content.[138] Again, however, much of the evidence remains circumstantial.

## VI. PHENYLPROPANOID PATHWAY

### A. Phenylpropanoid Metabolism
*1. Systemic Hosts*

During the course of productive, systemic infection, where virus multiplication is unrestricted, there is little if any perturbation in the normal activity of enzymes of the phenylpropanoid pathway. In TMV-infected 'Samsun' and in 'Xanthi nc' and 'Samsun NN' tobacco maintained at a temperature above which the hypersensitive response is not induced, activity of phenylalanine ammonia lyase (PAL), the first key enzyme of the pathway, was marginally less than in controls.[145-148] Similarly, there was only a slight, if detectable, change in the concentrations of the diverse polyphenolic products of the pathway. Concentration of chlorogenic acid, for example, began to rise above control levels only in the later stages of infection.[149]

*2. Hypersensitive Hosts*

In direct contrast to the situation outlined above, there are very significant changes in the activity of enzymes of the phenylpropanoid pathway and in the concentrations of the products of this pathway in plants reacting hypersensitively. Because of the invariable association of these alterations with hypersensitivity and their potential role in either lesion formation or restriction of virus transport or both, they have generated considerable interest, and many virus/host combinations have been studied. The early literature on these and related changes has recently been comprehensively reviewed and tabulated.[2]

In every case, necrosis of infected tissue is accompanied by a stimulation in PAL activity. Examples include the combination which has perhaps received most attention, N gene-containing tobacco infected with TMV[150-153] and bean infected with AlMV.[154] The degree of stimulation is positively correlated with lesion number and size, with the extent of necrosis rather than with virus accumulation. In those experiments employing adequate precision, rise in PAL activity was detectable just before the appearance of visible lesions, reached a

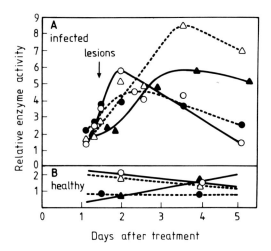

FIGURE 3. Changes in enzyme activities following inoculation of tobacco with TMV (A) compared to controls (B). ○-○, PAL; ●-●, CAH; △-△, OMT; ▲-▲, peroxidase. (Reprinted with permission from Legrand, M., Fritig, B., and Hirth, L., *Phytochemistry*, 15, 1353, 1976. Copyright 1976 by Pergamon Press, Oxford, U.K.)

maximum level as lesions expanded, and subsequently declined (Figure 3). In a continuation of their previous investigations on TMV-infected 'Samsun NN' tobacco, Fritig and colleagues[155,156] have also established that the activities of cinnamic acid 4-hydroxylase and the *S*-adenosyl-L-methionine and *O*-methyl transferase enzymes involved in the formation of lignin precursors are also enhanced (Figure 3). The stimulation of activity of these enzymes and also of PAL was accounted for by *de novo* synthesis, and occurred principally in cells immediately surrounding the lesions, spreading radially with lesion expansion. Although there may be little virus present in such cells, estimates have suggested that the stimulation in PAL activity in individual cells could be as high as 100- to 200-fold. Since these enzymes are involved in the biosynthesis of lignin precursors, and lignin deposition has been proposed as a contributory factor to the restriction of virus translocation, the stimulation of their activity could be critical to the development of hypersensitive resistance. This, and the question of their involvement in necrosis, are discussed more comprehensively in Section VIII.B.

Since there are substantial changes in the activities of enzymes involved in their biosynthesis, it might be expected that there would be changes in the concentrations of at least some of the many polyphenols present in plant tissue.[149,157] Indeed, there have been many reports of the accumulation of scopoletin either within or in cells immediately surrounding necrotic tissue. In TMV-infected tobacco, Fritig and colleagues[150] observed a stimulation in the rate of biosynthesis of both chlorogenic acid and scopoletin in the early stages of lesion development, with the concentration of scopoletin and its glycoside, scopolin, continuing to increase as lesions matured. However, concentrations of cinnamic, 4-caffeoylquinic, and 5-caffeoylquinic acids declined as the rate of biosynthesis increased, suggesting their utilization in subsequent metabolic processes. Although there is evidence for stimulation in concentration of some polyphenols in advance of necrosis, their role is uncertain (see also Section VIII).

## B. Peroxidases and Polyphenoloxidases

Virus infection appears to stimulate peroxidase activity in all hosts in which necrotic or chlorotic symptoms are induced, with a degree of stimulation which correlates with symptom severity. Stimulation is greatest in hosts reacting hypersensitively with increases observed in, for example, TMV-infected tobaccos[145,158,159] or AlMV-infected bean.[154] Stimulation has also been recorded in hosts permitting systemic infection, including CMV-infected cucumber[160] and tobacco[161] and TMV-infected tobaccos.[159,162] The enhancement of activity in uninoculated leaves of hypersensitive hosts has been suggested to contribute to the acquired resistance of these leaves.[163,164] There is, however, evidence to the contrary.

There are many isoenzymes exhibiting peroxidase activity which can be resolved either by gel electrophoresis or isoelectric focusing. A stimulation in activity arising from virus infection could conceivably reflect a perturbation in the regulation of the biosynthesis of an enzyme already present at a low level in the uninfected plant. Alternatively, it could represent biosynthesis of a "new" isoenzyme, the biosynthesis of which does not occur in healthy plants. Evidence on this is conflicting, and conclusions could depend on the sensitivity of the assay. In hosts reacting hypersensitively, including SBMV-infected bean and TMV-infected tobaccos, the presence of "new" isoenzymes in infected tissue has been reported by some,[161,165] but not others.[166] In systemically infected tissue, including tobacco[161] and bean[167] infected with TMV and cucumber infected with CMV,[168] only quantitative rather than qualitative changes have been reported.

Polyphenoloxidase activity seems, in most cases, to behave in a manner similar to that just described for the peroxidases. A significant stimulation in activity almost invariably accompanies the expression of hypersensitivity, with increases to occur in many virus/host combinations, including TMV-infected tobaccos,[158,159,162,169] TNV-infected cowpea,[170,171] and CCMV-infected soybean.[172] Stimulation, if it occurs at all, is much less in systemic hosts[159,162,173] and did not occur in AlMV-infected bean. Enhancement of activity of both peroxidases and polyphenoloxidases usually occurs shortly after the appearance of visible lesions.

## VII. GROWTH REGULATORS

Virus infection can lead to the formation of a diverse range of overt symptoms, depending on the combination of virus and host. Many, particularly those involving perturbation of growth habit, epinasty, mottling, and leaf distortion, can be partly if not completely simulated by altering the hormone balance of the plant. It is not surprising, therefore, that virus infection can lead to substantial changes in the concentrations of growth substances. The many investigations in these changes have been broadly of two types. In the first, endogenous concentrations of various growth hormones have been determined in both virus-containing and virus-free tissue. In the absence of more precise techniques, these investigations have usually employed biological assays, procedures which are inevitably imprecise and subject to interference. Only recently have more reliable, nonbiological assays been developed and used, particularly by Fraser and colleagues in their work on abscisic acid and cytokinins.[174] In the second, growth hormones have been added exogenously, before or after virus inoculation and before or after symptom development, and effects on either host response or virus multiplication (or both) have been assessed. The results from this sort of experiment, however, have often been difficult, if not impossible to interpret. Amounts used have sometimes been of a level which would produce nonphysiological, or even cytotoxic, concentrations in the treated tissue. The outcome may also be critically dependent upon other factors, including time of application and whether or not the treated tissue is detached from the plant. Some results from both types of experiment, which have been documented more comprehensively elsewhere,[2,174] are described below.

### A. Cytokinins
Although decreases in cytokinin content of cowpea and *N. glutinosa* tissue have been

reported to result from TobRSV infection,[124,175] the more usual observation is of an increase in concentration in virus-infected plants. For example, cytokinin concentration was stimulated in TMV-infected 'Samsun' tobacco leaves,[176] in BGMV-infected bean,[177] in TNV-infected *Chenopodium amaranticolor*,[178] and in both inoculated and uninoculated leaves of 'Xanthi nc' tobacco infected with TMV.[179] In hosts which permit systemic virus infection, preinoculation treatment with cytokinin often appears to depress the level of virus accumulation. TMV replication in tobacco, for example, was reduced by treatment with carbendazim (methyl benzimidazol-2-yl-carbamate)[180] or kinetin.[181] Postinoculation treatments often have the opposite effect, with a stimulation in virus accumulation reported in virus-infected *N. rustica* and tobacco.[182] There are discrepancies, however, with Reunov et al.[183] reporting a decreased accumulation of TMV in kinetin-treated tobacco.

Since a stimulation in cytokinin concentration is often positively correlated both with restriction of virus accumulation to areas in and around local lesions, and with the induction of acquired resistance, its involvement in the development of these responses has been considered to be a possibility. Generally, enhancement of cytokinin concentration by treatment of tissue prior to virus infection results in the apparent reduction in lesion size or number (or both). For example, treatment of tobaccos,[184] bean,[185] or *Datura stramonium* before TMV inoculation, or petunia or *N. rustica* prior to infection with TSWV[186] appeared to reduce the ensuing lesion number; where measured, sizes were also reduced. Postinoculation treatment can also produce similar results, apparently reducing lesion number when added up to 72 h after infection of petunia and *N. rustica* by TSWV. Postinoculation treatment of AlMV-infected bean with kinetin also reduced lesion size, an effect which could be at least partially reduced by including $Ca^{++}$ in the treatment.[187] Evidence suggests that the apparent effect of cytokinin on the numbers of lesions which develop is to influence lesion development, rather than the initial establishment of infection or virus replication. Microscopic examination of TMV-inoculated tobacco leaves with apparently reduced lesion numbers[188] and enhanced cytokinin content revealed the presence of "microlesions", which had failed to develop. Virus accumulation in lesions which did develop was greater than in lesions in untreated tissue. Cytokinin treatment also reduced $HgCl_2$-induced necrosis in Xanthi nc tobacco.[179] Finally, cytokinin treatment stimulated virus accumulation in TNV-infected *C. amaranticolor*, a result which led the authors to conclude that cytokinins were not directly involved in systemic acquired resistance.[178] It is of interest, however, that the dark green areas of tobacco leaves, which are systemically infected with TMV and which are resistant to virus challenge, also contain enhanced cytokinin levels.[176] It is not clear whether or not the two are related.

## B. Auxins

Reported changes in auxin concentration have often been contradictory, and in many cases are as difficult to interpret as the results of experiments on cytokinins. In host supporting systemic infection, the general observation is of a decrease in auxin concentration. Decreases were reported in, for example, potato infected with either PVX, PVX+PVY or PLRV,[189] BCTV-infected beet, bean, and tomato,[190] TMV-infected tobacco,[191] and TSWV-infected tomato.[192] However, BYDV appeared to have little effect on auxin concentration in barley,[193] while enhancement of auxin concentration was reported in TSWV-infected tobacco and tomato[194] and in virus-infected grapevine.[195] When there is a decrease it is not necessarily correlated with symptom severity.

A common feature of infected plants with auxin deficiency is stunting. Since auxins are required for growth, provision of exogenous auxin, in attempts to redress the deficit caused by virus infection, might be expected to at least partially restore the normal growth habit. This does not happen. Stunting was even enhanced by treatment of SBMV-infected beans with several auxins,[196] including β-naphthoxyacetic acid and *p*-chlorophenoxyacetic acid.

Their application did, however, restore apical dominance to SBMV-infected plants which, as a result of virus infection, appeared to have lost it and were developing axillary shoots. A similar effect was observed when peanut plants with "witch's broom" were treated with indolylacetic acid (IAA).[197] Generally, treatment with exogenous auxin has been found to decrease virus accumulation; TMV replication in tobacco, for example, is reduced by IAA application.[198] There are, of course, exceptions.[199]

In contrast to the commonly reported reduction in auxin concentration which accompanies systemic infection in plants responding hypersensitively, an increase is usually observed. For example, van Loon and Berbée[200] reported substantial increases in IAA concentration in TMV-inoculated 'Samsun NN' tobacco leaves, with smaller increases occurring in non-inoculated leaves. Treatment of young leaves with exogenous IAA either before or after infection decreased the final lesion size. Lesion size was greater, however, on leaves with a higher endogenous IAA concentration (young ones). There was no correlation, therefore, between IAA concentration and lesion development. IAA also reduced the size of AlMV lesions in bean.[187.]

The observed changes in auxin concentration may be related to a virus-induced change in efficiency of transport,[201] rate of biosynthesis, or metabolism. It has been suggested, for example, that ethylene, the biosynthesis of which is stimulated in plants reacting hypersensitively (see below), can in turn stimulate auxin biosynthesis. The activities of IAA oxidases and peroxidases, which have IAA oxidase activity, are also stimulated, to a degree which generally correlates with symptom severity. They might be expected to act to reduce IAA concentration. Stunting and reduced auxin levels are often correlated with stimulated IAA/oxidase and peroxidase activity.

### C. Abscisic Acid

Abscisic acid (ABA) is an inhibitor of growth, and in the relatively few virus/host combinations which have received attention its concentration seems usually to be stimulated. Reports of changes in ABA concentration in CMV-infected cucumber are conflicting,[202,203] and although Rajagopal[191] reported a reduction of ABA concentration in the early stages of TMV infection of tobacco, later work by Fraser,[3] Whenham and Fraser,[204] and Whenham et al.[205] demonstrated a substantial increase in tobacco systemically infected with TMV. There was also a substantial increase in the concentration of phaseic acid, an oxidation product of ABA, which suggested a stimulation of flux through the ABA pathway. In addition, the reduction in growth of these plants could be simulated by application of exogenous ABA to uninfected plants, at a level which led to concentration within the plant reaching those produced by virus infection. This would suggest that ABA at least makes a contribution to the reduction in growth in tobaccos, both ABA and virus infection reducing the rate of cell division. In tomato, however, which has a much higher endogenous ABA concentration than tobaccos, the situation is rather different. Infection by TMV had little effect on concentration of ABA or phaseic acid, and although plants became stunted, the reduction in growth could not be simulated in uninfected plants by application of ABA. When tobacco was infected with several TMV strains producing symptoms of differing severity and multiplying to different levels, there was a correlation between ABA stimulation and symptom severity. However, most of this ABA was not in chloroplasts, and it is not clear whether or not it has any direct relevance to the development of the yellow symptoms. Addition of exogenous ABA to infected plants can also stimulate both viral RNA synthesis[204] and virus accumulation[206] in TMV-infected tobacco.

ABA concentration may also be enhanced in plants reacting hypersensitively. In White Burley tobacco infected with the *flavum* strain of TMV, for example, the stimulation was much greater than in plants infected with the *vulgare* strain, which spread systemically. Since ABA concentration is normally enhanced by water stress, this stimulation could be a result of necrosis in infected leaves. As in systemically infected plants, the degree of stimulation was high enough to account for the observed reduction in growth.[2]

A reduction in endogenous ABA level in virus-infected stem cuttings of *Euphorbia pulcherrima* has been reported (Section VII.E).

Application of exogenous ABA to tobacco tissue caused a reduction in both size and number of TMV-induced lesions; the basis for this observation remains to be determined.

## D. Ethylene

The concentrations of ethylene seem either to remain unchanged or to increase on virus infection, the degree of stimulation being positively correlated with the severity of the symptoms produced. Although ethylene concentration did not appear to be enhanced in tobacco systemically infected with TMV[207-209] or in BYMV-infected *Tetragona expansa*, in which chlorotic lesions developed,[210] there was an enhancement in CMV-infected cucumber seedlings, which accompanied stunting, epinasty, and chlorosis. Enhancement of ethylene concentration was observed before the appearance of visible chlorosis, and removal from infected seedlings depressed its severity. Similarly, removal of ethylene also reduced epinasty, while application promoted it. It has been argued, therefore, that ethylene plays an important part in the induction of these symptoms and is also involved in the suppression of seedling growth.[211,212] In some cases, application of exogenous ethylene can influence the level of virus accumulation.[213]

Expression of hypersensitivity seems to be invariably accompanied by a stimulation in ethylene production. There are, for example, many reports of enhanced ethylene concentration in various hypersensitively reacting tobaccos infected with TMV.[208,209,214-216]

In 'Samsun NN' tobacco, stimulation in synthesis of both ethylene and its precursor, 1-amino-cyclopropan-1-carboxylic acid, began prior to the appearance of visible lesions (Figure 4). Enhancement of ethylene production, however, has usually been observed to occur coincidentally with lesion appearance. The degree of stimulation in TMV-infected 'Samsun NN' tobacco was correlated with the extent of necrosis, and its relevance or nonrelevance to the development of this response is discussed in Section VIII.

## E. Gibberellins

There is often, but not always, a change in gibberellin concentration in virus-infected hosts which become stunted; it is usually a decrease. There is, for example, a decrease in gibberellin content of cucumber seedlings infected with CMV,[202,213,217] which correlates with reduction in stem elongation, and in BYDV-infected barley. Gibberellins in CMV-infected cucumber may also differ qualitatively as well as quantitatively from those in uninfected plants,[218] although reports are to some extent conflicting.

In some instances, application of endogenous gibberellins may partially alleviate virus-induced stunting. Such is the case in tomato infected with TAV[219] and in tobacco and pepper infected with TEV.[220] In other cases, however, application of endogenous gibberellins has little effect, and even when an effect on stunting is observed, treated plants do not behave as healthy ones. This is not entirely unexpected, since the concentrations of other growth substances, including auxins, ethylene, and ABA, are also changing. Virus infection affects principally cell division rather than cell expansion, and gibberellins appear to mainly affect cell expansion, so that changes in the concentrations of this group of compounds would seem to play at most a minor role in the induction of disease.

Buds on virus-infected rooted stem cuttings of *Euphorbia pulcherrima* grow into leafy branches during the normal dormant period, while buds on similar cuttings obtained from healthy stock remain quiescent or sometimes give rise to cyathia. This break in dormancy of buds on infected stem cuttings has been correlated with low endogenous ABA level and high endogenous gibberellin level.[221]

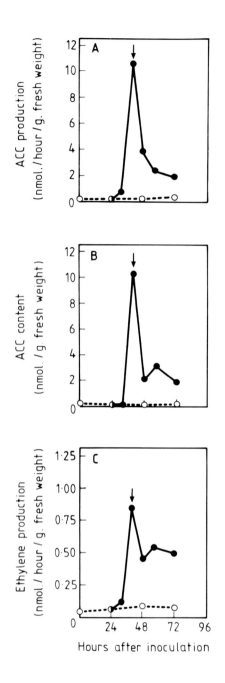

FIGURE 4.   Changes in (A) rate of production of 1-aminocyclopropane-1-car-
boxylic acid (ACC), (B) ACC content, and (C) ethylene production in 'Samsun
NN' tobacco leaves following inoculation with TNV (●-●); water control, ○-
○. The arrow indicates the time of lesion appearance. (From de Laat, A. M. M.
and van Loon, L. C., *Physiol. Plant Pathol.*, 22, 261, 1983. With permission.)

## VIII. SYMPTOMS

### A. Chlorosis, Yellowing, and Mosaics

The response of the plant resulting in the various types of mosaics and yellowing, often
accompanied by growth reduction or other abnormalities, is also a highly complex process.

Areas of the leaf which become lighter green or, at the extreme, yellow, are characterized by both cytological and metabolic perturbations. Many show similarities with changes occurring during senescence, and these symptoms may be regarded at least superficially, as a form of accelerated senescence. Commonly observed changes, many of which have been described at length in preceding sections, include a depression of cytokinin concentration and stimulation of ethylene and ABA production. Rate of photosynthesis and chlorophyll content are reduced, but not necessarily proportionately, and rate of chloroplast ribosomal RNA biosynthesis is also depressed. However, the events which initiate these changes are more difficult to define; it is to investigations which relate to this question that this section is devoted.

It is of interest, and indeed of importance when considering the primary causes of these perturbations, that symptoms can depend very much on both the route of infection and on the age of the leaf at the time it becomes infected. Cotyledons and leaves of cucumber seedlings inoculated with CMV often develop pale chlorotic spots or ringspots, but when young leaves become systemically infected via the vascular system, then pale green or yellow-green mosaics develop. Similarly, CMV-inoculated tobacco leaves develop chlorotic spots at the site of infection, sometimes accompanied by broken necrotic ringspots, while systemic invasion produces characteristic mosaic patterns. Some strains produce distinct yellow-green mosaics, but only in leaf tissue which becomes infected when young.

It is perhaps hardly surprising, therefore, that although many changes are characteristic of hormone imbalance, these symptoms, in common with others, can neither be fully alleviated nor reproduced precisely in uninfected plants by application of growth substances. Application of carbendazim, for example, to TMV-infected tobacco, modified the severity of the symptoms significantly, but at the same time reduced the level of virus accumulation. This confuses the interpretation of the results since, for a specific virus/host combination, it is usually found that the severity of symptoms is related to the extent of viral replication. In BYDV-infected barley and oats[222] and in CMV-infected cucumber,[160] for example, virus accumulation correlates with the severity of the systemic symptoms, and manipulation of the environment to change symptom severity also tends to concomitantly change the rate of virus replication. This might suggest that whatever the nature of the interaction between the virus and its host which initiates phenotypic expression, it is at least partially dose dependent. However, the severity of the host response is clearly not simply dependent upon the quantity of virus produced. As emphasized by several authors,[223] symptoms produced by viruses which replicate to a level which is several orders of magnitude less than that of TMV can still be just as severe, if not even more so than those produced by TMV. The only well-documented exception is TMV-infected tobacco. Fraser and colleagues[224] have demonstrated that here, where the amount of virus produced can be prodigious, there is at least a partial relationship between virus replication, growth reduction, and symptom severity. In other hosts, however, in which the amount of TMV produced is rather less, but still much greater than that of most other viruses, this relationship does not apply.

Since many changes are associated with the chloroplast, it might be inferred that this would be the site of initiation of virus-induced aberrations. Indeed, there have been reports of the occurrence of particles of TMV[225] and occasionally other viruses in chloroplasts, although it was not altogether clear how they got there and what, if anything, they were doing there. In some cases, they may have been in invaginations of cytoplasm.[226] However, sites of various aspects of virus replication have generally been considered to be outside the chloroplast. Any effects inside would normally be expected to be induced either via transport of a viral macromolecule through the plastid envelope, or possibly by transmission of appropriate signals across it. Replication of TYMV was one of the few viruses demonstrated to replicate in close association with the chloroplast, although particles still seem to remain on the cytoplasmic side of the envelope.[227] This is not to say that processes taking place inside could not be influenced.

Recent investigations are beginning to confirm that TMV and other viruses may indeed be in a position to directly influence plastid function. Rods, shorter than full-length viral particles and consisting of chloroplast RNA encapsidated with TMV capsid proteins, have been detected in infected chloroplasts.[228,229] It is not yet clear what their significance is. The association of both TMV RNA[230] and capsid protein[231] with chloroplasts has also been recently demonstrated. Tobacco chloroplasts also contain H protein, which is a minor component of TMV particles and contains amino acid sequences from both capsid protein and a host protein.[230] Evidence to suggest that BMV can replicate in plastids of infected barley has recently been presented,[232] and TEV RNA has been demonstrated to be associated with tobacco chloroplasts.[233] Finally, the presence of ss AbMV DNA in plastids of *Abutilon sellovianum* suggests that it can be transported through the plastid envelope.[234] Although there is therefore an increasing body of evidence to suggest that viruses may be able to influence chloroplast function directly from within, it is equally possible for the primary site of interaction to be outside (or even both). Many chloroplast proteins are, for example, encoded by nuclear DNA; their rate of biosynthesis is regulated outside the chloroplast.

We must now consider the nature of the viral macromolecule(s) which might be involved in the initiation process. Is symptom expression initiated by a genome-encoded polypeptide, or a virus-related nucleic acid?

In attempts to determine the section of the genome involved in symptom determination, many pseudorecombinants have been prepared from viruses with multicomponent genomes. Using this approach, it was determined, for example, that RNA2 of RRV controlled symptom severity in petunia.[235] In some cases, however, the situation seemed less clear, with several sections of the tripartite CMV genome seemingly participating in symptom production in different hosts.[236] The information to be gained from this type of experiment, however, is limited, and has only indicated rather imprecisely the parts of the genome which may be important.

Several authors have suggested that capsid protein may in some circumstances be involved in symptom induction. For example, temperature-sensitive TMV mutants which produced capsid protein having a defect in reversible assembly at the restrictive temperature induced more severe symptoms on tobacco than did the wild type. This symptom enhancement was correlated with alteration in the capsid protein charge. It was suggested that this altered charge could influence intracellular pH and in some way lead to chlorophyll breakdown. Alternatively, it was suggested that the defective protein could precipitate intracellularly, forming aggregates which might interact with membranes and interfere with their normal function.[237] Further evidence for the involvement of capsid protein has recently come from studies by Dawson and colleagues[238] on TMV mutants with insertions or deletions in the capsid protein region. Mutants which failed to produce capsid protein, or produced only small amounts, did not induce systemic symptoms in tobacco. Mutants which produced the carboxyl terminus of capsid protein induced yellow symptoms. In other cases, correlation of symptoms with capsid protein has not always been good. Some strains of TMV, for example, which produce mild symptoms on tobacco, have capsid proteins which are identical to those of the common strain. Virus accumulation, however, may be less. Also, symptoms produced by pseudorecombinant viruses are not always typical of those induced by the strain donating the capsid protein sequence. There is also evidence to suggest that capsid protein could have a direct influence on host DNA. Capsid protein of both TMV and CMV has been reported to be associated with chromatin in infected tobacco.[239]

If the evidence for involvement of capsid protein in symptom induction has until recently remained elusive, then that for the involvement of any other viral genome-encoded protein has been equally so. Experiments by van Loon and colleagues[240] suggested that the 126K protein encoded by TMV genomic RNA was associated with chromatin in infected tobacco leaves, suggesting the possibility of a regulatory role in addition to its putative function as a replicase. However, more recent observations employing immunoelectron microscopy indicated that this protein is associated not directly with chromatin, but with X-bodies.[241] If it does have a regulatory role, its site and mode of action is still in doubt.

Nishiguchi and colleagues,[242] addressing the question of the virulence of strains of TMV in tomato, have compared the nucleotide sequences of a common (L) strain with those of the attenuated LIIA and the strain LII, which is of intermediate virulence; both are derived from L. Although there were ten nucleotide differences between the sequences of LII and L, only three, at positions 1117, 2349, and 2754, resulted in amino acid changes. All three were in the 123K and 186K protein ORFs; only the change at 1117 was common to LII and LIIA. It caused a change from cysteine at amino acid position 348 in the virulent L strain to tyrosine in the attenuated strains. The authors suggested that the resulting change in conformation of these proteins, with an attendant modification in biological activity, was important in attenuation of virulence. They did not, however, rule out the possibility of contributions from other nucleotide changes. It may also be relevant to the interpretation of these results that the attenuated LIIA only accumulated to a final level of approximately one fifth that of the virulent L strain; this may well have contributed to the observed difference in symptoms. Since LIIA replicated normally in protoplasts, but produced a reduced amount of the 30K putative transport protein, it has been suggested that the reduced multiplication was the result of a deficiency in translocation.[243]

In their investigations on RDV, Kimura and colleagues[244] reported that changes in the translation product of the fourth largest genomic RNA could enhance symptom severity in rice. However, this enhancement was correlated with a stimulation in virus replication, and it was this the authors concluded to be responsible for the appearance of more severe symptoms.

Perhaps the most direct evidence to date for the involvement of a viral-encoded protein in symptom induction comes from the studies of Howell and colleagues[245] on CaMV. The translation product of gene VI, P66, is the major protein associated with inclusion bodies in infected host plants. Production of chimeric genomes consisting of DNA segments from isolates producing different symptoms indicated that it was this part of the DNA which controlled symptom phenotype.[246] Although tobacco is not normally a host for CaMV, production of transgenic tobaccos incorporating gene VI yielded plants expressing symptoms characteristic of virus infection. Symptom production was correlated with the presence of both P66 and 19S mRNA. Although present in all cells, their expression was not uniform, but produced a mosaic pattern reminiscent of, but not exactly reproducing, that typical of virus infection. Whether this is a reflection of different levels of production or differential effects in different cells remains to be determined.

The observation that a nucleotide change within an ORF which results in an amino acid change in the corresponding protein correlates with attenuation does not necessarily implicate the protein in pathogenicity. Interaction between a virus-related nucleic acid and a host component may be equally possible. Viroids, so far as is known, do not encode proteins, yet some are capable of inducing severe effects in their hosts.[247] In many cases these effects, such as leaf distortion and severe stunting, are reminiscent of hormone changes. It seems likely that, as with viruses, one of their primary effects is to disturb hormone balance. It is not yet clear how these effects are initiated, although some progress has been made. Sanger and colleagues,[248] for example, have compared the nucleotide sequences of several PSTV isolates of differing virulence in tomato. The PSTV genome, in common with that of other viroids, is a ss RNA circle, with extensive base pairing between two halves of the circle to produce a rigid, rodlike structure. They have discovered that although there are relatively few differences between the isolates, the changes which do occur affect the stability of this rodlike structure. The degree of destabilization correlates positively with virulence. It is suggested that such destabilization might influence the binding of the genome to a host

A                                                                          B

FIGURE 5.   Leaves from tomato plants 28 d after inoculation with the genomic RNAs of CMV-S with (A) or without (B) CARNA 5. (From Kaper, J. M. and Waterworth, H. E., *Science*, 196, 429, 1977. Copyright © 1977 by The American Association for the Advancement of Science. With permission.)

structure or component which so far remains unidentified. It could, for instance, affect transcription by interaction with either the host DNA or a regulatory protein. Alternatively, mechanisms involving interference with mRNA splicing can be envisaged.[249] However, the isolates can induce different symptoms in different hosts, a property also common to most viruses. The precise nature of the interaction and/or its consequences must therefore differ in detail between hosts. Nucleotide sequences important in the pathogenicity of CEV have also been identified.[250,251]

It is also of interest that symptom severity can be dramatically modified by the presence of satellite RNAs replicating in association with helper virus.[252] For example, some variants of the CARNA 5 satellite of CMV (e.g., n and D), when present in association with helper CMV, can induce a severe necrosis in tomato[253] (Figure 5), a symptom which is not characteristic of infection by CMV alone. Conversely, other variants, including Q, Y, and S, have the reverse effect; symptoms produced in tomato in the presence of these satellites are milder than those produced by helper alone. In this case, symptom attenuation is accompanied by a reduction in accumulation of helper CMV. This observation has been used to good effect in the construction of transgenic plants which are resistant to CMV.[254,255] Harrison, Baulcombe and colleagues have constructed transgenic tobacco plants which constitutively express satellite sequences; when challenged with CMV, symptom severity and virus accumulation are lower than in nontransformed plants. However, since symptom phenotype can depend on both satellite and helper,[256] it is conceivable that challenge with different CMV strains could produce different results.

The mechanism by which CMV satellites exert their effects is clearly of considerable interest and again, although we do not yet have the answers, it is being actively pursued. All CMV satellites for which the complete nucleotide sequence has been determined have one or more short putative ORFs. There is some evidence for the in vitro translation of one of them, ORF IIB, which begins around nucleotide 135, into small polypeptides. It is not clear, however, whether these putative ORFs are translated in vivo and whether they have any significance. Evidence from site-directed mutagenesis studies certainly suggests that any translation product of ORF I, which begins at nucleotides 11 to 13 and which is common to strains which produce necrosis, is not involved in pathogenicity.[257] In addition, although there seems to be a considerable degree of sequence consensus between necrogenic strains,[258] there do not appear to be features which correlate with pathogenicity. It may well be that sequences at several positions are important. It is also relevant that satellites which induce necrosis in tomato do not necessarily do so in other hosts, nor in all *Lycopersicon* species.[259]

In those situations where symptoms are alleviated, it may be of importance that at least some satellites have sequences which are complementary to viral RNA sequences[260] and have the potential capacity to regulate the expression of helper genes and helper replication. This could be a contributory factor to the reduction in CMV accumulation, although here, the amount of helper produced and symptom severity may not always be correlated. When plants are infected with another cucumovirus, TAV, for example, symptoms may be alleviated, although replication of helper is normal. In this case, satellite sequences appear to be influencing symptom phenotype independently of viral RNA synthesis.

Satellites are also associated with many other viruses (see Volume I) and, like the satellites of CMV, can markedly modify symptoms produced by the helper virus. Usually, symptoms are more pronounced, although in some cases they may be reduced. The satellite of TobRSV, for example, can attenuate the severity of symptoms induced by the helper virus, and transgenic plants expressing sequences of the satellite of TobRSV are resistant to challenge by helper TobRSV.[261] However, there is presently little information on these satellites which can illuminate our understanding of their ability to modify virus virulence.

## B. Hypersensitivity

When the genotype of virus and host and the environmental conditions are appropriate, the interaction results in a hypersensitive response, a severe, localized reaction involving necrosis and death of cells at and around the site of initiation of infection. Often, but not always, virus multiplication and translocation are limited to cells within and immediately surrounding the ensuing lesion. There is considerable diversity in both the timing of events and in lesion size. In some virus/host combinations, including 'Samsun NN' tobacco infected with TMV or cowpea infected with CMV, for example, visible lesions can appear soon after infection, within 1 to 2 d. In other instances, lesion development can take much longer, some taking up to 5 d to appear. In some cases, including CMV-infected cowpea, lesions may have attained almost their final size by the time they become visible, while in others, including 'White Burley' tobacco infected with some strains of TMV, lesions may continue to expand for several days after their first appearance. The final diameter of some lesions may be less than 1 mm, while others may reach several millimeters in diameter. In all cases, cells within the lesion area undergo dramatic changes, many occurring long before necrosis becomes visible. Several virus/host combinations have been examined by both light and electron microscopy, with *Nicotiana* reacting hypersensitively to TMV infection, inevitably having received by far the most attention.[262,263]

Early signs of impending cell degeneration, visible within 8 to 10 h after infection, often include increase in size of starch grains and increase in cytoplasmic vesiculation. Cytoplasmic membranes then begin to degenerate, a process which continues during the progressive deterioration in cellular organization, and culminates in complete breakdown of the plasmalemma. Membranes of nucleus, chloroplasts, and mitochondria suffer a similar fate.

Nuclei and nucleoli swell, and as the membranes disintegrate, contents merge with those of the cytoplasm. Chloroplast lamellae become distorted with vacuoles, membranes become disrupted, and again contents are released. Finally, cells collapse completely, breaking away from cell walls, and become opaque and featureless. While details and timing of this degenerative process may vary, the overall picture in other situations examined, including PVX- or TBSV-infected *G. globosa*[264,265] or bean infected with PVM,[266] is broadly the same. It is of interest that this process of cell degeneration appears to occur in many cells simultaneously,[150] perhaps implying a rapid rate of virus translocation from the initially infected cell.

The progressive deterioration in cellular form and function is, hardly surprisingly, accompanied by significant metabolic changes, many of which have been described in the preceding sections. Perturbations detected in advance of lesion appearance include stimulation in PAL activity, enhanced production of phenolics, and stimulation in ethylene production. The deterioration in membrane integrity revealed by microscopy is confirmed by an enhanced cell permeability, with loss of electrolytes. As lesions begin to appear, there is a continuing stimulation in PAL activity, now accompanied by enhancement in activity of peroxidases and polyphenoloxidases, stimulation in ABA and IAA production, the induction of synthesis of PR proteins, and deposition of callose and lignin as cells within the lesion area collapse.

The expression of this hypersensitive response, involving as it does such dramatic cellular and metabolic changes, poses a number of questions: How is the response initiated? Why does the necrotic zone normally reach a finite, usually small size, and not continue to enlarge? Which features of the response are responsible for, in most cases, the cessation of virus multiplication, and restriction of virus translocation and development of a systemic infection? We shall principally be concerned here with the first of these questions; the last and, to some extent, the second are dealt with elsewhere (Chapter 12).

In seeking to identify the primary events which initiate the hypersensitive or incompatible response, the timing of the changes, both cytological and biochemical, is critical and it is here that difficulties arise. It is necessary to identify the site at which infection has been initiated and, with leaves mechanically inoculated in the usual way, this is difficult. In a very few cases, the development of infection following, for example, inoculation with a pin, has been studied. This procedure does permit the earliest cytological changes to be determined. They need to be correlated with metabolic changes, and the determination of the timing of the beginning of these perturbations also poses problems. Much depends on the sensitivity of the assay used. An early event may be so designated because there is a sensitive procedure for detecting it, and a small early change in a relatively few cells can be identified. A less sensitive assay might require changes in many more cells. Histochemical detection of metabolic perturbations, or investigation of events taking place in tissue immediately surrounding a developing lesion can to some extent reduce these difficulties.

Of the many metabolic changes which are associated with necrosis, those relating to the activity of peroxidases and polyphenoloxidases have received more attention than most. Their activities are invariably enhanced when tissue responds hypersensitively, and they oxidize polyphenols to quinones which, in turn, can interact with proteins to produce cytotoxic melanins. Early membrane deterioration could lead to decompartmentalization of substrate and enzyme, and these oxidative enzymes were considered therefore to play a key role in the development of necrosis. However, the timing of the enhancement in their activity does not convincingly support such a view. It succeeds the appearance of visible lesions, and evidence suggests it to be a consequence rather than a cause of necrosis.[267] Evidence is stronger for the participation of these enzymes and PAL in limitation of virus movement. As indicated in Section VI, PAL is also stimulated, providing an increase in concentration of the substrates of peroxidases and polyphenoloxidases. This also provides an increased supply of the quinone precursors of lignin. Lignin deposition in cell walls around the lesion area is a common feature of the hypersensitive response, and has been suggested to contribute to the limitation of virus movement. This view is supported by experiments in which PAL activity has been inhibited using either α-aminooxyacetic acid (AOA) or α-aminooxy-β-phenylpropionic acid. In tobacco leaves treated with either inhibitor, the size of TMV lesions was significantly higher than in controls (Figure 6). It was greater, however, in leaves treated with AOA, the inhibitor which was the more efficient in reducing the production of lignin. There was therefore a correlation between stimulation of the phenylpropanoid pathway, production of lignin, and the development of resistance to virus movement.

The earliest visual evidence of impending necrosis is an alteration in the integrity of cell membranes and, in particular, a change in permeability of the plasmalemma which results in loss of osmotic control. Visual evidence is supported by observations of a stimulation in electrolyte leakage, in both TMV-infected 'Xanthi nc' tobacco and CMV-infected cowpea, which is apparent long before lesions become visible.[268,269] In TBSV-infected *G. globosa*, permeability changes were observed as long as 33 h before lesions became visible.[270] These early membrane changes therefore allow decompartmentalization of cellular constituents and are early signs of the more dramatic events to follow. Membrane permeability changes are not unique to virus infection; precisely similar effects are caused by other stresses including those of both chemical and physical injury, and indeed, the pattern of subsequent events is also closely parallel. In some cases, including TMV-infected 'Xanthi nc' tobacco maintained at 33°C and then reduced to 25°C[271] and in CMV-infected cowpea,[272] permeability changes have been demonstrated to be associated with changes in fatty acid composition. There was, for example, an increase in the saturated palmitic acid and a decrease in the unsaturated linolenic acid. Lipoxygenase was stimulated in advance of both changes in fatty acid composition and permeability change. There was also a significant use in adenylate cyclase activity very soon after TMV infection of 'Xanthi nc' tobacco;[273] its role in ensuing membrane changes remains unclear.

The early stimulation in ethylene production, sometimes detected in advance of lesion appearance, has inevitably led to suggestions that it may be involved in the development of necrosis. When ethylene activity was reduced with $CO_2$, lesions failed to develop,[210] and conversely, addition of ethylene stimulated necrosis. However, inhibition of ethylene stimulation with aminoethyoxyvinylglycine only partially prevented the development of necrosis-related metabolic changes. It is clear that membrane changes are an early event in the development of necrosis, and treatments which damage membranes cause necrosis.[274] It is of interest therefore that ethylene can also cause membrane damage. Ethylene production is stimulated at membranes, and may well be enhanced by membrane damage. To date, the suggestion that ethylene might play a key role in the initiation of necrosis is not well supported by the evidence. It has been suggested, however, that ethylene might play a role in the induction of PR protein synthesis, and indeed it has been shown to enhance glucanase and chitinase production in peas.[275] However, both chitinase and β-1,3-glucanase activity were stimulated by fungal elicitors when a stimulation in ethylene concentration was prevented by aminoethyoxyvinylglycine. In this case, therefore, the two are not necessarily related.

There is some evidence that oligosaccharides can act as elicitors for the induction of defense mechanisms which are activated in response to fungal infection, and their involvement in the early stages of virus-induced hypersensitivity has also been proposed.[150] It has been suggested, for example, that a stimulation in hydrolase activity by early events associated with membrane damage could be responsible for their release from neighboring cell walls. The released oligosaccharides may then be involved in subsequent perturbations to the normal regulatory controls of many metabolic pathways. Whereas fungi possess their own hydrolases, viruses of course do not, so that the proposed scheme requires an early stimulation of hydrolase activity, possibly via membrane damage. PR proteins (Chapter 3)

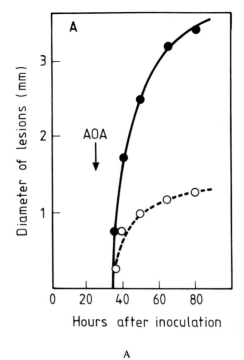

A

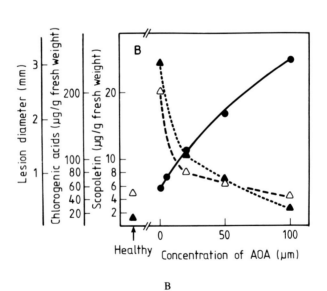

B

FIGURE 6.    Effect of α-aminooxyacetic acid (AOA) on size of TMV lesions in 'Samsun NN' tobacco and on flux through the phenylpropanoid pathway. (A) Lesion size in leaves detached 24 h after inoculation and maintained in medium with (●-●) or without (○-○) AOA. (B) Lesion size (●-●) and chlorogenic acid (△-△) and scopoletin (▲-▲) concentrations in leaves maintained in various concentrations of AOA; determinations were made 54 h after virus inoculation. (From Massala, R., Legrand, M., and Fritig, B., *Physiol. Plant Pathol.*, 16, 213, 1980. With permission.)

possess glucanase or chitinase activity, and some could therefore fulfill this requirement. A stimulation in their concentration would need to occur well in advance of visible necrosis, rather earlier than reports would suggest. Further studies are clearly needed.

It is of interest, however, that oligosaccharides can apparently stimulate the hypersensitive response. Infiltration of hypersensitive tobacco leaves with plant cell wall components significantly reduced the size of TMV lesions which developed.[276]

In addition, there have been reports of the production of necrosis-inducing factors in tobacco reacting hypersensitively to TMV.[277,278] It is not clear whether or not there is any relationship between these factors and oligosaccharide elicitors.

Another early event which is observed in cells destined to become necrotic is a stimulation in the production of superoxide anion $O_2^-$. Superoxide anion production was detectable almost immediately after TMV-inoculated 'Samsun NN' tobacco leaf discs were transferred to 20 to 26°C after prior incubation at 30°C; lesions appeared 6 to 7 h after transfer.[279] It is possible therefore that early production of toxic $O_2^-$, which is unlikely to be quickly removed by a modest increase in superoxide dismutase activity,[280] has a role in the induction of necrosis.

With techniques now available, adding precision to the timing of many of these events now becomes possible. Of equal importance, however, is the determination of the nature of the initial interaction between virus and host which initiates the events described. As with induction of other types of host response, genotype of host and virus are critical. While most strains of TMV induce a hypersensitive response in N gene-containing tobaccos, other viruses generally induce the same response in tobaccos, whether they contain the N gene or not. Most strains of CMV, for example, replicate systemically in both 'Xanthi' and 'Xanthi nc'. TobRSV induces a necrotic response in most tobaccos, although systemic spread is not prevented. Strains of TMV inducing hypersensitivity in N gene-containing tobaccos do not necessarily do so in tobaccos containing the allelic N' gene. Many pseudorecombinants have been prepared from viruses with multicomponent genomes establishing, for example, that production of local lesions in bean by AlMV was determined by RNA2,[281] and that in most cases RNA2 of CMV was responsible for inducing hypersensitivity in cowpea. As we have seen already, however, symptom production appeared to depend on interaction between several genome components. Perhaps the virus/host combination about which we have the most information is TMV and *N. sylvestris*. The common strain of TMV produces a systemic infection in this host. Early studies identified mutants which induced a hypersensitive response, although the location of the mutation was uncertain and imprecise.[282,283] Later work[284] established a correlation between lesion size in *N. sylvestris* and the temperature sensitivity of capsid protein of several strains. Additionally, the OM strain of TMV produces a systemic response, while the L strain induces hypersensitivity in *N. sylvestris*. Production of chimeric genomes indicated that the difference lay in the capsid protein region.[285] Finally, Knorr and Dawson[286] have demonstrated that the difference between a strain which produces a systemic infection and a mutant which induces a hypersensitive response is a single cytosine to uracil nucleotide change at position 6157, substituting phenylalanine for serine in capsid protein. The capsid protein sequence is therefore of crucial importance in inducing the hypersensitive response in this system. It is of interest that a mutant derived from TMV-L, which does not produce capsid protein, does not induce a hypersensitive response in tobacco containing the N' gene, but does in 'Xanthi nc', which contains the N gene. Dawson and colleagues also find that mutants lacking capsid protein can induce hypersensitivity in 'Xanthi nc'. Hypersensitivity appears to be initiated by different mechanisms in the two hosts.

Indeed, there may well be many diverse mechanisms by which hypersensitivity, or any other type of symptom, is induced. They may differ significantly, or only slightly, depending on host and virus. In the case of the N gene, the expression of hypersensitivity is temperature sensitive, implying interactions with a thermolabile structure; in other cases, it is not. Necrosis can be induced by many types of stress. In particular, lesions can be produced in hosts which normally become infected systemically by subjecting the infected plants to a stress in addition to that imposed by the normal course of virus infection.[287,288] Lesions can be induced, for example, in tobacco cultivars which normally support systemic TMV infection by hot- or cold-water treatment. When tobaccos which react hypersensitively are kept above the temperature at which lesions develop, lesions can be induced by ultraviolet irradiation. Lesions may, however, differ in appearance from those normally observed, and it remains to be seen whether or not they can contribute to our understanding of the mechanisms on initiation of the hypersensitive response.

The associated questions relating to limitation of lesion growth and virus translocation are equally important. In many cases, virus is not contained, while in others, virus may be present in cells around the periphery of a necrotic lesion, yet seems unable to replicate. In other cases, virus is contained near the site of initiation of infection in the absence of a necrotic response. These and other questions relating to resistance are addressed elsewhere (Chapter 12).

## IX. CONCLUSIONS

As I suggested in the Introduction, there is a lot of descriptive information on virus-induced changes which occur in plant tissue, but very little which enables the primary cause of these changes to be determined. Fortunately, this unsatisfactory state of affairs is now changing rapidly, and this area of plant virology is currently entering a particularly exciting and productive period. Techniques for nucleic acid sequencing, site-directed mutagenesis, the transcription of infective viral RNA genomes from cloned cDNAs, and plant transformation are already beginning to make possible the identification of regions of the viral genome which are important in pathogenesis. In addition, techniques for isolation and characterization of host messengers are beginning to provide information on changes in host regulation and control from the time of initiation of infection. Procedures which will allow the determination of the sites and nature of the interaction between the viral genome and/or its translation products are also becoming available. We should soon have some answers.

# ABBREVIATIONS

## Viruses

| | | | | |
|---|---|---|---|---|
| AbMV | Abutilon mosaic geminivirus | | PVM | Potato M carlavirus |
| AlMV | Alfalfa mosaic virus | | PVX | Potato X potexvirus |
| BCMV | Bean common mosaic potyvirus | | PVY | Potato Y potyvirus |
| BCTV | Beet curly top geminivirus | | RRV | Raspberry ringspot nepovirus |
| BGMV | Bean golden mosaic geminivirus | | RTV | Rice tungro virus |
| BMV | Brome mosaic bromovirus | | SBMV | Southern bean mosaic sobemovirus |
| BSMV | Barley stripe mosaic hordeivirus | | SCMV | Sugarcane mosaic potyvirus |
| BYDV | Barley yellow dwarf luteovirus | | SMV | Soybean mosaic potyvirus |
| BYMV | Barley yellow mosaic virus | | SqMV | Squash mosaic comovirus |
| BYV | Beet yellows closterovirus | | TAV | Tomato aspermy cucumovirus |
| CaMV | Cauliflower mosaic caulimovirus | | TBSV | Tomato bushy stunt tombusvirus |
| CCMV | Cowpea chlorotic mottle bromovirus | | TEV | Tobacco etch potyvirus |
| CMV | Cucumber mosaic cucumovirus | | TMV | Tobacco mosaic tobamovirus |
| CPMV | Cowpea mosaic comovirus | | TNV | Tobacco necrosis necrovirus |
| CuTV | Curly top | | ToYMV | Tomato yellow mosaic geminivirus |
| EMV | Euphorbia mosaic geminivirus | | TobRSV | Tobacco ringspot nepovirus |
| LMV | Lettuce mosaic potyvirus | | TSWV | Tomato spotted wilt virus |
| LNYV | Lettuce necrotic yellows rhabdovirus | | TYMV | Turnip yellow mosaic tymovirus |
| MDMV | Maize dwarf mosaic potyvirus | | WCMV | White clover mosaic potexvirus |
| PGMV | Peanut green mosaic potyvirus | | WSMV | Wheat streak mosaic potyvirus |
| PLRV | Potato leaf roll luteovirus | | | |
| PMV | Peanut mottle potyvirus | | | |

## Viroids

| | |
|---|---|
| CEV | Citrus exocortis |
| PSTV | Potato spindle tuber |

# REFERENCES

1. **Bos, L.,** *Symptoms of Virus Diseases in Plants,* 3rd ed., Centre for Agricultural Publication and Documentation, Wageningen, 1978.
2. **Goodman, R. N., Király, Z., and Wood, K. R.,** *Biochemistry and Physiology of Plant Disease,* University of Missouri Press, Columbia, 1986.
3. **Fraser, R. S. S.,** *Biochemistry of Virus-Infected Plants,* Research Studies Press, Letchworth, U.K., 1987.
4. **Martelli, G. P. and Russo, M.,** Virus-host relationships: symptomatological and ultrastructural aspects, in *The Plant Viruses,* Vol. 1, Francki, R. I. B., Ed., Plenum Press, New York, 1985, 163.
5. **van Loon, L. C.,** Disease induction by plant viruses, *Adv. Virus Res.,* 33, 205, 1987.
6. **Atchison, B. A.,** Division, expansion and DNA synthesis in meristematic cells of French bean (*Phaseolus vulgaris* L.) root-tip invaded by tobacco ringspot virus, *Physiol. Plant Pathol.,* 3, 1, 1973.
7. **Fraser, R. S. S.,** Effects of two strains of tobacco mosaic virus on growth and RNA content of tobacco leaves, *Virology,* 47, 261, 1972.
8. **Brakke, M. K., White, J. L., Samson, R. G., and Joshi, J.,** Chlorophyll, chloroplast ribosomal RNA, and DNA are reduced by barley stripe mosaic virus systemic infection, *Phytopathology,* 78, 570, 1988.
9. **Brakke, M. K.,** Mutations, the abberrant ratio phenomenon, and virus infection of maize, *Annu. Rev. Phytopathol.,* 22, 77, 1984.
10. **Mottinger, J. P. and Dellaporta, S. L.,** Stable and unstable mutations associated with virus infection in maize, in *Plant Infectious Agents: Viruses, Viroids, Virusoids and Satellites,* Robertson, H. D., Howell, S. H., Zaitlin, M., and Malmberg, R. L., Eds., Cold Spring Harbor Laboratory, Cold Spring Harbor, NY, 1983, 126.
11. **Fraser, R. S. S.,** Effects of two TMV strains on the synthesis and stability of chloroplast ribosomal RNA in tobacco leaves, *Mol. Gen. Genet.,* 106, 73, 1969.

12. **Fraser, R. S. S.,** The synthesis of tobacco mosaic virus RNA and ribosomal RNA in tobacco leaves, *J. Gen. Virol.,* 18, 267, 1973.
13. **Kubo, S.,** Chromatographic studies of RNA synthesis in tobacco mosaic virus, *Virology,* 28, 229, 1966.
14. **Oxelfelt, P.,** Development of systemic tobacco mosaic virus infection. II. RNA metabolism in systemically infected leaves, *Phytopathol. Z.,* 71, 247, 1971.
15. **Hirai, A. and Wildman, S. G.,** Effect of TMV multiplication on RNA and protein synthesis in tobacco chloroplasts, *Virology,* 38, 73, 1969.
16. **Pring, D. R.,** Viral and host RNA synthesis in BSMV-infected barley, *Virology,* 44, 54, 1971.
17. **White, J. and Brakke, M.,** Chloroplast RNA and proteins decrease as wheat streak and barley stripe mosaic viruses multiply in expanding, systemically infected leaves, *Phytopathology,* 72, 939, 1982.
18. **Randles, J. W. and Coleman, D. F.,** Loss of ribosomes in *Nicotiana glutinosa* L. infected with lettuce necrotic yellows virus, *Virology,* 41, 459, 1970.
19. **Reid, M. S. and Matthews, R. E. F.,** On the origin of mosaic induced by turnip yellow mosaic virus, *Virology,* 28, 563, 1966.
20. **Mohammed, N. A. and Randles, J. W.,** Effect of tomato spotted wilt virus on ribosomes, ribonucleic acids and fraction I protein in *Nicotiana tabacum* leaves, *Physiol. Plant Pathol.,* 2, 235, 1972.
21. **Roberts, P. L. and Wood, K. R.,** Decrease in ribosome levels in tobacco infected with a chlorotic strain of cucumber mosaic virus, *Physiol. Plant Pathol.,* 19, 99, 1981.
22. **Diener, T. O.,** Virus infection and other factors affecting ribonuclease activity of plant leaves, *Virology,* 14, 177, 1961.
23. **Randles, J. W.,** Ribonuclease isozymes in Chinese cabbage systemically infected with turnip yellow mosaic virus, *Virology,* 36, 556, 1968.
24. **Fraser, R. S. S. and Gerwitz, A.,** Tobacco mosaic virus infection does not alter the polyadenylated messenger RNA content of tobacco leaves, *J. Gen. Virol.,* 46, 139, 1980.
25. **Wildman, S. G., Cheo, C. C., and Bonner, J.,** The proteins of green leaves. III. Evidence for the formation of tobacco mosaic virus protein at the expense of a main protein component in tobacco leaf cytoplasm, *J. Biol. Chem.,* 180, 985, 1949.
26. **Crosbie, E. S. and Matthews, R. E. F.,** Effects of TYMV infection on growth of *Brassica pekinensis* Rupr., *Physiol. Plant Pathol.,* 4, 389, 1974.
27. **Randles, J. W. and Coleman, D.,** Changes in polysomes in *Nicotiana glutinosa* L. leaves infected with lettuce necrotic yellows virus, *Physiol. Plant Pathol.,* 2, 247, 1972.
28. **White, J. L. and Brakke, M. K.,** Protein changes in wheat infected with wheat-streak mosaic virus and barley infected with barley stripe mosaic virus, *Physiol. Plant Pathol.,* 22, 87, 1983.
29. **Singer, B. and Condit, C.,** Protein synthesis in virus-infected plants. III. Effects of tobacco mosaic virus mutants on protein synthesis in *Nicotiana tabacum, Virology,* 57, 42, 1974.
30. **Ziemiecki, A. and Wood, K. R.,** Proteins synthesised by cucumber cotyledons infected with two strains of cucumber mosaic virus, *J. Gen. Virol.,* 31, 373, 1976.
31. **Camacho Henriquez, A. and Sänger, H.-L.,** Analysis of acid-extractable tomato leaf proteins after infection with a viroid, two viruses and a fungus and partial purification of the pathogenesis-related protein p14, *Arch. Virol.,* 74, 181, 1982.
32. **van Loon, L. C., Gerritsen, Y. A. M., and Ritter, C. E.,** Identification, purification and characterization of pathogenesis-related proteins from virus-infected Samsun NN tobacco leaves, *Plant Mol. Biol.,* 9, 593, 1987.
33. **Kauffmann, S., Legrand, M., Geoffroy, P., and Fritig, B.,** Biological function of 'pathogenesis-related' proteins: four PR proteins of tobacco have 1,3-β-glucanase activity, *EMBO J.,* 6, 3209, 1987.
34. **Legrand, M., Kauffmann, S., Geoffroy, P., and Fritig, B.,** Biological function of pathogenesis-related proteins: four tobacco pathogenesis-related proteins are chitinases, *Proc. Natl. Acad. Sci. U.S.A.,* 84, 6750, 1987.
35. **Rao, G. P., Shukla, K., and Gupta, S. N.,** Effect of cucumber mosaic virus infection on nodulation, nodular physiology and nitrogen fixation of pea plants, *J. Plant Dis. Prot.,* 94, 606, 1987.
36. **Chowdhury, J. R., Srivastava, R. S., and Singh, R.,** Effect of common bean mosaic virus infection on nitrogenase activity in root nodules of mung bean, *J. Plant Dis. Prot.,* 94, 126, 1987.
37. **Khadhair, A. H., Sinha, R. C., and Peterson, J. F.,** Effect of white clover mosaic virus infection on various processes relevant to symbiotic $N_2$ fixation in red clover, *Can. J. Bot.,* 62, 38, 1984.
38. **Wongkaew, S. and Peterson, J. F.,** Effect of peanut mottle virus infection on peanut nodulation and nodule function, *Phytopathology,* 73, 377, 1983.
39. **Cooper, P. and Selman, I. W.,** An analysis of the effects of tomato mosaic virus on growth and the changes in the free amino compounds in young tomato plants, *Ann. Bot.,* 38, 625, 1974.
40. **Tu, J. C. and Ford, R. E.,** Maize dwarf mosaic virus infection in susceptible and resistant corn: virus multiplication, free amino acid concentrations and symptom severity, *Phytopathology,* 60, 1605, 1970.
41. **Tu, J. C. and Ford, R. E.,** Free amino acids in soybeans infected with soybean mosaic virus, bean pod mottle virus, or both, *Phytopathology,* 60, 660, 1970.

42. **Mohanty, S. K. and Sridhar, R.**, Physiology of rice tungro virus disease: proline accumulation due to infection, *Physiol. Plant.*, 56, 89, 1982.
43. **Welkie, G. W., Yang, S. F., and Miller, G. W.**, Metabolic changes induced by cucumber mosaic virus in resistant and susceptible strains of cowpea, *Phytopathology*, 57, 472, 1967.
44. **French, R. and Ahlquist, P.**, Characterisation and engineering of sequences controlling in vivo synthesis of brome mosaic virus subgenomic RNA, *J. Virol.*, 62, 2411, 1988.
45. **Siegel, A., Hari, V., and Kolacz, K.**, The effect of tobacco mosaic virus infection on host and virus-specific protein synthesis in protoplasts, *Virology*, 85, 494, 1978.
46. **Van Pelt-Heerschap, H., Verbeek, J., Huisman, M. J., Loesch-Fries, L. S., and Van Vloten-Doting, L.**, Non-structural proteins and RNAs of alfalfa mosaic virus synthesized in tobacco and cowpea protoplasts, *Virology*, 161, 190, 1987.
47. **Hills, G. J., Plaskitt, K. A., Young, N. D., Dunigan, D. D., Watts, J. W., Wilson, T. M. A., and Zaitlin, M.**, Immunogold localization of the intracellular sites of structural and non-structural tobacco mosaic virus proteins, *Virology*, 161, 488, 1987.
48. **Stussi-Garaud, C., Garaud, J.-C., Berna, A., and Godefroy-Colburn, Th.**, In situ location of an alfalfa mosaic virus non-structural protein in plant cell walls: correlation with virus transport, *J. Gen. Virol.*, 68, 1779, 1987.
49. **Fraenkel-Conrat, H.**, RNA-dependent RNA polymerases of plants, *Proc. Natl. Acad. Sci. U.S.A.*, 80, 422, 1983.
50. **Khan, Z. A., Hiriyanna, K. T., Chavez, F., and Fraenkel-Conrat, H.**, RNA-directed RNA polymerases from healthy and from virus-infected cucumber, *Proc. Natl. Acad. Sci. U.S.A.*, 83, 2383, 1986.
51. **Kamer, G. and Argos, P.**, Primary structural comparison of RNA-dependent polymerases from plant, animal and bacterial viruses, *Nucleic Acids Res.*, 12, 7269, 1984.
52. **Quadt, R., Verbeek, H. J. M., and Jaspars, E. M. J.**, Involvement of a nonstructural protein in the RNA synthesis of brome mosaic virus, *Virology*, 165, 256, 1988.
53. **Candresse, T., Mouches, C., and Bové, J.-M.**, Characterization of the virus encoded subunit of turnip yellow mosaic virus RNA replicase, *Virology*, 152, 322, 1986.
54. **Pietrzak, M. and Hohn, T.**, Translation products of cauliflower mosaic virus ORF V, the coding region corresponding to the retrovirus pol gene, *Virus Genes*, 1, 83, 1987.
55. **Armour, S. L., Melcher, U., Pirone, T. P., Lyttle, D. J., and Essenberg, R. C.**, Helper component for aphid transmission encoded by region II of cauliflower mosaic virus DNA, *Virology*, 129, 25, 1983.
56. **Wellink, J. and Van Kammen, A.**, Proteases involved in the processing of virus polyproteins, *Arch. Virol.*, 98, 1, 1988.
57. **Atabekov, J. G. and Dorokhov, Y. L.**, Plant virus-specific transport function and resistance of plants to viruses, *Adv. Virus Res.*, 29, 313, 1984.
58. **Deom, C. M., Oliver, M. J., and Beachy, R. N.**, The 30-kilodalton gene product of tobacco mosaic virus potentiates virus movement, *Science*, 237, 389, 1987.
59. **Meshi, T., Watanabe, Y., Saito, T., Sugimoto, A., Maeda, T., and Okada, Y.**, Function of the 30Kd protein of tobacco mosaic virus: involvement in cell-to-cell movement and dispensability for replication, *EMBO J.*, 6, 2557, 1987.
60. **Roberts, D. A. and Corbett, M. K.**, Reduced photosynthesis in tobacco plants infected with tobacco ringspot virus, *Phytopathology*, 55, 370, 1965.
61. **Owen, P. C.**, The effects of infection with tobacco mosaic virus on the photosynthesis of tobacco leaves, *Ann. Appl. Biol.*, 45, 456, 1957.
62. **Owen, P. C.**, The effect of infection with tobacco etch virus on the rates of respiration and photosynthesis of tobacco leaves, *Ann. Appl. Biol.*, 45, 327, 1957.
63. **Hopkins, D. L. and Hampton, R. E.**, Effects of tobacco etch virus upon the dark reactions of photosynthesis in tobacco leaf tissue, *Phytopathology*, 59, 1136, 1969.
64. **Hunter, C. S. and Peat, W. E.**, The effect of tomato aspermy virus on photosynthesis in the young tomato plant, *Physiol. Plant Pathol.*, 3, 517, 1973.
65. **Panopoulos, N. J., Faccioli, G., and Gold, A. H.**, Kinetics of carbohydrate metabolism in curly top virus-infected tomato plants, *Phytopathol. Mediterr.*, 11, 48, 1972.
66. **Leal, N. and Lastra, R.**, Altered metabolism of tomato plants infected with tomato yellow mosaic virus, *Physiol. Plant Pathol.*, 21, 1, 1984.
67. **Jensen, S. G.**, Photosynthesis, respiration and other physiological relationships in barley infected with barley yellow dwarf virus, *Phytopathology*, 58, 204, 1968.
68. **Jensen, S. G.**, Composition and metabolism of barley leaves infected with barley yellow dwarf virus, *Phytopathology*, 59, 1694, 1969.
69. **Jensen, S. G.**, Metabolism and carbohydrate composition in barley yellow dwarf virus-infected wheat, *Phytopathology*, 62, 587, 1972.
70. **Hall, A. E. and Loomis, R. S.**, Photosynthesis and respiration by healthy and beet yellows virus-infected sugar beets (*Beta vulgaris* L.), *Crop Sci.*, 12, 566, 1972.

71. **Tu, J. C., Ford, R. E., and Krass, C. J.,** Effect of maize dwarf mosaic virus infection on respiration and photosynthesis of corn, *Phytopathology,* 58, 282, 1968.

72. **Naidu, R. A., Krishnan, M., Ramanujam, P., Gnanam, A., and Nayudu, M. V.,** Studies on peanut green mosaic virus infected peanut (*Arachis hypogaea* L.) leaves. I. Photosynthesis and photochemical reactions, *Physiol. Plant Pathol.,* 25, 181, 1984.

73. **Owen, P. C.,** Photosynthesis and respiration rates of leaves of *Nicotiana glutinosa* infected with tobacco mosaic virus and of *N. tabacum* infected with potato virus X, *Ann. Appl. Biol.,* 46, 198, 1958.

74. **Magyarosy, A. C., Buchanan, B. B., and Schurmann, P.,** Effect of a systemic virus infection on chloroplast function and structure, *Virology,* 55, 426, 1973.

75. **Hampton, R. E., Hopkins, D. L., and Nye, T. G.,** Biochemical effects of tobacco etch virus infection on tobacco leaf tissue. I. Protein synthesis by isolated chloroplasts, *Phytochemistry,* 5, 1181, 1966.

76. **Hall, A. E. and Loomis, R. S.,** An explanation for the difference in photosynthetic capabilities of healthy and beet yellows infected sugar beets (*Beta vulgaris* L.), *Plant Physiol.,* 50, 576, 1972.

77. **Doke, N. and Hirai, T.,** Effects of tobacco mosaic virus infection on photosynthetic $CO_2$ fix₁⁺ion on $^{14}CO_2$ incorporation into protein in tobacco leaves, *Virology,* 42, 68, 1970.

78. **Makovcova, O. and Sindelar, L.,** Changes in phosphoenolpyruvate carboxylase and ribulosebisphosphate carboxylase activities in tobacco plants infected with tobacco mosaic virus, *Biol. Plant,* 20, 135, 1978.

79. **Bedbrook, J. R. and Matthews, R. E. F.,** Changes in the flow of early products of photosynthetic carbon fixation associated with replication of TYMV, *Virology,* 53, 84, 1973.

80. **Welkie, G. W., Yang, S. F., and Miller, G. W.,** Metabolic changes induced by cucumber mosaic virus in resistant and susceptible strains of cowpea, *Phytopathology,* 57, 472, 1967.

81. **Mohammed, N. A.,** Some effects of systemic infection by tomato spotted wilt virus on chloroplasts of *Nicotiana tabacum* leaves, *Physiol. Plant Pathol.,* 3, 509, 1973.

82. **Ghorpade, L. N. and Joshi, G. V.,** Development of photosynthesis in the sugarcane plant (Var. Co. 740) infected by mosaic virus, *Indian J. Exp. Biol.,* 18, 1202, 1980.

83. **Peterson, P. D. and McKinney, H. H.,** The influence of four mosaic diseases on the plastid pigments and chlorophyllase in tobacco leaves, *Phytopathology,* 28, 329, 1938.

84. **Roberts, P. L. and Wood, K. R.,** Effects of a severe (P6) and mild (W) strain of cucumber mosaic virus on tobacco leaf chlorophyll, starch and cell ultrastructure, *Physiol. Plant Pathol.,* 21, 31, 1982.

85. **Gates, D. W. and Gudauskas, R. T.,** Photosynthesis, respiration and evidence of a metabolic inhibitor in corn infected with maize dwarf mosaic virus, *Phytopathology,* 59, 575, 1969.

86. **Omar, R. A., Mehiar, F. F., Zayed, E. A., and Dief, A. A.,** Physiological and biochemical studies on soybean, bean and lettuce plants infected with seed-borne viruses, *Acta Phytopathol. Entomol. Hung.,* 21, 63, 1986.

87. **Sai Gopal, D. V. R., Satyanarayana, T., Gopinath, K., and Sreenivasulu, P.,** Effect of bavistin on chlorophylls and delta-aminolevulinic acid (ALA) in peanut green mosaic virus (PGMV)-infected groundnut leaves, *J. Plant Dis. Prot.,* 94, 600, 1987.

88. **Naidu, R. A., Krishnan, M., Nayudu, M. V., and Gnanam, A.,** Studies on peanut green mosaic virus infected peanut (*Arachis hypogaea* L.) leaves. II. Chlorophyll-protein complexes and polypeptide composition of thylakoid membranes, *Physiol. Plant Pathol.,* 25, 191, 1984.

89. **Naidu, R. A., Krishnan, M., Nayudu, M. V., and Gnanam, A.,** Studies on peanut green mosaic virus infected peanut (*Arachis hypogaea* L.). III. Changes in the polypeptides of photosystem II particles, *Physiol. Mol. Plant Pathol.,* 29, 53, 1986.

90. **Platt, S. G., Hendriques, F., and Rand, L.,** Effects of virus infection on the chlorophyll content, photosynthetic rate and carbon metabolism of *Tolmiea menziesii, Physiol. Plant Pathol.,* 15, 351, 1979.

91. **Kato, S. and Misawa, T.,** Studies on the infection and multiplication of plant viruses. VII. The breakdown of chlorophyll in tobacco leaves systemically infected with cucumber mosaic virus, *Ann. Phytopathol. Soc. Jpn.,* 40, 14, 1974.

92. **Montalbini, P., Koch, F., Burba, M., and Elstner, E.,** Increase in lipid-dependent carotene destruction as compared to ethylene formation and chlorophyllase activity following mixed infection of sugar beet (*Beta vulgaris* L.) with beet yellows virus and beet mild yellowing virus, *Physiol. Plant Pathol.,* 12, 211, 1978.

93. **Crosbie, E. S. and Matthews, R. E. F.,** Effects of TYMV infection on leaf pigments in *Brassica pekinensis* Rupr., *Physiol. Plant Pathol.,* 4, 379, 1974.

94. **Spikes, J. D. and Stout, M.,** Photochemical activity of chloroplasts isolated from sugar beet infected with virus yellows, *Science,* 122, 375, 1955.

95. **Zaitlin, M. and Jagendorf, A. T.,** Photosynthetic phosphorylation and Hill Reaction activities of chloroplasts isolated from plants infected with tobacco mosaic virus, *Virology,* 12, 477, 1960.

96. **Hopkins, D. L. and Hampton, R. E.,** Effects of tobacco etch virus infection upon the light reactions of photosynthesis in tobacco leaf tissue, *Phytopathology,* 59, 677, 1969.

97. **Goffeau, A. and Bové, J. M.,** Virus infection and photosynthesis. I. Increased photophosphorylation by chloroplasts from Chinese cabbage infected with turnip yellow mosaic virus, *Virology,* 27, 243, 1965.

98. **Gangulee, R., Singh, B. R., and Singh, H. C.,** Studies on metabolism of cowpea leaves infected with southern bean mosaic virus. I. Effect on carbohydrate metabolism, *Sci. Cult.,* 44, 226, 1978.

99. **Boninsegna, J. A. and Sayavedra, E.,** Starch metabolism in healthy and tomato bushy stunt virus-infected *Lycopersicon esculentum* plants, *Phytopathol. Z.,* 91, 163, 1978.

100. **Burroughs, R., Goss, J. A., and Sill, W. H., Jr.,** Alterations in respiration of barley plants infected with bromegrass mosaic virus, *Virology,* 29, 580, 1966.

101. **Bates, D. C. and Chant, S. R.,** The effects of virus infection on oxygen uptake and respiratory quotient of leaves of French bean and cowpea, *Physiol. Plant Pathol.,* 16, 199, 1975.

102. **Orlob, C. B. and Arny, D. C.,** Some metabolic changes accompanying infection by barley yellow dwarf virus, *Phytopathology,* 51, 768, 1961.

103. **Menke, G. H. and Walker, J. C.,** Metabolism of resistant and susceptible cucumber varieties infected with cucumber mosaic virus, *Phytopathology,* 53, 1349, 1963.

104. **Dwurazna, M. M. and Weintraub, M.,** Respiration of tobacco leaves infected with different strains of potato virus X, *Can. J. Bot.,* 47, 723, 1969.

105. **Bell, A. A.,** Respiratory metabolism of *Phaseolus vulgaris* infected with alfalfa mosaic and southern bean mosaic viruses, *Phytopathology,* 54, 914, 1964.

106. **Tu, J. C. and Ford, R. E.,** Effect of maize dwarf mosaic virus infection on respiration and photosynthesis of corn, *Phytopathology,* 58, 282, 1968.

107. **Weintraub, M., Kemp, W. G., and Ragetli, H. W. J.,** Studies on the metabolism of leaves with localised virus infection. I. Oxygen uptake, *Can. J. Microbiol.,* 6, 407, 1960.

108. **Chant, S. R.,** Respiration rates and peroxidase activity in virus infected, *Phaseolus vulgaris, Experientia,* 23, 676, 1967.

109. **Sunderland, D. W. and Merrett, M. J.,** The respiration of leaves showing necrotic local lesions following infection by tobacco mosaic virus, *Ann. Appl. Biol.,* 56, 477, 1965.

110. **Yamaguchi, A.,** Increased respiration of leaves bearing necrotic local lesions, *Virology,* 10, 287, 1960.

111. **Yamaguchi, A. and Hirai, T.,** The effect of local infection with tobacco mosaic virus on respiration in leaves of *Nicotiana glutinosa, Phytopathology,* 49, 447, 1959.

112. **Parish, C. L., Zaitlin, M., and Siegel, A.,** A study of necrotic lesion formation by tobacco mosaic virus, *Virology,* 26, 413, 1965.

113. **Takahashi, T. and Hirai, T.,** Respiratory increase in tobacco leaf epidermis in the early stage of tobacco mosaic virus infection, *Physiol. Plant Pathol.,* 17, 63, 1964.

114. **Takahashi, W. N.,** Respiration of virus infected plant tissue and effect of light on virus multiplication, *Am. J. Bot.,* 34, 496, 1947.

115. **Fernandez-Gonzalez, O., Renaudin, J., and Bové, J.,** Infection of chlorophyll-less protoplasts from etiolated Chinese cabbage hypocotyls by turnip yellow mosaic virus, *Virology,* 104, 262, 1980.

116. **Kano, H.,** Effects of light and inhibitors of photosynthesis and respiration on the multiplication of tobacco mosaic virus in tobacco protoplasts, *Plant Cell Physiol.,* 26, 1241, 1985.

117. **Weintraub, M., Ragetli, H. W. J., and Lo, E.,** Mitochondrial content and respiration in leaves with localized virus infections, *Virology,* 50, 841, 1972.

118. **Russo, M. and Martelli, G. P.,** Cytology of *Gomphrena globosa* L. plants infected by beet mosaic virus (BMV), *Phytopathol. Mediterr.,* 8, 65, 1969.

119. **Pierpoint, W. S.,** Cytochrome oxidase and mitochondrial protein in extracts of leaves of *Nicotiana glutinosa* L. infected with tobacco mosaic virus, *J. Exp. Bot.,* 19, 264, 1968.

120. **Dwurazna, M. M. and Weintraub, M.,** The respiratory pathways of tobacco leaves infected with potato virus X, *Can. J. Bot.,* 47, 731, 1969.

121. **Solymosy, F. and Farkas, G. L.,** Metabolic characteristics at the enzymatic level of tobacco tissues exhibiting localised acquired resistance to viral infection, *Virology,* 21, 210, 1963.

122. **Baur, J. R., Halliwell, R. S., and Langston, R.,** Effect of tobacco mosaic virus infection on glucose metabolism in *Nicotiana tabacum* L. var Samsun. I. Investigations with $^{14}$C-labelled sugars, *Virology,* 32, 406, 1967.

123. **Baur, J. R., Richardson, B., Halliwell, R. S., and Langston, R.,** Effect of tobacco mosaic virus infection on glucose metabolism in *Nicotiana tabacum* L. var. Samsun. III. Investigation of hexosemono-phosphate shunt enzymes and steroid concentration and biosynthesis, *Virology,* 32, 580, 1967.

124. **Kuriger, W. E. and Agrios, G. N.,** Cytokinin levels and kinetin-virus interactions in tobacco ringspot virus infected cowpea plants, *Phytopathology,* 67, 604, 1977.

125. **Owen, P. C.,** Some effects of virus infection on leaf water contents of *Nicotiana* species, *Ann. Appl. Biol.,* 46, 205, 1958.

126. **Tinklin, R.,** Effects of aspermy virus infection on the water status of tomato leaves, *New Phytol.,* 69, 515, 1970.

127. **Cooper, P. and Selman, I. W.,** An analysis of the effects of tobacco mosaic virus on growth and the changes in the free amino compounds in young tomato plants, *Ann. Bot.,* 38, 625, 1974.

128. **Wynd, F. L.,** Metabolic phenomena associated with virus infection in plants, *Bot. Rev.,* 9, 395, 1943.

129. **Ghabrial, S. A. and Pirone, T. P.,** Physiological changes which precede virus-induced wilt of tabasco pepper, *Phytopathology,* 54, 893, 1964.

130. **White, J. C. and Horn, N. L.,** The histology of tabasco peppers infected with tobacco etch virus, *Phytopathology,* 55, 267, 1965.

131. **Greenleaf, W. H.,** Effects of tobacco etch virus in peppers (*Capsicum* spp.), *Phytopathology,* 43, 564, 1953.

132. **Beute, M. K. and Lockwood, J. H.,** Mechanism of increased root rot in virus-infected peas, *Phytopathology,* 58, 1643, 1968.

133. **Hancock, J. G. and Magyarosy, A.,** The influence of squash mosaic on squash hypocotyl permeability, *Phytopathology,* 62, 762, 1972.

134. **Esau, K.,** Phloem degeneration in *Gramineae* affected by the barley yellow-dwarf virus, *Am. J. Bot.,* 44, 245, 1957.

135. **Lindsey, D. W. and Gudauskas, R. T.,** Effects of maize dwarf mosaic virus on water relations of corn, *Phytopathology,* 65, 434, 1975.

136. **Singh, R. and Singh, A. K.,** Distribution of photosynthetic assimilates between healthy and common bean mosaic virus infected sunn hemp plant parts, *Phytophylactica,* 16, 239, 1984.

137. **Watson, M. A. and Watson, D. J.,** The effect of infection with beet yellows and beet mosaic viruses on the carbohydrate content of sugar beet leaves, and on translocation, *Ann. Appl. Biol.,* 38, 276, 1951.

138. **Panopoulos, N. J., Faccioli, G., and Gold, A. H.,** Translocation of photosynthate in curly top virus-infected tomatoes, *Plant Physiol.,* 50, 266, 1972.

139. **Murphy, P. A.,** On the cause of rolling in potato foliage and on some further insect carriers of the disease, *Sci. Proc. R. Dublin Soc.,* 17, 163, 1923.

140. **Faccioli, G., Panopoulos, N. J., and Gold, A. H.,** Kinetics of the carbohydrate metabolism in potato leaf roll infected *Physalis floridana, Phytopathol. Mediterr.,* 10, 1, 1971.

141. **Lastra, R. and Gil, F.,** Ultrastructural host cell changes associated with tomato yellow mosaic virus, *Phytopathology,* 71, 524, 1981.

142. **Kim, K. S. and Flores, E. M.,** Nuclear changes associated with euphorbia mosaic virus transmitted by the whitefly, *Phytopathology,* 69, 980, 1979.

143. **Kim, K. S., Shock, T. L., and Goodman, R. M.,** Infection of *Phaseolus vulgaris* by bean golden mosaic virus: ultrastructural aspects, *Virology,* 89, 22, 1978.

144. **Rasa, E. A. and Esau, K.,** Anatomic effects of curly top and aster yellows viruses on tomato, *Hilgardia,* 30, 496, 1961.

145. **Legrand, M., Fritig, B., and Hirth, L.,** Enzymes of the phenyl-propanoid pathway and necrotic reaction of hypersensitive tobacco to tobacco mosaic virus, *Phytochemistry,* 15, 1353, 1976.

146. **Paynot, M., Martin, C., and Giraud, M.,** Activité phenylalanine ammoniac lyase du *Nicotiana tabacum* var. Xanthi n.c. et infection systemique par le virus du mosaïque du Tabac, *C. R. Acad. Sci.,* 277D, 1713, 1973.

147. **Paynot, M. and Martin, C.,** Effect d'un transfert de 20 a 32°C sur l'activité phenylalanine ammoniac lyase de *Nicotiana tabacum* var. Xanthi n.c. sains et inoculés par le virus de la mosaïque du Tabac, *C. R. Acad. Sci.,* 278D, 533, 1974.

148. **Paynot, M., Martin, C., and Javelle, F.,** Activité phenylalanine ammoniac lyase de divers *Nicotiana tabacum* et réaction nécrotique d'hypersensibilité au virus de la mosaïque du Tabac, *C. R. Acad. Sci.,* 280D, 1841, 1975.

149. **Tanguy, J. and Martin, C.,** Phenolic compounds and the hypersensitivity reaction in *Nicotiana tabacum* infected with tobacco mosaic virus, *Phytochemistry,* 11, 19, 1972.

150. **Fritig, B., Kauffmann, S., Dumas, B., Geoffroy, P., Kopp, M., and Legrand, M.,** Mechanisms of the hypersensitivity reaction in plants, in *Plant Resistance to Viruses,* CIBA Found. Symp. No. 133, John Wiley & Sons, Chichester, U.K., 1987, 92.

151. **Fritig, B., Gosse, J., Legrand, M., and Hirth, L.,** Changes in phenylalanine ammonia lyase during the hypersensitive reaction of tobacco to TMV, *Virology,* 55, 371, 1973.

152. **Massala, R., Legrand, M., and Fritig, B.,** Effect of α-aminoacetate, a competitive inhibitor of phenylalanine ammonia lyase, on the hypersensitive resistance of tobacco to tobacco mosaic virus, *Physiol. Plant Pathol.,* 16, 213, 1980.

153. **Massala, R., Legrand, M., and Fritig, B.,** Comparative effects of two competitive inhibitors of phenylalanine ammonia lyase on the hypersensitive resistance of tobacco to tobacco mosaic virus, *Plant Physiol. Biochem.,* 25, 217, 1987.

154. **Vegetti, G., Conti, G. G., and Pesci, P.,** Changes in phenylalanine ammonia lyase, peroxidase and polyphenoloxidase during development of local necrotic lesions in pinto bean leaves infected with alfalfa mosaic virus, *Phytopathol. Z.,* 84, 153, 1975.

155. **Collendavelloo, J., Legrand, M., and Fritig, B.,** Plant disease and the regulation of enzymes involved in lignification. *De novo* synthesis controls O-methyltransferase activity in hypersensitive tobacco leaves infected by tobacco mosaic virus, *Physiol. Plant Pathol.,* 21, 271, 1982.

156. **Collendavelloo, J., Legrand, M., and Fritig, B.,** Plant disease and the regulation of enzymes involved in lignification. Increased rate of *de novo* synthesis of the three tobacco O-methyltransferases during the hypersensitive response to infection by tobacco mosaic virus, *Plant Physiol.,* 73, 550, 1983.

157. **Best, R. J.,** Studies on a fluorescent substance present in plants. 3. The distribution of scopoletin in tobacco plants and some hypotheses on its part in metabolism, *Aust. J. Exp. Biol. Med. Sci.,* 26, 225, 1948.

158. **Cabanne, F., Scalla, R., and Martin, C.,** Oxidase activities during the hypersensitive reaction of *Nicotiana Xanthi* to tobacco mosaic virus, *J. Gen. Virol.,* 11, 119, 1971.

159. **Suseno, H. and Hampton, R. E.,** The effect of three strains of tobacco mosaic virus on peroxidase and polyphenoloxidase activity in *Nicotiana tabacum, Phytochemistry,* 5, 819, 1966.

160. **Wood, K. R. and Barbara, D. J.,** Virus multiplication and peroxidase activity in leaves of cucumber (*Cucumis sativus* L.) cultivars systemically infected with the *W* strain of cucumber mosaic virus, *Physiol. Plant Pathol.,* 1, 73, 1971.

161. **Gáborjányi, R., Sagi, F., and Balász, E.,** Growth inhibition of virus infected plants: alterations of peroxidase enzymes in compatible and incompatible host-parasite relations, *Acta Phytopathol. Acad. Sci. Hung.,* 8, 81, 1973.

162. **van Loon, L. C. and Geelen, J. L. M. C.,** The relation of polyphenol-oxidase and peroxidase to symptom expression in tobacco var. "Samsun NN" after infection with tobacco mosaic virus, *Acta Phytopathol. Acad. Sci. Hung.,* 6, 9, 1971.

163. **Simons, T. J. and Ross, A. F.,** Metabolic changes associated with systemic induced resistance to tobacco mosaic virus in Samsun NN tobacco, *Phytopathology,* 61, 293, 1971.

164. **Simons, T. J. and Ross, A. F.,** Changes in phenol metabolism associated with induced systemic resistance to tobacco mosaic virus in Samsun NN tobacco, *Phytopathology,* 61, 1261, 1971.

165. **Farkas, G. L. and Stahmann, M. A.,** On the nature of changes in peroxidase isoenzymes in bean leaves infected by southern bean mosaic virus, *Phytopathology,* 56, 669, 1966.

166. **Novacky, A. and Hampton, R. E.,** Peroxidase isoenzymes in virus-infected plants, *Phytopathology,* 58, 301, 1968.

167. **Bates, D. C. and Chant, S. R.,** Alterations in peroxidase activity and peroxidase isoenzymes in virus-infected plants, *Ann. Appl. Biol.,* 65, 105, 1970.

168. **Wood, K. R.,** Peroxidase isoenzymes in leaves of cucumber (*Cucumis sativus* L.) cultivars systemically infected with the *W* strain of cucumber mosaic virus, *Physiol. Plant Pathol.,* 1, 133, 1971.

169. **Solymosy, F., Farkas, G. L., and Király, Z.,** Biochemical mechanism of lesion formation in virus-infected plant tissues, *Nature (London),* 184, 706, 1969.

170. **Wagih, E. E. and Coutts, R. H. A.,** Peroxidase, polyphenoloxidase and ribonuclease in tobacco necrosis virus infected or mannitol osmotically stressed cowpea and cucumber tissue. I. Quantitative alterations, *Phytopathol. Z.,* 104, 1, 1982.

171. **Wagih, E. E. and Coutts, R. H. A.,** Peroxidase, polyphenoloxidase and ribonuclease in tobacco necrosis virus infected or mannitol osmotically stressed cowpea and cucumber tissue. II. Qualitative alterations, *Phytopathol. Z.,* 104, 124, 1982.

172. **Batra, G. K. and Kuhn, C. W.,** Polyphenoloxidase and peroxidase activities associated with acquired resistance and its inhibition by 2-thiouracil in virus-infected soybean, *Physiol. Plant Pathol.,* 5, 239, 1975.

173. **Farkas, G. L. and Solymosy, F.,** Host metabolism and symptom production in virus-infected plants, *Phytopathol. Z.,* 53, 85, 1965.

174. **Fraser, R. S. S. and Whenham, R. J.,** Plant growth regulators and virus infection: a critical review, *Plant Growth Reg.,* 1, 37, 1982.

175. **Tavantzis, S. M., Smith, S. H., and Witham, F. H.,** The influence of kinetin on tobacco ringspot virus infectivity and the effect of virus infection on the cytokinin activity in intact leaves of *Nicotiana glutinosa* L., *Physiol. Plant Pathol.,* 14, 227, 1979.

176. **Sziráki, I. and Balász, E.,** The effect of infection by TMV on cytokinin level of tobacco plants, and cytokinins in TMV-RNA, in *Current Topics in Plant Pathology,* Király, Z., Ed., Akademia Kiadó, Budapest, 1977, 345.

177. **De Fazio, G.,** Cytokinin levels in healthy and bean golden mosaic virus (BGMV) infected bean plants (*Phaseolus vulgaris* L.), *Rev. Bras. Bot.,* 4, 57, 1981.

178. **Faccioli, G., Rubies-Autonell, C., and Albertini, R.,** Role of cytokinins in the acquired resistance of *Chenopodium amaranticolor* towards an infection of tobacco necrosis virus, *Phytopathol. Mediterr.,* 23, 15, 1984.

179. **Sziráki, I., Balázs, E., and Király, Z.,** Role of different stresses in inducing systemic acquired resistance to TMV and increasing cytokinin levels in tobacco, *Physiol. Plant Pathol.,* 16, 277, 1980.

180. **Fraser, R. S. S. and Whenham, R. J.,** Inhibition of the multiplication of tobacco mosaic virus by methyl benzimidazol-2-yl-carbamate, *J. Gen. Virol.,* 39, 191, 1978.

181. **Király, Z. and Pozsár, B. I.,** On the inhibition of TMV production by kinetin and adenine in intact tobacco leaves, in *Host-Parasite Relations in Plant Pathology,* Király, Z. and Ubrizsy, G., Eds., Plant Profection Institute, Budapest, 1964, 61.

182. **Milo, G. E. and Srivastava, B. I. S.**, Effect of cytokinins on tobacco mosaic virus production in local-lesion and systemic hosts, *Virology*, 38, 26, 1969.

183. **Reunov, A. V., Reunova, G. D., Vasilyeva, L. A., and Reifman, G.**, Effect of kinetin on tobacco mosaic virus and potato virus X replication in leaves of systemic hosts, *Phytopathol. Z.*, 90, 342, 1977.

184. **Király, Z., El Hammady, M., and Pozsár, B. I.**, Susceptibility to tobacco mosaic virus in relation to RNA and protein synthesis in tobacco and bean plants, *Phytopathol. Z.*, 63, 47, 1968.

185. **Nakagaki, Y.**, Effect of kinetin on local lesion formation on detached bean leaves inoculated with tobacco mosaic virus or its nucleic acid, *Ann. Phytopathol. Soc. Jpn.*, 37, 307, 1971.

186. **Aldwinckle, H. S. and Selman, I. W.**, Some effects of supplying benzyladenine to leaves and plants inoculated with viruses, *Ann. Appl. Biol.*, 60, 49, 1967.

187. **Tu, J. C.**, Interaction of calcium with indole-3-acetic acid and kinetin during the formation of local lesions in bean (*Phaseolus vulgaris*) by alfalfa mosaic virus, *Can. J. Bot.*, 64, 1097, 1986.

188. **Balázs, E., Barna, B., and Király, Z.**, Effect of kinetin on lesion development and infection sites in Xanthi-nc tobacco infected by TMV: single-cell local lesions, *Acta Phytopathol. Acad. Sci. Hung.*, 11, 1, 1976.

189. **Pavillard, J.**, La teneur en auxine des pommes de terre virosées, in *Proc. 2nd Conference Potato Virus Diseases*, Pudoc, Wageningen, 178, 1955.

190. **Smith, S. H., McCall, S. R., and Harris, J. H.**, Alterations in the auxin levels of resistant and susceptible hosts induced by the curly top virus, *Phytopathology*, 58, 575, 1968.

191. **Rajagopal, R.**, Effect of tobacco mosaic virus infection on the endogenous levels of indoleacetic, phenylacetic and abscisic acids of tobacco leaves at various stages of development, *Z. Pflanzenphysiol.*, 83, 403, 1977.

192. **Grieve, B. J.**, Studies in the physiology of host-parasite relations. 4. Some effects of tomato spotted wilt virus on growth, *Aust. J. Exp. Biol. Med. Sci.*, 21, 89, 1943.

193. **Russell, S. L. and Kimmins, W. C.**, Growth regulators and the effect of BYDV on barley (*Hordeum vulgare* L.), *Ann. Bot. (London)*, 35, 1037, 1971.

194. **Jones, J. P.**, Studies on the auxin levels of healthy and virus infected plants, *Diss. Abstr.*, 16, 1567, 1956.

195. **Ochs, G.**, Untersuchungen über den Einfluss eines phytopathogen Virus auf den Wuchsstoff-haushalt der Rebe, *Naturwissenschaften*, 14, 343, 1958.

196. **Hartman, R. T. and Price, W. C.**, Synergistic effect of plant growth substances and southern bean mosaic virus, *Am. J. Bot.*, 37, 820, 1950.

197. **Thung, T. H. and Hadiwidjaja, T.**, Growth substances in relation to virus diseases: experiments with *Arachis hypogea, Tijdschr. Plantenziekten*, 57, 95, 1951.

198. **van Loon, L. C.**, Regulation of changes in proteins and enzymes associated with active defence against virus infection, in *Active Defence Mechanisms in Plants*, NATO Advanced Study Institute Series, Vol. 37, Wood, R. K. S., Ed., Plenum Press, New York, 1982, 247.

199. **Cheo, P. C.**, Effect of plant hormones on virus-replicating capacity of cotton infected with tobacco mosaic virus, *Phytopathology*, 61, 869, 1971.

200. **van Loon, L. C. and Berbée, A. T.**, Endogenous levels of indoleacetic acid in leaves of tobacco reacting hypersensitively to TMV, *Z. Pflanzenphysiol.*, 89, 373, 1978.

201. **Lockhart, B. E. L. and Semancik, J. S.**, Growth inhibition, peroxidase and 3-indole acetic acid oxidase activity, and ethylene production in cowpea mosaic virus-infected cowpea seedlings, *Phytopathology*, 60, 553, 1970.

202. **Bailiss, K. W.**, Gibberellins, abscisic acid and virus-induced stunting, in *Current Topics in Plant Pathology*, Király, Z., Ed., Akademiai Kadó, Budapest, 1977, 361.

203. **Aharoni, N., Marco, S., and Levy, D.**, Involvement of gibberellins and abscisic acid in suppression of hypocotyl elongation in CMV-infected cucumbers, *Physiol. Plant Pathol.*, 11, 189, 1977.

204. **Whenham, R. J. and Fraser, R. S. S.**, Effect of systemic and local-lesion-forming strains of tobacco mosaic virus on abscisic acid concentration in tobacco leaves: consequences for the control of leaf growth, *Physiol. Plant Pathol.*, 18, 267, 1981.

205. **Whenham, R. J., Fraser, R. S. S., and Snow, A.**, Tobacco mosaic virus-induced increased in abscisic acid concentration in tobacco leaves: intracellular location and relationship to symptom severity and to extent of virus multiplication, *Physiol. Plant Pathol.*, 26, 379, 1985.

206. **Balázs, E., Gáborjányi, R., and Király, Z.**, Leaf senescence and increased virus susceptibility in tobacco: the effect of abscisic acid, *Physiol. Plant Pathol.*, 3, 341, 1973.

207. **Balázs, E., Gáborjányi, R., Toth, A., and Király, Z.**, Ethylene production in Xanthi tobacco after systemic and local virus infections, *Acta Phytopathol. Acad. Sci. Hung.*, 4, 355, 1969.

208. **Gáborjányi, R., Balázs, E., and Király, Z.**, Ethylene production, tissue senescence and local virus infections, *Acta Phytopathol. Acad. Sci. Hung.*, 6, 51, 1971.

209. **Nakagaki, Y., Hirai, T., and Stahmann, M. R.**, Ethylene production by detached leaves infected with tobacco mosaic virus, *Virology*, 40, 1, 1970.

210. **Bailiss, K. W., Balázs, E., and Király, Z.**, The role of ethylene and abscisic acid in TMV-induced symptoms in tobacco, *Acta Phytopathol. Acad. Sci. Hung.*, 12, 133, 1977.

211. **Levy, D. and Marco, S.,** Involvement of ethylene in epinasty of CMV-infected cucumber cotyledons which exhibit increased resistance to gaseous diffusion, *Physiol. Plant Pathol.,* 9, 121, 1976.
212. **Marco, S. and Levy, D.,** Involvement of ethylene in the development of cucumber mosaic virus-induced chlorotic lesions in cucumber cotyledons, *Plant Pathol.,* 14, 235, 1979.
213. **Balázs, E. and Gáborjányi, R.,** Ethrel-induced leaf senescence and increased TMV susceptibility in tobacco, *Z. Pflanzenkr. Pflanzenschutz,* 81, 389, 1974.
214. **Pritchard, D. W. and Ross, A. F.,** The relationship of ethylene to formation of tobacco mosaic virus lesions in hypersensitive responding tobacco leaves with and without induced resistance, *Virology,* 64, 295, 1975.
215. **De Laat, A. M. M. and van Loon, L. C.,** Regulation of ethylene biosynthesis in virus-infected tobacco leaves. II. Time course of level, of intermediates and in vivo conversion rates, *Plant Physiol.,* 69, 240, 1982.
216. **De Laat, A. M. M. and van Loon, L. C.,** The relationship between stimulated ethylene production and symptom expression in virus-infected tobacco leaves, *Physiol. Plant Pathol.,* 22, 261, 1983.
217. **Bailiss, K. W.,** The relationship of gibberellin content to cucumber mosaic virus infection of cucumber, *Physiol. Plant Pathol.,* 4, 73, 1974.
218. **Ben-Tal, Y. and Marco, S.,** Qualitative changes in cucumber gibberellins following cucumber mosaic virus infection, *Physiol. Plant Pathol.,* 16, 327, 1980.
219. **Bailiss, K. W.,** Gibberellins and the early disease syndrome of aspermy virus in tomato (*Lycopersicon esculentum* Mill.), *Ann. Bot.,* 32, 543, 1968.
220. **Fernandez, T. F. and Gáborjányi, R.,** Reversion of dwarfing induced by virus infection: effect of polyacrylic acid and gibberellic acid, *Acta Phytopathol. Acad. Sci. Hung.,* 11, 271, 1976.
221. **Nath, S. and Mandahar, C. L.,** Involvement of gibberellins in breaking bud dormancy in *Euphorbia* crinkle mosaic virus-infected stem cuttings of *Euphorbia pulcherrima* Wild., *Biol. Plant.,* 30, 260, 1988.
222. **Skaria, M., Lister, R. M., Forster, J. E., and Shaner, G.,** Virus content as an index of symptomatic resistance to barley yellow dwarf virus in cereals, *Phytopathology,* 75, 212, 1985.
223. **Zaitlin, M.,** How viruses and viroids induce disease, in *Plant Disease: An Advanced Treatise,* Vol. 4, Horsfall, J. G. and Cowling, E. B., Eds., Academic Press, New York, 1979, 257.
224. **Fraser, R. S. S., Gerwitz, A., and Morris, G. E. L.,** Multiple regression analysis of the relationships between tobacco mosaic virus multiplication, the severity of mosaic symptoms, and the growth of tobacco and tomato, *Physiol. Mol. Plant Pathol.,* 29, 239, 1986.
225. **Esau, K. and Cronshaw, J.,** Relation of tobacco mosaic virus to the host cells, *J. Cell. Biol.,* 33, 665, 1967.
226. **Tu, J. C.,** Temperature-induced variations in cytoplasmic inclusions in clover yellow mosaic virus-infected alsike clover, *Physiol. Plant Pathol.,* 14, 113, 1979.
227. **Matthews, R. E. F.,** *Plant Virology,* 2nd ed., Academic Press, New York, 1981, chap. 7.
228. **Shalla, T. A., Petersen, L. J., and Giunchedi, L.,** Partial characterisation of virus-like particles in chloroplasts of plants infected with the U.5. strain of TMV, *Virology,* 66, 94, 1975.
229. **Rochon, D. and Siegel, A.,** Chloroplast DNA transcripts are encapsidated by tobacco mosaic virus coat protein, *Proc. Natl. Acad. Sci. U.S.A.,* 81, 1719, 1984.
230. **Zaitlin, M. and Hull, R.,** Plant virus host interactions, *Annu. Rev. Plant Physiol.,* 38, 291, 1987.
231. **Reinero, A. and Beachy, R. N.,** Association of TMV coat protein with chloroplast membranes in virus-infected leaves, *Plant Mol. Biol.,* 6, 291, 1986.
232. **Nakayama, M., Horikoshi, M., Mise, K., Yamaoka, N., Park, P., Furusawa, I., and Shishiyama, J.,** Replication of brome mosaic virus RNA in chloroplasts, *Ann. Phytopathol. Soc. Jpn.,* 53, 301, 1987.
233. **Gadh, I. P. S. and Hari, V.,** Association of tobacco etch virus related RNA with chloroplasts in extracts of infected plants, *Virology,* 150, 304, 1986.
234. **Groening, B. R., Abouzid, A., and Jeske, H.,** Single-stranded DNA from abutilon mosaic virus is present in the plastids of infected *Abutilon sellovianum, Proc. Natl. Acad. Sci. U.S.A.,* 84, 8996, 1987.
235. **Harrison, B. D., Murant, A. F., Mayo, M. A., and Roberts, I. M.,** Distribution of determinants for symptom production, host range and nematode transmissibility between the two RNA components of raspberry ringspot virus, *J. Gen. Virol.,* 22, 233, 1974.
236. **Rao, A. L. N. and Francki, R. I. B.,** Distribution of determinants for symptom production and host range in the three RNA components of cucumber mosaic virus, *J. Gen. Virol.,* 61, 197, 1982.
237. **Jockusch, H. and Jockusch, B.,** Early cell death caused by TMV-mutants with defective coat proteins, *Mol. Gen. Genet.,* 102, 204, 1968.
238. **Dawson, W. O., Bubrick, P., and Grantham, G. L.,** Modification of the tobacco mosaic virus coat protein gene affecting replication, movement, and symptomatology, *Phytopathology,* 78, 783, 1988.
239. **Van Telgen, H. J., Van der Zaal, E. J., and van Loon, L. C.,** Evidence for an association between viral coat protein and host chromatin in mosaic-diseased tobacco leaves, *Physiol. Plant Pathol.,* 26, 83, 1985.

240. **Van Telgen, H. J., Goldbach, R. W., and van Loon, L. C.,** The 126,000 molecular weight protein of tobacco mosaic virus is associated with host chromatin in mosaic-diseased tobacco plants, *Virology,* 143, 612, 1985.

241. **Wijdeveld, M. M. G., Goldbach, R. W., Verduin, B. J. M., and van Loon, L. C.,** Association of viral 126kDa protein-containing X-bodies with nuclei in mosaic-diseased tobacco leaves, *Arch. Virol.,* 104, 225, 1989.

242. **Nishiguchi, M., Kikuchi, S., Kiho, Y., Ohno, T., Meshi, T., and Okada, Y.,** Molecular basis of plant viral virulence; the complete nucleotide sequence of an attenuated strain of tobacco mosaic virus, *Nucleic Acids Res.,* 13, 5585, 1985.

243. **Watanabe, Y., Morita, N., Nishiguchi, M., and Okada, Y.,** Attenuated strains of tobacco mosaic virus. Reduced synthesis of a viral protein with a cell-to-cell movement function, *J. Mol. Biol.,* 194, 699, 1987.

244. **Kimura, I., Minobe, Y., and Omura, T.,** Changes in a nucleic acid and a protein component of rice dwarf virus particles associated with an increase in symptom severity, *J. Gen. Virol.,* 68, 3211, 1987.

245. **Baughman, G. A., Jacobs, J. D., and Howell, S. H.,** Cauliflower mosaic virus gene VI produces a symptomatic phenotype in transgenic tobacco plants, *Proc. Natl. Acad. Sci. U.S.A.,* 85, 733, 1988.

246. **Daubert, S. D., Schoelz, J., Debao, L., and Shepherd, R. J.,** Expression of disease symptoms in cauliflower mosaic virus genomic hybrids, *J. Mol. Appl. Genet.,* 2, 537, 1984.

247. **Diener, T. O.,** Viroids and their interactions with host cells, *Annu. Rev. Microbiol.,* 36, 239, 1982.

248. **Schnölzer, M., Haas, B., Ramm, K., Hofmann, H., and Sänger, H. L.,** Correlation between structure and pathogenicity of potato spindle tuber viroid, *EMBO J.,* 4, 2181, 1985.

249. **Gross, H. J.,** Viroids: their structure and possible origin, in *Subviral Pathogens of Plants and Animals: Viroids and Prions,* Maramorosch, K. and McKelvey, J. J., Eds., Academic Press, New York, 1985, 165.

250. **Visvader, J. E. and Symons, R. H.,** Eleven new sequence variants of citrus exocortis viroid and the correlation of sequence with pathogenicity, *Nucleic Acids Res.,* 13, 2907, 1985.

251. **Visvader, J. E. and Symons, R. H.,** Replication of in vitro constructed viroid mutants: location of the pathogenicity-modulating domain of citrus exocortis viroid, *EMBO J.,* 5, 2051, 1986.

252. **Kaper, J. M. and Collmer, C. W.,** Modulation of plant viral diseases by secondary RNA agents, in *RNA Genetics,* Vol. 3, Domingo, E., Holland, J., and Ahlquist, P., Eds., CRC Press, Boca Raton, FL, 1988.

253. **Kaper, J. M. and Waterworth, H. E.,** Cucumber mosaic virus associated RNA 5: causal agent for tomato necrosis, *Science,* 196, 429, 1977.

254. **Harrison, B. D., Mayo, M. A., and Baulcombe, D. C.,** Virus resistance in transgenic plants that express cucumber mosaic virus satellite RNA, *Nature (London),* 328, 799, 1987.

255. **Baulcombe, D. C., Hamilton, W. D. O., Mayo, M. A., and Harrison, B. D.,** Resistance to viral disease through expression of viral genetic material from the plant genome, in *Plant Resistance to Viruses,* CIBA Found. Symp. 133, John Wiley & Sons, Chichester, U.K., 1987, 170.

256. **García-Arenal, F., Zaitlin, M., and Palukaitis, P.,** Nucleotide sequence analysis of six satellite RNAs of cucumber mosaic virus: primary sequence and secondary structure alterations do not correlate with differences in pathogenicity, *Virology,* 158, 339, 1987.

257. **Collmer, C. W. and Kaper, J. M.,** Site-directed mutagenesis of potential protein-coding regions in expressible cloned cDNAs of cucumber mosaic viral satellites, *Virology,* 163, 293, 1988.

258. **Kaper, J. M., Tousignant, M. E., and Steen, M. T.,** Cucumber mosaic virus-associated RNA 5. XI. Comparison of 14 CARNA 5 variants relates ability to induce tomato necrosis to a conserved nucleotide sequence, *Virology,* 163, 284, 1988.

259. **White, J. L. and Kaper, J. M.,** Absence of lethal stem necrosis in select *Lycopersicon* spp. infected by cucumber mosaic virus strain D and its necrogenic satellite CARNA 5, *Phytopathology,* 77, 808, 1987.

260. **Rezaian, M. A. and Symons, R. H.,** Anti-sense regions in satellite RNA of cucumber mosaic virus form stable complexes with the viral coat protein gene, *Nucleic Acids Res.,* 14, 3229, 1986.

261. **Gerlach, W. L., Llewellyn, D., and Haseloff, J.,** Construction of a plant disease resistance gene from the satellite RNA of tobacco ringspot virus, *Nature (London),* 328, 802, 1987.

262. **Israel, H. W. and Ragetli, H. W. J.,** An electron microscope study of tobacco mosaic virus lesions in *Nicotiana glutinosa, J. Cell Biol.,* 23, 499, 1964.

263. **Israel, H. W. and Ross, A. F.,** Fine structure of local lesions induced by tobacco mosaic virus in tobacco, *Virology,* 33, 272, 1967.

264. **Allison, A. V. and Shalla, T. A.,** The ultrastructure of local lesions induced by potato virus X: a sequence of cytological events in the course of infection, *Phytopathology,* 64, 784, 1974.

265. **Appiano, A., Pennazio, S., D'Agostino, G., and Redolfi, P.,** Fine structure of necrotic local lesions induced by tomato bushy stunt virus in *Gomphrena globosa* leaves, *Physiol. Plant Pathol.,* 11, 327, 1977.

266. **Hiruki, C. and Tu, J. C.,** Light and electron microscopy of potato virus M lesions and marginal tissue in red kidney bean, *Phytopathology,* 62, 77, 1972.

267. **Farkas, G. L., Király, Z., and Solymosy, F.,** Role of oxidative metabolism in the localization of plant viruses, *Virology,* 12, 408, 1960.

268. **Weststeijn, M.,** Permeability changes in the hypersensitive reaction of *Nicotiana tabacum* cv. Xanthi-nc after infection with tobacco mosaic virus, *Physiol. Plant Pathol.,* 13, 253, 1978.

269. **Pennazio, S. and Sapetti, C.,** Electrolyte leakage in relation to viral and abiotic stresses inducing necrosis in cowpea leaves, *Biol. Plant.,* 24, 218, 1982.

270. **Pennazio, S., Appiano, A., and Redolfi, P.,** Changes occurring in *Gomphrena globosa* leaves in advance of the appearance of tomato bushy stunt virus necrotic local lesions, *Physiol. Plant Pathol.,* 15, 177, 1979.

271. **Ruzicska, P., Gombos, Z., and Farkas, G. L.,** Modification of the fatty acid composition of phospholipids during the hypersensitive reaction in tobacco, *Virology,* 128, 60, 1983.

272. **Kato, S. and Misawa, T.,** Lipid peroxidation during the appearance of hypersensitive reaction in cowpea leaves infected with cucumber mosaic virus, *Ann. Phytopathol. Soc. Jpn.,* 42, 472, 1976.

273. **Abad, P., Guibbolini, M., Poupet, A., and Lahlou, B.,** Occurrence and involvement of adenylate cyclase activity in the first steps of tobacco mosaic virus infection of *Nicotiana tabacum* cv. Xanthi-nc leaves, *Biochim. Biophys. Acta,* 882, 44, 1986.

274. **Ohashi, Y. and Shimomura, T.,** Modification of cell membranes of leaves systemically infected with tobacco mosaic virus, *Physiol. Plant Pathol.,* 20, 125, 1982.

275. **Mauch, F., Hadwiger, L. A., and Boller, T.,** Ethylene: symptom, not signal for the induction of chitinase and β-1,3-glucanase in pea pods by pathogens and elicitors, *Plant Physiol.,* 76, 607, 1984.

276. **Modderman, P. W., Schot, C. P., Klis, F. M., and Wieringa-Brants, D. H.,** Acquired resistance in hypersensitive tobacco against tobacco mosaic virus induced by plant cell wall components, *Phytopathol. Z.,* 113, 165, 1985.

277. **Weststeijn, E. A.,** Evidence for a necrosis-inducing factor in tobacco mosaic virus-infected *Nicotiana tabacum* cv. Xanthi-nc grown at 22°C but not at 32°C, *Physiol. Plant Pathol.,* 25, 83, 1984.

278. **Hooley, R. and McCarthy, D.,** Extracts from virus infected hypersensitive tobacco leaves are detrimental to protoplast survival, *Physiol. Plant Pathol.,* 16, 25, 1980.

279. **Doke, N. and Ohashi, Y.,** Involvement of an $O_2^-$ generating system in the induction of necrotic lesions on tobacco leaves infected with tobacco mosaic virus, *Physiol. Mol. Plant Pathol.,* 32, 163, 1988.

280. **Montalbini, P. and Buonaurio, R.,** Effect of tobacco mosaic virus infection on levels of soluble superoxide dismutase (SOD) in *Nicotiana tabacum* and *Nicotiana glutinosa* leaves, *Plant Sci.,* 47, 135, 1986.

281. **Roosien, J. and Van Vloten-Doting, L.,** Complementation and interference of ultraviolet-induced Mts mutants of alfalfa mosaic virus, *J. Gen. Virol.,* 63, 189, 1982.

282. **Kado, C. I. and Knight, C. A.,** Location of a local lesion gene in tobacco mosaic virus RNA, *Proc. Natl. Acad. Sci. U.S.A.,* 55, 1276, 1966.

283. **Wilson, T. M. A., Perham, R. N., Finch, J. T., and Butler, P. J. G.,** Polarity of the RNA in the tobacco mosaic virus particle and the direction of protein stripping in sodium dodecyl sulphate, *FEBS Lett.,* 64, 285, 1976.

284. **Fraser, R. S. S.,** Varying effectiveness of the *N'* gene for resistance to tobacco mosaic virus in tobacco infected with virus strains differing in coat protein properties, *Physiol. Plant Pathol.,* 22, 109, 1983.

285. **Saito, T., Meshi, T., Takamatsu, N., and Okada, Y.,** Coat protein gene sequence of tobacco mosaic virus encodes a host response determinant, *Proc. Natl. Acad. Sci. U.S.A.,* 84, 6074, 1987.

286. **Knorr, D. A. and Dawson, W. O.,** A point mutation in the tobacco mosaic virus capsid protein gene induces hypersensitivity in *Nicotiana sylvestris, Proc. Natl. Acad. Sci. U.S.A.,* 85, 170, 1988.

287. **Foster, J. A. and Ross, A. F.,** The detection of symptomless virus-infected tissue in inoculated tobacco leaves, *Phytopathology,* 65, 600, 1975.

288. **Ohashi, Y. and Shimomura, T.,** Induction of local lesion formation on leaves systemically infected with virus by a brief heat or cold treatment, *Ann. Phytopathol. Soc. Jpn.,* 37, 211, 1971.

Chapter 3

PATHOGENESIS-RELATED PROTEINS

**O. P. Sehgal and F. Mohamed**

TABLE OF CONTENTS

# I. INTRODUCTION

The development of symptoms, especially that of necrosis, following invasion of many pathogens is accompanied by an increased synthesis and/or accumulation of a class of plant proteins termed pathogenesis-related proteins, or PRs.[1] The production of PRs is not an exclusive pathogen-specific response because similar proteins are synthesized under conditions of abiotic stress such as chemical injury,[2-5] osmotic shock,[6] or aging.[7] PR proteins are induced also if plant cells cultivated in vitro are exposed to an elicitor of phytoalexin production[8] or upon spraying leaves with the culture filtrate of a plant pathogenic fungus.[9] Apparently, activation of genes coding for these proteins is the response of plants to conditions of parasitic or physiological stress. Under normal growth conditions these genes are maintained in a repressed state or are expressed at low levels. The term plant "stress proteins" has also been proposed[10] for these proteins because of their induction under diverse stressful situations. Several PRs accumulate in the intercellular space;[2,11-14] consequently, the term "extracellular stress-related proteins"[12] has also been applied. PRs, however, are not induced upon heat shock treatment. Further, of the several different metal salts tested for PR induction in tobacco, only those of barium and manganese proved effective.[15] Obviously, some sort of a specificity exists in the recognition of a stress by plants and the subsequent synthesis of these proteins. In certain specialized cases, PRs are produced constitutively, as in the reciprocal hybrids of *Nicotiana glutinosa* and *N. debneyi*.[2,16] These hybrids develop normally and show no signs of karyological or physiological dysfunction. Apparently, a variety of biotic and abiotic stimuli cause induction of PRs. Gene derepression, mediated by a chemical "messenger", may be involved.[2] In *N. glutinosa* × *N. debneyi* amphidiploids, this messenger is expressed constitutively.[16]

Synthesis or accumulation of PRs is occasionally accompanied by the development of local or systemic resistance (acquired resistance) to infection by related or unrelated pathogens.[2,4] Several PR proteins possess chitinase or glucanase activity.[17-20] Consequently, PR proteins have attracted considerable attention from the viewpoint of the applicability of the recombinant DNA technology for engineering plant defense molecules in combating pathogens and pests. PR proteins also are desirable models for examining gene regulatory processes in higher plants.

# II. NOMENCLATURE AND CLASSIFICATION

PR proteins were described independently by van Loon and Van Kammen[21] and Gianinazzi et al.[22] in tobacco *N. tabacum* leaves responding hypersensitively to infection by tobacco mosaic virus (TMV). Much of the information concerning the properties of PRs has been derived from these proteins, which have been studied most extensively. PR proteins have been detected in approximately 20 plant species following viral infection or chemical treatment.

There is some evidence that PR proteins from divergent plant species are chemically similar or serologically related[23-29] (see also Section V). This implies a conservation of the structural traits of certain PRs and underscores their functional commonalities during evolution. Additionally, a spectrum of related or unrelated PRs is synthesized within a plant species.[2,11,24,25] A need consequently exists to standardize a unified system of nomenclature for these proteins.

Generally, PRs are identified following their separation by anionic or alkaline (pH 8.3) nondenaturing polyacrylamide gel electrophoresis (PAGE), and a classification system based upon their relative mobilities was proposed.[1] This system has considerable merit, but has not been applied or used consistently. According to this proposal, the fastest migrating protein in a plant, such as tobacco cv. Xanthi-nc, should be designated as Xanthi-nc PR-1. Other proteins are designated as Xanthi-nc PR-2, Xanthi-nc PR-3, etc., according to their decreasing mobilities. Those proteins which resemble Xanthi-nc PR-1 in electrophoretic mobility, possess the same mass, and are related serologically should be regarded as members of the PR-1 group, and designated PR-1a, PR-1b, PR-1c, etc., according to their decreasing mobilities. It is necessary to include the name of the cultivar of a plant species in cases where PR induction is examined in more than one cultivar (e.g., tobacco cvs. Samsun, Xanthi) and qualitative differences are indicated; otherwise, the cultivar name may be omitted. It is worth considering that the first letters of the name of the genus and the species be incorporated as a part of the PR nomenclature system, rather than the common name of the plant. Thus, tobacco Xanthi-nc PR-1 becomes NtXanthi-nc PR-1. This is to allow a distinction between the PR proteins in other plant species, such as *N. glutinosa* or *N. sylvestris*, both of which are commonly referred to as tobacco. The fastest migrating proteins in these respective species should be designated as NgPR-1 and NsPR-1. Similarly, in bean (*Phaseolus vulgaris*) cv. Pinto, a comparable protein should be named PvPinto PR-1. The basic PRs are detected and identified upon their separation by cationic or acidic (pH 4.3) PAGE. For these proteins, the Roman numerals may be used. Thus, the fastest migrating basic protein in bean cv. Pinto may be designated PvPinto PR-I, the next as PvPinto PR-II, etc.

Adoption of a uniform system of nomenclature for PR proteins is highly desirable and needed to alleviate confusion which exists in the literature. The system proposed by Antoniw et al.,[1] along with the suggestions made herein, is flexible enough to designate and differentiate the various PRs. It can be subjected to appropriate modifications or refinements as the need arises.

## III. ISOLATION AND PURIFICATION

Extraction of fresh or frozen leaf tissue with 0.1 *M* citrate-phosphate buffer (pH 2.8), or 0.1 *M* sodium acetate-acetic acid buffer (pH 5.0), containing ascorbic acid and β-mercaptoethanol, is a preferred method[2,3,10,23,25] for isolating most PRs. Clarification of the extract by low-speed centrifugation is followed by an ultracentrifugation step to remove virions.[23] The soluble proteins are precipitated with ammonium sulfate (30 to 75% saturation) and then subjected to preparative nondenaturing PAGE.[23,30] The protease-resistant nature[2,11] of several PRs and their tolerance to heat and/or acetone[23,24] are features which may be exploited in rigorous purification. Contaminating leaf glycoproteins may be removed from a PR preparation by passage through a column of immobilized concanavalin A.[30] A final purification step of several tobacco PRs involves high-pressure gel permeation chromatography or high-performance chromatofocusing.[31]

Several PRs have been identified as chitinases or glucanases, and antibodies against many have been prepared. Consequently, techniques incorporating affinity chromatography offer excellent opportunities for obtaining these proteins in a highly purified form. Intercellular fluid recovered from leaves has served as the starting source material in purifying some PRs.[30,32]

## IV. MODEL SYSTEMS

### A. Solanaceous PRs
*1. Tobacco*
At least nine acidic PRs can be detected (Figure 1) in tobacco leaves that produce local lesions upon TMV infection.[15] PR proteins b1 (PR-1a), b2 (PR-1b), and b3 (PR-1c) of 'Xanthi-nc' and proteins IV, III, and II of 'Samsun NN' are related serologically and possess the same mass (15 kDa); these constitute the PR-1 group of proteins. The Xanthi-nc b4 (PR-2), b5, and b6 are comparable to Samsun NN components I, N, and O, respectively, and possess glucanase activity;[29] these proteins are serologically similar (molecular weight 40

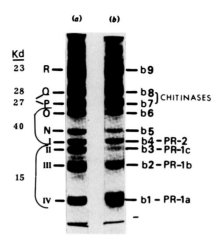

FIGURE 1. Nondenaturing PAGE of extracts of TMV-infected leaves of *Nicotiana tabacum* 'Samsun NN' (a) and 'Xanthi-nc' (b). (Modified from Antoniw, J. F. and White, R. F., in *Plant Resistance to Viruses,* Evered, D. and Harnnet, S., Eds., John Wiley & Sons, Chichester, U.K., 1987, 57. Copyright © 1987. Reprinted by permission of John Wiley & Sons, Ltd.)

kDa) and constitute another group of related PRs. The Xanthi-nc b7 (27 kDa) and b8 (28 kDa) correspond to Samsun NN components P and Q, respectively, are related serologically, and possess chitinase activity. Finally, Xanthi-nc b9, which corresponds to Samsun NN PR-R, is a thaumatin-like protein[30] with a molecular weight of 23 kDa. Two basic chitinases (32 and 34 kDa) are also induced in Samsun NN following TMV infection and are related serologically with the acidic chitinases PR-P and PR-Q.[17,18]

cDNA clones, representing seven classes of mRNAs (designated as clusters A to G) which were induced in TMV-infected Samsun NN leaves, were used to characterize the corresponding mRNAs and proteins.[33] These mRNAs occurred in low levels in healthy leaves, but were greatly stimulated (some more than a hundredfold) in the diseased tissue. Hybrid-selected translation with clusters B, D, E, and F cDNAs gave products that were precipitated with an antiserum to a mixture of PR proteins.[34] Subsequent tests showed that PRs 1a, 1b, and 1c corresponded to cluster B; PRs P and Q to cluster D; and PR-R to cluster E. The clusters G and F corresponded to the basic proteins homologous to those coded by the clusters B and D.[33]

Figure 2 is a schematic representation of the 'Samsun NN' PR-1b mRNA as well as of the amino acid sequence encoded in its open reading frame (ORF).[35] A leader sequence of 29 nucleotides precedes an ORF of 504 nucleotides, which is in turn followed by a 235 nucleotide noncoding sequence. The ORF encodes for a 30 amino acid signal peptide and the mature PR-1b protein of 138 amino acids (15 kDa). Preliminary observations indicate a homology of more than 90% among the partial PR-1a and PR-1b sequences and that of the PR-1c sequence. A similar level of homology also exists in the corresponding signal peptides. Isolation and nucleotide sequence of the genomic clones of Xanthi-nc PR-1a and PR-R have been reported recently.[36,37]

Several chemicals cause induction of PR proteins in tobacco leaves.[3,11,15] 2-Chloroethylphosphonic acid (ethephon) and 1-aminocyclopropane-1-carboxylic acid, which are ethylene-releasing compounds, are excellent inducers of some PRs.[2] Similarly, benzoic acid and its derivatives are effective in inducing several PR proteins.[2,11] Salicylic acid induces only the PR-1 group of proteins and PR-N in tobacco, indicating that it is a more specific inducer of acidic PRs compared to, for example, TMV.[33]

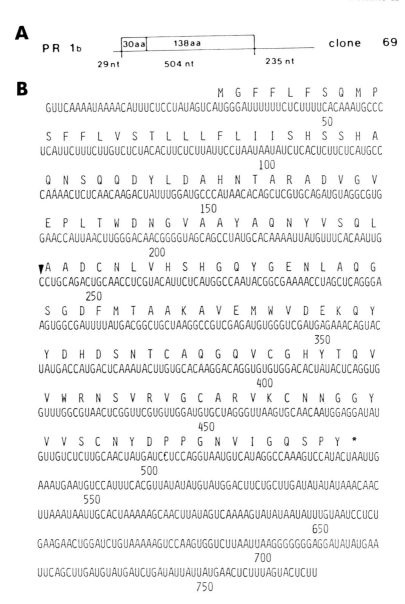

FIGURE 2. Diagrammatic representation (A) and complete nucleotide sequence (B) of the mRNA of tobacco PR-1b protein. (Modified from Cornelissen, B. J. C., Hooft van Huijsduijnen, R. A. M., van Loon, R. C., and Bol, J. F., *EMBO J.*, 5, 37, 1986. With permission.)

### 2. Tomato

A highly basic (pI 10.7) protein of molecular weight 14.2 kDa (P14) is induced in tomato (*Lycopersicon esculentum*) leaves upon infection with viroids, viruses, a fungus (*Cladosporium fulvum*), or upon treatment with ethephon or heavy metals.[38] P14 accumulates in the intercellular space around the mesophyll cells, and a close correlation exists between the symptom severity and P14 synthesis.[39]

P14 has been the subject of intensive study,[39] and its primary structure is shown in Figure 3. It consists of 130 amino acids. The five randomly distributed hydrophobic domains (residues 15 to 20, 42 to 48, 66 to 72, 94 to 101, and 109 to 114) render P14 suitable for interaction with cellular membranes. It is not a glycoprotein. P14 shows no relationship,

```
1                               10                              20
PCA-Asn-Ser-Pro-Gln-Asp-Tyr-Leu-Ala-Val-His-Asn-Asp-Ala-Arg-Ala-Gln-Val-Gly-Val-

21                              30                              40
Gly-Pro-Met-Ser-Trp-Asp-Ala-Asn-Leu-Ala-Ser-Arg-Ala-Gln-Asn-Tyr-Ala-Asn-Ser-Arg-

41                              50                              60
Ala-Gly-Asp-Cys-Asn-Leu-Ile-His-Ser-Gly-Ala-Gly-Glu-Asn-Leu-Ala-Lys-Gly-Gly-Gly-

61                              70                              80
Asp-Phe-Thr-Gly-Arg-Ala-Ala-Val-Gln-Leu-Trp-Val-Ser-Glu-Arg-Pro-Ser-Tyr-Asn-Tyr-

81                              90                              100
Ala-Thr-Asn-Gln-Cys-Val-Gly-Gly-Lys-Lys-Cys-Arg-His-Tyr-Thr-Gln-Val-Val-Arg-Leu-

101                             110                             120
Gly-Cys-Gly-Arg-Ala-Arg-Cys-Asn-Asn-Gly-Trp-Trp-Phe-Ile-Ser-Cys-Asn-Tyr-Asp-Pro-

121                             130
Val-Gly-Asn-Trp-Ile-Gly-Gln-Arg-Pro-Tyr
```

FIGURE 3.    Amino acid sequence of tomato P14 protein. (From Lucas, J., Henriquez, A. C., Lottspeich, F., Henschen, A., and Sanger, H. L., *EMBO J.*, 4, 2745, 1985. With permission.)

based upon sequence homologies, with a large variety of plant proteins, including cytochromes, lectins, peroxidases, storage proteins, and histones. The homology between P14 and tobacco PR-1b is approximately 60%.[39]

### 3. Potato

Nine acidic (pI 3.95 to 5.60; molecular weight range 28 to 41 kDa) and six basic (pI 7.2 to 9.2; molecular weight range 14 to 33 kDa) proteins accumulate in potato leaves, developing local lesions following inoculation with the U2 strain of TMV.[40] None of the acidic potato PRs resemble the tobacco PR-1 group of proteins.

Kombrink et al.[19] reported induction of two glucanases (36 and 36.2 kDa) and six chitinases (molecular weight ranging from 32.6 to 38.7 kDa) in potato leaves upon infection with the late blight fungus. These enzymes accumulate in the intercellular fluid of the leaves.

## B. Legume PRs
### 1. Vigna Species

Several new proteins are synthesized in cowpea (*Vigna unguiculata*) leaves reacting hypersensitively to infection by tobacco necrosis virus (TNV).[41] The concentration of these proteins was correlated with the symptom severity, with larger amounts contained in the necrotic area. Wilson[42] reported the induction of a 26-kDa protein in cowpea following infection by southern bean mosaic virus (SBMV), but no additional details were provided.

An acidic, 38-kDa protein (VuPR-1) is induced in cowpea plants upon infection with tobacco ringspot virus (TRSV) or cowpea chlorotic mottle virus (CCMV), or treatment of leaves with 0.1% salicylic acid.[23] VuPR-1 is also induced in *V. marina* and *V. sesquipedalis* (asparagus bean) upon TRSV infection. It lacks carbohydrate and possesses 1, 3-β-glucanase activity.

Pennazio et al.[43] reported synthesis of five new proteins in the primary leaves of *V. sesquipedalis* upon infection with TNV or tobacco rattle virus. One of these proteins was also induced upon treatment with sodium salicylate. Information on the properties of these proteins is lacking.

### 2. Phaseolus Species

Infection with TNV or alfalfa mosaic virus (AMV) or treatment with mercuric chloride causes induction of several new proteins in *P. vulgaris* cv. Saxa.[44-46] De Tapia et al.[47]

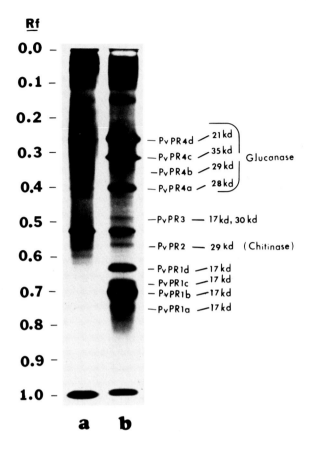

FIGURE 4. Electrophoresis of pH 5.0 extracts of (a) healthy and (b) SBMV-infected primary leaves of *Phaseolus vulgaris* 'Pinto' in Tris-borate-EDTA buffer, pH 8.3.

analyzed comparatively the types of proteins synthesized in response to AMV infection vs. treatment with mercuric chloride. Four acid-soluble proteins, PR-1 and PR-2 (22 kDa), PR-3 (29 kDa), and PR-4 (33.5 kDa), were synthesized upon mercuric chloride treatment. In the AMV-infected leaves only, three proteins were detected which corresponded to PR-1, PR-3, and PR-4 induced with mercuric chloride. All of these proteins were also present in the intercellular fluid. No serologic cross-reactivity was observed between PR-4 and PR-1, or PR-2 and PR-3 proteins. Some evidence exists[48] that Saxa PR-4 is synthesized as a 35-kDa nascent product and is then processed to yield a mature protein of 33.5 kDa.

Exposure to ethylene stimulates the synthesis of an endochitinase (35.4 kDa) in Saxa bean leaves.[49] A full-length copy of its mRNA has been cloned. From the nucleotide sequence of this clone, the primary structure of this enzyme has been deduced. There is a 27-residue amino signal sequence followed by 301 residues of the mature protein. It will be of interest to establish whether this protein is related to the Saxa PR-4 protein.

Nondenaturing electrophoretic analysis (Figure 4) of the soluble proteins from the primary leaves of *P. vulgaris* cv. Pinto reacting hypersensitively to SBMV infection shows the presence of ten acidic proteins which are not detected in healthy leaves.[24,50] The PvPinto PR-1 group contains four proteins with an apparent molecular weight of 17 kDa; these proteins are serologically interrelated. PvPinto PR-2 is a 29-kDa protein and possesses chitinase activity. Two proteins (17 and 30 kDa) comprise the PvPinto PR-3 component, but it is unclear whether these are distinctive proteins or are subunits of a single protein.

The PvPR-4 group consists of four serologically related 1,3-β-glucanases (21, 28, 29, and 35 kDa). The intensely staining protein (Rf 0.53) present in the healthy and diseased leaves is a phosphatase.

Infection of Pinto primary leaves with three other viruses (TMV, TRSV, and bean pod mottle virus [BPMV]) which produce lesions of divergent phenotypes causes protein changes similar to those induced by SBMV.[24,50,51]

Other than 'Pinto', 16 SBMV-susceptible *P. vulgaris* cultivars, reacting with either lesion formation or production of chlorosis on the primary leaves, were screened for the presence of PvPinto PR-1c and PvPinto PR-4d proteins.[52] Both of the proteins were synthesized in cultivars reacting with lesion formation, whereas, among those showing chlorosis, only PvPinto PR-1 was present. This observation indicates that induction of PvPR-4 protein is correlated with tissue necrotization. SBMV-induced necrotic lesion formation on *P. acutifolius* (tepary bean), *P. abrogineus*, *P. angustifolius*, *P. anisotrichus*, *P. dumosus*, and *P. microcarpus* is accompanied by the induction of PvPinto PR-1c and PR-4d proteins.[52]

Infection of the primary leaves of lima bean (*P. lunatus* cv. Nemagreen) with TRSV causes production of purple or brown lesions and a marked accumulation of an 18 kDa protein (P1PR-1).[27] A similar protein is also induced in *P. vulgaris* 'Pinto' infected with SBMV, TRSV, BPMV, or TMV; soybean (*Glycine max* cv. Avery) infected with CCMV; or cowpea infected with TRSV. P1PR-1 shows no serological relationship with either the cowpea PR-1 (38 kDa) or tobacco PR-1 or PR-2 proteins. P1PR-1 protein is neither a chitinase nor a glucanase.

### C. Cucumber PRs

Several PR proteins (ranging from 16 to 28 kDa) accumulate in cucumber leaves infected with viruses, bacteria, or fungi.[53,54] One of these proteins has been identified as an acidic endochitinase.[55,56] This protein, irrespective of the pathogen involved in its induction, accumulates in the intercellular fluid of the directly inoculated (first true leaf) as well as in the noninfected upper leaves. It is also induced following ethylene treatment of the leaves.

### D. Maize PRs

Infection of maize (*Zea mays* cv. INRA 258) with brome mosaic virus or treatment with 0.2% mercuric chloride stimulates the production of eight acidic PR proteins.[25] The estimated molecular weights for these proteins are PR-1, 14.2 kDa; PR-2, 16.5 kDa; PR-3 and PR-4, 25 kDa; PR-5, 29 kDa; PR-6a, 32 kDa; PR-6b, 30.5 kDa; and PR-7, 34.5 kDa. Maize PR-3 and PR-4 are serologically related, as are PR-6a and PR-6b, or PR-5 and PR-7. Maize PR-3, PR-4, PR-5, and PR-7 are endochitinases of comparable specific activities (50 to 80 nKat/mg protein). The multiple isoelectric forms of chitinases may result from posttranslational modifications, such as glycosylation, phosphorylation, methylation, etc.[49]

Table 1 compares the amino acid composition of selected maize PRs with those from the other plant species. A rather common feature of these proteins is a fairly high content of glycine.

## V. IMMUNOLOGICAL RELATIONSHIPS

A coordinated study was designed by van Loon et al.[57] to ascertain whether or not PR proteins from different plant species were serologically related. It was concluded that PRs from different genera were unrelated, but within a genus, such as *Nicotiana*, some PRs were interrelated. Ahl et al.[58] examined the serological relationship of PR proteins induced in *N. tabacum* cv. Xanthi-nc, *N. glutinosa*, and *N. sylvestris* by TMV; in *N. debneyi* by TNV; and in *N. tomentosiformis* treated with acetylsalicylic acid. The various PRs could be classified into three serological groups, irrespective of the species source. Furthermore, it was

## Table 1
## AMINO ACID COMPOSITION OF SELECTED PATHOGENESIS-RELATED PROTEINS

Number of residues (tentative)

| Amino acid | Bean cv. Pinto[a] | | Bean cv. Saxa[b] | | | Cowpea[c] | Tobacco[d] | Maize[e] | | | |
|---|---|---|---|---|---|---|---|---|---|---|---|
| | PvPR-1c (17 kDa) | PvPR4d (21 kDa) | PR-2 (22 kDa) | PR-3 (29 kDa) | PR-4 (33.5 kDa) | VuPR-1 (38 kDa) | PR-R (23 kDa) | PR-1 (14.2 kDa) | PR-2 (16.5 kDa) | PR-4 (25 kDa) | PR-5 (29 kDa) |
| Asp/Asn | 23 | 24 | 20 | 22 | 18 | 43 | 31 | 8 | 21 | 30 | 39 |
| Thr | 3 | 14 | 13 | 14 | 21 | 15 | 16 | 9 | 5 | 11 | 13 |
| Ser | 12 | 15 | 16 | 23 | 23 | 24 | 9 | 6 | 16 | 17 | 20 |
| Glu/Gln | 15 | 12 | 11 | 16 | 31 | 35 | 13 | 13 | 13 | 11 | 14 |
| Pro | 5 | 7 | 7 | 17 | 15 | 20 | 17 | 9 | 9 | 11 | 14 |
| Gly | 22 | 22 | 28 | 36 | 46 | 45 | 30 | 25 | 27 | 53 | 53 |
| Ala | 9 | 13 | 24 | 28 | 34 | 26 | 12 | 11 | 21 | 25 | 34 |
| Cys | 2 | 4 | ND | ND | 15 | 2 | 16 | 1 | 1 | 3 | ND |
| Val | 10 | 7 | 12 | 10 | 11 | 21 | 8 | 12 | 12 | 11 | 15 |
| Met | 1 | 1 | ND | 1 | 2 | 1 | 2 | 1 | 1 | 1 | 1 |
| Ileu | 3 | 5 | 11 | 19 | 9 | 15 | 7 | 5 | 1 | 10 | 13 |
| Leu | 6 | 7 | 13 | 22 | 14 | 23 | 9 | 6 | 5 | 16 | 21 |
| Tyr | 7 | 7 | 4 | 6 | 7 | 3 | 5 | 3 | 4 | 5 | 8 |
| Phe | 3 | 9 | 8 | 9 | 7 | 20 | 10 | 4 | 1 | 4 | 5 |
| His | 3 | 1 | 7 | 5 | 4 | 4 | Trace | 2 | 2 | 2 | 2 |
| Lys | 4 | 6 | 11 | 13 | 8 | 7 | 6 | 4 | 3 | 3 | 2 |
| Arg | 5 | 8 | 2 | 4 | 6 | 11 | 9 | 1 | 2 | 2 | 2 |
| Trp | ND[f] | ND | ND | ND | 5 | ND | 3 | ND | ND | ND | ND |

a Mohamed.[24]
b Calculated from De Tapia et al.[47]
c Dallali and Sehgal.[23]
d Pierpoint et al.[30]
e Calculated from Nasser et al.[25]
f Not determined.

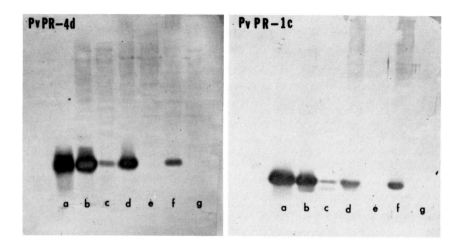

FIGURE 5.   Western blots for detecting the presence of PvPR-4d and PvPR-1c proteins in TRSV-infected plants. In both immunoblots the arrangement of the samples was as follows: (a) purified protein, PvPR-4d or PvPR-1c; (b) diseased 'Pinto' extract; (c) healthy 'Pinto'; (d) diseased Lima bean; (e) healthy Lima bean; (f) diseased cowpea; and (g) healthy cowpea.

observed that PRs within a group, but from different *Nicotiana* species, were more closely related to one another than those belonging to different groups, but present in the same species. The application of immunoblotting and the use of highly specific antibodies have permitted identification of related PR proteins in divergent plants. Thus, tomato P14 protein has been shown to be related to PR-1 group of tobacco PRs and a protein induced in cowpea leaves infected with TNV.[28] The presence of serologically related acidic PRs in *N. clevelandii*, *N. rustica*, and *Chenopodium amaranticolor* infected with the U1 strain of TMV, and in *N. sylvestris* and potato infected with the U2-TMV strain, also has been indicated.[14] Similarly, PRs related immunologically to Xanthi-nc PR-1a have been detected in virus-infected or salicylic acid-treated plants of corn, potato, tomato, and *C. amaranticolor*.[26] Additionally, untreated *Gomphrena globosa* plants contain a similar protein in small quantities, but its concentration increases upon treatment with salicylic acid.[26] Nasser et al.[25] observed that maize PR-2 protein was serologically related to tobacco PR-1b, with which it shows similarity in the amino acid composition.

Figure 5 illustrates that PvPR-1c and PvPR-4d, which are produced in SBMV-infected 'Pinto' leaves also are synthesized in cowpea, Lima bean, and 'Pinto' bean following infection with TRSV.

No serological interrelationship, however, has been indicated between Saxa bean PR-2 or PR-4 and tobacco PR-1a;[47] between cowpea VuPR-1 and tobacco PR-1 or PR-2 proteins;[23] and between lima PR-1 and tobacco PR-1 or PR-2 proteins.[27]

## VI. INDUCTION

Matsuoka and Ohashi[59] examined the induction of PR-1 group of proteins in 'Samsun NN' tobacco leaf discs infected with TMV or treated with salicylic acid by pulse labeling (Figure 6). In the salicylate-treated samples, PR synthesis began after a lag period of 8 h, whereas in the virus-infected tissue it occurred after a period of 18 h. PR synthesis declined rapidly in both cases after 50 h. Quantitatively more PRs were detected in the virus-infected tissue, with the most dramatic difference obvious in the case of PR-1b. Jamet et al.[60] observed that a significant amount of radioactivity was incorporated into tobacco PR-1a, PR-1c, PR-2, and PR-N when [14]C-labeled amino acids were injected into TMV-infected leaves. On the

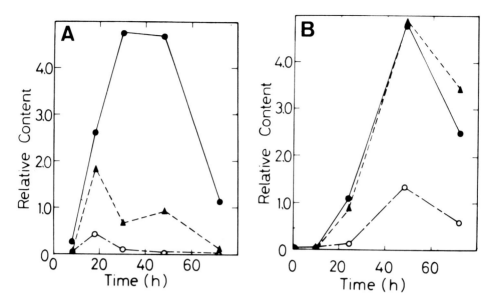

FIGURE 6. Time course of the synthesis of tobacco PR-1 proteins following (A) treatment with salicylic acid; (B) infection with TMV. PR-1a (●), PR-1b (▲), and PR-1c (○). (From Matsuoka, M. and Ohashi, Y., *Plant Physiol.*, 80, 505, 1985. With permission.)

basis of the comparative levels of the specific radioactivities of various PRs and of other host-coded proteins, it was concluded that PR proteins are synthesized de novo and are not products of proteolytic cleavage of preformed proteins. Similar conclusions were drawn concerning the synthesis of tomato P14 protein.[28] The synthesis of PR proteins is inhibited by cycloheximide, but not by chloramphenicol, indicating that production of cytoplasmic proteins is necessary for PR induction.[2]

The temporal induction of tobacco PR-1 mRNA in directly inoculated Samsun NN lower leaves, and the noninoculated upper leaves (Figure 7) was monitored by Northern blotting. In the lower leaves, PR-1 mRNA was first detected 2 d after inoculation and persisted for at least 12 additional days. In the upper leaves, which contain no virus, mRNA was detected between 8 and 11 d. Apparently, mRNAs coding for PR proteins are synthesized in the distant plant parts not invaded by a virus upon transport of a signal or the mobile "messenger" which emanates from the infected tissue.[2]

Neither actinomycin D nor cordycepin, which are inhibitors of DNA-dependent RNA synthesis, prevent PR induction.[2] Further, poly(A) mRNA from healthy or TMV-infected 'Xanthi-nc' leaves are equally efficient in directing PR synthesis in the in vitro translation system.[61] These observations indicate that synthesis of Xanthi-nc PR proteins is modulated at the translational level. On the other hand, in TMV-infected 'Samsun NN' tobacco, PR-1a mRNA concentration is more than a hundredfold greater than that in noninfected tissue. This suggests that PR-1a synthesis is regulated at the transcriptional level.[35] Additional studies are needed to obtain an improved understanding of the regulatory mechanism(s) involved in the synthesis of PR proteins.

The time course of induction of PvPinto PR-1 and PvPinto PR-4 in relation to SBMV virion synthesis is shown in Figure 8. Virions first become detectable at the time of lesion appearance (approximately 30 h postinoculation), increase in concentration to reach a maximum level between 72 to 100 h postinoculation, and then decline somewhat. The production of PR proteins begins at 60 h postinoculation and increases rapidly, reaching maximum values between 180 to 200 h postinoculation. The fact that the PR proteins are produced in abundant quantities when SBMV increase is restricted indicates that PRs may be involved, directly or indirectly, in influencing virus production.

FIGURE 7.    Induction of tobacco PR-1 mRNA in Samsun NN leaves following inoculation (0 to 14 d) of the lower leaves with TMV. RNA extracted from lower inoculated leaves (A—I), upper noninoculated leaves (J—M), or mock-infected plants (N—P) was electrophoresed, blotted onto nitrocellulose membrane, and hybridized with a $^{32}$P-labeled cDNA clone. (From Cornelissen, B. J. C., Hooft van Huijsduijnen, R. A. M., van Loon, R. C., and Bol, J. F., *EMBO J.*, 5, 37, 1986. With permission.)

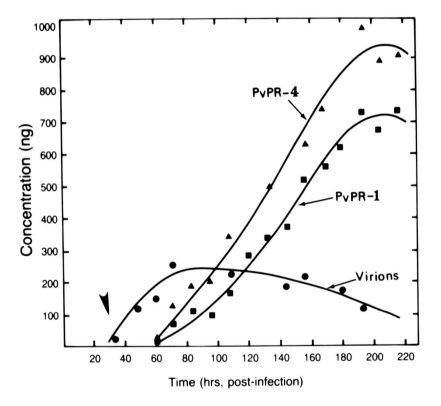

FIGURE 8.    Time course of the synthesis of SBMV (●), PvPR-1 (■), and PvPR-4 (▲) in the primary leaves of *Phaseolus vulgaris* 'Pinto'. Quantitative rocket immunoelectrophoresis was used for determining the concentration (nanograms per 2 mg leaf tissue) of virions and the PR proteins. Arrow indicates the time of the appearance of local lesions.

## VII. ENDOGENOUS CONCENTRATION AND DISTRIBUTION

PR proteins may reach a fairly high concentration in plant tissues. For example, in TMV-infected tobacco, PRs constitute approximately 10% of the total soluble leaf proteins.[11] The concentration of the PR-1 group of proteins in TMV-infected tobacco leaves may vary from 20 to 50 μg/g tissue,[62] while in the salicylate-treated tissue it may be as high as 90 μg/g tissue.[10] TMV-induced chitinases and glucanases may comprise as much as 4% (approximately 450 μg/g tissue) of the total soluble leaf protein fraction in tobacco 'Samsun NN'.[20] Other estimates of the endogenous concentrations of the PR proteins include 290 μg/g tissue for VuPR-1 in TRSV-infected cowpea leaves,[23] and 250 to 300 μg/g for PvPR-1 and PvPR-4 in SBMV-infected primary leaves of 'Pinto' bean.[24]

The concentration of PR-1a protein in a single TMV-induced lesion on 'Xanthi-nc' tobacco is approximately 1800 ng.[15,62] In an individual SBMV-induced lesion on 'Pinto' primary leaves, the concentrations of PvPR-1 and PvPR-4 proteins are 115 and 296 ng, respectively.[24]

The relative distribution of TMV virions and PR-1a in the different sections of an individual lesion was evaluated by Antoniw and White.[62] TMV concentrations were highest near the center and decreased progressively toward the periphery. Contrastingly, little PR-1 was detected in the center of the lesion; highest concentrations were just outside this area. This observation indicates, but does not prove, that PR-1a might actively be involved in viral localization.

Immunocytochemical studies[63] show that tomato P14 protein is localized mostly within the disorganized cytosol of the leaf cells infected with citrus exocortis viroid and in the intercellular space. Additionally, this protein was also observed in those cells of the non-infected leaves in which the cytosol was undergoing disorganization, as well as in the intercellular space of such leaves. There was no indication of a preferential P14 association with any specific cellular organelle.

The distribution of tobacco PR proteins examined by immunofluorescence microscopy[64] revealed their presence around epidermal, palisade, and lacunar cells, and occasionally in large amounts in the intercellular spaces. With immunogold labeling, these proteins were localized in the cytosol, particularly in association with endoplasmic reticulum, and also in the apoplast.[64] Figure 9 shows the association of PR-b1 protein with the middle lamella of tobacco leaves treated with polyacrylic acid.[33]

The time course of the accumulation of 'Pinto' PR proteins in the intercellular space of the SBMV-infected leaves is shown in Figure 10. The following conclusions may be drawn from these observations: (1) accumulation of PRs in the intercellular fluid follows lesion formation; (2) the most prominent PRs in the intercellular space are PvPR-4c and PvPR-4d, which have been identified as 1,3-β-glucanases; (3) PvPR-1b, PvPR-4a, and PvPR-4b are either excreted or not into the intercellular fluid or occur in extremely low concentrations. which have been identified as 1,3-β-glucanases; (3) PvPR-1b, PvPR-4a, and PvPR-4b are either not excreted into the intercellular fluid or occur in extremely low concentrations.

Figure 11 shows the relative concentration of VuPR-1 in the various parts, including roots, of a cowpea plant infected with TRSV; it occurs in the highest concentration in plant tissue showing necrosis.

## VIII. PROBABLE FUNCTIONS AND FUTURE PERSPECTIVES

Considerable evidence exists[2-4,11] that PR protein synthesis increases plant resistance to viruses, but the exact mechanism is unclear. No viral-inactivating activity appears to be associated per se with the PR proteins. Further, it is difficult to perceive how the extracellular PRs may affect multiplication or spread of viruses. Perhaps only those PRs contained within the cell are effective.[2] Gianinazzi et al.[65] postulated that PR proteins act in a manner similar

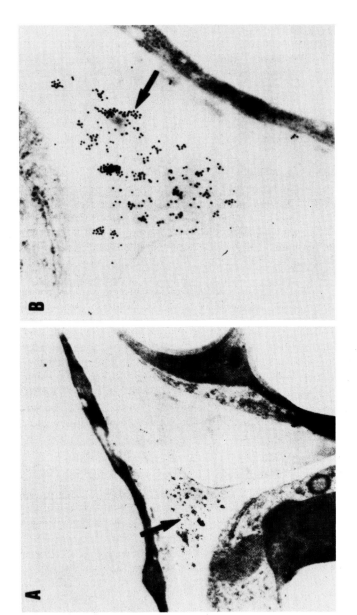

FIGURE 9.   Immunocytochemical localization of the PR-b1 protein in the middle lamella of tobacco leaves following induction with polyacrylic acid. (Reproduced from Bol, J. F., Hooft van Huijsduijnen, R. A. M., Cornelissen, B. J. C., and van Kan, J. A. L., in *Plant Resistance to Viruses*, Evered, D. and Harnett, S., Eds., John Wiley & Sons, Chichester, U.K., 1987, 72. Copyright © 1987. Reprinted by permission of John Wiley & Sons, Ltd.)

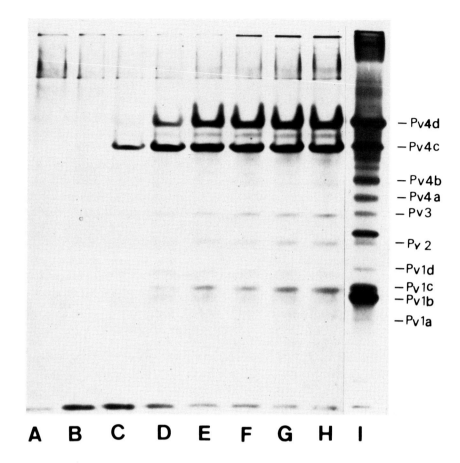

FIGURE 10.   Temporal accumulation of PR proteins in the intercellular fluid of the primary leaves of *Phaseolus vulgaris* 'Pinto' inoculated with SBMV. Samples were obtained at: lane (A) 0 h; (B) 36 h; (C) 60 h; (D) 84 h; (E) 108 h; (F) 132 h; (G) 156 h; and (H) 180 h. Lane I was charged with pH 5.0 extract of the virus-infected leaves (6 d postinoculation). Electrophoresis was in Tris-borate-EDTA buffer, pH 8.3.

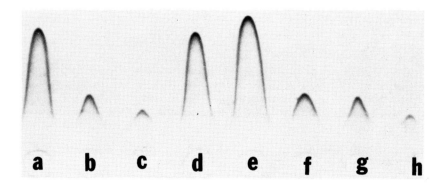

FIGURE 11.   Detecting VuPR-1 protein in the various parts of the cowpea plant infected with TRSV. Five microliter samples (1 g fr wt tissue per milliliter of 0.02 $M$ phosphate buffer, pH 7.0) were analyzed with rocket immunoelectrophoresis in 0.8% agarose. (a) Purified VuPR-1 protein, 600 ng; (b) stem; (c) roots; (d) directly inoculated primary leaf, with approximately 65% of the area showing necrosis, 25% chlorosis, and 10% normal green; (e) necrotic area only of the primary leaf; (f) chlorotic area; (g) green area; and (h) noninoculated trifoliate leaf.

to that of interferon in the animal cells. Alternatively, PR proteins may influence viral multiplication indirectly by affecting cellular constituents or structures necessary for viral life cycle. There is no actual evidence that PR proteins are toxic and cause cell death and necrotization, thereby arresting viral multiplication or spread. Vera et al.,[63] however, suggested such a mode of action for the tomato P14 protein. The involvement of any elicitors in the virus-induced hypersensitive reaction and the role of PR proteins in this process are conjectural.[66]

That ethylene plays an important role in the induction of PR proteins in the virus-infected and the noninfected plant parts is well substantiated.[2,4,11,67] Ethylene-stimulated transcriptional and/or translational processes may lead to the synthesis or accumulation of PRs.[2,49] The recognition of some PRs as chitinases or glucanases[17,20] underscores their importance as plant defense molecules against pathogens and pests. However, no biological function has been ascribed to the PR-1 group of proteins which are the predominant tobacco PRs. The 'Pinto' PR-1 group of proteins was tested for DNA-binding attributes, but none were found.[24]

The availability of complete cDNA clones for several PR proteins[32,33,36,37,49] offers opportunities for engineering plants in which one or more of these proteins is expressed constitutively. By using appropriate regulatory sequences and suitable promoters, these genes can be conditioned to express at preferred sites of pathogen invasion or establishment.[33]

## ACKNOWLEDGMENTS

We wish to thank Drs. L. C. van Loon and H. L. Sanger for permitting us to reproduce some of their illustrative materials, and Dr. S. G. Pueppke for helpful suggestions in preparation of this chapter. This is a contribution from the Missouri Agricultural Experiment Station, Journal Series No. 10,724.

## REFERENCES

1. **Antoniw, J. F., Ritter, C. E., Pierpoint, W. S., and van Loon, L. C.,** Comparison of three pathogenesis-related proteins from plants of two cultivars of tobacco infected with TMV, *J. Gen. Virol.,* 47, 79, 1980.
2. **van Loon, L. C.,** Pathogenesis-related proteins, *Plant Mol. Biol.,* 4, 111, 1985.
3. **White, R. F.,** Acetylsalicylic acid (aspirin) induces resistance to tobacco mosaic virus in tobacco, *Virology,* 9, 410, 1979.
4. **van Loon, L. C.,** The induction of pathogenesis-related proteins by pathogens and specific chemicals, *Neth. J. Plant Pathol.,* 89, 265, 1983.
5. **Barker, H.,** Effects of virus infection and polyacrylic acid on leaf proteins, *J. Gen. Virol.,* 28, 155, 1975.
6. **Wagih, E. E. and Coutts, R. H. A.,** Similarities in the soluble protein profiles of leaf tissue following either a hypersensitive reaction to virus infection or plasmolysis, *Plant Sci. Lett.,* 21, 61, 1981.
7. **Fraser, R. S. S.,** Evidence for the occurrence of the pathogenesis-related proteins in leaves of healthy tobacco plants during flowering, *Physiol. Plant Pathol.,* 19, 69, 1981.
8. **Somssich, I. E., Schmeltzer, E., Bollmann, J., and Hahlbrock, K.,** Rapid activation by fungal elicitor of genes encoding pathogenesis-related proteins in cultured parsley cells, *Proc. Natl. Acad. Sci. U.S.A.,* 83, 2427, 1986.
9. **Maiss, E. and Poehling, H. M.,** Resistance against plant viruses induced by culture filtrates of the fungus *Stachybotrys chartarium, Neth. J. Plant Pathol.,* 89, 323, 1983.
10. **Ohashi, Y. and Matsuoka, M.,** Synthesis of stress proteins in tobacco leaves, *Plant Cell Physiol.,* 26, 473, 1985.
11. **van Loon, L. C.,** Transcription and translation in the diseased plant, in *Biochemical Plant Pathology,* Callow, J. A., Ed., John Wiley & Sons, New York, 1983, chap. 18.
12. **Parent, J.-G. and Asselin, A.,** Detection of pathogenesis-related proteins (PR or B) and of other proteins in the intercellular fluid of hypersensitive plants infected with tobacco mosaic virus, *Can. J. Bot.,* 62, 564, 1984.

13. **Parent, J.-G., Hogue, R., and Asselin, A.,** Glycoprotein, enzymatic activities and b proteins in intercellular fluid extracts from hypersensitive *Nicotiana* species infected with tobacco mosaic virus, *Can. J. Bot.,* 63, 928, 1984.

14. **Parent, J.-G., Hogue, R., and Asselin, A.,** Serological relationships between pathogenesis-related leaf proteins from four *Nicotiana* species, *Solanum tuberosum,* and *Chenopodium amaranticolor, Can. J. Bot.,* 66, 199, 1988.

15. **Antoniw, J. F. and White, R. F.,** The role of pathogenesis-related proteins, in *Plant Resistance to Viruses,* Evered, D. and Harnett, D., Eds., John Wiley & Sons, Chichester, U.K., 1987, 57.

16. **Ahl, P. and Gianinazzi, S.,** b proteins as a constitutive component in highly (TMV) resistant interspecific hybrids of *Nicotiana glutinosa* × *Nicotiana debneyi, Plant Sci. Lett.,* 26, 173, 1982.

17. **Hooft van Huijsduijnen, R. A. M., Kauffmann, S., Brederode, F. T., Cornelissen, B. J. C., Legrand, M., Fritig, B., and Bol, J. F.,** Homology between chitinases that are induced by TMV infection of tobacco, *Plant Mol. Biol.,* 9, 411, 1987.

18. **Legrand, M., Kauffmann, S., Geoffroy, P., and Fritig, B.,** Biological functions of pathogenesis-related proteins: four tobacco pathogenesis related proteins are chitinases, *Proc. Natl. Acad. Sci. U.S.A.,* 84, 6750, 1987.

19. **Kombrink, E., Schroder, M., and Hahlbrock, K.,** Several "pathogenesis-related" proteins in potato are 1, 3-β-glucanases and chitinases, *Proc. Natl. Acad. Sci. U.S.A.,* 85, 782, 1988.

20. **Vogeli-Lange, R., Hansen-Gehri, A., Boller, T., and Meins, F.,** Induction of the defense-related glucanohydrolases, 1, 3-β-glucanase and chitinase, by tobacco mosaic virus infection of tobacco leaves, *Plant Sci.,* 54, 171, 1988.

21. **van Loon, L. C. and Van Kammen, A.,** Polyacrylamide disc electrophoresis of the soluble leaf proteins from *Nicotiana tabacum* var. Samsun and Samsun NN. Changes in protein constitution after infection with tobacco mosaic virus, *Virology,* 40, 199, 1970.

22. **Gianinazzi, S., Martin, C., and Vallee, J. C.,** Hypersensibilite aux virus, temperature et proteines solubles chez le *Nicotiana* Xanthi n.c. Apparition de nouvelles macromolecules lors de la repression de la synthese virale, *C.R. Acad. Sci.,* Paris, D270, 2383, 1970.

23. **Dallali, A. and Sehgal, O. P.,** Induction of a pathogenesis-related protein in selected *Vigna* species in response to virus infection, *Curr. Top. Plant Biochem. Physiol.,* 7, 232, 1988.

24. **Mohamed, F.,** unpublished observations, 1988.

25. **Nasser, W., De Tapia, M., Kauffmann, S., Montasser-Kouhsari, S., and Burkhard, G.,** Identification and characterization of maize pathogenesis-related proteins. Four maize PR proteins are chitinases, *Plant Mol. Biol.,* 11, 529, 1988.

26. **White, R. F., Rybicki, E. P., Von Wechmar, M. B., Dekker, J. L., and Antoniw, J. F.,** Detection of PR-1 type proteins in Amaranthaceae, Chenopodiaceae, Gramineae and Solanaceae by immunoelectroblotting, *J. Gen. Virol.,* 68, 2043, 1987.

27. **Sehgal, O. P. and Wilson, R. R.,** Stimulation of a 18 kd leaf protein in several legumes following viral infection, *Curr. Top. Plant Biochem. Physiol.,* 7, 222, 1988.

28. **Nassuth, A. and Sanger, H. L.,** Immunological relationships between "pathogenesis-related" leaf proteins from tomato, tobacco, and cowpea, *Virus Res.,* 4, 229, 1986.

29. **Kauffmann, S., Legrand, M., Geoffroy, P., and Fritig, B.,** Biological function of pathogenesis-related proteins: four PR proteins of tobacco have 1, 3-β-glucanase activity, *EMBO J.,* 6, 3209, 1987.

30. **Pierpont, W. S., Tatham, A. S., and Pappin, D. J. C.,** Identification of the virus-induced protein of tobacco leaves that resemble the sweet-protein thaumatin, *Physiol. Mol. Plant Pathol.,* 31, 291, 1987.

31. **Jamet, E. and Fritig, B.,** Purification and characterization of 8 of the pathogenesis-related proteins in tobacco reacting hypersensitively to tobacco mosaic virus, *Plant Mol. Biol.,* 6, 69, 1986.

32. **Payne, G., Dawn-Parks, T., Burkhart, W., Dincher, S., Ahl, P., Metraux, J. P., and Ryals, J.,** Isolation of the genomic clones for pathogenesis-related protein 1a from *Nicotiana tabacum* cv. Xanthinc., *Plant Mol. Biol.,* 11, 89, 1988.

33. **Bol, J. F., Hooft van Huijsduijnen, R. A. M., Cornelissen, B. J. C., and Van Kan, J. A. L.,** Characterization of pathogenesis-related proteins and genes, in *Plant Resistance to Viruses,* Evered, D. and Harnett, S., Eds., John Wiley & Sons, Chichester, U.K., 1987, 72.

34. **Hooft van Huijsduijnen, R. A. M., Cornelissen, B. J. C., van Loon, L. C., Van Boom, J. H., Tromp, M., and Bol., J. F.,** Virus induced synthesis of messenger RNAs for precursors of pathogenesis-related proteins in tobacco, *EMBO J.,* 9, 2167, 1985.

35. **Cornelissen, B. J. C., Hooft Van Huijsduijnen, R. A. M., van Loon, L. C., and Bol., J. F.,** Molecular characterization of messenger RNAs for 'pathogenesis-related' proteins 1a, 1b and 1c, induced by TMV infection of tobacco, *EMBO J.,* 5, 37, 1986.

36. **Payne, G., Middlesteadt, W., Williams, S., Desai, N., Dawn-Parks, T., Dincher, S., Carnes, M., and Ryals, J.,** Isolation and nucleotide sequence of novel cDNA encoding the major form of pathogenesis-related protein R, *Plant Mol. Biol.,* 11, 223, 1988.

37. **De Witt, P. G. M. and van Der Meer, F. E.,** Accumulation of pathogenesis-related tomato leaf protein P14 as an early indicator of incompatibility in the interaction between *Cladosporium fulvum* (Syn. *Fulva fulva*) and tomato, *Physiol. Mol. Plant Pathol.,* 28, 203, 1986.
38. **Camacho, H. and Sanger, H. L.,** Analysis of acid-extractable tomato leaf proteins after infection with a viroid, two viruses, and a fungus, and partial purification of the "pathogenesis-related" protein, *Arch. Virol.,* 74, 181, 1982.
39. **Lucas, J., Henriquez, A. C., Lottspeich, F., Henschen, A., and Sanger, H. L.,** Amino acid sequence of the 'pathogenesis-related' leaf protein P14 from viroid-infected tomato reveals a new type of structurally unfamiliar proteins, *EMBO J.,* 4, 2745, 1985.
40. **Parent, J.-G. and Asselin, A.,** Acidic and basic extracellular pathogenesis-related leaf proteins from fifteen potato cultivars, *Phytopathology,* 77, 1125, 1987.
41. **Coutts, R. H. A.,** Alterations in the soluble protein patterns of tobacco and cowpea leaves following inoculation with tobacco necrosis virus, *Plant Sci. Lett.,* 12, 189, 1978.
42. **Wilson, T. M. A.,** Pathogenesis-related protein synthesis in selected cultivars of beans and cowpeas following leaf damage by carborundum, treatment with aspirin, infection with tobacco mosaic virus or with bean or cowpea strain of southern bean mosaic virus, *Neth. J. Plant Pathol.,* 89, 313, 1983.
43. **Pennazio, S., Colariccio, D., Roggero, P., and Lenzi, R.,** Effect of salicylate stress on the hypersensitive reaction of asparagus bean to tobacco necrosis virus, *Physiol. Mol. Plant Pathol.,* 30, 347, 1987.
44. **Szczepanski, M. and Redolfi, P.,** Changes in the proteins of bean leaves infected with tobacco necrosis or alfalfa mosaic virus, *Phytopathol. Z.,* 113, 57, 1985.
45. **Redolfi, P.,** Protein changes and hypersensitive reaction in virus-infected bean leaves, *Riv. Patol. Veg.,* 19, 7, 1983.
46. **Redolfi, P. and Cantisani, A.,** Preliminary characterization of new soluble proteins in *Phaseolus vulgaris* cv Saxa reacting hypersensitively to viral infection, *Physiol. Plant Pathol.,* 25, 9, 1984.
47. **De Tapia, M., Bergmann, P., Awade, A., and Burkhard, G.,** Analysis of acid extractable bean leaf proteins induced by mercuric chloride treatment and alfalfa mosaic virus infection. Partial purification and characterization, *Plant Sci.,* 45, 167, 1987.
48. **De Tapia, M., Dietrich, A., and Burkhard, G.,** In vitro processing of a bean pathogenesis-related (PR 4) protein, *Eur. J. Biochem.,* 166, 559, 1987.
49. **Broglie, K., Gaynor, J. J., and Broglie, R. M.,** Ethylene-regulated gene expression: molecular cloning of the genes encoding an endochitinase from *Phaseolus vulgaris, Proc. Natl. Acad. Sci. U.S.A.,* 83, 6820, 1986.
50. **Mohamed, F. and Sehgal, O. P.,** Alterations in leaf proteins accompanying southern bean mosaic virus-induced necrosis in *Phaseolus vulgaris* L. cv. 'Pinto', *Phytopathology,* 75, 965, 1985.
51. **Mohamed, F. and Sehgal, O. P.,** Immunoreactivity of viral-induced pathogenesis-related proteins, *Curr. Top. Plant Biochem. Physiol.,* 6, 172, 1987.
52. **Sehgal, O. P.,** unpublished observations.
53. **Andebrhan, T., Coutts, R. H. A., Wagih, E. E., and Wood, R. K. S.,** Induced resistance and changes in the soluble protein fraction of cucumber leaves locally infected with *Colletotrichum lagenarium* or tobacco necrosis virus, *Phytopathol. Z.,* 98, 47, 1980.
54. **Gessler, C. and Kuc, J.,** Appearance of a host protein in cucumber plants infected with viruses, bacteria and fungi, *J. Exp. Bot.,* 33, 58, 1982.
55. **Metraux, J. P. and Boller, T.,** Local and systemic induction of chitinase in cucumber plants in response to viral, bacterial and fungal infections, *Physiol. Mol. Plant Pathol.,* 28, 161, 1986.
56. **Metraux, J. P., Streit, L., and Staub, T. H.,** A pathogenesis-related protein in cucumber is a chitinase, *Physiol. Mol. Plant Pathol.,* 33, 1, 1988.
57. **van Loon, L. C., Gianinazzi, S., White, R. F., Abu-Jawdah, Y., Ahl, P., Antoniw, J. F., Boller, T., Camacho-Henriquez, A., Conjero, V., Coussirat, J. C., Goodman, R. N., Maiss, E., Redolfi, P., and Wilson, T. M. A.,** Electrophoretic and serological comparison of pathogenesis-related (b) proteins from different plant species, *Neth. J. Plant Pathol.,* 89, 293, 1983.
58. **Ahl, P., Antoniw, J. F., White, R. F., and Gianinazzi, S.,** Biochemical and serological characterization of b-proteins from *Nicotiana* species, *Plant Mol. Biol.,* 4, 31, 1985.
59. **Matsuoka, M. and Ohashi, Y.,** Induction of pathogenesis-related proteins in tobacco leaves, *Plant Physiol.,* 80, 505, 1986.
60. **Jamet, E., Kopp, M., and Fritig, B.,** The pathogenesis-related proteins of tobacco; their labelling from $^{14}C$ amino acids in leaves reacting hypersensitively to infection by tobacco mosaic virus, *Physiol. Plant Pathol.,* 27, 29, 1985.
61. **Carr, J. P., Antoniw, J. F., White, R. F., and Wilson, T. M. A.,** Latent messenger RNA in tobacco (*Nicotiana tabacum*), *Biochem. Soc. Trans.,* 10, 353, 1982.
62. **Antoniw, J. F. and White, R. F.,** Changes with time in the distribution of virus and PR protein around single local lesions of TMV infected tobacco, *Plant Mol. Biol.,* 6, 145, 1986.

63. **Vera, P., Hernandez Yago, J., and Conjero, V.,** Immunocytochemical localization of the major "pathogenesis-related" (PR) protein of tomato plants, *Plant Sci.,* 55, 223, 1988.
64. **Dumas, E., Lherminier, J., Gianinazzi, S., White, R. F., and Antoniw, J. F.,** Immunocytochemical location of pathogenesis-related b1 protein induced in tobacco mosaic virus-infected or polyacrylic acid-treated tobacco plants, *J. Gen. Virol.,* 69, 2687, 1988.
65. **Gianinazzi, S., Pratt, H. M., Shewry, P. R., and Miflin, B. J.,** Partial purification and preliminary characterization of soluble leaf proteins specific to virus infected tobacco plants, *J. Gen. Virol.,* 34, 345, 1977.
66. **Fritig, B., Kaufmann, S., Dumas, B., Geoffroy, P., Kopp, M., and Legrand, M.,** Mechanism of hypersensitivity reaction of plants, in *Plant Resistance to Viruses,* Evered, D. and Harnett, D., Eds., John Wiley & Sons, Chichester, U.K., 1987, 92.
67. **Roggero, P. and Pennazio, S.,** Biochemical changes during the necrotic systemic infection of tobacco by potato virus Y, necrotic strain, *Physiol. Mol. Plant Pathol.,* 32, 105, 1988.

Chapter 4

# VIRUS TRANSPORT IN PLANTS

## C. L. Mandahar and I. D. Garg

## TABLE OF CONTENTS

# I. INTRODUCTION

Viruses enter their hosts through one or more of the following ways: mechanical inoculation through abrasion, and introduction through vectors, dodder, or graft union. Entry is through the epidermis in the case of mechanical inoculation and chiefly by nonpersistent transmission through vectors. Transmission through graft union, dodder, or vectors transmitting the virus in a persistent manner, generally places the virus in the deeper tissue layers of the host. The common sites of virus entry in mechanical inoculation are supposed to be the broken epidermal hairs and/or protoplasmic membranes exposed in the wounded epidermal cells. Abundant plasmodesmata usually occur in cell walls of hairs and between hair cells and the underlying epidermal cells. Virions coming directly into contact with the exposed plasmalemma may enter the cell by pinocytosis. Once inside the cell, virions undergo deproteinization. The released viral nucleic acid then produces one or more polypeptides through translation by using the host protein-synthesis machinery. The viral RNA translation products then take part in replication of viral genomic RNA, which in turn encodes viral coat protein besides one or more species of nonstructural virus proteins.

Upon mechanical inoculation, only a few host cells are infected directly by the virus. Thus, the number of primarily infected cowpea cells upon mechanical inoculation with tobacco mosaic virus (TMV) was estimated to be one in 50,000 to 1,50,000 mesophyll cells.[1] For an infection leading to virus disease symptoms, it is essential that virus infection spreads from the initially infected cells to the neighboring cells. This is the cell-to-cell spread of the virus, called slow or short distance transport, and is assumed to take place through plasmodesmata. Systemic infection of a plant can occur only when virus is able to become distributed throughout the plant. This is the long-distance or rapid movement of virus and takes place in phloem. When long-distance transport of a virus is combined with its slow cell-to-cell movement, the two together result in a thorough invasion of the host and lead to rapid appearance of systemic symptoms in infected plants. Thus, cell-to-cell movement of tobacco ringspot virus from inoculated areas to sieve tubes and its long-distance transport in phloem results in systemic infection of soybean plants within 3 to 10 d after inoculation of cotyledons or leaves.[2] A third type of virus transport which is intermediate in speed has been suggested by Schneider.[3]

Virus transport has been reviewed.[3-6] Much new and fundamental information has recently become available which has completely transformed our understanding of the various aspects of virus movement in plants.

# II. VIRUS TRANSPORT

## A. Cell-to-Cell Transport
*1. General*
Cell-to-cell transport occurs between parenchyma cells and between parenchyma and vascular tissue. Virus particles moving from cell to cell are not the ones originally introduced into the cell(s) by inoculation, but are the new generation of particles produced by multiplication of the original inoculum. The process is presumably repeated for every new cell infected, which explains the slow cell-to-cell movement of a virus. The bulk of the particles seem to remain in the cell in which they are synthesized, while only a limited, but an unknown number, moves on to the adjacent cells. Thus, cucumber mosaic virus (CMV) appeared to be transported from an infected cell to an adjacent cell after considerable accumulation of antigen in infected cells has taken place.[7]

Virus particles, infectious nucleic acid, and protein coat are nonmobile, and their movement within a cell or to the contiguous cells has obviously been thought to be dependent upon the transport systems operating within the host cell. Cytoplasmic streaming is an

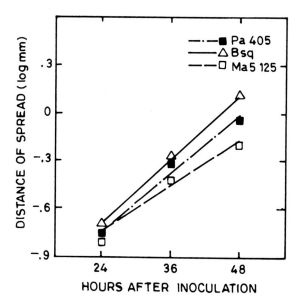

FIGURE 1.    Rate of cell-to-cell transport of maize dwarf mosaic virus strain B in mechanically inoculated leaves of Ma5125 (susceptible), Pa405 (resistant), and Bsq (resistant) corn inbreds. (From Lei, J. D. and Agrios, G. N., *Phytopathology*, 76, 1034, 1986. With permission.)

important force that brings about the intracellular movement of cytoplasmic entities. Viruses until quite recently were, therefore, also assumed to be translocated passively within the cell, to plasmodesmata, and through plasmodesmata to the adjoining cells with cytoplasmic streaming.[3,8]

As already mentioned, cell-to-cell transport is a slow process and its rate is significantly lower than that of long-distance transport. The rate of cell-to-cell movement is thought to be determined by velocity of cyclosis, length of cells, number of plasmodesmata connecting adjacent cells, and rate of virus multiplication.[9] The rate of cell-to-cell virus movement is determined by measuring the radial spread of a local lesion or by employing immunofluorescent antibodies to trace the spread of viral antigen. The former method is the one commonly employed in plants. Various estimates have been given of the cell-to-cell virus movement on this basis. It is estimated to be about 5 to 15 $\mu$m/h or one cell every 4 to 10 h;[3] 17 $\mu$m h$^{-1}$ of potato aucuba mosaic virus spread in Xanthi-nc tobacco at 15°C but 27 $\mu$m h$^{-1}$ at 25°C.[10] In compact callus tissue TMV spread at 0.8 mm/d,[11] and TMV infectious material took 8 and 10 h at 20 to 22° and 17 to 19°C, respectively, for its spread from epidermal to mesophyll cells.[12] Tobacco necrosis virus-induced local lesions increased in diameter from 0.2 mm at 48 h postinoculation (hpi) to 1.8 mm at 120 hpi.[13] The concentric spread of CMV from primarily infected cells of tobacco leaves was 8.4, 20.3, and 26.3 $\mu$m/h at 20, 25, and 30°C, respectively.[7] It is clear from some of the data given above that spread of viral infectious material from primarily infected epidermal cells to the adjacent mesophyll cells is influenced by temperature. The rate of cell-to-cell transport of maize dwarf mosaic virus in susceptible and resistant corn inbreds is shown in Figure 1.

## 2. Role of Plasmodesmata

The plant viruses are generally held to move from cell to cell through plasmodesmata. This is based on the common observation that particles of many spherical and several elongated viruses have been seen to occur within plasmodesmata and that particles of some

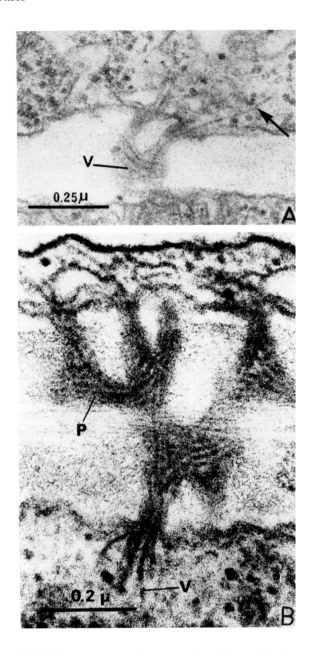

FIGURE 2.   (A) A plasmodesma-containing tobacco etch virus
particles out of which one particle (arrow) is emerging into the
cytoplasm from the plasmodesmatal pore. (From Weintraub, M.,
Ragetli, H. W. J., and Leung, E., *J. Ultrastruct. Res.,* 56, 351,
1976. With permission.) (B) A group of potato virus Y particles
(V) emerging into cytoplasm from the pore of a plasmodesma (P).
(From Weintraub, M., Ragetli, H. W. J., and Lo, E., *J. Ultrastruct.
Res.,* 46, 131, 1974. With permission.)

elongated viruses protrude from the plasmodesmatal channel into the cytoplasm of a cell
(Figure 2). Some such viruses are bean pod mottle, beet western yellows, beet yellows,
black raspberry necrosis, carnation etched ringspot, carrot mottle, carrot red leaf, cauliflower
mosaic, cherry leaf roll, chrysanthemum aspermy, cowpea mosaic, dahlia mosaic, grapevine
fanleaf, maize rough dwarf, potato leaf roll, potato virus X, potato virus Y, radish mosaic,

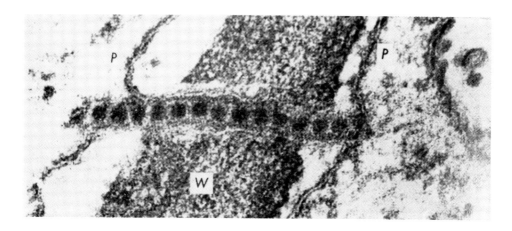

FIGURE 3. A tubule containing a linear row of strawberry latent ringspot virus particles passing through a plasmodesma of *Chenopodium amaranticolor* leaf cells and continues into the cytoplasm of the cell at the right. P = Plasmalemma; W = cell wall. (From Roberts, I. M. and Harrison, B. D., *J. Gen. Virol.*, 7, 47, 1970. With permission.)

raspberry ringspot, rice dwarf, strawberry latent ringspot, tobacco etch, tobacco mosaic, tobacco ringspot, tomato ringspot, and wound tumor.[14-16] These viruses belong to various virus groups with particle diameter shown in parenthesis: nepoviruses (20 to 25 nm), comoviruses (25 to 30 nm), caulimoviruses (45 nm), plant reoviruses (70 to 90 nm), luteoviruses (25 to 27 nm), closteroviruses (10 to 12 nm), potyviruses (11 to 15 nm), and some unassigned spherical and elongated viruses whose diameter falls in the above ranges.

Potato virus Y and tobacco etch virus occurred between the desmotubule and plasmalemma lining plasmodesmata. Particles of spherical viruses (nepo-, como-, caulimo-, and reoviruses) are often arranged in single files within double-walled tubules (Figure 3). These tubules are in or pass through and/or protrude from plasmodesmata and are 30 to 40 nm in diameter (Figure 4). The tubules have been suggested to be modified plasmodesmata[17] or part of the cellular microtubules, since virus-containing tubules have been suggested to be continuous with cellular microtubules.[18] Probably, tubules help in aligning virus particles along the opening of plasmodesmata for their free transit. The beet western yellows virus particles present within and outside plasmodesmata are spatially closely related, so that all particles appear to be moving from cell to cell.[19]

Plasmodesmata[15,20,21] are thin protoplasmic connections between adjacent cells — but not all cells. A "simple" plasmodesma is a plasmalemma-lined pore through which an unbranched desmotubule passes, with each of its ends being closely associated with endoplasmic reticulum and with a central rod being located within the desmotubule (Figure 5). The plasmalemma closely encompasses the desmotubule in the collar region at both ends of a plasmodesma. The desmotubule is considered to be composed of protein subunits and is thought to be similar in dimension and structure to a microtubule. However, even the structure of this simple plasmodesma has not been resolved unequivocally. For example, the existence of the central rod is doubted by several workers. It is considered to be an artifact and, if present, will profoundly affect the carrying capacity of plasmodesmata. Moreover, this simple plasmodesma appears to have different structures in different plant groups. More complicated plasmodesmata are also found. Many plasmodesmata have a median nodule, plasmodesmata/desmotubule of other plasmodesmata are branched, while the desmotubule is absent in some cases.

The diameter of plasmodesmata observed by optical microscope has usually been given in the range of 0.1 to 0.5 µm, which is practically the lower limit of resolution of such a

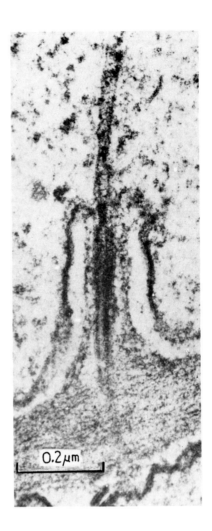

0.2 μm

FIGURE 4.    A tubule protruding from a mod-
ified plasmodesma in a carrot mottle virus in-
fected palisade cell of *Nicotiana clevelandii.*
(From Murant, A. F., Roberts, I. M., and
Goold, R. A., *J. Gen. Virol.*, 21, 269, 1973.
With permission.)

microscope. Reported diameters of plasmodesmata by electron microscope are given in Table 1. The outer diameter of plasmodesmata of some other plants generally varies from 40 to 60 nm; their internal diameters are still smaller so that size of plasmodesmatal canal is very small indeed. The distribution and frequency patterns of plasmodesmata in different types of cells suggest a specific predetermined function for them. This specific function is the translocation of materials like sugars and amino acids. Organelles and large macromolecules are much larger and cannot normally pass through plasmodesmata whether the desmotubule is present or not. Normal plasmodesmata have never been observed to contain ribosomes.[22]

Most of the plant viruses as well as their nucleic acid and coat protein are macromolecules of sizes too large to pass freely through the desmotubule of a normal plasmodesma. The majority of spherical viruses are in the range of 25 to 30 nm diameter, while particles of some spherical viruses are in the 40 to 80 nm diameter range. There is hardly any question of the ability of these viruses to pass through normal plasmodesmata. Free viral RNA of 2

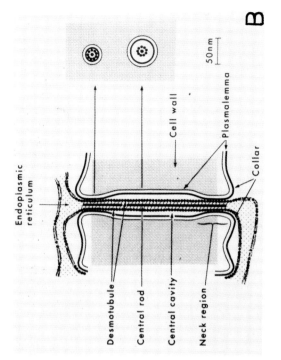

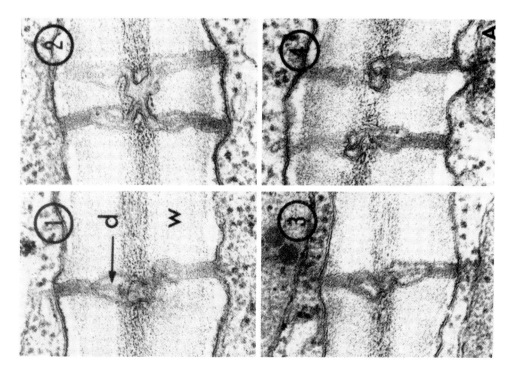

FIGURE 5. (A) Longitudinal sections of plasmodesmata in cells of uninoculated zinnia leaf (1 and 2). Longitudinal and cross-sections of plasmodesmata having normal structure in cells of zinnia leaf inoculated with dahlia mosaic virus (3 and 4). W = Cell wall; d = desmotubule. (From Kitajima, E. W. and Lauritis, J. A., *Virology*, 37, 681, 1969. With permission.) (B) Model of a simple plasmodesma showing various plasmodesmatal features. (From Robards, A. W., in *Intercellular Communication in Plants: Studies on Plasmodesmata*, Gunning, B. E. S. and Robards, A. W., Eds., Springer-Verlag, Berlin, 1976, 15. With permission.)

**Table 1**
**DIMENSIONS OF PLASMODESMATA (nm)**

| | Outer diameter of plasmalemma | Inner diameter of plasmalemma | Outer diameter of desmotubule | Inner diameter of desmotubule | Central rod |
|---|---|---|---|---|---|
| *Azolla*, young root cortical cells | 35 | 25 | 16 | 7 | 3 |
| *Hordeum*, young root (4 mm from tip) endodermal cells | 46 | 33 | 20 | 9 | 3 |
| *Hordeum*, older root (120 mm from tip) endodermal cells | 60 | 44 | 20 | 10 | 4 |
| *Abutilon*, distal cross-wall of stalk cell of nectary hair | 44 | 29 | 16 | 10 | 3 |

From: Robards, A. W., Plasmodesmata in higher plants, in *Intercellular Communication in Plants: Studies on Plasmodesmata,* Gunning, B. E. S. and Robards, A. W., Eds., Springer-Verlag, Berlin, 1976, 15. With permission.

$\times$ $10^6$ mol wt presumably occurs as an irregular spheroid because of extensive base-pairing and is estimated to have an average diameter of 10 nm.[15] Many elongated viruses have a diameter of 10 to 15 nm. The free nucleic acid and elongated viruses in their lengthwise direction may just be able to pass through plasmodesmata if the desmotubule is absent.

Still, viruses occur within plasmodesmata and move surprisingly freely and rapidly in parenchymatous tissue connected by plasmodesmata. Thus, plasmodesmata in virus-infected cells must undergo structural changes to allow the passage of virions from one cell to the other, or some virus-coded function is necessary, as it were, to open the gates to infection.[5] Several reports show structural (qualitative) and quantitative changes in plasmodesmata in virus-infected cells. About 1% of the plasmodesmata in dahlia mosaic virus-infected zinnia leaf cells were observed to have an altered fine structure.[23] They were characterized by the absence of a desmotubule and had an opening of 60 to 80 nm in diameter, as against 25 to 35 nm in the normal ones, and they contained virus-like particles (Figure 6). Similarly, the tubular core (desmotubule) was reported to be absent in plasmodesmata which contained beet western yellows virus particles in a single tubular profile in infected beet leaf cells.[23a] Desmotubule was also absent in plasmodesmata containing potato leaf roll virus particles.[24] Fine structural change in the form of pushing of the desmotubule to one side in plasmodesmata containing elongated virus particles of tobacco mosaic virus was observed by Weintraub et al.[25] Particles of the elongated beet yellows virus become so oriented as to pass plasmo-desmatal pores lengthwise and the plasmodesmata are wide enough to accommodate a group of particles that way.[26] In contrast, there was no difference in structure or size of plasmodesmata in healthy, mosaic TMV L strain (type strain), or LS1 (TMV temperature-sensitive mutant of the L strain) infected tobacco leaves at 22 or 32°C.[27]

A relation between the number of plasmodesmata connecting neighboring cells and the rate of movement of viral infection has been demonstrated by Wieringa-Brants[28] in cowpea and tobacco leaves inoculated with TMV and tobacco necrosis virus (TNV) and by Shalla et al.[27] in the case of a temperature-sensitive (*ts*) strain of TMV. The *ts* TMV isolate LS1 both replicates and moves from cell to cell in intact tobacco leaves at 22°C, while it only replicates without showing any cell-to-cell transport in leaves at the restrictive temperature of 32°C. This was due to a significantly decreased number of plasmodesmata between the healthy and LS1 infected cells at 32°C compared with in similar leaves kept at 22°C or between the healthy and type TMV isolate L (which can move from cell to cell at restrictive temperature)-infected cells in leaves kept at 32°C.[27] Thus the LS1-induced reduction in plasmodesmata at 32°C explains in part the lack of cell-to-cell movement of this ToMV isolate at restrictive temperature.[27] This also supports the view that cell-to-cell transport of

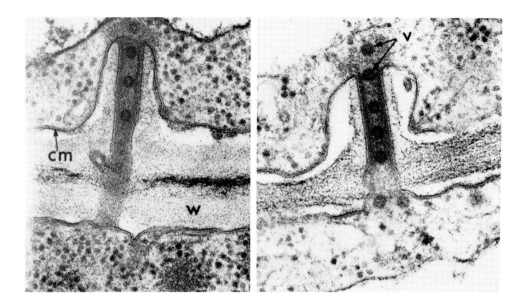

FIGURE 6. Longitudinal sections of transformed plasmodesmata in cells of zinnia leaf infected with dahlia mosaic virus. Virions (V) can be seen in the transformed plasmodesmata and cell wall (W) is projecting into the cytoplasm around an end of a plasmodesma. Compare these modified plasmodesmata with the normal plasmodesmata in uninfected and dahlia mosaic virus-infected zinnia leaf shown in Figure 5A. cm = Cytoplasmic membrane. (From Kitajima, E. W. and Lauritis, J. A., *Virology*, 37, 681, 1969. With permission.)

viruses occurs through plasmodesmata. Wieringa-Brants found that TMV and TNV in inoculated Xanthi tobacco and cowpea, respectively, took 6 to 8.5 h to travel from inoculated epidermis to mesophyll of plants kept in light, but only 1.0 to 1.5 h to travel from epidermis to mesophyll of plants kept 24 h in complete darkness.[28] This has been correlated with the increased number of plasmodesmata formed between epidermis and mesophyll after dark treatment: the average number of plasmodesmata per mm of cell was 0.8 and 3.2, respectively, in nondarkened and darkened cowpea leaves and 1.6 and 7.6, respectively, in nondarkened and darkened Xanthi tobacco leaves.[28] (See Figure 7.)

## B. Long-Distance Transport

Systemic spread of a virus within a host is the result of long-distance transport of virus through phloem tissue. Long-distance movement starts once the viral infection leaves the parenchyma and enters the phloem tissue. Some viruses may be introduced directly into this transport system by the vectors or through graft union. However, some vector-transmitted viruses show a lag period before their rapid translocation through the phloem.[3] Lag is considerably longer in mechanically transmitted viruses. Markedly different periods have been estimated for the egress of TMV infection from parenchyma and entry into phloem after mechanical inoculation. Thus, Kondo recorded entry of TMV infection into phloem in *Nicotiana glutinosa* cv. Xanthi 6 h after inoculation of an attached leaf in the midrib region.[29] Other workers have reported a period of 18 h to 4 or 5 d for the first detection of long-distance transport of TMV after inoculation.[3,30] It usually takes less than 24 h for TMV to enter the long-distance transport system after inoculation, but the lag period may be appreciably more in the case of other viruses.[3]

The considerable reported differences (6 h to 5 days) in the time taken for TMV to egress into the phloem from mechanically inoculated leaves have been attributed by Helms and Wardlaw[30] to differences in conditions under which studies were made and to the techniques

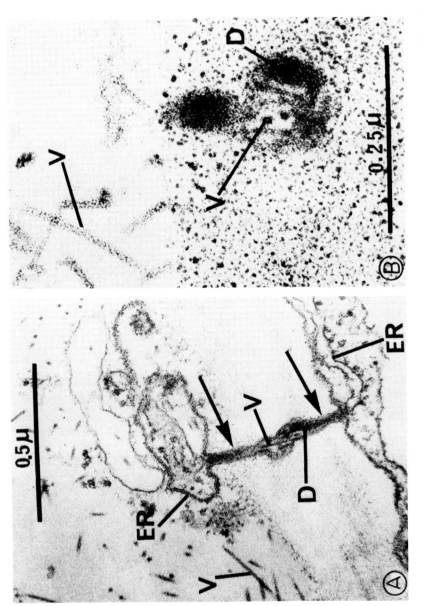

FIGURE 7.   (A) Longitudinal section of a plasmodesma. A TMV particle (V) occurs within the plasmodesma and the desmotubule (D) is pushed to one side of the plasmodesmatal channel. The desmotubule is in the center of the plasmodesma at places where no TMV particles occur (arrow). Plasmodesma is associated in normal fashion with cisternae of endoplasmic reticulum (ER) of the two adjacent cells. (B) Transverse sections of TMV particles (V) present in a plasmodesma. The electron-dense structure present at one side is suggested to be the desmotubule (D). The central hole of TMV virions is visible as the translucent area. (From Weintraub, M., Ragetli, H. W. J., and Leung, E., *J. Ultrastruct. Res.*, 56, 351, 1976. With permission.)

employed for detection of viral infection. Besides, many other factors such temperature, point of inoculation, position and the age of organ (leaf) inoculated, concentration of virus in inoculum, and rate of multiplication of virus in primarily infected cells influence the time required for entry of a virus into the long-distance transport system. In addition, it is possible that egress of the virus from parenchyma and entry into phloem may actually take more time in some viruses. The effect of age on entry of potato virus X (PVX) from inoculated leaves into tubers has been studied by Beemster.[31] There is a marked delay in entry of the virus into tubers if an old leaf is inoculated. This phenomenon has been termed mature plant resistance and is of considerable importance in the production of virus-free seed potatoes.

The rate of long-distance transport of the virus has been found to lie between 1.27 cm $h^{-1}$ [32] and 17.8 cm $h^{-1}$.[33] Helms and Wardlaw have reported a rate of 0.11 to 2.33 cm $h^{-1}$ for TMV in *N. glutinosa*.[30] The rate of virus transport in vascular tissues is around 15 to 80 mm $h^{-1}$ according to Gibbs,[15] while transport of vector-borne viruses upon entry into phloem was reported to be 20 mm $h^{-1}$ in the case of maize streak virus and up to 1500 mm $h^{-1}$ in the case of beet curly top virus. The rate of long-distance transport is many-fold higher than that of the cell-to-cell spread.

There is strong evidence to show that long-distance transport of a virus occurs through the phloem along with the photosynthetic assimilate.[3] Though rates of translocation of the photosynthetic assimilate (43 to 92 cm $h^{-1}$) and of long-distance virus transport (0.11 to 2.31 cm $h^{-1}$) are quite dissimilar, there is a strong relation between the two to support the view that viruses move in the phloem along with the photosynthate.[30,34] The much lower rate of long-distance movement of TMV in phloem may be due to the relatively large size of virions. A similarity between the direction of movement of the photosynthate and of virus has been reported by Bennett in his classical studies on beet curly top virus and triple-crown beet plants.[35] One of the crowns was inoculated with the virus and kept unshaded, the second was shaded but left uninoculated, while the third was neither shaded nor inoculated. Only the inoculated unshaded and the shaded uninoculated crowns developed the virus disease symptoms, demonstrating thereby the movement of the virus along with the photosynthate from the former to the latter. The spread of maize dwarf mosaic virus strain B in a susceptible inbred line correlated with metabolic flow in the vascular system, but this was not so in the resistant genotypes.[36] In potato, virus moves from the lower inoculated leaves towards the shoot tip before tuber initiation but towards the tubers after their initiation and thus follows the direction of photosynthetic assimilate towards the food sink.[31]

Long-distance transport of a virus is discontinuous in the initial stages of long-distance movement.[3,30] Thus, the first detection of TMV after 18 h of inoculation was restricted to some but not all of the adjacent segments of petiole and stem. It can be explained by postulating the reentry of virus from the phloem sieve tube to the adjacent parenchyma cells for multiplication at some irregular intervals.

The rate of transport of some viruses is intermediate between the slow cell-to-cell transport and the rapid long-distance transport. The path of intermediate transport is speculated to be the highly elongated immature fiber cells of vascular tissue. These cells possess a high velocity of directional cyclosis, and consequently movement of virus in these cells is supposed to be faster than that in the cell-to-cell transport.

Great importance was attached to the length of cells in virus transport by Zech.[37] By employing cytological abnormalities as a guide, he estimated the movement of TMV as 0.18 mm $h^{-1}$ in interveinal areas, 0.4 mm $h^{-1}$ in elongated epidermal cells along the veins, and 3.6 mm $h^{-1}$ in elongated bundle sheath cells.[37] In bean hypocotyl, the 2.00 mm average length of a phloem fiber is about 25 times that of a parenchyma cell. Spread of southern bean mosaic virus (SBMV) was 30 to 40 times faster in the region of the pericycle phloem than that in the inner cortical cells when the hypocotyl of young 'Pinto' bean was invaded by the virus.[38] Similar results were reported by Smith and McWhorter with respect to primary symptom development in *Vicia faba* inoculated with a strain of tomato ringspot virus.[39]

Necrosis progressed rapidly along the stem down to the roots; progression was along the parenchyma tissue external to the phloem, i.e., in the immature elongated primary phloem fibers.

Other probable candidates for the intermediate transport are transfer cells, which are characterized by the presence of specific protrusions in their walls and by numerous plasmodesmatic connections with adjacent sieve elements on one side and parenchyma on the other. Aggregates of clover yellow mosaic virus particles were observed in transfer cells and adjacent parenchyma in *Pisum sativum* 3 weeks after inoculation.[40] Due to their intermediary position, Hiruki et al.[40] suggested that transfer cells may play an intermediary role between cell-to-cell transport and long-distance transport of the virus.

## C. Transport Form of the Virus

Translocation of viral infection in plants has been variously suggested to occur as mature virions, as free viral nucleic acid, as viral replicative complex associated with cell membranes,[8] or as virus-specific informosomes, which are the ribonucleoprotein particles of nonribosomal nature.[41,42] Evidences in favor of each exist.

### 1. Translocation as Complete Virus Particles

TMV cannot multiply in the nonhost plant dodder. Nevertheless, trace amounts of TMV particles were found in dodder when the latter was grown on TMV-infected *Nicotiana* spp. These complete virus particles could only have moved as such from *Nicotiana* spp. to dodder. Observation by electron microscope of virions in plasmodesmata and phloem is considered to be the best proof of translocation of viruses as complete particles.

### 2. Translocation as Free Viral Nucleic Acid

Kinetic and radiation inactivation experiments show that the cells adjoining the primary infected cell become infectious before intact TMV particles can be detected in them. Thus, TMV-inoculated epidermal cells acquired the ability to infect the underlying mesophyll cells 3 h before the appearance of virions in epidermal cells.[43] TMV infection in case of coat protein-defective mutants produce lesions that expand at normal speed in inoculated leaves. Some *ts* coat protein mutants spread from cell to cell at the nonpermissive temperatures (30 to 35°C).[44,45] Similarly, a TMV mutant lacking coat protein at normal temperature does move from cell to cell.[5] Long tobacco rattle virus (TRV) particles do not carry the information for coat protein, but their RNA can replicate by itself and can spread through an infected plant and cause disease symptoms independently of the short rod particles. Deletion variants of TRV, which are incapable of forming complete nucleoprotein particles but induce lesion formation at normal speed, also appear to be transported as naked infective viral RNA. Powell et al.[46] implicated viral RNA as the more likely means of intercellular movement of pea enation mosaic virus. Lower molecular weight infectious RNAs, that is viroids, spread through plants. If a virus uncoats at a lipid-virus interface outside the cell, the infection must be travelling into a cell as naked RNA. Several of the common and extensively studied viruses like CMV, plant rhabdoviruses, and other viruses have not so far been observed in plasmodesmata.[15]

Differential temperature treatment (DTT) studies with TMV by Dawson et al.[47] also indicate that the infection travels as naked RNA. In these studies, lower leaves were mechanically inoculated and maintained at 25°C, whereas the upper uninoculated leaves were maintained at 5°C. In this system, the upper young leaves maintained at low temperature become infected with TMV from the older lower infected leaves of the same plant that are maintained at a high temperature (25°C) which is permissive for viral replication. After this treatment for several days, part of the upper leaves were immediately analyzed for TMV particles and the rest were incubated at 25°C for a couple of days and were then analyzed.

Results showed that the upper leaves lacked TMV virions immediately after DTT but had many of them after incubation. Viral infection thus reached the upper uninoculated leaves in a form other than virus. This form could be the viral RNA, since free viral RNA made up most of the infectivity of the upper leaves kept at low temperature.

### 3. Translocation as Viral Replicative Complex Associated with Cell Membranes

Vesicular cytopathological structures have been observed in phloem and plasmodesmata of pea enation mosaic virus-infected tissue.[46] It has been suggested on this basis that long-distance transport of this virus occurs as RNA associated with vesicle membrane which protects the viral RNA from RNases present in phloem tissue.[46]

### 4. Translocation as Informosomes

TMV mutants now commonly employed in the studies on virus transport in plants fall into two categories. TMV mutants Ni 118 and *flavum* have a temperature-sensitive *ts* coat protein gene, but are temperature-resistant (*tr*) with respect to their cell-to-cell movement. At restrictive temperature when coat protein is not produced, these mutants can still replicate and systemically spread in tobacco plants. The TMV mutants Ni 2519 and LS1 form the second category. They are *ts* with regard to their cell-to-cell spread, but *tr* with regard to coat protein. Being *ts* for transport function, they multiply normally in infected host cells but fail to spread systemically in tobacco plants at nonpermissive temperature at 32°C.[45,48,49]

Dorokhov et al.[50,51] analyzed the transport form of Ni 118 mutant at restrictive temperature of 35°C when it spreads in a form other than virions, since it fails to produce coat protein at the restrictive temperature. They found and implicated a new type of ribonucleoprotein particles called virus-specific informosome-like ribonucleoproteins (vRNP). The vRNP is comprised of genomic RNA ($2.1 \times 10^6$ mol wt), two subgenomic RNAs (RNAs $1_1$ and $1_2$), coat protein (17,000 mol wt), and other TMV-specific proteins (including the 30,000 mol wt protein).[51] The characteristic properties of TMV vRNP are 1.36 to 1.45 g/cm$^3$ buoyant density in cesium chloride density gradients, susceptibility to RNase, and filamentous structure different from the virion.[52] vRNP does not act as precursor of virions.[42] Subgenomic RNAs of vRNP guide the synthesis of TMV-specific proteins, one of which is the nonstructural transportation 30K protein which may be a structural component of vRNP.[41,42] vRNP has also been demonstrated in host cells infected with various TMV strains like *vulgare* and *ts* mutants Ni 2519 and LS1 and with potato virus X.[42,51]

vRNP in LS1 infected leaves is produced only at the permissive temperature (24°C) and not at the restrictive temperature (33°C).[51] The correlation between vRNP production and virus transport suggest that vRNP is involved in the cell-to-cell spread of viral infection. RNA 3 of alfalfa mosaic virus (AMV)[53] and brome mosaic virus[54] is necessary both for cell-to-cell spread and for coat protein synthesis of the respective virus. This correlation is significant in view of the fact that coat protein is an integral part of vRNP in TMV.

Dorokhov et al.[41] have also shown the possible involvement of vRNP even in the long-distance transport of viral infection. Studies were conducted by employing DTT and TMV mutants LS1 and Ni 118. In one experiment, lower leaves were inoculated with the common strain of TMV, while in the other experiment they were inoculated simultaneously with LS1 (*ts* in cell-to-cell transport function) and Ni 118 (*ts* coat protein) mutants. Lower inoculated leaves were maintained at 25°C in the first experiment and at 35°C in the second experiment, but the upper uninoculated leaves were kept at 5°C. In both the cases, vRNP was observed in the upper uninoculated leaves at the end of the DTT, indicating the involvement of the coat protein in long-distance transport. The presence of coat protein in vRNP could possibly protect viral RNAs during long-distance transport.

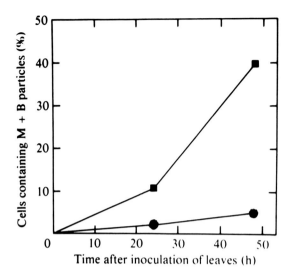

FIGURE 8. Cowpea mosaic virus (CPMV) is transported throughout an inoculated cowpea leaf only if inoculated with both the bottom (B) and middle (M) components together (■). The B component, when inoculated alone, can replicate in inoculated cells but fails to spread to other cells (●). The spread of the virus (percentages of infected cells) was detected by staining with fluorescent anti-CPMV serum the protoplasts prepared from inoculated leaves at times indicated. (Modified from Rezelman, G., Franssen, H. J., Goldbach, R. W., Ie, T. S., and van Kammen, A., *J. Gen. Virol.*, 60, 335, 1982. With permission.)

## III. THE ROLE OF VIRAL GENOME/GENETIC BASIS OF VIRUS TRANSPORT

A sizeable amount of evidence has become available within the last decade or so, which demonstrates that virus movement in a plant is a viral genome-controlled function.[5,6] The isolation of three TMV mutants (Ni 2519, II-27, and LS1), that are defective in their ability to move from cell to cell at restrictive temperature.[45,48,49,55] is clear proof of the genetic basis of the transport function. The postulated involvement of a virus-encoded transport protein in the transport function[56,57] is again indicative of the genetic control by virus of its transport in plants.

The next evidence favoring a genetic basis of virus transport came from multipartite viruses whose spread in host plants is controlled by one of the RNA segments of virus genome. The bottom (B) component of cowpea mosaic virus (CPMV) codes for functions related to viral RNA replication. However, even though B RNA is capable of infecting cells and replicating in them, the CPMV infection cannot spread to neighboring cells without the middle (M) component[58] (Figure 8). Cowpea chlorotic mottle virus RNA 1 is required for systemic spread of this virus in cowpea.[59] The RNA 2 (or M RNA) of red clover mosaic virus determines the morphology of lesions induced in cowpea and the ability of the virus to invade the plants systemically.[60] Only one of the two DNA molecules of tomato golden mosaic virus codes for cell-to-cell spread of the virus and expression of disease symptoms in transgenic plants.[61] The DNA 1 of cassava latent virus can replicate autonomously, but the virus can spread systemically in plants only if DNA 2 is also present.[62]

Even the long-distance spread of a virus in a susceptible plant is determined by viral

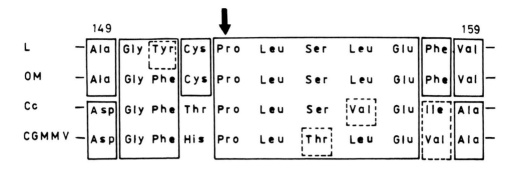

FIGURE 9.   The conserved amino acid sequences (residues 149 to 159) of the 30K proteins of TMV strains L (tomato strain), OM (common strain), and Cc (cowpea strain) and CGMMV (cucumber green mottle mosaic virus) around amino acid residue [proline (Pro)] (black arrow) which undergoes substitution to amino acid residue serine in TMV LS1 mutant (see text). The TMV LS1 mutant is defective in cell-to-cell movement at restrictive temperature and is derived from the wild-type tomato TMV (L strain). Solid lines enclose the sets of identical residues while dashed lines enclose the sets of residues that are regarded as frequent amino acid substitution. (From Ohno, T., Takamatsu, N., Meshi, T., Okada, Y., Nishiguchi, M., and Kiho, Y., *Virology*, 131, 255, 1983. With permission.)

genome. One of the most common resistance mechanisms of a plant is the inability of a virus strain to spread systemically in that plant. On the other hand, a resistance-breaking strain (formed by mutation in viral genome) will do so successfully.

## A. Transport Protein

Atabekov and Morozov[56] suggested that a TMV subgenomic RNA-coded nonstructural 30K protein, called the transport protein, controlled cell-to-cell spread of TMV infection. This suggestion was supported by Taliansky et al.[63] and by Leonard and Zaitlin.[57] Occurrence of 30K protein and its mRNA was detected in TMV-infected protoplasts first by Joshi et al.[64] and Ooshika et al.[65] Watanabe et al.,[66] while studying the time course of in vivo production in TMV-infected tobacco protoplasts of 30K protein, found that the protein and its mRNA were short-lived and transiently synthesized. They were first detected 3 to 4 h after inoculation, reached a maximum at 8 to 9 h post-inoculation, then decreased, and gradually disappeared thereafter. The maximum amount of the 30K protein was detected in intact leaves at 24 h after inoculation.[64] Various types of evidence have demonstrated that involvement of 30K protein in cell-to-cell spread of TMV.

Ohno et al.[67] and Leonard and Zaitlin[57] compared the polypeptide maps of 30K protein produced by TMV *ts* mutant LS1 and by its normal parent strain L. They observed a change in this protein produced by LS1. The sequence analysis of clones of 30K protein gene from LS1 and L revealed that a single base substitution at residue 1019 of C to U in the 30K protein cistron results in a serine replacement in the LS1 strain for a proline residue of the L strain at amino acid residue 153.[67] Thus the inability of LS1 strain to spread cell-to-cell seems to be due to a single amino acid substitution in the 30K protein.[67] It is significant in this connection that some highly conserved sequences occur in the amino acid sequence of 30K protein of several TMV strains (Figure 9).[67-69]

More direct evidence for the involvement of 30K protein in cell-to-cell transport of TMV has been obtained by Meshi et al.[70] They constructed several mutants of TMV with a modified 30K protein gene using an in vitro system to produce an infectious TMV RNA from a cloned cDNA copy and analyzed their ability for cell-to-cell spread and replication using tobacco plants and protoplasts. A mutant having a single base substitution in the 30K protein gene was found. This mutant was identical to the LS1 mutant in its genomic composition and in its sensitivity to temperature for cell-to-cell spread. Still more evidence in its favor was obtained by transforming a tomato plant by incorporating the 30K gene from L strain and

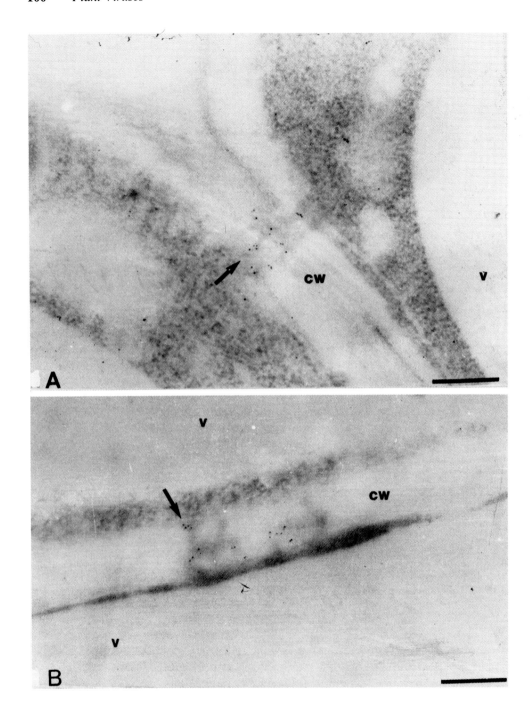

FIGURE 10.    (A and B)    Plasmodesmata in TMV infected cells showing localization of the 30K protein label
at 24 hours postinoculation. Position of gold-labeled 30K protein is indicated by arrows. CW = Cell wall; V
= vacuole; bars = 300 nm. (From Tomenius, K., Clapham, D., and Meshi, T., *Virology*, 160, 363, 1987.
With permission.)

demonstrating the ability of the LS1 TMV mutant to spread in the transgenic plant at restrictive temperature.[71a]

A stable association of the 30K protein with cell walls, particularly with plasmodesmata of systemically infected cells of *Nicotiana tabacum* cv. Samsun, has been demonstrated by Godefroy-Colburn et al.[71] and Tomenius et al.[72] Tomenius et al[72] studied the distribution and localization of 30K protein in infected cells by employing immunogold probing. They found that the 30K protein became localized in plasmodesmata rather than nuclei. The 30K protein accumulated inside plasmodesmata prior to the passage of any significant amounts of virus through them (Figure 10). It was detected in plasmodesmata at least by 16 hours postinoculation (hpi), reached maximum accumulation at 24 hpi, and number of plasmodesmata as well as intensity of label began decreasing from 25 hpi.[72] Watanabe et al.[73] employed pulse-labeling and pulse-chase experiments with ($^{35}$S) methionine and observed that the 30K protein, after production in cytoplasm, were translocated to the crude nuclear fraction. But this nuclear localization of the 30K protein may be due to their studies being conducted on a system which lacks cell walls, i.e., protoplasts.

Watanabe et al.[74] conducted studies on the site of mutation causing attenuation in TMV. The attenuation of TMV was found to be associated with reduced capability of cell-to-cell spread coupled with reduced synthesis of the 30K protein.

All the above experimental evidence strongly supports the view that the 30K protein mediates the cell-to-cell spread of TMV. Other virus-coded proteins shown to play role in cell-to-cell spread of viral infection are 48K protein produced by CPMV, and 32K protein produced by alfalfa mosaic virus (AMV). The RNA of B component of CPMV can replicate in cells independently of RNA of the M component particles, but fails to spread throughout leaves because presumably M component RNA encodes the transport protein.[58] It was further suggested that a 48K protein is involved in transport of this virus throughout leaves.[58] Wellink et al.[75] detected the association of this protein with membrane fraction of infected cells and protoplasts. The 48K protein is derived by proteolytic cleavage of a polyprotein, which is the translation product of M RNA. On the basis of association of 48K protein with the membrane fraction[75] and its limited homology with the 30K protein of TMV,[76] Wellink et al.[75] supported its postulated role in cell-to-cell spread of CPMV or its RNA.

Occurrence and localization of 32K (=35K) nonstructural protein encoded by AMV in the infected cells has been studied by immunochemical technique by Godefroy-Colburn et al.[77] The 32K protein is found associated with the membrane fraction shortly after inoculation,[78] but disappears from this fraction rapidly and is translocated to cell walls. Consequently, the 32K protein was detected by specific antiserum in a cell wall fraction of AMV-infected tobacco plants.[77] Translocation of 32K to the cell walls was demonstrated by employing immunoblotting with antibodies directed against a synthetic peptide corresponding to the C-terminus of the 32K protein.[77] The 32K protein undergoes slight modification after its translocation to cell walls. It has been speculated that the 32K protein modified the cell walls and plasmodesmata in such a way as to allow the cell-to-cell passage of viral infection. The modification of 32K protein reported in AMV-infected tobacco leaves[77] was also observed by Van Pelt-Heerschap et al.[79] in AMV-infected bean leaves but not in the protoplasts. In infected protoplasts, unmodified 32K protein was present only in the 100 g fraction as also reported by Watanabe et al.[73] in the case of TMV-infected tobacco protoplasts.

In further studies, Stussi-Garaud et al.[80] have demonstrated the localization immunocytochemically of 32K protein in the middle lamella of walls of those parenchyma or epidermal cells of tobacco leaves which have just been reached by the infection front and in which AMV multiplication has just begun. The 32K protein was not found when AMV had accumulated to high levels in infected cells. Since 32K protein is localized in middle lamella of those cells where virus multiplication has just begun, this supports the speculation that it modifies plasmodesmata and renders them permeable to viral infection.

Other viruses in which movement of viral infection is controlled by the viral genome and where tentative involvement of nonstructural virus-coded proteins with the virus transport is supposed to exist are tobacco streak virus,[81] brome mosaic virus (BMV), cauliflower mosaic virus,[82] tomato golden mosaic virus,[61] red clover necrotic mosaic virus,[60] and tomato black ring virus.[76] A region on the 150K polyprotein immediately N-terminal to the putative coat protein sequence of tomato black ring virus may perform the transport function.[76] It is suggested to be so because the amino acid sequence of this region resembles parts of the 30K transport protein of tobamoviruses (tobacco, tomato, and cowpea strains of TMV and CGMMV) as well as to, but to a lesser extent, the analogous part of the 105K polypeptide encoded by the RNA 2 of cowpea mosaic virus.[76]

Comparison of amino acid sequences of the putative movement proteins revealed a limited number of matches between AMV and BMV but no matches with that of TMV 30K protein.[83] Antibodies directed against 35K ( = 32K) protein encoded by RNA 3 of AMV failed to react with the corresponding proteins directed by RNAs from three other viruses with tripartite genomes, namely, tobacco streak, CMV, and BMV.[84] However, dot matrix comparisons of sequences revealed a strong homology between the TMV, tobacco streak virus, and TRV putative movement proteins.[85] Dot matrix comparison also revealed homologies between the movement protein of TMV and the open reading frame 1 protein of the DNA viruses, namely, cauliflower mosaic virus (CaMV) and carnation etched ring virus.[82,86]

Different TMV strains differ markedly in the amino acid sequences of their 30K proteins. These sequences have been derived from the nucleotide sequence data of TMV RNA.[68,69,87] They may markedly differ in the content of a particular amino acid residue.

The potyvirus-induced cylindrical inclusions are composed of virus-coded nonstructural proteins. It is highly probable that these inclusions or their constituent protein(s) play some significant role in cell-to-cell transport of these viruses. Three types of evidences suggest this. Cylindrical inclusions have close relationship to plasmodesmata and are even initially formed and deposited on plasmodesmata.[25,88,89] Second, homologous capsid proteins or virus particles were found associated with cylindrical inclusions of wheat streak mosaic (WSMV) and wheat spindle streak mosaic (WSSMV) viruses in vivo by immunoelectron microscopy employing gold-labeled antibodies.[90] Third, cylindrical inclusions were only formed on cell walls shared by two adjacent cells and were always first found to be attached perpendicularly to the walls of cells adjoining infected cells prior to the appearance of WSMV particles.[90] It was propounded on these bases that cylindrical inclusions provide an active mechanism of virus transport which ensures alignment of virions along plasmodesmata resulting in the location of a large number of virions with plasmodesmata.[90]

Little is known about the mode of action of transport proteins. One of the probable mechanisms is that movement of transport protein somehow brings about change(s) in plasmodesmata rendering them permeable to virions, viral nucleic acid and/or vRNP.

## B. Complementation of Viral Transport Function

Complementation, when products of one virus may be used by another virus to overcome some of the latter's deficiencies, of viral transport function, is exhibited by TMV mutants.[91-93] The cell-to-cell transport *ts* property of TMV Ni 2519 mutant was overcome when co-inoculated with TMV *vulgare* strain. This resulted in systemic spread of Ni 2519 at the restrictive temperature of 33°C.[94-95] Similarly, the cell-to-cell transport *ts* property of TMV LS1 strain was complemented by the unrelated potato virus X (PVX).[63] The experiment was based on the fact that TMV, when imbibed into a plant through a cut stem end, cannot cause infection. Preinfection of the upper leaves with PVX, followed by imbibition of TMV LS1, resulted in infection of upper leaves with TMV LS1, showing thereby that PVX complemented the function of TMV LS1 movement from tracheal elements into parenchyma. Complementation of the cell-to-cell transport function of red clover mottle virus in tobacco is done by dolichos enation mosaic virus.[96] Barley stripe mosaic virus complements the replication and systemic spread of TMV, but not of any other virus among several other viruses tested, in barley plants.[97] Since complementation of transport function occurs even between unrelated viruses, the function does not seem to be virus specific. It is possible that transport protein of a virus brings in some alterations in structure of plasmodesmata which open them, resulting in movement of an unrelated virus.

Some other instances where mixed virus infection leads to change in distribution of a virus are also known. Some variants of barley yellow dwarf virus remain confined to phloem in single infections but become capable of entering xylem in paired virus infections.[98] Potato leaf roll virus (PLRV), a phloem-restricted virus, enters parenchyma of *Datura stramonium* leaves in the presence of PVX.[99] This virus also enters parenchyma cells, with remarkable increase in its concentration, in *Nicotiana clevelandii* leaves doubly infected with PLRV and potato virus Y (PVY).[100] But, in potato leaves doubly infected with PLRV and PVY, PLRV remained confined to the phloem. Bean golden mosaic virus, another phloem-restricted virus, spreads systemically in beans when co-inoculated with the bean strain of TMV.[101]

Another form of complementation of virus movement function is the rendering of a virus-resistant plant susceptible to a given virus. Tomato plants are resistant to the Russian strain of BMV but become susceptible to BMV in the presence of the tomato strain of TMV. Bean plants resistant to BMV become susceptible to it in the presence of the bean strain of TMV or dolichos enation mosaic virus. TMV was able to spread in barley plants in the presence of barley stripe mosaic virus (BSMV)[102] or BMV,[103] and TMV *vulgare* was able to spread in wheat plants in the presence of BSMV.[104]

Spread of the virus in a resistant host was halted the moment the helper virus spread was stopped. Thus the spread of BMV in the tomato plant, which was made possible by the *ts* mutant LS1 of tomato strain of TMV at 24°C, came to a stop after the temperature was raised to 32°C.

## IV. TRANSPORT IN XYLEM

Southern bean mosaic virus is the only virus known to be translocated naturally in xylem tissue.[3] SBMV can infect the plant through the xylem and can move through steamed sections of stem.[105] It causes infection in foliage of 'Pinto' bean plants after its introduction directly into the tracheary elements by stem puncture or through graft union.[105,106] It has been hypothesized that SBMV enters parenchyma of young leaves through only those xylem elements which are undergoing transition from living to dead cells.[3]

Other viruses, like TMV and turnip yellow mosaic virus, fail to cause infection in leaves if experimentally introduced into the vascular tissue through a cut stem end. Perhaps some barrier checks their entry from xylem vessels into living parenchyma tissue.

## V. TRANSPORT OF VIROIDS

Viroids are known to move in phloem. Thus, potato spindle tuber viroid moved to the growing tip of tomato plant via the phloem after 6 to 7 days of inoculation of the lower leaves.[107] It was detected in tomato shoot tip about 1 week before the appearance of disease symptoms. Viroids do not code for any virus-specific protein in infected plant.[108] Free nucleic acids are vulnerable to attack by the host RNases, particularly so in the phloem. It is therefore still a mystery how viroids are spread in the long-distance transport system.

# REFERENCES

1. **Sulzinski, M. A. and Zaitlin, M.,** Tobacco mosaic virus replication in resistant and susceptible plants: in some resistant species virus is confined to a small number of initially infected cells, *Virology,* 121, 12, 1982.
2. **Halk, E. L. and McGuire, J. M.,** Translocation of tobacco ringspot virus in soybean, *Phytopathology,* 63, 1291, 1973.
3. **Schneider, I. R.,** Introduction, translocation and distribution of viruses in plants, *Adv. Virus Res.,* 11, 163, 1965.
4. **Bennett, C. W.,** Biological relation of plant viruses, *Annu. Rev. Plant Physiol.,* 7, 143, 1956.
5. **Atabekov, J. G. and Dorokhov, Yu. L.,** Plant virus-specific transport function and resistance of plants to viruses, *Adv. Virus Res.,* 29, 313, 1984.
6. **Zaitlin, M. and Hull, R.,** Plant virus-host interactions, *Annu. Rev. Plant Physiol.,* 38, 291, 1987.
7. **Hosokava, D. and Mori, K.,** Studies on multiplication and distribution of viruses in plants by the use of fluorescent antibody technique. II. Multiplication and distribution of the virus in tobacco leaves inoculated with cucumber mosaic virus, *Ann. Phytopathol. Soc. Japan,* 48, 444, 1982.
8. **de Zoeten, G. A.,** Early events in plant virus infections, in *Plant Diseases and Vectors: Ecology and Epidemiology,* Maramorosch, K. and Harris, K. F., Eds., Academic Press, New York, 1981, 221.
9. **Rappaport, I. and Wildman, S. G.,** A kinetic study of local lesion growth on *Nicotiana glutinosa* resulting from tobacco mosaic virus infection, *Virology,* 4, 265, 1957.
10. **Henderson, H. M. and Cooper, J. I.,** Effect of thermal shock treatment on symptom expression in test plants inoculated with potato aucuba mosaic virus, *Ann. Appl. Biol.,* 86, 389, 1977.
11. **Omura, T. and Wakimoto, S.,** Effect of plant hormones on tobacco mosaic virus concentration in tobacco tissue culture, *J. Fac. Agric. Kyushu Univ.,* 22, 211, 1978.
12. **Dijkstra, J.,** On the early stages of infection by tobacco mosaic virus in *Nicotiana glutinosa* L., *Virology,* 18, 142, 1962.
13. **Kimmins, W. C. and Wuddah, D.,** Hypersensitive resistance: determination of lignin in leaves with a localized virus infection, *Phytopathology,* 67, 1012, 1977.
14. **Weintraub, M., Ragetli, H. W. J., and Lo, E.,** Potato virus Y particles in plasmodesmata of tobacco leaf cells, *J. Ultrastruct. Res.,* 46, 131, 1974.
15. **Gibbs, A.,** Viruses and plasmodesmata, in *Intercellular Communication in Plants: Studies on Plasmodesmata,* Gunning, B. E. S. and Robards, A. W., Eds., Springer-Verlag, Berlin, 1976, 150.
16. **Puffinberger, C. W. and Corbett, M. K.,** Euonymus chlorotic ringspot disease caused by tomato ringspot virus, *Phytopathology,* 75, 423, 1985.
17. **Roberts, I. M. and Harrison, B. D.,** Inclusion bodies and tubular structures in *Chenopodium amaranticolor* plants infected with strawberry latent ringspot virus, *J. Gen. Virol.,* 7, 47, 1970.
18. **Kim, K. S. and Fulton, J. P.,** Tubules with viruslike particles in leaf cells infected with bean pod mottle virus, *Virology,* 43, 329, 1971.
19. **Esau, K. and Hoefert, L. L.,** Ultrastructure of sugarbeet leaves infected with beet western yellows virus, *J. Ultrastruct. Res.,* 40, 556, 1972.
20. **Robards, A. W.,** Plasmodesmata, *Annu. Rev. Plant Physiol.,* 26, 13, 1975.
21. **Robards, A. W.,** Plasmodesmata in higher plants, in *Intercellular Communication in Plants: Studies on Plasmodesmata,* Gunning, B. E. S. and Robards, A. W., Eds., Springer-Verlag, Berlin, 1976, 15.
22. **Gunning, B. E. S. and Overall, R. L.,** Plasmodesmata and cell-to-cell transport in plants, *Bioscience,* 33, 260, 1983.
23. **Kitajima, E. W. and Lauritis, J. A.,** Plant virions in plasmodesmata, *Virology,* 37, 681, 1969.
23a. **Esau, K. and Hoefert, L. L.,** Development of infection with beet western yellows virus in sugarbeet, *Virology,* 48, 724, 1972.
24. **Shepardson, S., Esau, K., and McCrum, R.,** Ultrastructure of potato leaf phloem infected with potato leaf roll virus, *Virology,* 105, 379, 1980.
25. **Weintraub, M., Ragetli, H. W. J., and Leung, E.,** Elongated virus particles in plasmodesmata, *J. Ultrastruct. Res.,* 56, 351, 1976.
26. **Esau, K., Cronshaw, J., and Hoefert, L. L.,** Relation of beet yellows virus to the phloem and to movement in the sieve tube, *J. Cell Biol.,* 32, 71, 1967.
27. **Shalla, T. A., Peterson, L. J., and Zaitlin, M.,** Restrictive movement of a temperature-sensitive virus in tobacco leaves is associated with a reduction in number of plasmodesmata, *J. Gen. Virol.,* 60, 355, 1982.
28. **Wieringa-Brants, D. H.,** The role of epidermis in virus-induced local lesions on cowpea and tobacco leaves, *J. Gen. Virol.,* 54, 209, 1981.
29. **Kondo, A.,** Long-distance movement of tobacco mosaic virus within the tobacco plant, *Ann. Phytopathol. Soc. Japan,* 40, 299, 1974.

30. **Helms, K. and Wardlaw, I. F.,** Translocation of tobacco mosaic virus and photosynthetic assimilate in *Nicotiana glutinosa, Physiol. Plant Pathol.,* 13, 23, 1978.
31. **Beemster, A. B. R.,** *Translocation of Virus X in the Potato (Solanum tuberosum* L.) in *Primarily Infected Plants,* Veenman and Zonen, Wageningen, 1958, 1.
32. **Bennett, C. W.,** Plant tissue relations of sugar beet curly top virus, *J. Agric. Res.,* 48, 665, 1934.
33. **Kunkel, L. O.,** Movement of tobacco mosaic virus in tomato plants, *Phytopathology,* 29, 684, 1939.
34. **Helms, K. and Wardlaw, I. F.,** Movement of viruses in plants: long distance movement of tobacco mosaic virus in *Nicotiana glutinosa,* in *Transport and Transfer Processes in Plants,* Wardlaw, I. F. and Passioura, J. B., Eds., Academic Press, New York, 1976, 283.
35. **Bennett, C. W.,** Correlation between movement of the curly top virus and translocation of food in tobacco and sugarbeet, *J. Agric. Res.,* 54, 479, 1937.
36. **Lei, J. D. and Agrios, G. N.,** Mechanisms of resistance in corn to maize dwarf mosaic virus, *Phytopathology,* 76, 1034, 1986.
37. **Zech, H.,** Untersuchungen über den Infektionsvorgang und die Wanderung des Tabakmosaikvirus im Pflanzenkörper, *Planta,* 40, 461, 1952.
38. **Mitchell, J. W., Preston, W. H., Jr., and Beal, J. M.,** Stem inoculation of Pinto bean with southern bean mosaic virus, a promising method for use in screening chemicals for antiviral activity, *Phytopathology,* 46, 479, 1956.
39. **Smith, F. H. and McWhorter, F. P.,** Anatomical effects of tomato ringspot virus in *Vicia faba, Am. J. Bot.,* 44, 470, 1957.
40. **Hiruki, C., Shukla, P., and Rao, D. V.,** Occurrence of clover yellow mosaic virus aggregates in transfer cells of *Pisum sativum, Phytopathology,* 66, 594, 1976.
41. **Dorokhov, Yu. L., Alexandrova, N. M., Miroshnichenko, N. A., and Atabekov, J. G.,** The informosome-like virus-specific ribonucleoprotein may be involved in the transport of tobacco mosaic virus infection, *Virology,* 137, 127, 1984.
42. **Dorokhov, Yu. L., Miroshnichenko, N. A., Alexandrova, N. M., and Atabekov, J. G.,** Polypeptide analysis of virus-specific informosomes induced by tobacco mosaic virus and potato virus X, *Mol. Biol. (Moscow),* 18, 1001, 1984.
43. **Fry, P. R. and Matthews, R. E. F.,** Timing of some early events following inoculation with tobacco mosaic virus, *Virology,* 19, 461, 1963.
44. **Jockusch, H.,** In vivo- und in vitro-Verhalten temperature-sensitive Mutanten des Tabakmosaikvirus, *Z. Vererbungsl.,* 95, 379, 1964.
45. **Jockusch, H.,** Temperatursensitive Mutanten des Tabakmosaicvirus. I. In vivo-Verhalten, *Z. Vererbungsl.,* 98, 320, 1966.
46. **Powell, C. A., de Zoeten, G. A., and Gaard, G.,** The localization of pea enation mosaic virus-induced RNA-dependent RNA polymerase in infected peas, *Virology,* 78, 135, 1977.
47. **Dawson, W. O., Schlegel, D. E., and Lung, M. C. Y.,** Synthesis of tobacco mosaic virus in intact tobacco leaves systemically inoculated by differential temperature treatment, *Virology,* 65, 565, 1975.
48. **Nishiguchi, M., Motoyoshi, F., and Oshima, N.,** Behaviour of a temperature sensitive strain of tobacco mosaic virus in tomato leaves and protoplasts, *J. Gen. Virol.,* 39, 53, 1978.
49. **Nishiguchi, M., Motoyoshi, F., and Oshima, N.,** Further investigation of temperature-sensitive strain of tobacco mosaic virus: its behaviour in tomato leaf epidermis, *J. Gen. Virol.,* 46, 497, 1980.
50. **Dorokhov, Yu. L., Alexandrova, N. M., Miroshnichenko, N. A., and Atabekov, J. G.,** Role of virus-specific ribonucleoprotein particles in systemic spreading of tobacco mosaic virus, *Biol. Nauki (Moscow),* 9, 23, 1980.
51. **Dorokhov, Yu. L., Alexandrova, N. M., Miroshnichenko, N. A., and Atabekov, J. G.,** The formation of virus-specific informosome-like ribonucleoprotein in tobacco cells infected by *ts* mutants of tobacco mosaic virus, *Biol. Nauki (Moscow),* 12, 17, 1983.
52. **Dorokhov, Yu. L., Alexandrova, N. M., Miroshnichenko, N. A., and Atabekov, J. G.,** Isolation and analysis of virus-specific ribonucleoprotein of tobacco mosaic virus-infected tobacco, *Virology,* 127, 237, 1983.
53. **Nassuth, A. and Bol, J. F.,** Altered balance of the synthesis of plus- and minus-strand RNAs induced by RNAs 1 and 2 of alfalfa mosaic virus in the absence of RNA 3, *Virology,* 124, 75, 1983.
54. **Kiberstis, P. A., Loesch-Fries, L. S., and Hall, T. C.,** Viral protein synthesis in barley protoplasts inoculated with native and fractionated brome mosaic virus RNA, *Virology,* 112, 804, 1981.
55. **Peters, D. L. and Murphy, T. M.,** Selection of temperature-sensitive mutants of tobacco mosaic virus by lesion morphology, *Virology,* 65, 595, 1975.
56. **Atabekov, J. G. and Morozov, Yu. S.,** Translation of plant virus messenger RNAs, *Adv. Virus Res.,* 25, 1, 1979.
57. **Leonard, D. A. and Zaitlin, M.,** A temperature-sensitive strain of tobacco mosaic virus defective in cell-to-cell movement generates an altered viral-coded protein, *Virology,* 117, 416, 1982.

58. **Rezelman, G., Franssen, H. J., Goldbach, R. W., Ie, T. S., and Van Kammen, A.,** Limits to the independence of bottom component RNA of cowpea mosaic virus, *J. Gen. Virol.,* 60, 335, 1982.

59. **Wyatt, S. D. and Kuhn, C. W.,** Derivation of a new strain of cowpea chlorotic mottle virus from resistant cowpeas, *J. Gen. Virol.,* 49, 289, 1980.

60. **Osman, T. A. M., Dodds, S. M., and Buck, K. W.,** RNA 2 of red clover necrotic mosaic virus determines lesion morphology and systemic invasion in cowpea, *J. Gen. Virol.,* 67, 203, 1986.

61. **Sunter, G., Gardiner, W. E., Rushing, A. E., Rogers, S. G., and Bisaro, D. M.,** Independent encapsulation of tomato golden mosaic virus A component DNA in transgenic plants, *Plant Mol. Biol.,* 8, 477, 1987.

62. **Ward, A., Stanley, J., Townsend, R., Ettesami, P., and Kunert, K. J.,** Genomic organisation of geminiviruses infecting dicotyledonous hosts, *VII Int. Congr. Virol.,* Edmonton, Abstr. No. S23.1, 1987.

63. **Taliansky, M. E., Malyshenko, S. I., Pshennikova, E. S., Kaplan, I. B., Ulanova, E. F., and Atabekov, G. J.,** Plant virus-specific transport function. I. Virus genetic control required for systemic spread, *Virology,* 122, 318, 1982.

64. **Joshi, S., Pleij, C. W. A., Haenni, A. L., Chapeville, F., and Bosch, L.,** Properties of the tobacco mosaic virus intermediate length RNA 2 and its translation, *Virology,* 127, 100, 1983.

65. **Ooshika, I., Watanabe, Y., Meshi, T., Okada, Y., Igano, K., Inouye, K., and Yoshida, N.,** Identification of the 30K protein of TMV by immunoprecipitation with antibodies directed against a synthetic peptide, *Virology,* 132, 71, 1984.

66. **Watanabe, Y., Emori, Y., Ooshika, I., Meshi, T., Ohno, T., and Okada, Y.,** Synthesis of TMV-specific RNAs and proteins at the early stage of infection in tobacco protoplasts: transient expression of the 30K protein and its mRNA, *Virology,* 133, 18, 1984.

67. **Ohno, T., Takamatsu, N., Meshi, T., Okada, Y., Nishiguchi, M., and Kiho, Y.,** Single amino acid substitution in 30K protein of TMV defective in virus transport function, *Virology,* 131, 255, 1983.

68. **Goelet, P., Lomonossoff, G. P., Butler, P. J. G., Akam, M. E., Gait, M. J., and Karn, J.,** Nucleotide sequence of tobacco mosaic virus RNA, *Proc. Nat. Acad. Sci. U.S.A.,* 79, 5818, 1982.

69. **Meshi, T., Ohno, T., and Okada, Y.,** Nucleotide sequence of the 30 K protein cistron of cowpea strain of tobacco mosaic virus, *Nucleic Acids Res.,* 10, 6111, 1982.

70. **Meshi, T., Watanabe, Y., Saito, T., Maeda, T., and Okada, Y.,** Function of the 30K protein of TMV, *VII Int. Congr. Virol,* Edmonton, Abstr. No. OP5.10, 1987.

71. **Godefroy-Colburn, T., Gagey, M.-J., Moser, O., Ellwart-Tschurtz, M., Nitschko, H., and Stussi-Garaud, C.,** Immunodetection of the transport factor (30K) of tobacco mosaic virus in infected plants, *VII Int. Congr.* Virol., Edmonton, Abstr. No. OP5.26, 1987.

71a. **Deom, C. M., Oliver, M. J., and Beachy, R. N.,** unpublished results, cited in Reference 6.

72. **Tomenius, K., Clapham, D., and Meshi, T.,** Localization of immunogold cytochemistry of the virus-coded 30K protein in plasmodesmata of leaves infected with tobacco mosaic virus, *Virology,* 160, 363, 1987.

73. **Watanabe, Y., Ooshika, I., Meshi, T., and Okada, Y.,** Subcellular localization of the 30K protein in TMV-inoculated tobacco protoplasts, *Virology,* 152, 414, 1986.

74. **Watanabe, Y., Morita, N., Nishiguchi, M., and Okada, Y.,** Attenuated strains of tobacco mosaic virus: reduced synthesis of a viral protein with a cell-to-cell movement function, *J. Mol. Biol.,* 194, 699, 1987.

75. **Wellink, J., Jaegle, M., Prinz, H., Van Kammen, A., and Goldbach, R.,** Expression of middle component RNA of cowpea mosaic virus in vivo, *J. Gen. Virol.,* 68, 2577, 1987.

76. **Meyer, M., Hemmer, O., Mayo, M. A., and Fritsch, C.,** The nucleotide sequence of tomato black ring virus RNA 2, *J. Gen. Virol.,* 67, 1257, 1986.

77. **Godefroy-Colburn, T., Gagey, M.-J., Berna, A., and Stussi-Garaud, C.,** A non-structural protein of alfalfa mosaic virus in the walls of infected tobacco cells, *J. Gen. Virol.,* 67, 2233, 1986.

78. **Berna, A., Briand, J.-P., Stussi-Garaud, C., and Godefroy-Colburn, T.,** Kinetics of accumulation of the three non-structural proteins of alfalfa mosaic virus in tobacco (*Nicotiana tabacum* cv. *Xanthi-NC*) plants, *J. Gen. Virol.,* 67, 1135, 1986.

79. **Van Pelt-Heerschap, H. H., Verbeek, H. H., Huisman, M. J., Loesch-Fries, L. S., and Van Vloten-Doting, L.,** Nonstructural proteins and RNAs of alfalfa mosaic virus synthesized in tobacco and cowpea protoplasts, *Virology,* 161, 190, 1987.

80. **Stussi-Garaud, C., Garaud, J.-C., Brema, A., and Godefroy-Colburn, T.,** In situ localization of an alfalfa mosaic virus non-structural protein in plant cell walls: correlation with virus transport, *J. Gen. Virol.,* 68, 1779, 1987.

81. **Cornelissen, B. J. C., Janssen, H., Zuidema, D., and Bol, J. F.,** Complete nucleotide sequence of tobacco streak virus RNA 3, *Nucleic Acids Res.,* 12, 2427, 1984.

82. **Hull, R. and Covey, S. N.,** Cauliflower mosaic virus: pathways of infection, *Bioassays,* 3, 160, 1985.

83. **Haseloff, J., Goelet, P., Zimmern, D., Ahlquist, P., Dasgupta, R., and Kaesberg, P.,** Striking similarities in amino acid sequence among non-structural proteins encoded by RNA viruses that have dissimilar genomic organisation, *Proc. Natl. Acad. Sci. U.S.A.,* 81, 4358, 1984.

84. **van Tol, R. G. L. and Van Vloten-Doting, L.,** Lack of serological relationship between the 35K non-structural protein of alfalfa mosaic virus and the corresponding protein of three other plant viruses with a tripartite genome, *Virology,* 109, 444, 1981.

85. **Boccara, M., Hamilton, W. D. O., and Baulcombe, D. C.,** The organisation and interviral homologies of genes at the 3' end of tobacco rattle virus RNA 1, *EMBO J.,* 5, 223, 1986.

86. **Hull, R., Sadler, J., and Longstaff, M.,** The sequence of carnation etched ring virus DNA: comparison with cauliflower mosaic virus and retroviruses, *EMBO J.,* 5, 3083, 1986.

87. **Takamatsu, N., Ohno, T., Meshi, T., and Okada, Y.,** Molecular cloning and nucleotide sequence of the 30K and the coat protein cistron of TMV (tomato strain) genome, *Nucleic Acids Res.,* 11, 3667, 1983.

88. **Lawson, R. H., Hearon, S. S., and Smith, F. F.,** Development of pinwheel inclusions associated with sweet potato russet crack virus, *Virology,* 46, 453, 1971.

89. **McMullen, C. R. and Gardner, W. S.,** Cytoplasmic inclusions induced by wheat streak mosaic virus, *J. Ultrastruct. Res.,* 72, 65, 1980.

90. **Langenberg, W. G.,** Virus protein associated with cylindrical inclusions of two viruses that infect wheat, *J. Gen. Virol.,* 67, 1161, 1986.

91. **Atabekov, J. G., Schaskolskaya, N. D., Atabekova, T. I., and Sacharovskaya, G. A.,** Reproduction of temperature-sensitive strains of TMV under restrictive conditions in the presence of temperature-resistant helper strain, *Virology,* 41, 397, 1970.

92. **Atabekov, J. G., Novikov, V. K., Vishnichenko, V. K., and Kaftanova, A. S.,** Some properties of hybrid viruses reassembled in vitro, *Virology,* 41, 519, 1970.

93. **Kassanis, B. and Conti, M.,** Defective strains and phenotypic mixing, *J. Gen. Virol.,* 13, 361, 1971.

94. **Taliansky, M. E., Atabekova, T. I., Kaplan, I. B., Morozov, S. Yu., Maeyshenko, S. I., and Atabekov, J. G.,** A study of TMV *ts* mutant Ni 2519. I. Complementation experiments, *Virology,* 118, 301, 1982.

95. **Taliansky, M. E., Kaplan, I. B., Yarvekulg, L. V., Atabekova, T. I., Agranovsky, A. A., and Atabekov, J. G.,** A study of TMV *ts* mutant Ni 2519. II. Temperature sensitive behaviour of Ni 2519 RNA upon reassembly, *Virology,* 118, 309, 1982.

96. **Lapchik, L. G., Malyshenko, S. I., Kuznetsova, L. L., and Talianskii, M. E.,** Accumulation of red clover mottle virus in tobacco plants under conditions of mixed infection, *Mikrobiol. Zh. (Kiev),* 49, 81, 1987.

97. **Ouchi, S., Muraoka, A., Shimazu, Y., Fukaya, S., and Takanami, Y.,** Interaction of barley stripe mosaic virus and tobacco mosaic virus in barley leaves, *VII Int. Congr. Virol.,* Edmonton, Abstr. No. OP5.27, 1987.

98. **Gill, C. C. and Chong, J.,** Vascular cell alterations and predisposed xylem infection in oats by inoculation with paired barley yellow dwarf viruses, *Virology,* 114, 405, 1981.

99. **Atabekov, J. G., Taliansky, M. E., Drampyan, A. H., Kaplan, I. B., and Tupka, I. E.,** Systemic infection by a phloem-restricted virus in parenchyma cells in a mixed infection, *Biol. Nauk,* 10, 28, 1984.

100. **Barker, H.,** Invasion of non-phloem tissue in *Nicotiana clevelandii* by potato leafroll luteovirus is enhanced in plants also infected with potato Y potyvirus, *J. Gen. Virol.,* 68, 1223, 1987.

101. **Carr, R. J. and Kim, K. S.,** Evidence that bean golden mosaic virus invades non-phloem tissue in double infections with tobacco mosaic virus, *J. Gen. Virol.,* 64, 2489, 1983.

102. **Dodds, J. A. and Hamilton, R. I.,** The influence of barley stripe mosaic virus on the replication of tobacco mosaic virus in *Hordeum vulgare* L., *Virology,* 50, 404, 1972.

103. **Hamilton, R. I. and Nichols, C.,** The influence of barley stripe mosaic virus on the replication of tobacco mosaic virus in *Hordeum vulgare, Phytopathology,* 67, 484, 1977.

104. **Taliansky, M. E., Malyshenko, S. I., Pshennikova, E. S., and Atabekov, J. G.,** Plant virus-specific transport function. II. A factor controlling virus host range, *Virology,* 122, 327, 1982.

105. **Schneider, I. R. and Worley, J. F.,** Upward and downward transport of infectious particles of southern bean mosaic virus through steamed portions of bean stems, *Virology,* 8, 230, 1959.

106. **Schneider, I. R. and Worley, J. F.,** Rapid entry of infectious particles of southern bean mosaic virus into living cells following transport of the particles in the water stream, *Virology,* 8, 243, 1959.

107. **Palukaitis, P.,** Potato spindle tuber viroid: investigation of the long distance, intra-plant transport route, *Virology,* 158, 239, 1987.

108. **Dickson, E.,** Viroids: infectious RNA in plants, in *Nucleic Acids in Plants,* Vol. 2, Hall, T. C. and Davies, J. W., Eds., CRC Press, Boca Raton, U.S., 1979, 153.

Chapter 5

# VARIABILITY OF PLANT VIRUSES

## C. L. Mandahar

## TABLE OF CONTENTS

# I. VARIABILITY

Several of the plant viruses are highly variable, but a few show only little variation. Alfalfa mosaic (AMV), cucumber mosaic (CMV), tobacco mosaic (TMV), tobacco ringspot, and turnip mosaic viruses belong to the former category, while cauliflower mosaic (CaMV), maize dwarf mosaic, and squash mosaic viruses belong to the latter category. TMV is the most highly variable virus.

Many of the variants of a virus have been characterized by experimentally isolating them in laboratory from hosts showing the typical disease syndrome. A TMV strain was first characterized in this way.[1] Later, as many as 51[2,3] and 130[4] such TMV strains were isolated. McKinney selected bright yellow areas on TMV-infected tobacco leaves showing typical disease symptoms, used them as inoculum, and found that TMV from those areas induced bright yellow mottling clearly different from the typical mosaic symptoms.[1] Jensen used yellow spots as sources of inoculum and isolated variants by merely making pin pricks through these yellow spots into healthy leaves of tobacco plants.[2,3] A number of CMV variants were similarly located by merely cutting out yellow spots from leaves showing mottle symptoms.[5] Since these pioneering studies, variants have been isolated experimentally by some such or other methods in many of the plant viruses.

Consecutive transfer of a virus isolate through a series of single necrotic local lesions on a host is the passage of a virus. Such local lesions are taken to have completely separated each major strain, particularly if dilute inoculum is used and results in the formation of well separated local necrotic lesions. Still, variants do appear in a strain preparation purified and obtained by transfer from a single necrotic or chlorotic lesion. Garcia-Arenal et al.[6] therefore concluded that the local lesion method for obtaining biologically pure virus is of doubtful value, since some contaminants can also be passaged along with the major virus strain. The contaminant may differ widely from the parent strain and would need unexpected base changes for their origin from the parent as, for example, an atypical TMV mutant.[7]

"Variant" is a general term used for any isolate that differs in any one or more characters from the parent strain. Variants that arise under experimental conditions and whose descent can be traced back are considered mutants, while naturally occurring variants isolated from field and whose genetic descent cannot be known are termed strains.[8] However, no such distinction is generally made by investigators, and the three terms are often used interchangeably, as is done here. No effort has been made to evaluate these terms.

Viruses have often been diagnosed and identified in the past based on insufficient experimentation so that either an already known virus was redescribed wrongly as a new virus or the description was so inadequate that it was difficult to describe whether a so-called new virus was really new or a strain of an old, well-established virus. The Plant Virus Subcommittee of the International Committee on Taxonomy of Viruses (ICTV) at the Fourth International Congress for Virology held in 1978 at the Hague decided that it is necessary to have a detailed paper on procedurs to be adopted for proper identification and characterization of plant viruses. Such a detailed paper has since been published by a subgroup of the Plant Virus Subcommittee in 1981.[9] A subsidiary paper, aimed at the less equipped laboratories of the developing countries, has also been published.[10] Both papers offer guidelines for the type and standards of data to be acquired, the equipment needed, and the basic techniques to be employed for this purpose, and the standards of presentation of these data for publishing descriptions and for proper identification and characterization of a new plant virus. The guidelines for establishing the taxonomic status of new virus isolates as suggested by Hamilton et al.[9] are shown in Table 1. Bos had also earlier tabulated various steps in diagnosing plant virus diseases and the properties to be used for plant virus description and characterization.[11]

The VIDE (Virus Identification Data Exchange) project, based at Australian National

**Table 1**
**GUIDELINES FOR ESTABLISHING THE TAXONOMIC STATUS OF NEW VIRUS ISOLATES**

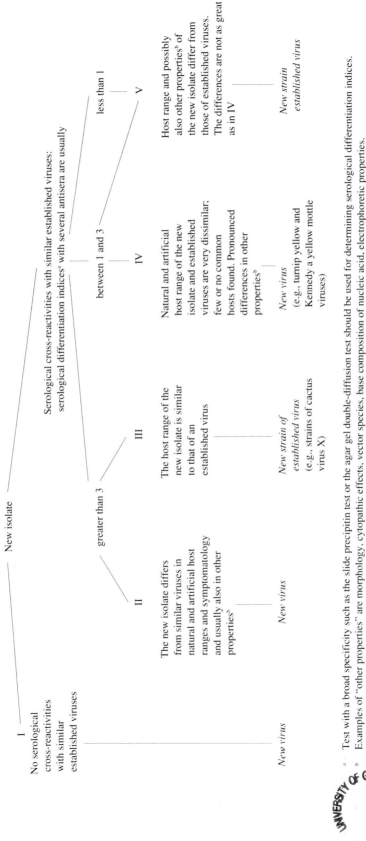

New isolate

**I** — No serological cross-reactivities with similar established viruses

*New virus*

Serological cross-reactivities with similar established viruses: serological differentiation indices[a] with several antisera are usually

greater than 3

**II** — The new isolate differs from similar viruses in natural and artificial host ranges and symptomatology and usually also in other properties[b]

*New virus*

**III** — The host range of the new isolate is similar to that of an established virus

*New strain of established virus* (e.g., strains of cactus virus X)

between 1 and 3

**IV** — Natural and artificial host range of the new isolate and established viruses are very dissimilar; few or no common hosts found. Pronounced differences in other properties[b]

*New virus* (e.g., turnip yellow and Kennedy a yellow mottle viruses)

less than 1

**V** — Host range and possibly also other properties[b] of the new isolate differ from those of established viruses. The differences are not as great as in IV

*New strain established virus*

[a] Test with a broad specificity such as the slide precipitin test or the agar gel double-diffusion test should be used for determining serological differentiation indices.

[b] Examples of "other properties" are morphology, cytopathic effects, vector species, base composition of nucleic acid, electrophoretic properties.

From Hamilton, R. I., Edwardson, J. R., Francki, R. I. B., Hsu, H. T., Hull, R., Koenig, R., and Milne, R. G., *J. Gen. Virol.*, 54, 223, 1981. With permission.

University, Canberra, Australia, collected data for providing up-to-date information on plant viruses. The data were organized, stored and manipulated using the DELTA system (*DE*scription *L*anguage for *TA*xonomy) for computer-generated identification aids.[12] The third microfiche edition of the database was produced in 1985 and carries detailed descriptions of all the plant virus groups and most of the viruses that infect leguminous plants.

## A. Inoculum Is an Heterogenous Mixture of Strains

In nature each virus exists in an infected plant more often as a conglomerate of strains differing slightly from each other in various characteristics but with one dominant genotype. Each of the four citrus tristeza virus (CTV)-infected orange plants harbored several virus strains differing in aphid transmissibility from poor (less than 5%) to high (above 30%).[13] A poorly transmissible CTV isolate seemingly quantitatively dominates over the highly transmissible isolates. A TMV substrain virulent to resistant tomato plants was present within the original nonvirulent parent inoculum.[14] Most of the wound tumor virus isolates that evolved in the greenhouse by continued maintenance in sweet clover plants were mixtures of vectorial virus and deletion mutants of the virus, including subvectorial and exvectorial isolates,[15] and vectorial virus could be recovered from subvectorial isolates.[16] This indicates that vectorial and subvectorial isolates occur in a mixed population. Mixtures of citrus exocortis viroid species within an isolate are known to occur.[17] Such a mixture of strains, comprising three mild and one severe isolates of this viroid, appear to occur in chrysanthemum.[18] Several other examples of mixtures of strains present in an inoculum obtained from natural infected plants are known.

These variants could have originated either by mutation of the parent virus or could have been contaminants in the original inoculum. Garcia-Arenal et al.[6] investigated this problem by employing nucleic acid hybridization using cDNA and by two-dimensional RNA fingerprinting. They analyzed the sequence relationship of genomic RNA of variants found in a preparation of TMV U1 strain obtained after the passage of inoculum through a necrotic local lesion from minimum dose inoculation. The variants thus obtained fell into two classes: members of one class possessed close sequence homology to the parent U1 strain and were therefore considered to have arisen by mutation of the parent. Members of the second class were distantly related TMV strains and were considered to have passaged as contaminants along with the U1 strain. This was in spite of the fact that the virus from which the inoculum was prepared was itself obtained from a single necrotic lesion from dilute inoculum. In addition, two strains of a virus can infect and multiply in the same cell and may lead to recombination (Section II.B). Moreover, a lesion from which the original inoculum was obtained may contain two strains of a virus, since these strains can infect and multiply in adjacent cells as happens in case of AMV.[19] Thus recombination, mutation, and the presence of more than one strain in the original inoculum can account for the presence of a mixture of strains in an inoculum.

Variants thus eventually appear even in the most rigorously cloned virus isolates as well as in purified virus preparations. All these are clonal variants, are closely related to the parent strain, and are probably its spontaneous mutants.[20] This is true of TMV.[8,21] However, one particular genotype has the greatest selective advantage in all such cases and may constitute up to 90% of the clone.[21,22] It is known that the single lesion isolates from stock cultures of turnip yellow mosaic virus rapidly revert to a mixture of strains.[23] Three variants were isolated from the same culture of tobacco rattle virus.

## B. Species Concept

A species is defined as groups of interbreeding natural populations which are reproductively isolated from other such natural interbreeding groups or, in other words, a virus species is a population of viruses having a common gene pool which is distinct from the gene pools of other viruses. A common gene pool is formed by regular genetic interchange between a population of viruses and is distinct from such a gene pool of another population of viruses.[24] No genetic interchange is possible between the different gene pools because of the barriers between them.

The above species concept cannot be applied to all viruses. Viruses of some major groups do not have interbreeding natural populations and so neither have regular gene exchange nor possess common gene pools that are distinguished from other gene pools. All viruses that reproduce entirely or mainly by clonal means will accumulate in strains which will result in almost continuous variation with no genetic exchange and with no clear breaks. This can be illustrated with respect to the potyviruses, which is a large virus group with several known strains of the better studied viruses. Serological relationship of members of a virus group range from close to distant or have no serological relationship at all. Thus all definite potyviruses are serologically related to at least one other potyvirus and different strains of a potyvirus may be closely to distantly related and, in the case of distantly related strains, the relationship between two strains of a potyvirus may be as distant as between two potyviruses. It is thus difficult to define the natural boundaries of a strain or a virus and the greater the number of known strains of a potyvirus, the more difficult it becomes to do so.[25,26] In some cases, a continuum of strains exists between two (poty)viruses. Similarly, each carlavirus is serologically related to at least one other carlavirus, while seven of the better known carlaviruses are serologically interrelated.[27] Tobacco rattle virus isolates belonging to serotype I—II intergrade between serotypes I and II. Tobacco necrosis virus contains many serologically distinct but intergrading isolates and their apparent serological relationships are variable and depend upon the serological complexity of the isolates as well as upon the test antiserum used.[28] Most luteoviruses are serologically related so that these viruses may form an interacting system in nature.[28a]

All the above observations are possibly valid for other single-stranded positive-sense plant viruses. Another conclusion flows from the above observations. Different strains can be taken as a range of variant clones that have been selected by the host plant(s) for biological fitness out of a much larger number of mutants that naturally keep on arising in an inoculum. Consequently, there may not be consistent sharp discontinuities in variation of strains of a virus or viruses of a group. This makes it difficult to delineate separate viruses of a group.

Three specific instances can be cited in this connection. Bean yellow mosaic (BYMV) and pea necrosis (PNV) viruses are placed in BYMV subgroup of potyviruses on the basis of amino acid composition. Clover yellow mosaic virus (CYMV) is placed in this group on the basis of serology. In fact, BYMV, CYMV, and PNV are serologically related. BYMV and PNV are considered to be separate viruses by some but are regarded as synonymous by many.[29,30] Similarly, BYMV and CYMV are considered as synonymous by some but as separate viruses by others.[29,31] Reddick and Barnett[32] and Barnett et al.[33] studied the relationships between isolates of these three viruses by comparing nucleotide sequence homologies by molecular hybridization and ELISA. They concluded that possibly all three viruses form an evolutionary continuum in relationship to each other since each of them has many strains. This supports the assumption of Jones and Diachun who, on the basis of host range, symptomatology, and serological relationships, placed all these viruses under BYMV with three distinct serotypes.[29] Serotype I contains CYMV strains, serotype II contains BYMV strains, and serotype III contains PNV strains.,

The second example pertains to tymoviruses. Andean potato latent virus (APLV) and eggplant mosaic virus (EMV) have many strains and isolates and several of the strains of both the viruses can be regarded as strains of either one of the two viruses. The same holds true for belladonna mottle (BMV) and dulcamara mottle (DMV) viruses. The serological properties of "physalis mottle strain of BMV" and physalis mottle virus (PhyMV) are intermediate between the EMV-APLV pair on one side and the BMV-DMV pair on the other. Hence, all these viruses of solanaceous plants (in the series EMV-APLV-PhyMV-

DMV-BMV) can be regarded as a large cluster of strains of one virus.[34] The third example pertains to prunus necrotic ringspot, apple mosaic, and rose mosaic ilarviruses. Rose mosaic virus and apple mosaic virus strain P are serologically identical and cross-react weakly with prunus necrotic ringspot virus. Strains intermediate between apple mosaic and prunus necrotic ringspot viruses occur in rose and hops. It is therefore possible that a continuum of strains exist between these two ilarviruses.

Similarly, variation in several other plant virus groups (tombus-, luteo-, tobamo-, carla-, and potexviruses) is continuous primarily because of clonal propagation of these viruses. There are no clear breaks which could help us to differentiate between viruses and virus strains and to define and delineate a virus "species". Harrison concluded that it is not prudent to adopt the concept of biological species for viruses, because this implies a degree of distinctness, fixity, and uniformity which is not present in viruses.[35]

The situation with multipartite nepoviruses is somewhat different. Classification of nepoviruses based on serological relationships, that based on the ability to form pseudo-recombinants, and that based on nucleic acid homology determined by nucleic acid hybridization are nearly identical. Harrison suggested, therefore, that each of the six nepoviruses (tomato black ring, raspberry ringspot, tomato ringspot, grapevine chrome mosaic, cocoa necrosis, and artichoke Italian latent viruses) can be considered to closely approach a biological species.[35] Nevertheless, even here well-defined edges of strain clusters based on antigenic affinity or genome homology are missing. Discovery of 16 and N5 isolates of tobraviruses shows that evolution of tobraviruses has involved true genetic recombination between genomes of tobacco rattle and pea early browning viruses (Section II.B). This demolishes the boundaries between these two tobraviruses.

However, if some viruses do have interbreeding reproductive methods and establish separate gene pools, then the species concept may be applied to them. This may as yet be true only of the Reoviridae.[24]

Milne,[24,36] Harrison,[35] and Gibbs[37] discuss the species problem in plant viruses.

## II. ORIGIN OF STRAINS

Genetic variants in plant viruses are produced in four different ways: by mutation, mutation followed by specific selection of strains by host, recombination, and pseudorecombination in the case of multipartite viruses. The presence of satellites, if not recognized, can be another factor. Loss or acquisition of satellite nucleic acids changes the characteristics of the helper virus leading to the apparent production of its new variants. Genetic masking may also apparently appear to produce new variants differing in vector transmissibility and/or host range.

### A. Mutation

Single TMV-infected cells may contain more than $10^6$ virions, while more than $10^9$ virions may be located in a local lesion, and about a thousand mutant particles could be present in a local lesion even if the mutation rate is as little as one per million.[20] In fact, virus preparations may contain about 0.5 to 2% of their total particles as variants distinguishable from the main strain.[4,20] According to another estimate, about $10^7$ TMV virions are contained in a single local lesion, while up to $10^{14}$ virus particles are present in a single systemically infected leaf,[14] so that a large number of mutants may be present in it. Garcia-Arenal et al.[6] also concluded that the rate of transcriptional infidelity during RNA synthesis is very high. It has also been estimated that there is a 3 to 1 probability that each copy of an RNA genome of 4500 nucleotides will have at least one nucleotide different from the parent RNA.[37a] Bawden therefore suggested that there is no need to postulate any mechanism other than random mutation to explain variability in plant viruses because virion populations are so large that, even with mutation rates much lower than in other systems, a very large number of mutants will be produced.[20] Mutation is thus the most common and primary cause of variation in plant viruses, and the spontaneous mutation can be very high because of the large populations of virus particles. The majority of the present day strains of a virus must have arisen at one time or other by mutation. However, it is difficult to establish whether a newly isolated strain arose by mutation or was originally present in mixture with other strains and subsequently became selected due to some type of selection pressure.

All mutations breed true and an initial mutation generally remains unaltered by subsequent mutations. Both Tsugita,[38] who studied more than 80 naturally occurring strains and induced TMV mutants, and Wittmann and Wittmann-Liebold,[39] who studied about 200 induced and spontaneous TMV mutants, found that no more than three exchanges per mutant ever occurred. Recently, a case of double mutation in AMV has been reported in which mutations occurred in both RNAs 1 and 2. The mutation rate of a virus is the probability that a particular nucleotide position of a virus genome is altered during a single replication cycle. Mutation rates of different regions of a virus genome are different, primarily due to the effects of secondary structure on the fidelity of replication.

No evidence is available showing that mutation rate of a virus determines its rate of field variation. In fact, ultimately the apparent mutation rate as borne out of the rate of field variation may most likely be determined by some factors that alter the intensity of selection. These factors will determine the rate of field variation of a virus by inhibiting or accelerating the spread of variants in a population. It is possible that the different rates of variation of different viruses reflect differences in the efficiency of selection processes rather than differences in the mutation rate. It is, of course, certain that the relatively high mutation rate of viruses is a very important factor in generating a pool of mutant genomes on which various selection pressures can work, resulting in the emergence of new strains.

A few examples of mutations have already been given in Section I.A. More are mentioned here. Two AMV *ts* mutants, Bts4 and Mts3, showed *ts* defects in RNAs 1 and 2, and so behave as double mutants in cowpea protoplasts.[40] Such multiple mutations may be expressed in different hosts. An RNA3 *ts* mutant defect in the Mts3 mutant was expressed in tobacco leaf discs and bean plants but not in cowpea protoplasts, while the reverse was true for *ts* mutations in RNAs 1 and 2 of this mutant. Jones concluded that potato virus X(PVX) strain group 4 spontaneously appears in stock cultures of PVX group 2 strains maintained by serial subculture in *Nicotiana glutinosa*.[41] A spontaneous AMV mutant arose and was easily recognized phenotypically because it produced spherical particles, besides producing a small proportion of very long bacilliform particles.[42] This change in shape was due to mutations that occurred in RNA3 which brought forth some alteration(s) in both the proteins (35K and capsid protein) encoded by it. Cowpea mosaic virus strain Sb underwent spontaneous mutation(s) to generate new strains S1 and S8, which overcome hypersensitive resistance in cowpea cv. Early Red and induced systemic mosaic symptoms.[43] A spontaneous mutant of tobacco rattle virus (TRV) strain CAM was isolated from a plant showing a single yellow spot in a systemically infected leaf among otherwise symptomless *N. clevelandii* plants.[44] Southern bean mosaic virus (SBMV) strain A appears to have great potential for variability by mutation. It rapidly overcame resistance of 'Black Turtle Soup' bean plants against its systemic spread by undergoing mutation.[45] Various types of SBMV strains have possibly arisen by mutation of GH(Ghana) strain, due to its continuous propagation in a single bean or cowpea cultivar under high temperature conditions.[46] In fact, the origin of resistance-breaking strains of various viruses by mutation has been commonly reported.

Two types of changes in viral nucleic acids that seem to be the major causes of mutation are nucleotide substitution and addition or deletion of one or more nucleotides. Nucleotide substitution is also called base substitution, since a mutational change in a nucleotide is usually accomplished by a change in the purine or pyrimidine base of nucleotide. Substitution of one purine for another or of one pyrimidine for another is called transition, whereas substitution of one pyrimidine with a purine or vice versa is called transversion. Transitions and transversions affect only one nucleotide and are hence said to cause point mutations.

## 1. Deletion Mutation

Work on the deletion mutants of various viruses is better documented and is discussed in detail.

Wound tumor virus (WTV) maintained in sweet clover plants for a number of years without passage through a vector caused the differentiation of WTV isolates which, after various intervals, lost their ability to be transmitted by the vector either partially (subvectorial isolates) or completely (exvectorial isolates). Reddy and Black used electrophoresis in polyacrylamide gel to determine if the RNA genomes of different WTV isolates have different or identical segments.[15] The electrophorograms of the 12 RNA segments of each WTV isolate that was maintained in sweet clover for more than 2 years was different. This was indicative of the mutated nature of the genome. In all, 28 such mutations were detected. Segments 5, 1, 2, and 7 underwent 10, 9, 4 and 1 mutations, respectively, while segments 3, 4, 6, 8, 9, 10, 11, and 12 did not produce any detectable mutants. Genomes of such WTV mutant isolates lacked certain RNA segments or parts of segment and showed many different zonal patterns in polyacrylamide gel electrophoresis (PAGE) columns. Such isolates are therefore in reality deletion mutants, and it is reasonable to assume that the changes in genome segments could be the cause of loss of transmissibility or infectivity of the isolates.

As already mentioned, the deletion mutations were restricted to only four of the WTV genome segments, that is, segments 1, 2, 5, and 7.[15] Segment 5 in three of the isolates was not complete, but existed only as a remnant of the wild segment, while it was entirely absent from one isolate. A later study also found that segment 2 or 5 was entirely absent in certain isolates.[16] In contrast, the remnant of segment 7 completely replaced the original segment.[16] Remnants of segments which had undergone deletion mutation did not encode any polypeptide. The genome segments generated by the deletion mutation are replicated and encapsidated in normal fashion to produce virus particles in systemically infected plants and are also transcribed in vitro by the virion-associated transcriptase.[15,16,47]

Several WTV deletion mutant RNAs are derived from specific genome segments by deletion of an internal portion so that their 5' and 3' termini are identical to the termini of the parent segment.[47] Mutants, in this way, are considered to be the terminally conserved remnants of the genome segments. Genome segment 2 undergoes an internal deletion by 15% while genome segment 5 shows an internal deletion of 85% of some portion. Deletion boundaries are located more than 40 base pairs (bp) from either terminus.[47] The ribosome-binding sites, initiation codons, and the assembly origin are conceivably conserved in the mutant RNAs since they are functional with regard to transcription, replication, and encapsidation.

Soil-borne wheat mosaic furovirus (SBWMV) genome also undergoes deletion mutations.[48,49] Virus particles of the bipartite wild-type (WT) SBWMV purified in spring from infected wheat contained two types of particles: large 281 to 300 nm long (1.0L) particles and small 138 to 160 nm long (0.5L) particles. Sometimes still smaller particles 92 to 110 nm long (0.35L) were present, in addition to the 1.0L and 0.5L particles.[50] Repeated manual transfer of SBWMV with three particle types (1.0L, 0.5L, and 0.35L) ultimately produced an isolate with only 1.0L and 0.35L particles,[50] which is called Lab 1 mutant. Inoculation of wheat plants with 1.0L and 0.35L RNAs gave progeny of 1.0L and 0.35L particles only. However, inoculation of wheat with 1.0L and 0.5L RNAs produced particles of 1.0L, 0.5L, and 0.35L and in one case a new isolate with 1.0L and 0.4L RNAs (Lab 2 isolate) was produced.

In wheat plants that had been inoculated manually with unpurified WT isolate several months earlier or when the naturally infected plants were grown at 17°C, the short rod RNA (RNA II or 0.5L RNA) underwent spontaneous deletion mutation to delete as many as 1000

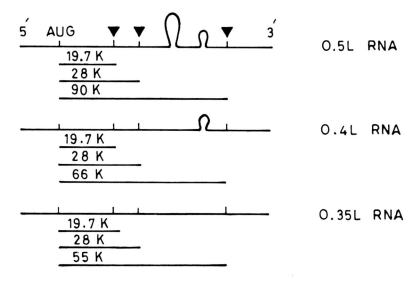

FIGURE 1. The tentative genetic maps of RNA IIs of soil-borne wheat mosaic virus (SBWMV). The vertical alignments of different RNAs indicate the correspondences among their nucleotide sequences. The loop structure(s) represent the nucleic acid area(s) that are deleted to produce sequentially the next smaller deletion mutants 0.4L and 0.35L RNA. The translation products and their estimated mol wt are shown below each RNA. ▼, Termination site. (From Hsu, Y. H. and Brakke, M. K., *Virology*, 143, 272, 1985. With permission.)

bases to produce 0.35L RNA, while nothing happened to long-rod RNA (RNA I or 1.0L RNA).[48,49] The Lab 1 deletion mutant is composed of long rod (1.0L) RNA and deleted RNA II (0.35L) and causes more severe disease symptoms in wheat plants that the WT SBWMV. The RNA combination of 1.0L RNA and 0.5L RNA or 1.0L RNA and 0.35L RNA was equally infectious. The molecular weights of 1.0L, 0.5L, 0.4L, and 0.35L RNAs under denaturing conditions were determined to be $2.28 \times 10^6$ (6500 bases), $1.23 \times 10^6$ (3500 bases), $0.97 \times 10^6$ (2800 bases), and $0.86 \times 10^6$ (2450 bases), respectively. Thus, 0.5L RNA lost approximately 700 bases to produce 0.4L RNA and about 1050 bases to yield the 0.35L mutant RNA (Figure 1).

In vitro translation studies of WT and mutant SBWMV demonstrated that differences between the translation products of the two were confined only to the polypeptides encoded by the respective 0.5L RNA.[51] RNAs II (0.5L RNAs) of all strains encoded 19.7K (coat protein) and 28K proteins. The differences in the RNAs II of mutants were in the synthesis of additional proteins: 90K, 66K, and 55K by 0.5L RNA, 0.4L RNA, and 0.35L RNA, respectively.[52] The pattern of translation products by 0.5L RNA of different strains was consistent with the formation of 0.4L and 0.35L RNA by internal deletion in the 3' region of 0.5L RNA. The deleted sequences seem to be located internally between the termination codons of 28K and 90K polypeptides[53] (Figure 1).

Beet necrotic yellow vein furovirus (BNYVV) is in nature restricted to beet roots and causes sugar beet rhizomania disease; it contains four RNA species, viral RNAs 1 to 3 and RNA4, which is a satellite RNA.[54] The extrapolated length of RNA4 of BNYVV isolate F2 is 1431 nucleotides and contains an open reading frame (ORF) of 282 codons, while this ORF of RNA4 of BNYVV isolate G1 has undergone an internal deletion mutation of 324 nucleotides.[55] The RNA3 of BNYVV isolate F2 is 1775 nucleotides long and contains a coding region of 219 codons. However, this coding region has undergone an internal deletion of 354 nucleotides in BNYVV isolate G1.[55] The RNA3 of F2 and G1 isolates shows great homology except simple base substitutions, 30 transitions, and 10 transversions, and an internal deletion of 354 nucleotides in RNA3 of the F2 isolate between positions 730 and

1083, which portion is absent from RNA3 of isolate G1 (Figure 2). It seems, therefore, that RNA3 of G1 may have originated from RNA3 of F2 by this internal deletion. Koenig et al.[56] also suggested the formation of deletion mutants in this virus because BNYVV isolates isolated from mechanically inoculated leaves of *Chenopodium* failed to infect sugar beet roots.

Some isolates of tomato spotted wilt virus (TSWV) do not produce virus particles, but instead produce only amorphous masses in infected cells. One such defective form of TSWV is TSWV-P, which has lost the capacity to produce membrane proteins so that the defective form consists of naked nucleocapsids and is infectious. Serological analysis of protein components carried out by Verkleij and Peters[57] showed that the defective form cannot produce TSWV protein 4. Nucleic acid hybridization demonstrated that the defective form contains normal RNA segments 1 and 3 but has lost RNA segment 2 and instead contains a new RNA segment with a molecular weight between that of RNAs 2 and 3 of the normal form. The molecular weight of RNA2 of the defective form is $1.4 \times 10^6$ and of the normal TSWV is $1.7 \times 10^6$, which indicates that the latter has undergone a deletion mutation to produce the former, and that the deleted RNA2 part had the genetic information for synthesis of protein 4.[57]

When the TRV strain OR was maintained in the greenhouse by transfer to fresh plants of *N. clevelandii* at approximately monthly intervals, the capacity to induce yellows symptoms decreased slowly until it produced scarcely any yellow symptoms after about 2 years.[58] The RNA2 of the OR strain contains the genetic information for inducing yellow symptoms in *Nicotiana* spp., while RNA2 of the above OR mutant does not. Moreover, RNAs 2 of the two strains have extensive sequence homology, except that RNA2 of the mutant is somewhat smaller in size than the OR RNA2. A deletion of about 150 nucleotides in the OR RNA2 is suggested to produce the small RNA2 of the mutant.[58]

The vector nontransmissible isolates of pea enation mosaic virus have lost a minor protein,[59] as well as a portion of the larger of the two RNA segments.[60] The deleted RNA sequence obviously encoded the minor protein responsible for vector transmissibility of this virus. Deletion mutations in vector-transmitted isolates of cauliflower mosaic virus lead to the origin of the mutants that have lost the capacity to be insect transmitted.[61] The aphid nontransmissible strain CM4-184 is possibly a natural deletion mutant of CM-1841 and it arose during successive mechanical transfers over a period of years. The CM4-184 deletion consists of 421 bp from ORF II of its genome.[61] Thus, the specific region where deletion occurs has been mapped in the genome of CM4-184. Size variations in RNA3 of the type, ND 18, and AM strains of barley stripe mosaic virus (BSMV) indicate that BSMV strains with different numbers of RNAs could have originated by specific deletions having occurred in RNA3.

The coat protein of TMV strain HR has 156 instead of the usual 158 amino acid residues. This must have come about due to the deletion of two codons in RNA of the parent strain. This explanation should also be valid in all those cases where coat protein of mutants contains a lesser number of amino acid residues than the parent strain.

### 2. Base Substitution

Two unequivocal examples of base substitution as the cause of mutation are known. One pertains to cauliflower mosaic virus and the second to TMV. Several naturally occurring aphid-nontransmissible isolates of cauliflower mosaic virus (CAMV) are known. Three of them are CM4-184, CM 1841, and Campbell. Analysis of cauliflower mosaic virus deletion mutants[61,61a] and the formation of hybrid genomes between aphid-transmissible (AT$^+$) and aphid-nontransmissible (AT$^-$) isolates[62,62a] correlated aphid transmissibility to the ORF II encoded 18,000 mol wt protein product (p 18) present in inclusion body preparations from CaMV-infected tissue.[62,62b] Woolston et al.[62a] compared the nucleotide sequences between the *Bst*EII site at position 126 and the *Xho* I site at position 1643 of an AT$^+$ CaMV isolate

```
F2  AGAAATTCAAAATTTACCATTACATATTGGTATTTATTTACCCTCAGTTGGTGATATATG   60
G1  .GAAATTCAAAATTTACCATTACATATTGGTATTTATTTACCCTCAGTTGGTGATATATG


F2  TGAGGACGCTAGCCTGTTGGGTTTCCTGACCGACCAAATCCAAGCGAGCTTAATCCAAGT  120
G1  TGAGGACGCTAGCCTGTTGGGTTTCCTGACCGACCAAATCCAAGCGAGCTTAATCCAAGT


F2  ACCTCGTCTCAAATTGAGTGTCAAGTGAATAAGCATAGTGACTCCATCGTTTCAGGGTAG  180
G1  ACCTCGTCTCAAATTGAGTGTCAAGTGAATAAGCATAGTGACCCCATCGTTTCAGGGTAG
                                               *

F2  TTAACGGCTATTAATAGACATATTACGAACGCTTCTCTTTATTTATCACCAACATGGGAT  240
G1  TTGACGGCTATTAATAGACATATTACAAACGCTTCTCTTTATTTATCACCAACATGGGAT
      *                       *

F2  GTAATGTTTATGCGTGAGCTACGGCCGCATTGTAAAATTAGTGGTTTTGAATTTCTATTC  300
G1  GTAATGTTTATGCGTGAGCTACGGCCGCATTGTAAAATTAGTGGTTTTGAATTTCTATTC


F2  TTCGGAATATCCAAGGTTTAAAAGACCAGCATTTGGGTTAAAAATTTTTAAACCTTACTA  360
G1  TTCGGAATATACAAGGTTTAAAAGACCAGCATTTGGGTTAAAAATTTTTAAACCTTACTA
               *

F2  TCTTTAACTAGTAACTTGAATTCGATTTATATTCAGATTTTGCATATCAAGTTGTTGTGT  420
G1  TCTTCAACTAGTAACTCGAACTTGATTTATATTCAGATTTTAAATATCAAGTTGTTGTGT
        *              *    *   *            **

                            M  G  D  I  L  G  A  V  Y  D  L  G
F2  TTTCTGATCATCATTAAGTGGCCGTCATGGGTGATATATTAGGCGCAGTTTATGATTTAG  480
G1  TTTCTGATCATCATTAAGTGACCGTCATGGGTGATATATTAGGCGCAGTTTATGATTTAG
                        *

     H  R  P  Y  L  A  R  R  T  V  Y  E  D  R  L  I  L  S  T  N
F2  GGCACAGACCTTACCTAGCACGGCGTACGGTTTATGAGGATCGTTTGATTCTTAGCACAA  540
G1  GGCACAGACCTTACCTAGCACGGCGTACGGTTTATGAGGATCGTTTGATTCTTAGCACAC
                                                                *

     G  N  I  C  R  A  I  N  L  L  T  H  D  N  R  T  S  L  V  Y
F2  ATGGTAATATCTGTCGAGCTATTAACTTGTTAACTCACGATAATCGTACTTCACTGGTGT  600
G1  ATGGTAATATCTGTCGGGCTATTAACTTGTTAACTCACGATAATCGTACTACACTGGTGT
                    *                                    *

     H  N  N  T  K  R  I  R  F  R  G  L  L  C  A  Y  H  R  P  Y
F2  ATCACAATAACACTAAACGCATAAGGTTTCGTGGTTTATTGTGTGCTTATCATAGGCCTT  660
G1  ATCACAATAACTACTAAACGCATAAGGTTTCGTGGATTATTGTGTGCTCTTCATGGGCCTT
               *                       *               **   *

     C  G  F  R  A  L  C  R  V  M  L  C  S  L  P  R  L  C  D  I
F2  ATTGTGGGTTTCGTGCCTTATGTAGAGTAATGTTATGTTCTTTACCTCGTTTGTGTGACA  720
G1  ATTGTGGGTTTCGTGCCTTATGTAGAGTAATGTTATGTTCTCTACCTCGTTTGTGTGACA
                                                 *

     P  I  N  G  S  R  D  F  V  A  D  P  T  R  L  D  S  S  V  N
F2  TCCCTATCAATGGATCTCGCGACTTTGTTGCGGATCCTACCAGACTCGACAGCTCTGTTA  780
G1  TCCCTATCA...................................................

     E  L  L  V  S  T  G  L  V  I  H  Y  D  R  V  H  D  V  P  I
F2  ATGAGTTGTTGGTTTCTACCGGTCTCGTCATCCACTATGATCGTGTTCATGATGTTCCCA  840
G1  ...........................................................

     H  T  D  G  F  E  V  V  D  F  T  T  V  F  R  G  P  G  N  F
F2  TACACACTGATGGTTTTGAAGTTGTAGATTTTACGACTGTCTTTCGTGGTCCTGGAAATT  900
G1  ...........................................................

     L  L  P  N  A  T  N  F  P  R  P  T  T  T  D  Q  V  Y  M  V
F2  TTCTTTTGCCTAATGCAACAAATTTCCCTCGGCCGACCACAACCGATCAGGTTTACATGG  960
G1  ...........................................................

     C  L  V  N  T  V  D  C  V  L  R  F  E  S  E  L  T  V  W  I
F2  TGTGTTTGGTAAACACGGTTGATTGTGTGTTACGTTTTGAGTCTGAACTTACAGTGTGGA 1020
G1  ...........................................................
```

FIGURE 2. The cDNA sequences of BNYVV RNA3 of isolates F2 and G1. Asterisks beneath the sequences indicate point mutation differences between the F2 and G1 sequences, while asterisks above sequences indicate the termination codon at the end of the long ORF. The F2 and G1 sequences are from clone pBF11 and pBA4, respectively. (From Bouzoubaa, S., Guilley, H., Richards, K., and Putz, C., *J. Gen. Virol.*, 66, 1553, 1985. With permission.) (Continued on page 120.)

```
          H   S   G   L   Y   T   G   D   V   L   D   V   D   N   N   V   I   Q   A   P
F2    TTCACTCTGGTTTGTATACAGGTGATGTTTTAGATGTGGATAATAATGTTATTCAAGCCC       1080
G1    . . . . . . . . . . . . . . . . . . . . . . . . . . . . . . . . . . . . . . . . . .


          D   G   V   D   D   D   D   *
F2    CTGACGGTGTTGATGATGATGATTAGAGTTATCACAATTTCAACAACACACTTATTGGTG       1140
G1    . . . GACTCGTTGATGATGATGATTAGAGTTATCACAATTTCAACAACACACTTATTGGTG
              * * * *

F2    TGTTGTTCTGTTACACCATTTGAAAGTTTAATAATTATCTCAATCCGATTGTTGATCTGG       1200
G1    TGTTGTTCTGTTACACCATTTGAAAGTTTAATAATTGTCTCAATTCGATTGTTGATCTGG
                                          *              *

F2    TTGGGACAATTATTTTATTTTCTTTTGGTGTAATCGTCCGAAGACGTTAAACTACACGTG       1260
G1    TTGGGACAATTATTTTATTTTCTTTTGGTGTAATCGTCCGAAGACGTTAAACTACACGTG

F2    ATTTCACGGTGTTCTATGAGAAGATTGTTTAACGGTGTCACGTTGTGCATTTTTAAGCCT       1320
G1    ATTTCACGGTGTTCGATGAGAAGATTGTTTAACGGTGTTTACGTTGTGTACCTTTAAGCTT
                        *                          *       * **         *

F2    TCTTCTCATTTAACCACATGTGATGATTGTAGCCTGTGGGTTGTTATGTGGACAATTATG       1380
G1    TCTTCTCATTTAACCACATGTGATGATTGTAGCCTGTGGGTTGTTATGTGGACAATTATG

F2    GTTACTTATTTGTAAATGGTAAAGAGTGTGCAGTAGCGACTTTATGCGAGTGGGAGTAGT       1440
G1    GTTACTTATTTGTAAATGATAAAGAGTGTGCGGTAGCGACTTTATGCGAGTGGGAGTAGT
                          *                *

F2    TGTGTTATTACTACTATTCTGGTTCGTATAAAGATCCTTGACGGCGGCATCGTGGGTTCC       1500
G1    TGTGTTATTACTACTATTCTGGTTCGTATAAAGATCCTTGACGGCGGCATCGTGGGTTCC

F2    ACAGCCGGTTACATGGTGTTCCCGTCCGTTTACGAAGGTTTAACTGTGAGCCTTGTATTT       1560
G1    ACAGCCGGTTACATGGTGTTCCCGTCCGTTTACGAAGGTTTAACTGTGAGCCTTGTATTT

F2    TACGAATACACAGTTTTTATCCTAACAGGCTCGTTCACAAGCCTCCTTTTACATTAAGTT       1620
G1    TACAAATACACAGTTTTTATCTTAACAGGCTCGTTCACAAGCCTCCTTTTACATTAAGTT
          *                    *

F2    TAAAGGTTTATGTGGACACAAAAATATGGCTTATTGGTTATGCTAAACCTCATATCATGT       1680
G1    TAAAGGTTTATGTGGACACAAAAATATGGCTTATTGGTTATGCTAAACCTCATATCATGT

F2    TATAGTATTTGTTTTATATTATAATTAAGGTTAAGATGTACTGACTGGGTGTGAAATGTA       1740
G1    TATAATATTCGTTTCATATTATAATTAAGGTTAAGATGTACTGACTGGGTGTGAAATGTA
          *        *         *

F2    CCAGTCCTTGTAGGGTTCTTTGTCAGTATATTGAC      1775
G1    CCAGTCCTTGTAGGGTTCTTTGTCAGTATATTGAC
```

FIGURE 2 (continued)

Cabb B-S and an AT⁻ isolate Campbell. They found only a variation of about 3.4% nucleotide sequence between the two and it was entirely due to base substitutions of the type A-G or C-T. They found only five base changes within ORF II from Cabb B-S to Campbell isolates. Two of these could be responsible for the changes in the predicted amino acid sequence: amino acid 33 is altered from serine in Cabb B-S to asparagine in Campbell, and amino acid 94 is altered from glycine in Cabb B-S to arginine in Campbell. The predicted amino acid change at site 33 is silent, while amino acid change 94 from glycine to arginine is an active change which alters the properties of the ORF II product p 18, resulting in loss of its function for aphid transmissibility. Thus change from AT⁺ to AT⁻ phenotype is due to a single amino and change in p 18 from glycine to arginine at amino acid position 94.

It is significant in this connection that glycine at position 94 is conserved in p 18 of AT⁺ isolates PV 147,[63] Cabb B-S,[63a] and XJ,[63b] and that the glycine to arginine change occurs in p 18 of AT⁻ CaMV isolates CM-1841 and Campbell.[62a] Hull (unpublished results, quoted in Woolston et al.[62a]) suggested that computer prediction of the secondary structure of p 18 (the ORF II product) indicates the presence of a possible interdomain turn near position 94 and that transition of glycine to arginine at position 94 in Campbell weakens this interdomain turn. It must be inhibiting the p 18 to perform its function of aphid transmissibility. Thus aphid transmissibility or nontransmissibility is not directly correlated with the presence or absence of ORF II product p 18[62b] as was suggested earlier. The earlier suggestion that a mutation in control sequences distant to ORF II was responsible for aphid-nontransmissibility of the CM-1841 isolate[63] is also no longer valid.

The 30K TMV-encoded protein is involved in a virus transport function within the infected plant (Chapter 4). The nucleotide sequences of the 30K and coat protein cistrons of a temperature-sensitive (*ts*) mutant TMV LS1, which is defective in cell-to-cell transport at the restrictive temperature, and of the wild-type TMV L (tomato strain) were worked out and compared.[64] A single base substitution leading to the exchange of proline codon in the L strain by a serine codon in the LS1 mutant occurs in a highly conserved amino acid sequence (Figure 4 of Chapter 4). A conserved proline residue is predicted to be present in the region of a β turn of the secondary structure in the 30K protein. Its substitution by serine could lead to decreased rigidity of the 30K protein of the LS1 mutant. This change may make it temperature sensitive, resulting in the failure of this *ts* mutant to spread at the restrictive temperature.[64] The attenuated TMV strain $L_{11}A$ is derived through spontaneous mutation of strain $L_{11}$, which in turn is derived from the virulent TMV strain L. The determined nucleotide sequence shows that the strain $L_{11}A$ is derived from strain L by ten base substitutions which result in three amino acid changes in TMV-encoded protein of 126K (or its readthrough product of 184K) molecular weight.[64a]

## B. Recombination

Recombination implies the involvement of genomes of at least two parents and a realignment/transfer/combination of genes of one parent to the other. The new genome of the progeny is constructed by deriving the majority of genes from one parent and some genes from the other parent. The progeny are the recombinants or hybrids and can arise only when nucleic acids of the two parents multiply in the same cell. The presence of two types of inclusions, each characteristic of a strain, in the same cell is a definite indication of the replication of two strains in that cell. Particles of some of the AMV strains form four definite strain-specific and distinct groups of aggregation bodies. Tobacco leaves inoculated with mixed inoculum containing two such AMV strains (Caldy and 15/64) show two important characters:[19] in some cases cytoplasm of a single cell is divided into two regions: one region contains aggregates specific for one strain and the other contains aggregates specific for the second strain, while in other cases the two types of aggregates occur side by side within the same cell and even seem to merge with each other. In some other cases the two types of aggregates occur separately in adjacent cells or groups of cells. Since the aggregation form is determined by the coat protein and the coat protein is in turn coded by top *a* RNA component, it is clear that top *a* RNA components of the two strains occur and multiply in the same cell in cases where the two aggregate forms occur within the same cell, but they multiply in adjacent cells in cases where the two aggregates occur within neighboring cells. It was shown earlier that two TMV strains could infect the same cell.

Later work involving inoculation of plant protoplasts with plant viruses has established beyond any doubt that two related viruses or strains of a virus or even unrelated viruses can infect and replicate simultaneously in doubly infected protoplasts. Genomes of two cowpea chlorotic mottle virus (CCMV) strains appeared to coexist in mixed infections of tobacco protoplasts.[65,66] A *ts* and a *tr* CCMV mutant were employed. The *ts* strain was dominant over the *tr* strain at all temperatures so that most of the progeny had the *ts* genome, but *tr* remained available to rescue the *ts* genome by providing its coat protein under restrictive conditions (35°C). On simultaneous inoculation of tobacco protoplasts with S and E strains of raspberry ringspot virus, more than half of the protoplasts contained both the strain-specific antigens.[67] Change in the ratio of particles of the two strains in inoculum resulted in a change in the proportion of protoplasts containing antigen aggregates of both the strains. Tobacco protoplasts could be doubly infected by two TMV strains, the common and tomato strains, and both strains replicated.[68] Mixed inoculum containing nearly the same concentrations of the two TMV strains caused double infection of about 80% protoplasts but only 50% of the protoplasts were infected by the two strains after sequential infection.

Thus there is every possibility that new strains can arise by recombination of genetic material of two different strains of a virus when multiplying in the same cell. However, it may be mentioned that, in almost all cases of recombination reported below, the actual infection of cells by the two strains was not specifically investigated by the workers concerned, but was only implied.

Possibly the first report about recombination concerns tomato spotted wilt virus. New TSWV strains were produced after the two TSWV strains in a host cell of tomato or *N. glutinosa* exchanged determinants.[69] The new strains were stable, bred true over many years, and exhibited some of the characteristic symptoms of both the parental strains. Bean plants exhibited new combinations of symptoms and even qualitatively new symptoms after inoculation with mixtures of different AMV components. This was regarded to be due to genetic recombinations between RNAs of the heterologous components during replication rather than mere complementation.[70]

More convincing and direct evidence of recombination between strains of a plant virus as an important method of variation has become available only recently. An isolate of CMV is possibly a natural hybrid of two other strains, because competition hybridization of a large number of CMV isolates showed that one isolate was partially homologous to two of the nonhomologous CMV strains.[71] Studies on nucleotide sequence homology of the four genomic RNAs of Q strain of CMV shed new light on the possible mode of origin of new strains of multicomponent viruses.[72] The four RNAs possess nearly identical sequences up to residue 138 from the 3′ terminus. However, RNAs 1 and 2 did not possess any sequence homology with RNAs 3 and 4 from residue 139 to 169, but all four RNAs again showed complete sequence homology from residue 170 to 209. This homology-nonhomology-homology nucleotide arrangement between the four RNAs of this strain suggests that it possibly originated as a new recombinant strain after mixed infection of a cell by two CMV strains such that RNAs 1 and 2 came from one strain and RNAs 3 and 4 from the other strain.[72] This process is parallel to the formation of pseudorecombinants in vitro. In fact, pseudorecombinant formation clearly envisages the possibility of origin of new strains in vivo by an identical phenomenon. Possibly the in vivo origin of Q-CMV is the first such example to be explained on this basis.

Two naturally occurring tobravirus isolates, namely, 16 (an Italian tobravirus) and N5 (a tobravirus isolate from Scotland), induce tobacco rattle virus-like symptoms in herbaceous plants but react serologically with antisera to and are related to pea early browning virus (PEBV) isolates, 16 to the PEBV British serotype and N5 to the PEBV Dutch serotype. Nucleic acid hybridization studies of Robinson et al.[73] showed that nucleotide sequences of RNA1 of 16, N5, and TRV strains is similar suggesting that RNA1 in all these tobraviruses are identical. However, the situation in RNA2 of 16 and N5 was shown to be complicated:[73] the RNA2 of 16 contains internal sequences similar to the British serotype of PEBV but the 3′ and 5′ ends bear TRV-like sequences (Figure 3). Similarly, the internal region of the N5 RNA2 contains sequences related to the Dutch serotype of PEBV while its 3′ and 5′ ends bear more extensive (as compared to those of the 16) TRV-like sequences. This clearly establishes that the RNA2 of 16 and N5 are truly recombinant molecules formed by recombination of RNA2 sequences derived from PEBV and TRV. Additionally, 16 and N5 are true natural hybrids possessing the pathogenicity (and RNA1 which determines the pathogenicity) of TRV and serological properties (and RNA2 which determines the serological properties) of PEBV. The origin of N5 and 16 tobraviruses can occur only if, in a double infection of TRV and PEBV, the 3′ and 5′ end nucleotide sequences of TRV were erroneously substituted onto the somewhat similar PEBV RNA2 instead of TRV RNA2.[73]

The internal region of RNA2 of the Italian tobravirus (isolate 16) is constituted by the

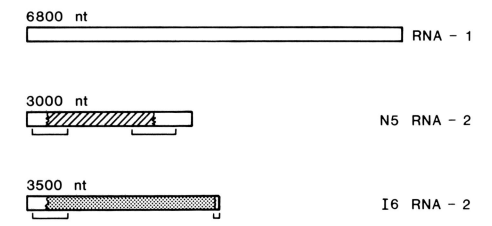

FIGURE 3. Structure of the recombinant (hybrid) RNA2 genomes of N5 and 16 tobravirus isolates as deduced from hybridization experiments. TRV-like structures are unshaded, sequences similar to those of PEBV-Dutch serotype are represented by the hatched region, sequences similar to those of PEBV-British serotype are represented by the speckled region, braces indicate the uncertainties in the positions of the boundaries of the various types of sequences, and sizes of the RNA molecules estimated to the nearest hundred nucleotides (nt) are shown. (From Robinson, D. J., Hamilton, W. D. O., Harrison, B. D., and Baulcombe, D. C., *J. Gen. Virol.*, 68, 2551, 1987. With permission.)

British isolate of PEBV, while that of the Scotch tobravirus (isolate N5) is constituted by the Dutch isolate of PEBV, while their 3′ and 5′ ends are identical to the TRV. This shows that these two isolates must have originated independently.[73]

RNA recombination has also been shown to occur in brome mosaic virus (BrMV).[74] Recombination between BrMV RNAs 1,2, and 3 was studied in barley plant inoculated with viral RNAs 1 and 2 and an RNA3 deletion mutant designated m4 RNA3. The RNA3 of the emergent virus in three barley plants on day 15 postinoculation was subjected to sequence analysis. It was found to have undergone recombination with one of the other genomic RNAs and was thus different from both the wild-type RNA3 and m4 RNA3. Consequently, the emergent viruses are the mutants and were designated mutants A (produced in plant I), B,C,D, and E (all produced in plants II and III).

Mutant A RNA3 contained RNA2 sequences, and its deleted portion was restored and contained 39 more bases than the wild-type RNA3, and 59 more bases than the m4 RNA3 ancestor. Mutant E RNA3 also had its deleted portion restored, and the last 206 bases of the wild-type RNA3 were found to be replaced by 215 3′-terminal nucleotides of RNA2. Both A RNA3 and E RNA3 were larger, because of the increased number of bases added, than the wild-type RNA3 (Figure 4) and the recombination in both cases occurred outside the 3′-end homologous region. The RNA3 of the B,C and D mutants was formed by recombination between RNA3 and RNA1 or RNA2 sequences at a site within their about-200-nucleotide-long 3′-terminal homologous regions (shown as open box in Figure 4). The mutant B RNA3 is formed by recombination of RNA3 and RNA1 sequences somewhere between 130 and 101 bases, mutant C RNA3 is a combination of RNA and RNA1 sequences somewhere between 175 and 131 bases, and mutant D RNA3 is a combination of RNA3 and RNA2 sequences somewhere between 205 and 101 bases. It is clear from the above that recombination can occur at various locations on viral RNA.

Dixon et al.[75] found that DNA of CaMV strain CM4-184 contains regions, some of which are related to CaMV strain S and some to CaMV strain CM-1841. It is suggested that this happens because of the intergenomic switching between the RNA templates of the two virus strains concerned during formation of nascent DNA minus strand by reverse transcription.

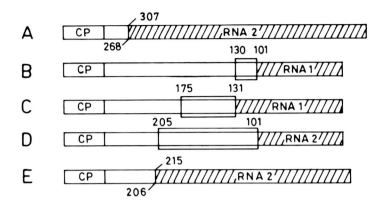

FIGURE 4.    The RNA3 of various brome mosaic virus (BrMV) mutants A,B,C,D, and E formed by recombination of wild-type RNA3 and RNA1 or RNA2 sequences outside or somewhere within the 200-nucleotide long 3′-terminal homologous sequences (shown as open box). Only the 3′ regions of RNA3 downstream from coat protein open reading frame (ORF) have been shown. Wild-type BrMV RNA3 sequences are shown as open spaces adjacent to the coat protein (CP) ORF. Cross-hatched areas represent the regions identical to wild-type RNA1 or RNA2, open boxes enclose areas in which recombination took place, and the numerical numbers represent the nucleotide numbers. (Reprinted by permission from *Nature*, 321, 528—531. Copyright © 1986. Macmillan Magazines Ltd.)

CaMV strain CM4-184 appears to be a product of a recombination between the viral DNA of the other two strains: a majority of the CM4-184 nucleotide sequences are derived from the CM1841 parent, while the remaining nucleotide sequences come from a CaMV strain closely related to strain S.[75] Also, the 401 to 1930 nucleotide sequences are similar to those of the CM-1841 except for a deletion in ORF II and two base exchanges (at positions 541 and 961) shared with CaMV-S.[75] It is significant in this connection that the formation of hybrid CaMV genomes by in vitro recombination between cloned viral DNA of Cabb B-JI and Campbell isolates[62,62a] has been accomplished (Figure 5). Schoelz et al.[76] also constructed some recombinant viruses in vitro from CaMV strains D4 and CM-1841. Thus template switching during reverse transcription causes recombination of viral DNA genomes.

It is evident from above that true recombination can play an active role in generating new well-adapted variants in nature.

## C. Selection Pressures and Selection of New Strains

Inoculum, as already shown, is often a mixture of strains, one of which is dominant in the original host under the particular environmental conditions. Many of the mutants fail to multiply and express themselves in the presence of the dominant parent strain which, over the years, has preferentially been selected to become the dominant strain under those particular environmental conditions. The new strains, with a change in selection pressure, are likely to prove superior to and more adaptable than the parent strain and may become stabilized and dominant under the changed environments. These, then, will be the new natural strains and may sweep through an area. Viruses are thus also governed by the processes of natural selection and survival of the fittest. Various types of selection pressures are responsible for preferential selection of new strains under new stresses. Some of the more important of these selection pressures whose role in sieving out strains has been clearly established by now are: passage through hosts different from the original host (host-selection), continued mechanical transfer of insect-transmitted strains which selects vector nontransmissible strains, selection by vectors for transmissibility, and rise in the ambient temperature which selects strains that can replicate at raised temperature (*tr* mutants). These factors can also result in a change in the dominant strain of a locality over years.

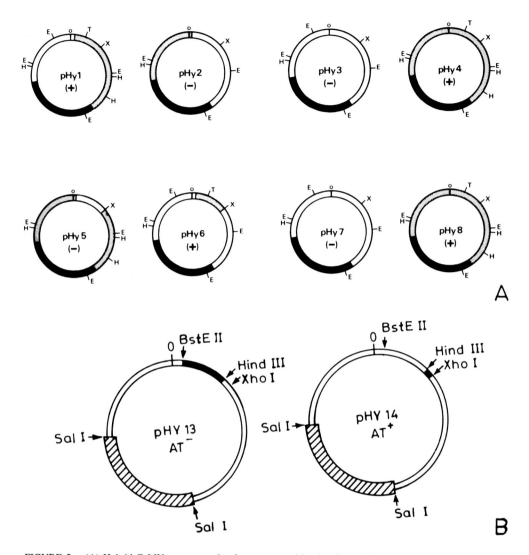

FIGURE 5. (A) Hybrid CaMV genome molecules constructed in vitro from DNA of AT⁺ Cabb B-JI (half-tone shading) and DNA of AT⁻ Campbell (unshaded). Plasmid vector pAT153 DNA is shown as black segments. Names of restriction enzymes are abbreviated to their first letter in the pHy1-pHy8 hybrid genomes. Aphid-transmissible progeny is shown as (+), aphid-nontransmissible progeny as (−), and the site of α-strand gap is represented by the symbol "O". (From Woolston, C. J., Covey, S. N., Penswick, J. R., and Davies, J. W., *Gene*, 23, 15, 1983. With permission.) (B) Formation of hybrid CaMV genomes pHy 13 and phy 14 by in vitro recombination of cloned Cabb B-JI (black) and Campbell (white) DNA isolates. Plasmid vector pAT153 is depicted as hatched. For clarity, other *Hind* III restriction sites are not shown. CaMV genome hybrid pHy 13 is composed of the Campbell genome and *Bst*E II to *Hind* III fragment of Cabb B-JI genome. The pHy 14 genome hybrid is composed of the Campbell genome and *Hind* III and *Xho* I Cabb B-JI DNA fragment. (From Woolston, C. J., Czaplewski, L. G., Markham, P. G., Goad, A. S., Hull, R., and Davies, J. W., *Virology*, 160, 246, 1987. With permission.)

## 1. Strain Selection by Host

The new host, when an inoculum is passaged through it, favors the selection and therefore the rapid multiplication and symptom expression of a new strain from the preexisting mixture of strains. The result is the so-called attenuation of the old strain, while a new strain, which was not able to express itself in the original host, becomes dominant and makes its appearance in the other host.

Virulence of beet curly top virus towards its sugar beet host decreased after passage through *Chenopodium murale* and it then induced only mild symptoms in the host.[77] A similar decrease in virulence towards the host also occurs in a TMV strain after passage through sea holly (*Eryngium aquaticum*).[78] The term "attenuation" was used for the decrease or loss in virulence of viruses after passage through nonhosts as exemplified above, and all the above-mentioned viruses were said to have become attenuated. Decrease in virulence of tobacco necrosis virus after being maintained in bean, and of potato virus Y and AMV after growth in nonhosts, was reported later. Similarly, severe ryegrass mosaic virus obtained from Italian ryegrass (*Lolium multiflorum*) became milder after one passage through the resistant ryegrass (*L. perenne*) cultivar Mascot.[79] Yarwood reviews the host passage effects on plant viruses.[80] Atabekov also discusses this topic.[81]

Physicochemical differences (like specific infectivity, sensitivity to inactivation by ultraviolet irradiation, electrophoretic mobility, protein composition, serology, and symptoms induced in host plants) exist between particles of a cowpea strain of TMV when purified from French beans and when purified from tobacco. Of the four possibilities (presence of both forms in inoculum and selection by bean or tobacco of the respective strain, selection of randomly occurring mutants, selection of host-induced mutants, and formation of phenotypic variants), the selection of randomly occurring mutants as the cause for these physicochemical differences between the same TMV strain when isolated from two different hosts was favored. Host selection of a TMV substrain, which already existed as a small proportion of the parent culture, explains its increased virulence.[14]

CMV, when inoculated to primary leaves of cowpea, gave rise to mutants which were less injurious to cucumber but more injurious to cowpea than the parent culture. Passage of these "cowpea mutants" through cucumber generally made them less virulent to cowpea and in several cases made them revert to the "original" CMV form which produces local lesions in cowpea. Thus passage of CMV through cowpea-cucumber series selects variants suitable to the host concerned and the mutants more or less return to the original form after passage through cucumber.[82] Barley stripe mosaic virus isolates from barley could not be easily mechanically transmitted to oat plants but, after passage through oats, could be easily transmitted from infected to healthy oat plants.[83] It was further investigated by Chiko by employing the C4 BSMV strain-barley-wild oats system.[84] To begin with, only a low proportion of wild oats could be systemically infected by C4 BSMV inoculum obtained from barley, but subsequently (when the C4 inoculum was obtained from infected wild oats), a very high proportion of healthy wild oats could be systemically infected by the isolate C4 BSMV. This was suggested to be due to strain selection during systemic passage of the isolate through wild oats.[84]

Variants of CCMV can be obtained by passage through specific hosts. Weekly serial propagation of type (T) CCMV strain in susceptible cowpea, followed by passage through susceptible beans, led to the isolation of a mild (M) strain of CCMV.[85] Propagation of T in susceptible cowpeas causes the appearance of a new spontaneous mutant M. The new variant co-exists with T in mixed infections, but later (during continued propagation in susceptible cowpeas) becomes dominant over T under greenhouse conditions and then is selected during the later passage through susceptible beans while T is seemingly excluded. Only a minor genetic change influencing systemic symptoms results in the appearance of strain M. Consequently, these two strains are closely related in specific infectivity, replication, several biophysical properties and serology but differ in symptoms induced in cowpea, in relative competitive ability in cowpeas and beans, and in the nature of RNA3.[85] A new and distinct CCMV strain R was isolated from the resistant cowpea introduction 186465 after inoculation with CCMV T strain.[86] This strain could systemically invade and replicate in T strain-resistant cowpea plants and was serologically distinct. At least two changes are envisaged to occur in the genome of CCMV-T for the origin of strain R: change in strain T to an intermediate form T3d which has less RNA3 than T, followed by a change of T3d to strain R.[86]

Valverde and Fulton[45] found that southern bean mosaic virus strain A gave rise to three strains depending upon the host through which it was passaged. SBMV-A gave rise to SBMV-C on inoculation to 'Pinto' bean but gave rise to SBMV-B on inoculation to 'Black Turtle Soup' bean, while inoculation of SBMV-C to 'Black Turtle Soup' beans gave rise to SBMV-D. Maintenance of PVX strains in tobacco sometimes results in the loss of their ability to infect potato. Inoculation of small lesion isolates of CMV on cowpea cv. Catjang produces large lesions at a rate of 0.53%. Large lesion CMV mutants arise by mutation of the parent strain and are later specifically host-selected by the Catjang cowpea.[87,88]

Hurtt conducted detailed studies on hibiscus chlorotic ringspot virus (HCRSV) and observed the following:[89]

1.  Wild-type HCRSV, isolated from *Hibiscus rosa-sinensis* and after serial passage in *Chenopodium quinoa*, reacted with antiserum to HCRSV but could not infect kenaf (*H. cannabinus*), which is a systemic host of the wild-type virus.
2.  Virus particles isolated from *H. rosa-sinensis*, kenaf, and *C. quinoa* had the same buoyant density in cesium chloride, the same particle size, the same genome size, and the same coat protein size. However, purified preparations of the virus from each plant, when subjected to electrophoresis in agarose slab gels, gave electrophoretic zones which differed in number, migration rate, and serological properties from zones of the virus from other two plants. These migrating components/zones are the electrophorotypes. Wild-type HCRSV contained three electrophorotypes. Virus from kenaf migrated as a single electrophorotype designated HCRSV-K. The virus serially passed in *C. quinoa* was a mixture of several unique electrophorotypes: a fast electrophorotype (HCRSV-F), a slow electrophorotype (HCRSV-S), and minor electrophorotypes with intermediate migration rates. The electrophorotypes were consistently differentiated in three experiments conducted more than a year apart.
3.  HCRSV-F and -S fail to infect kenaf and *H. rosa-sinesis*. Immunological studies showed that the HCRSV-F and -S were serologically related but were nevertheless unique, and their aspartic acid and threonine content differed.

Hurtt arrived at the following conclusions on the basis of the above observations.[89] The variants arise as a consequence of host passage effects. The origin of the same electrophorotypes after passage in *C. quinoa* even after a year (i.e., repetitive nature of the phenomenon) suggests that the origin of the same particular electrophorotypes cannot be due to random mutation or environmentally directed mutation but is instead most probably due to host selection or host-directed mutation. Inability of the *C. quinoa* electrophorotypes (HCRSV-F and -S) to infect hibiscus suggests that the host-directed mutation, rather than host selection, is the possible explanation of the origin of these electrophorotypes. If true, this could be an example of Lamarckism operative in plant viruses, i.e., host-induced, adaptive mutation and the inheritance of acquired characters. The similar size and density of the HCRSV electrophorotypes suggest that differences in surface charge of virus particles could be responsible for the differences in electrophoretic mobility. Replacement of some neutral amino acids in HCRSV-S by acidic amino acids may give rise to HCRSV-F.

In short, host-directed mutation is the reason for the origin of these electrophorotypes and the mutation involves some amino acid changes in capsid protein.

Certain earlier claims on the influence of host on strain selection have not been confirmed by later work. Bawden claimed that reversible changes between a tobacco form and a bean form of sunn hemp mosaic virus were possible.[90] However, this observation has not been confirmed by any other investigator[80,91] and may have been due to contamination. Host-induced transition of TMV U1 strain to TMV U5-like M5 strain by several transfers of the U1 strain to *N. glauca*, reported by Bald et al.,[92] has not been supported by Wetter.[91] It is not unexpected, since the transition of U1 to the U5 strain involves fundamental changes in coat protein cistron, as well as in genes controlling host range specificity and host reactions. Thus, this transition cannot represent a single mutation but depends on considerable changes in the viral genome — an unusual thing.

A few published reports suggest the possible mechanism(s) that can be responsible for the selection and/or replacement of a particular virus strain by a particular host. The severe TMV strain, which produces mild symptoms in tobacco after passage through sea holly, moves slowly in sea holly and may have been filtered out. In contrast, TMV strain U5 moves closer to the apex of cultured stem apices of infected plants of *N. glauca* than TMV strain U1. In a mixed infection, therefore, strain U5 may exclude strain U1 from reaching young meristematic tissues and thus ultimately from the plant itself. Similarly, the severe strain of citrus tristeza virus was transported from phloem cells to neighboring cells more rapidly than the mild strain.[93] and may ultimately replace the mild strain. The relative rate of transport of different virus strains in particular host plants seems to explain the host-selection in the above cases (see Chapter 4), and virus transport in plants is genetically controlled by the virus itself and is possibly mediated by some transport protein encoded by viral genome. Thus, the mutations in the above cases may be occurring in the cistron responsible for synthesis of transport protein.

Another factor controlling host-selection may be the relative specific infectivity of the concerned strains. A strain with greater specific infectivity may displace in a host a strain with lesser specific infectivity and may be selected. Out of the two raspberry ringspot virus strains S and E in mixed inoculum, the strain S has greater specific infectivity and tends to dominate so that, in presence of strain S, strain E infected a lesser number of protoplasts.[67]

Thermosensitivity of strains may be another factor. Cucumber mosaic virus strain C is thermoresistant and multiples faster in tobacco plants kept at 22°C than the thermosensitive strain B. The latter consequently tended to be replaced by the former. This tendency was greatly amplified at 32°C at which the thermoresistant strain C replaced the thermosensitive strain B in tobacco plants inoculated with a mixture of the two strains at a much quicker rate.[94] An AMV *ts* mutant, *Tbts7 (UV)*, carries a *ts* defect in the early function of capsid protein. It was found that in a mixed infection of wild type (WT) AMV and *Tbts 7(UV)* symptoms of the mutant did not appear at 30°C, that no mutant capsid protein and no mutant RNA were present in the progeny of these mixed infections at 30°C, and that only the WT RNA and predominantly WT coat protein were present in the progeny. It was concluded that mutant RNA3 is completely outcompeted by WT RNA3 at 30°C.[95] Another case of such competition between WT and a *ts* derivative occurs in CCMV: a *ts* mutant of CCMV RNA3 was dominant over the WT at all tested temperatures in mixedly infected protoplasts.[66] An avirulent masked TMV strain manifests itself when infected tobacco plants are kept at 30°C.[96]

Continued mechanical transfer of some vector-transmitted viruses leads to the development of strains which can no longer be transmitted by the insect vectors. The case of WTV, due to the formation of deletion mutations, has already been discussed. Repeated mechanical transfer of the leafhopper-transmitted potato yellow dwarf virus in *N. rustica* gave strains of the virus that could not be transmitted by leafhoppers. An aphid-transmitted potato virus C strain often became nontransmissible after passage through potato. Similarly, continuous passage of sowthistle yellow vein virus by mechanical inoculation of the aphid vector yielded strains that are no longer transmissible to plants. These changes in the mode of transmission of these virus strains was explained on the basis of selection by the host but may also be due to, after the proven work on WTV, mutation. However, experimental proof is still lacking.

## 2. Strain Selection by Vector

There is some evidence that vectors may also select a virus strain out of a mixture of strains present in a host. In Israel, citrus tristeza virus had been restricted for over two decades in several introduction plots. There had been no natural spread of the virus from these plots to other trees and several aphids, including *Aphis gossypii*, failed to transmit the introduced CTV strains or transmitted them only at low rates. However, a limited natural spread was found to have occurred since 1970 and this was correlated with the appearance of CTV isolates highly transmissible by *A. gossypii*.[97] Similarly, a highly transmissible CTV isolate, designated ST4, was obtained through *A. gossypii* from sweet orange trees infected with the low-transmissible ST isolates.[13,98] Possibly, these transmissible isolates had been cross-protected all these years by the nontransmissible isolates against transmission by *A. gossypii*. However, aging of infected source plants may have resulted in breakdown of cross-protection so that *A. gossypii* may have acquired and transmitted the transmissible isolates to healthy plants which in turn may have served as principal sources of natural spread of CTV after 1970.[97] Thus, the vector also possibly selects transmissible isolates out of a mixture of isolates. This may also be true of all viruses that are specifically transmitted by vectors.

A clear case of strain selection by vector has been reported by Kimura et al.[98a] in rice dwarf virus (RDV). The leaves of a rice plant showing severe symptoms were macerated in $0.1\ M$ sodium phosphate buffer, pH 7.3; centrifuged; the supernatant diluted and used as inoculum for injection into second instar nymphs which were then employed for transmitting the virus to healthy plants which later developed symptoms, and were used in turn as inoculum sources. This selection process was repeated 10 more times. Severe symptoms (severely stunted plants) were present in 85% of the plants after the first selection, in 93 to 96% plants after the second to sixth passage, and in all plants after the 10th selection, i.e., passage in vector. Thus, the vector selected a severe (S) RDV strain. Moreover, the S strain was more readily transmitted by vectors than the present O strain. The fourth largest RNA segment of the 12 RNA segments of RDV strain S has an apparent molecular weight of about 20,000 more than that of the corresponding segments of strain O.[98a] The protein encoded by segment 4 of the S strain has a molecular weight of 44,000 compared to the 43,000 mol wt protein encoded by segment 4 of the O strain.[98a]

## 3. Strain Flora of a Locality

Certain isolates of a virus are predominant in certain geographic areas. Australian bean yellow mosaic virus S strain-like isolates are dominant in *Vicia faba* in a region in South Australia and western Victoria, while this strain does not exist in *V. faba* or in any other host in other parts of Australia.[99] Barley yellow dwarf virus isolates causing severe symptoms in cereals, which are usually B isolates, are dominant in the south and west of England, while isolates causing mild symptoms, usually F isolates, are dominant in the north and east parts of England.[100] Argentina, Brazil, and the U.S. have their characteristic strains of beet curly top virus, each with a different species of leafhopper as vector. Isolates of Andean potato latent virus have been divided by Koenig et al.[101] into three major serological strain groups restricted to different geographical regions: the Hu group in the southern Andes (from Central Peru to Bolivia), the CCC group in the northern Andes (Bolivia), and the Col-Caj group throughout the Andean region, but more common in the north than in the south.[101]

However, some strains may have become widely distributed over the years. Yeh et al.[102] investigated the relationships of nine papaya ringspot virus isolates obtained from Taiwan, Hawaii, Florida, and Ecuador and concluded that these virus isolates from widely separated geographic areas of the world had very similar biological and serological properties and appeared to belong to the same biological group. A strain of beet curly top virus similar to the North American strain occurs in Turkey, and it is postulated that the virus originated in

**Table 2**
## DISTRIBUTION OF VARIANTS OF BARLEY YELLOW DWARF VIRUS RECOVERED FROM OATS COLLECTED IN NEW YORK STATE OVER A PERIOD OF 20 YEARS

| Year | Isolate identified (No.) | Isolate (%) similar to: | | | | |
|---|---|---|---|---|---|---|
| | | MAV | RPV | RMV | PAV | SGV |
| 1957 | 29 | 90 | 7 | — | 3 | — |
| 1958 | 42 | 88 | 0 | — | 12 | — |
| 1959 | 42 | 86 | 2 | 2 | 10 | 0 |
| 1960 | 72 | 89 | 1 | 1 | 8 | 0 |
| 1961 | 41 | 93 | 5 | 0 | 2 | 0 |
| 1962 | 34 | 85 | 9 | 0 | 6 | 0 |
| 1963 | 10 | 55 | 18 | 9 | 18 | 0 |
| 1964 | 28 | 29 | 11 | 35 | 25 | 0 |
| 1965 | 36 | 31 | 19 | 8 | 42 | 0 |
| 1966 | 67 | 19 | 27 | 4 | 49 | 0 |
| 1967 | 76 | 36 | 10 | 8 | 46 | 0 |
| 1968 | 54 | 61 | 6 | 6 | 28 | 0 |
| 1969 | 90 | 8 | 16 | 17 | 57 | 2 |
| 1970 | 47 | 6 | 26 | 9 | 59 | 0 |
| 1971 | 53 | 4 | 28 | 8 | 60 | 0 |
| 1972 | 81 | 3 | 17 | 3 | 76 | 1 |
| 1973 | 63 | 3 | 16 | 3 | 78 | 0 |
| 1974 | 75 | 0 | 7 | 9 | 84 | 0 |
| 1975 | 56 | 2 | 9 | 9 | 80 | 0 |
| 1976 | 59 | 0 | 0 | 0 | 98 | 2 |

From Rochow, W. F., *Phytopathology,* 69, 655, 1979. With permission.

Europe and was carried with the beet to several parts of the New World where it acquired different vectors.[103]

Strain flora of a virus in particular areas may show wide variations due to any one or more of several stresses and selection pressures. Changes in host cultivar, in temperature and in other environmental conditions, and in vector species are the major selection pressures. The predominant strain of tomato mosaic virus can rapidly change in response to new selection pressures. Strain O was the dominant greenhouse strain in England prior to 1966. In that year, two new tomato varieties (Eurocross and Virocross) were introduced. Both these varieties contain the gene *Tm-1* which imparted resistance against the then prevailing strain O but not against TMV strain 1. By 1968, strain 1 had become more common than strain O in greenhouses where new tomato varieties had been grown for 3 successive years. However, this strain declined when some other varieties were grown instead of Eurocross and Virocross.[104] This is a clear example of the selection pressure exerted by the host plant and it selects certain viral mutants depending upon the "resistance genes" present in that cultivar. Several cases are known where resistance-breaking strains appeared and then became dominant in an area in response to the cultivation of new cultivars.

Considerable fluctuations have occurred in the dominant strain of barley yellow dwarf virus (BYDV) in the U.S. over a period of 20 years. In 1957, 90% of the isolates were similar to MAV, and only 3% of the isolates were similar to PAV strain. This was followed by a gradual decline of MAV isolates and an almost parallel increase of PAV isolates over the 20-year period so that, by 1976, 98% of the isolates were of the PAV type while not even a single isolate of MAV could be detected[105] (Table 2). Some as yet unknown factors are responsible for this change, since change in the predominant BYDV strain could not be correlated to the changes in vector populations. No particularly significant change in vector population was detected in this duration. Similarly, considerable annual fluctuations occurred in BYDV race flora in Ontairo and Quebec.[106] Similarly, a change in the distribution pattern of BYDV isolates seems to have occurred over the years in England. BYDV isolates causing severe symptoms in cereals were originally dominant in the south and west parts of England, while the isolates causing mild symptoms were dominant in the northern parts of England.[100] However, a change in the above distribution pattern seems to have occurred over the years. BYDV isolates causing severe symptoms in cereals in the north of England appear to have become more common in 1984 than in earlier years.[107]

The two major barley yellow mosaic virus strains occurring in West Germany are the M and NM strains. The NM is the wild type which infects plants in the early spring but is replaced by the M strain late in the season as the temperature rises[108] and may be another example where strain selection is mediated by high ambient temperature. Strains of beet curly top virus occurring in the 1970s in beet fields and weeds in the foothills of the San Joaquin Valley in California are far more virulent than those found in the 1950s and the 1960s.[109]

## D. Satellite RNAs

A complicating factor in delineating strains of a virus is the presence of satellite RNA(s) in that virus. Satellite RNAs change the electrophoretic mobility, virulence, host range, symptoms, sedimentation coefficients, etc., of the virus. This seemingly produces new virus strains unless the role of satellites is established.[110-112]

## E. Pseudorecombination

Pseudorecombination formation, an experimental approach to the study of genetics and variation in multicomponent viruses, establishes that plant viruses can undergo "hybridization". "Hybrids" can be constructed between two related strains of a multicomponent virus by exchanging RNA or nucleoprotein species of one strain with the corresponding RNA or nucleoprotein species of another strain to "create" a viable new "hybrid". However, this hybrid does not entail any recombination of genetic elements in the conventional manner operative in higher forms of life. Hence a "hybrid" between two related strains of a multicomponent virus formed by exchange of an RNA species is, in fact, a pseudorecombinant[113] or a reassortment. Such functional and viable pseudorecombinants have been created in the laboratory in a large number of viruses belonging to nearly all the multicomponent virus groups.

Viable pseudorecombinants possess characteristic of each parent. The RNAs of a pseudorecombinant, although completely dependent upon each other for their replication, maintain their identity and individuality and are replicated true to type. The genetic information of an individual RNA in such cases is not detectably altered during replication of the pseudorecombinant genome. This means that it should be possible to reconstruct the parental strains from the reassortants. Pseudorecombinants were formed among four strains of raspberry ringspot virus, namely, E, LG, S, and D, and the original parental strains were regenerated on "back-crossing" of LG and S pseudorecombinants.

The type and number of pseudorecombinants formed between any two viruses or virus isolates depends upon the degree of their serological affinity. Serologically closely related strains produce pseudorecombinants with all combinations of RNA1 and RNA2 as in raspberry ringspot virus.[114] Viruses with an intermediate degree of serological relationship produce at least one of the possible kinds of pseudorecombinants but not all or may not produce any readily. Such a situation occurs in Brazilian Z and Oregon Y strains of tobacco rattle virus,[115] and strains of A and G12 of tomato black ring virus (TBRV).[116] Infectious, stable and viable pseudorecombinants between strains A and G12 of TBRV could only be prepared from middle component particles of isolate A and bottom component particles of isolate

G12 or from RNA2 of isolate A and RNA1 of isolate G12, but not with converse combinations. No pseudorecombinants can be produced from different components of serologically unrelated or very distantly related viruses. This is true of raspberry ringspot virus and TBRV as well as of CAM and PRN strains of TRV. Similarly, strains of distantly related comoviruses fail to complement each other while the closely related strains of a comovirus do so. Thus heterologous mixtures of M and B components of cowpea yellow mosaic virus and bean pod mottle virus cannot complement each other while the heterologous mixtures of M and B components of the 'Arkansas' and 'Puerto Rico' strains of cowpea severe mosaic virus can do so.[117] Several other such cases are known.

Thus, degree of serological relationship can indicate the probability of obtaining pseudorecombinants between any two viruses or virus strains. However, this is not absolutely true since a few exceptions are also known. Cucumber mosaic virus and tomato aspermy virus (TAV) are serologically distantly related and their RNAs share only little sequence homology. Yet viable recombinants can be constructed out of their RNAs. A pseudorecombinant was constructed from these two distinct cucumoviruses from RNAs 1 and 2 of TAV and RNA 3 of CMV.[118] Many naturally occurring strains of AMV and tobacco streak virus are compatible with each other. Heterologous mixtures of RNAs and coat protein from AMV, tobacco streak, citrus leaf rugose, and citrus variegation viruses are compatible and infectious. Thus, interstrain compatibility between these ilarviruses seems common. Tobravirus isolates 16 and N5 form reassortants with TRV by reassortment of the two genome segments in all combinations, but generally not that easily with PEBV segments.[73] The 3' and 5' ends of RNAs 1 and 2 of TRV, 16, and N5 are homologous, but not of PEBV. This suggests that terminal nucleotide sequences appear to determine compatibility between genome parts during pseudorecombinant formation.[73]

Biological properties of some pseudorecombinants suggest that pseudorecombination may be playing some role in bringing about variation in plant viruses. The reassortant prepared from middle component particles of tomato black ring virus isolate A and bottom component particles of its isolate G12 produced abnormally small lesions.[116] Isolates obtained from some of these lesions could induce, like the parent isolates, systemic symptoms on *C. quinoa* and the serological properties of these pseudorecombinants were stable on subculturing. All this suggests that this pseudorecombinant can successfully multiply to produce more of its type and in this way may produce a new virus strain. Leaves inoculated with a mixture of nucleoprotein components obtained from different AMV strains gave rise to certain single-lesion isolates which excited symptom pattern different from either parent.[70] Thus pseudorecombinants can be very important in nature in increasing the range of variability flowing from a given number of mutations. The separated RNA components of genomes of peanut stunt and cucumber mosaic cucumoviruses are interchangeable in pseudorecombinant experiments. Because such events probably also take place in vivo, it is usually difficult to make a sharp distinction between the naturally occurring isolates of these two cucumoviruses.

## F. Genomic Masking

Heterologous encapsidation of viral RNA in the capsid of another virus in a mixedly infected cell is genomic masking, known erroneously earlier as phenotypic mixing. It is by now known to be a fairly common phenomenon and can occur in vivo between serologically related or unrelated viruses. This process can also occur between structurally different viruses as encapsulation of RNA of rod-shaped BSMV in capsids of icosahedral brome mosaic virus.[119] Genomic masking can also be considered to result in variability of a virus since some properties of such an ''hybrid'' virus are different from the virus contributing the genome but are instead identical to the virus contributing the capsid. Some of these properties are: heterologous encapsidation of RNA of a defective virus, increased host range, and vector specificity and transmissibility.

When the strains of TMV, one of which is defective for assembly, interact in vivo in the same host, encapsidation of the genome of one into a capsid of the second strain commonly occurs.[120,121] Similarly, in mixed infections, TMV *ts* coat protein mutants could be complemented by different TMV *tr* helper strain.[81,122,123] The TMV defective mutant Ni 118 was coated with capsid protein of the heterologous legume TMV strain in mixedly infected plants kept at 35°C to yield complete virus particles with Ni 118 infectivity.

Heterologous encapsidation of TMV RNA in BSMV capsid protein in barley seedlings simultaneously inoculated with both the viruses and the consequent systemic invasion of TMV in the nonhost barley seedling has been reported.[124-126] This shows the increased host range of TMV. Additional examples of increased host range of a virus due to genomic masking are discussed in Section III.C.2.

Two serologically unrelated isolates of barley yellow dwarf virus, which are transmitted by different aphid vectors, became transmissible by a single aphid species from the mixedly infected plants.[127] This was so because the genome of the nontransmissible isolate was encapsidated within the capsid of the transmissible isolate. Carrot mottle virus (CMotV) is not transmitted by aphid vector *Caveriella aegopodii*, while carrot red leaf virus (CLRV) is transmissible by this vector. However, CMotV is persistently transmitted by the aphid vector from plants also infected by CLRV.[128] This dependence of CMotV on CLRV for transmission was suggested to be due to the CMotV genome becoming encapsidated in CLRV coat protein.[129] Later work showed that plants mixedly infected by these two viruses did contain some virus particles in which CMotV RNA was partially or entirely coated with CLRV coat protein.[130-131] Virus particles purified from such mixedly infected plants possessed the infectivity of both the viruses but were indistinguishable from CLRV particles, indicating that the CLRV capsid of some virus particles encapsidated CMotV RNA. The molecular weight of RNA, isolated from preparations with which CMotV infectivity was associated, had the same ($1.5 \times 10^6$) molecular weight as the infective RNA isolated from CMotV infected *N. clevelandii*.[131]

Similarly, mechanically transmissible lettuce speckles mottle virus can be transmitted by *Myzus persicae* only from plants mixedly infected with beet western yellows virus and was due to genomic masking of RNA of the former virus by coat protein of the latter virus.[132] Possibly, aphid transmission of bean yellow vein-banding virus depended on the coating of its nucleic acid with coat protein of the helper pea enation mosaic virus.[133]

Infection and multiplication of two related or unrelated viruses in the same cell is a prerequisite for genomic masking to occur. Several related viruses can do so, and their examples have already been given in Section B, "Recombination". Several unrelated viruses can also do so, and some examples of unrelated viruses infecting and multiplying within the same cell are: TMV and tobacco etch virus,[134-135] soybean mosaic virus and bean pod mottle virus,[136] TMV and BSMV,[126] TMV and brome mosaic virus,[137] and potato viruses X and Y.[138] Raspberry ringspot virus and TRV CAM strain occur in the same cells of doubly infected leaves of *N. benthamiana* as well as in tobacco mesophyll protoplasts inoculated simultaneously with both these viruses.[139] Three unrelated viruses, namely, TMV, CMV, and PVX can infect and replicate in the same protoplast.[140]

## III. DIFFERENCES BETWEEN STRAINS

### A. Introduction

Biological, serological, biophysical, biochemical, or electrophoretic heterogenetic variation among different isolates of plant viruses is a very common feature. Consequently, some of the accepted criteria employed for establishing strains of a virus are: host range, symptomatology, severity of symptoms, nucleotide sequence, particle morphology, amino acid composition, peptide maps, composition of coat protein, hydrodynamic properties, molecular architecture, physical properties, mode of transmission, serological cross-reactivity, vector species, virus-vector relationships, vector specificity, pathogenecity, virus-induced inclusions, differences in virus titer, cross-protection, and electrophoretic mobility. Hamilton et al.[9] concluded that at present antigenic properties and host ranges of the isolates are the most acceptable criteria on which to base the decision whether two viruses are distinct viruses or strains of the same virus (Table 1).

Variants of a virus isolated and determined or characterized on the basis of one character may differ from the variants designated by another character. Moreover, the relationship determined between various isolates by one criterion may or may not agree with the relationship determined by some other criterion. Thus, the conclusions arrived at on the basis of serology may or may not agree with the conclusions arrived at on the basis of nucleic acid hybridization or on the basis of any other criterion. Hence, the number and designation of strains of a virus as reported by different workers may keep on changing. There may not be any fixity about it. It is therefore advisable to specify the criterion employed to differentiate strains.

Host range or symptom characteristics have mainly been employed to differentiate members of sugarcane mosaic virus (SCMV) subdivision. Jensen et al.[141] used the properties of virus-induced proteins to differentiate 12 isolates of maize dwarf mosaic virus (MDMV) and four isolates of SCMV. Two virus-specific proteins in infected plants could be identified: the capsid protein whose size among isolates varied from 34.4 to 39.7 kilodaltons (kDa), and cytoplasmic cylindric inclusion protein whose size varied from 64.2 to 67.5 kDa among the 16 isolates. Certain new strains could be identified on the basis of the inclusion protein. Thus host reaction and serological typing showed that a naturally infected sorghum plant in field carried a MDMV-B-like strain, but electrophoresis of isolated inclusion proteins produced two distinct bands. This suggests that the plants were in fact infected with two virus strains having identical biological properties.[141] Thus, host range, host reaction, and serology, commonly used to classify isolates and strains of SCMV subdivisions, fail to reveal all the diversity. This shows that the use of new characteristics reveals the existence of variability within groups of isolates shown to be homogeneous by other criteria. It is necessary therefore to employ as many criteria as possible to know the full extent of diversity of a virus.

## B. Morphological Differences

There are hardly any morphological differences between virus strains primarily because the various morphological features such as size, shape, and geometrical arrangement of subunits of virus particles are too constant and too few to be of any practical value for distinguishing between strains of a virus. Bawden's[90] early claim that the particles of cowpea strain of TMV from bean had shorter particles than those from tobacco were not confirmed by Brandes and Wetter.[142] However, a few possible exceptions have been reported. The particles of *Odontoglossum* ringspot virus, a TMV strain, are 20 nm longer than those of the typical TMV strain.[143] Bean yellow mosaic virus isolates from different hosts have different lengths: 742 to 756 nm in leguminous hosts, 794 to 800 nm in *C. amaranticolor* (834 nm in *C. amaranticolor* according to another estimate), and 771 nm in pea. A spontaneous AMV mutant produces spherical particles besides a few very long bacilliform particles.[42]

## C. Chemical Differences
### 1. Capsid Protein
### a. General

The amino acid composition of several TMV strains and tobamoviruses has been worked out (Table 3). Some of the generalizations are amino acid composition of capsid protein of different strains of a virus varies; differences are confined mainly to the proportion of different amino acids present while the proportion of certain specific amino acids (like cysteine, leucine, and proline in the case of TMV) may be constant in all cases; sometimes differences may exist with regard to the presence or absence of certain amino acids — as histidine and methionine in case of TMV; amino acid exchanges in different strains are distributed at random over the total length of a polypeptide; strains differ in one to several amino acid exchanges; and the more closely the two strains are related, the more identical is their amino acid composition. These conclusions more or less hold true for different strains of all viruses investigated to date.

Per cent amino acid differences between different tobamoviruses are shown in Table 4. Meshi et al.[147] determined the amino acid sequence of cucumber green mottle mosaic virus (CGMMV) strain W from its nucleotide sequence and found the coat protein to be composed of 160 residues. They compared its amino acid sequence with the already reported amino acid sequences of TMV strains *vulgare* and cowpea (Cc) (Figure 6). All three of these strains are distantly related to each other and the homologies between the amino acid sequences of CGMMV-W and *vulgare,* CGMMV-W and Cc, and *vulgare* and Cc were 36, 44, and 44%, respectively.

Because of the relatively small amount of genetic information reflected in viral coat protein, the limitations of serology with antiserum to coat protein of a virus have been recognized.[37] Moreover, only amino acids that are on or near the surface of the protein subunit and are exposed are serologically active. Thus, only a limited part of the protein subunit accounts for the serological reactions. It was found that certain TMV mutants which had even one change in amino acid residue in the antigenically active regions of capsid protein were serologically distinct from the TMV type strain, while others with even several exchanges in other regions of the protein subunit were not altered serologically.[148,149] Serological tests, therefore, compare indirectly only a limited extent of the nucleotide sequence out of the whole length of the nucleic acid. It is possible, therefore, that viruses known to be related on the basis of other characters, because of the homology of the nucleotide sequence on other parts of nucleic acid, may appear unrelated serologically. Despite this, the importance of the amino acid composition of coat protein and serology in identification of plant viruses as well as the quantitative estimation of natural relationships between viruses and strains is great, since practical experience has shown that only those viruses which possess a large number of common characteristics are serologically related.

### b. Serological Differentiation Index

A numerical concept has been induced in the serological relationships between different strains of a virus by expressing the serological relationships in terms of the serological differentiation index (SDI).[150,151] The SDI corresponds to the number of two-fold dilution steps separating homologous and heterologous titers of antisera in agar gel double-diffusion tests. A SDI of reciprocal tests of 1 to 3 indicates fairly close serological relationship between the two concerned viruses, while the relationship is considered progressively distant with the increasing SDI from 4 onwards to 9. The SDI varies from 1 to 7 between pairs of TMV strains,[152] varies from 3.7 to 6.4 between various carlaviruses,[27] and varies from 1 to 3 between strains of CMV, but from 3 to 8 between different cucumoviruses.[153] Viruses with different host ranges and average SDIs of reciprocal tests of more than 3 should be regarded as separate viruses.[9] A correlation between the mean SDI and amino acid sequence homology as well as between SDI and amino acid composition of tobamoviruses has been established[145a,146a,151a] (Figure 7). Expression of serological relationships as SDI provides a reliable yardstick of serological relationship and enabled some authors to construct models based on SDI of reciprocal tests and serological interrelationships between different viruses of a group and/or between different strains of a virus[91,151,154] (Figures 8 and 9). Clement and Converse used SDI values to differentiate and group isolates of tobacco streak virus.[155]

## Table 3
## AMINO ACID COMPOSITION OF TMV STRAINS IN MOLES OF AMINO ACID RESIDUE PER MOLE OF PROTEIN SUBUNIT (RELATIVE MOLAR RATIO)

| Amino acid | U1 | J14D1 | YA | GA | D | Y-TAMV | G-TAMV | HR | U2 | Cc | ORSV | ToMV |
|---|---|---|---|---|---|---|---|---|---|---|---|---|
| Alanine | 14 | 14 | 14 | 14 | 11 | 11 | 18 | 18 | 17 | 12 | 11 | 11 |
| Arginine | 11 | 11 | 12 | 12 | 9 | 9 | 8 | 11 | 8 | 11 | 9 | 9 |
| Aspartic acid | 18 | 17 | 19 | 19 | 17 | 18 | 22 | 17 | 22 | 17 | 20 | 18 |
| Cysteine | 1 | 1 | 1 | 1 | 1 | 1 | 1 | 1 | 1 | — | 1 | 1 |
| Glutamic acid | 16 | 15 | 16 | 16 | 19 | 19 | 16 | 22 | 16 | 16 | 15 | 19 |
| Glycine | 6 | 6 | 6 | 5 | 6 | 6 | 4 | 4 | 5 | 4 | 7 | 6 |
| Histidine | 0 | 0 | 0 | 0 | 0 | 0 | 0 | 0 | 0 | 1 | 0 | 0 |
| Isoleucine | 9 | 9 | 8 | 8 | 7 | 7 | 8 | 8 | 8 | 10 | 8 | 7 |
| Leucine | 12 | 12 | 12 | 12 | 13 | 13 | 11 | 11 | 11 | 15 | 14 | 13 |
| Lysine | 2 | 3 | 2 | 2 | 2 | 2 | 1 | 2 | 1 | 1 | 1 | 2 |
| Methionine | 0 | 0 | 0 | 0 | 1 | 1 | 2 | 3 | 2 | — | 3 | 1 |
| Phenylalanine | 8 | 8 | 8 | 8 | 8 | 8 | 8 | 6 | 8 | 6 | 7 | 8 |
| Proline | 8 | 8 | 8 | 8 | 8 | 8 | 10 | 8 | 10 | 8 | 9 | 8 |
| Serine | 16 | 17 | 14 | 15 | 16 | 15 | 10 | 13 | 10 | 18 | 12 | 15 |
| Threonine | 16 | 16 | 17 | 17 | 17 | 17 | 19 | 14 | 19 | 19 | 21 | 16 |
| Tryptophane | 3 | 3 | 3 | 3 | 3 | 3 | 2 | 2 | 2 | 1 | 3 | 3 |
| Tyrosine | 4 | 4 | 4 | 4 | 5 | 5 | 6 | 7 | 6 | 8 | 7 | 5 |
| Valine | 14 | 14 | 14 | 14 | 15 | 15 | 12 | 10 | 12 | 12 | 10 | 15 |
| Total | 158 | 158 | 158 | 158 | 158 | 158 | 158 | 158 | 158 | 159 | 158 | 158 |

YA = Yellow aucuba; GA = green aucuba; D = dahlemense; HR = Holme's ribgrass; Y-TAMV and G-TAMV = yellow and green tomato atypical mosaic; Cc = cowpea; ORSV = odontoglossum ringspot; and ToMV = tomato mosaic.

Based on References 144, 144a, 145, and 145a.

## Table 4
## PERCENT DIFFERENCES BETWEEN AMINO ACID SEQUENCES OF CAPSID PROTEIN SUBUNIT OF SOME TOBAMOVIRUSES

| Strain | TMV (U1) | OM | Dahl | ORS | U2 | HR | ToMV | ORSV | SHMV | CGMMV |
|---|---|---|---|---|---|---|---|---|---|---|
| TMV (U1) | 0 | 2 | 18 | 26 | 30 | 55 | 18 | 25 | 59 | 62 |
| OM (tobacco) | 2 | 0 | 18 | 26 | 27 | 56 | | | | |
| Dahl (dahlmense) | 18 | 18 | 0 | 27 | 30 | 54 | | | | |
| ORS (orchid) | 26 | 26 | 27 | 0 | 30 | 53 | | | | |
| U2 (tobacco) | 30 | 27 | 30 | 30 | 0 | 55 | 32 | 28 | 62 | 68 |
| HR (Holme's ribgrass) | 55 | 56 | 54 | 53 | 55 | 0 | 54 | 53 | 67 | 63 |
| ToMV (tomato mosaic) | 18 | | | | 32 | 54 | 0 | 26 | 50 | 60 |
| ORSV (odontoglossum ringspot) | 25 | | | | 28 | 53 | 26 | 0 | 58 | 59 |
| SHMV (sunhemp mosaic) | 59 | | | | 62 | 67 | 50 | 58 | 0 | 53 |
| CGMMV (cucumber green mottle mosaic) | 62 | | | | 68 | 63 | 62 | 59 | 53 | 0 |

Adapted from References 146 and 146a.

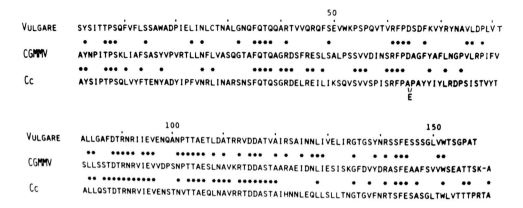

FIGURE 6.   Comparison of the vertically aligned amino acid sequences of coat proteins of CGMMV-W, TMV-*vulgare* and TMV-Cc strains. Asterisks indicate the common residues. (From Meshi, T., Kiyama, R., Ohno, T., and Okada, Y., *Virology,* 127, 54, 1983. With permission.)

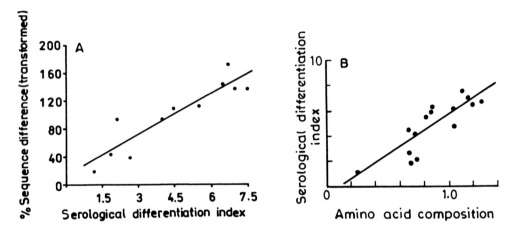

FIGURE 7.   (A) Relationships between transformed capsid protein sequence differences and serological similarity (shown as SDI values) of tobamoviruses. A slope of 19.31 occurs in the linear regression line which intercepts the Y-axis at 14.02. The points have a correlation coefficient of 0.908 ($p < 0.001$, 10 d). (From Gibbs, A., *Intervirology*, 14, 101, 1980. With permission.) (B) Relationship between amino acid composition and serological similarity (SDI) for different tobamoviruses. The linear regression is statistically significant ($p < 0.001$), with values of $y = 6.30x + 0.63$ and a correlation coefficient of 0.833 (15 d.f.). (From Paul, H. L., Gibbs, A., and Wittmann-Liebold, B., *Intervirology*, 13, 99, 1980. With permission.)

Isolates R and NS had SDIs predominantly <1 and were placed in one group while isolates WC and RNA had SDIs predominantly >1 and were placed in a second subgroup. SDI values of tombusviruses varied from 1 to 9.[156] Thus, the correlation between SDI of reciprocal tests and differences in amino acid composition/sequence is an important criterion for establishing relationships between viruses.

## 2. Nucleic Acid

All strains of a virus possess nearly the same amount of nucleic acid. However, this fact is of hardly any importance since viruses of different groups may also have the same percentage of nucleic acid. Nevertheless, base composition of a nucleic acid is an important criterion, since base composition is essentially the same in all strains of a virus but is different from that of all other viruses. Six turnip yellow mosaic virus strains could be placed into

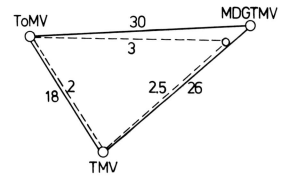

FIGURE 8. Comparison of interrelationships between tobacco mosaic virus (TMV), tomato mosaic (ToMV), and mild dark-green tobacco mosaic virus (MDGTMV), based on amino acid sequences of their coat protein and their SDI value. Percentages of amino acid exchanges are shown outside the triangle (solid lines (Henning and Wittmann[8]), while SDI values in reciprocal tests are shown inside the triangle (dashed lines) (Wetter[91]). The relationship is shown in length units with TMV-ToMV as the base line. (From Wetter, C., *Phytopathology*, 74, 1308, 1984. With permission.)

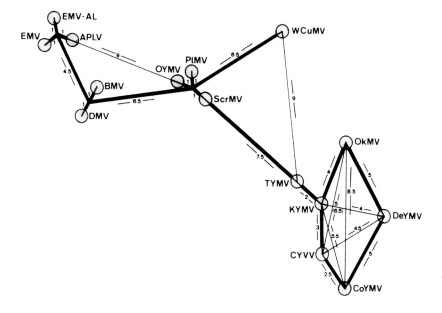

FIGURE 9. Serological relationship and serological classification of tymoviruses based on average SDIs of reciprocal tests that have been shown as length units. Andean potato latent virus (APLV), belladona mottle virus (BMV), cocoa yellow mosaic virus (CoYMV), clitoria yellow vein virus (CYVV), dulcamara mottle virus (DMV), desmodium yellow mottle virus (DeYMV), eggplant mosaic virus (EMV), erysimum latent virus (ELV), kennedya yellow mosaic virus (KYMV), okra mosaic virus (OKMV), ononis yellow mosaic virus (OYMV), plantago mottle virus (PlMV), scrophularia mottle virus (ScrMV), turnip yellow mosaic virus (TYMV), and wild cucumber mosaic virus (WCuMV). (From Koenig, R., *Virology*, 72, 1, 1976. With permission.)

two groups on the basis of their cytosine content: average cytosine content of strains of one group was 38.2%, while that of the strains of the other group was 41.6%.[157] These two groups are the same as indicated by their amino acid composition.

Nucleic acid hybridization techniques are now commonly employed to discover nucleotide sequence homology between strains of viruses. Van De Walle and Siegel employed competition hybridization and excess RNA-cDNA hybridization techniques for studying sequence homologies between various TMV strains.[158,159] They found that the TMV strains could be grouped on this basis into the same three subgroups as previously recognized on the basis of capsid protein characteristics. Further, the strains of a subgroup could not be distinguished on the basis of nucleotide sequence since they exhibited a high degree of homology, while strains of different subgroups had easily detectable differences in their genome homologies. A fourth group of TMV was also reported by them. The U1 and dahlemense strains exhibited only a limited amount, only 15%, of nucleotide sequence homology which complements capsid amino acid sequence data that show that the two strains differ in only 18% of their amino acid positions. Palukaitis and Symons[160] also used essentially the same methods as Van De Walle and Siegel to compare nucleotide sequences of 13 tobamovirus strains. The results were also the same in both cases except that no homology existed between RNA of cucumber virus 3 (CV3) and cucumber virus 4 (CV4) according to Van De Walle and Siegel[159] but possess 50% homology according to Palukaitis and Symons.[160]

The four subgroups of TMV strains erected by Van De Walle and Siegel are as follows:[159] subgroup U1 includes the strains U1, U6, YA, JI4D1, 06-67, and Ash; subgroup dahlemense contains the strains dahlemense, YTAMV (yellow tomato atypical mosaic) and HR (Holme's ribgrass); subgroup U2 includes the strains U2 and green tomato atypical mosaic virus (GATMV); and the fourth subgroup includes the Berkley and Czech isolates of cucumber virus 4, and probably cucumber green mottle mosaic virus.

Competition hybridization studies on strains of a number of other viruses lead to the following conclusions: 18 isolates of CMV fall into three subgroups and members of a subgroup had completely homologous RNAs, but no homology existed with members of any other group;[71] four strains of AMV were completely homologous;[161] and so were the five strains of BSMV.[162] Molecular hybridization showed various homologies between CMV strains: the four RNAs of CMV-Q could not be distinguished from the four RNAs of CMV-P, while the RNAs of CMV-M gave only partial sequence homologies, only 15 to 30%, with RNAs of CMV-Q.[163]

Cowpea mosaic virus (CPMV) strains DG and CPMV-SB and Echtes Ackerbohnenmosaic Virus (EAMV) show some serological cross-reactivity and have often been considered strains of a single virus. However, there are no sequence homologies between their genomes so that all three viruses are suggested to be considered as different comoviruses.[159] There is almost complete nucleotide sequence homology between the 189 residues at the 3' end of tomato aspermy virus strains V and N, indicating that the two are possibly minor variants of each other or of the original parent strain.[163a]

## D. Biological Differences
### 1. Symptoms and Degree of Virulence

Symptoms of a virus are extremely variable depending upon host plants, virus strains, and environmental conditions like temperature, light, and soil conditions. Symptom development can hardly be taken as an effective criterion for differentiating between different strains of a virus. However, degree of virulence has often been employed in distinguishing between strains of a virus. A virulent and a mild strain have in many cases been considered different from each other. Two strains of a virus can have identical coat protein cistrons but different genes controlling symptom expression and cytological changes in the host. Thus, strains of a virus can induce different pathological responses even in the same host and may have different specific infectivities. The tobacco form of cowpea TMV strain is about 10 times more infectious in *N. glutinosa* than the bean form.

## 2. *Host Range*

One of the major characteristics for identifying a new virus is its host range and host specificity. The host range of a virus in general is a fixed and relatively stable character. Moreover, species of a family from which a virus is first isolated are more likely to be susceptible to that virus than the plants of other families. Different strains of a virus usually have similar host ranges or, conversely, viruses having similar host ranges are likely to be related to each other. RNA-cDNA hybridization studies show that cherry leaf roll virus (CLRV) isolates from plants of the same genus possess a high degree of homology while isolates from plants of different genera have lower homology. Thus CLRV isolates from birch and cherry had 85 to 90% homologous sequences, while genome homology of birch or cherry CLRV isolates with that of rhubarb or dogwood isolates was only 46 to 48%.[164] Similarly, tobamoviruses restricted to different host families are distantly serologically related but have no nucleic acid homology while tobamoviruses with overlapping natural host ranges have very close to very distant relationship as indicated by serology and nucleotide homology.[150,160] However, exceptions are also known where natural host ranges of even closely related viruses may vary greatly, as in the case of TMV and cucumber virus 3 or where unrelated viruses may have overlapping host ranges as is true of TMV and tobacco etch virus. Nevertheless, host ranges or infections of particular host(s) by viruses have often been used to differentiate strains of a virus.

The viral genome largely specifies the host range of a virus, which explains some of the observations given above. Thus host range of a virus reassembled from protein and RNA from two different TMV strains exclusively depends upon the RNA and not upon the coat protein. Similarly, the "hybrid" virus particles, prepared from CCMV RNA and brome mosaic virus (BrMV) coat protein, could infect cowpea which is the CCMV host, but failed to infect barley, which is the BrMV host. Atabekov et al.[165,166] made more extensive studies in this connection by employing "hybrid" viruses reassembled in vitro from different combinations of coat protein and RNA from TMV, BrMV, CV4, or BSMV. Heterologous coat protein failed to broaden the host range of a viral RNA. On the contrary, coat protein in some cases restricted the host range of viral RNA since the "hybrid" virus prepared from TMV coat protein and BrMV RNA failed to infect *Zea mays* and *Hordeum vulgare* plants which can be infected by BrMV as well as free BrMV RNA isolated from the "hybrid" virus. The cross-protection phenomenon also indicates that coat protein most likely does play some specific role in controlling host range. Thus, both viral genome and coat protein can influence host range of a virus.

Hamilton and co-workers[124,125,137] found that TMV, which remains restricted to the inoculated barley leaves, can readily infect barley plants systemically when co-inoculated with a virus like BSMV or BrMV which normally spreads systemically in barley plants. This indicated a helper-dependent infection of a nonhost and was investigated further by Taliansky et al.[167] by employing several virus-nonhost combinations. They found that the PVX helper virus-infected resistant *Tm-2* tomato lines became susceptible to TMV; that BrMV was transported and replicated in the nonhost tomato plants preinfected with the helper virus TMV; that BrMV was also transported and replicated in the nonhost bean plants preinfected with the helper virus *Dolichos* enation mosaic virus; and that TMV was also transported and replicated in the nonhost wheat plants preinfected with BSMV (Table 5). It was concluded that a virus can systemically infect a nonhost due to the transport function being complemented and coded by the helper virus. This permits a dependent virus to spread systemically in a nonhost whose resistance is due to blocking of the transport function. In other words, virus-specific transport function controls the host range/host specificity of a plant virus.

**Table 5**
**EFFECT OF PREINFECTION BY "HELPER" VIRUSES ON INFECTION OF NONHOST PLANTS WITH THE NONSPREADING VIRUSES**

| Inoculated plant | Helper virus used for preinfection | Nonspreading virus (helper-dependent virus) | Accumulation of nonspreading virus in the upper leaves of nonhost plants (days after introduction of foreign virus) | |
|---|---|---|---|---|
| | | | 1 day | 10 days |
| Mayak tomato, *(L. esculen-tum)* (nonhost for BrMV) | TMV strain L | BrMV | (−) | (+) |
| | No TMV L (control) | BrMV | (−) | (−) |
| Pinto bean | DEMV | BrMV | (−) | (+) |
| (*Phaseolus vulgaris*) | No DEMV (control) | BrMV | (−) | (−) |
| Wheat | BSMV | TMV *vulgare* | (−) | (+) |
| (*Triticum vulgare*) | No BSMV (control) | TMV *vulgare* | (−) | (−) |
| Moperou tomato | PVX | | (+) | (+) |
| (TMV resistant) | No PVX (control) | | (−) | (−) |

TMV L = Tobacco mosaic virus strain L; DEMV = dolichos enation mosaic virus (serotype of TMV systemically spreading in beans); BrMV = brome mosaic virus; BSMV = barley stripe mosaic virus; and PVX = potato virus X.

Modified from Taliansky, M. E., Malyshenko, S. I., Pshennikova, E. S., and Atabekov, J. G., *Virology,* 122, 327, 1982. With permission.

There are several examples in which a virus may replicate in the initially infected cells of a resistant plant but fail to cause systemic infection due to restriction of its cell-to-cell movement. TMV-bean plants is one such system.[168] Several other examples of this phenomenon are known (Chapter 12, "Disease Resistance Mechanisms"). This clearly shows that helper-dependent broadening of the plant virus host range occurs in complementation of the transport function in the transport-deficient (dependent) virus by the helper even if the viruses are unrelated (Table 5). In contrast, host range of a virus is not controlled by the first stages of infection like virus adsorption and uncoating,[81,169] both of which largely are nonspecific phenomena.

### 3. Transmission
Different strains of a virus may be transmitted in different ways and by different insect vectors (Chapter 8, "Virus Transmission").

### E. Differences in Physical Properties: Electrophoretic Mobility
Electrophoretic mobility of a virus largely depends on its surface characteristics and hence on amino acid composition of the capsid protein. Consequently, strains with different amino acid composition are likely to have different electrophoretic mobilities. This is so in SBMV, TMV, and tobacco rattle virus. Strains U1 and U2 of TMV can be differentiated electrophoretically in *N. tabacum.* Similarly, tobacco rattle virus strains could be placed into three groups on the basis of their electrophoretic properties, and these three groups were the same as based on the serological properties and amino acid composition of capsid protein.

### F. Importance of Various Criteria in Assessing Relationships between Strains of a Virus or Different Viruses of a Plant Virus Group
Extensive work has been conducted on differentiation of various CMV isolates by different criteria: by host range and symptomatology,[170] by serological studies,[171] by ELISA,[172] by physical and chemical properties,[173] by peptide mapping,[172] and by nucleotide sequence

homologies.[71,163] All these methods placed the CMV isolates into the same two main groups; there were not many dissenting conclusions.

Similarly, the relationships of tobamoviruses as determined by complementary hybridization technique,[159,160] amino acid composition of coat protein and their serological behavior,[145] and on biological and physical attributes[174,175] were essentially the same. The TMV strains fall into three natural subgroups, namely, U1, U2, and dahlemense, and a fourth subgroup erected by Van de Walle and Siegel[159] with few or no differences detectable between strains of a subgroup except for symptomatology, while easily detectable differences exist between subgroups. However, one dissenting note appears. Competition molecular hybridization studies on cowpea and tobacco forms of cowpea strain of TMV and U1 TMV strain showed that RNAs of the tobacco form of cowpea TMV and U1 TMV were homologous but had no detectable relationship with the cowpea form of cowpea TMV.[176] Still the two cowpea TMV forms and U1 TMV possess many common physical, chemical, and biological properties and show common replication strategy so that all three viruses are regarded as TMV strains.[176] The distinct nucleotide sequence between U1 TMV and the cowpea form of cowpea TMV could be due to either extensive separate evolutionary development from a common ancestor or to convergent evolution from two distinct ancestors.[176]

The role of different criteria in recognizing strains of a virus can be illustrated with reference to blackeye cowpea mosaic (BCMV), bean yellow mosaic (BYMV), and cowpea aphid-borne (CAMV) viruses. BCMV was earlier commonly regarded as a strain of BYMV but these two viruses were regarded as two distinct potyviruses on the basis of cytological inclusions[177] and host range.[178] BCMV and CAMV were also earlier regarded as closely related viruses or even synonyms, while host range and nonreciprocal SDS-immunodiffusion tests with antiserum to BCMV suggested the two viruses to be distinct entities.[179] This was confirmed by reciprocal serological tests and differences in host range/reactions on selected cowpea lines and the six BCMV and CAMV isolates were distributed as follows: Florida and New York isolates of BCMV and Kenya and Nigeria isolates of CAMV were placed in BCMV group, while the Morocco and Cyperus isolates of CAMV were placed in CAMV group.[180,181] It may be mentioned that particle length, sedimentation rate of nucleic acid, and capsid protein size of all the six isolates were identical and were reclassified into the two above-mentioned groups entirely on the basis of host range and reciprocal serological tests.

The bean, cowpea, Ghana, severe bean mosaic, and type strains of SBMV possess similar chemical, physical and some biological characteristics but differ in host range, electrophoretic mobility, and serology. Mang et al.[182] further investigated the B (bean) and C (cowpea) strains. They found that molecular weights of their genome-linked proteins as well as of their four respective major proteins synthesized in vitro were significantly different and that little sequence homology existed between the noncoding regions of their genomes, but that extensive homology existed among the 400 bases from the 3' ends of RNAs of both the strains. Thus host range and serology are the methods of choice for delineating strains in SBMV.

CTV isolates showing considerable biological differences, however, do not show serological differences, i.e., were serologically similar. Rosner and Bar-Joseph studied nucleic acid homology of ten serologically identical CTV isolates by nucleic acid hybridization by employing cloned cDNA sequences derived from a severe CTV isolate.[183] Of the nine CTV isolates, six exhibited positive hybridization and three exhibited differential hybridization with the cDNA clones. Thus estimates of homology between different CTV isolates by serology and by nucleic acid hybridization did not agree with each other. Similarly, estimates of serological relatedness of tombusviruses[156] have no correlation with estimates of relatedness based on genome homology determined by hybridization tests.[184] Serologically the most distantly related virus to other tombusviruses is cymbidium ringspot virus, but nucleic acid hybridization did not indicate that distant relationship. In contrast, genome homology between the serologically closely related artichoke mottle crinkle and petunia asteroid mosaic virus was only 23%. There was also no correlation between SDI values and estimates of genome homology of tymoviruses. The above situation in CTV, tombusviruses, and tymoviruses is unlike the situation that obtains in CMV and tobamoviruses where the virus/isolates are placed in the same groups by serological and nucleic acid hybridization studies.

The four Australian sugarcane mosaic virus isolates have been placed into two groups on the basis of molecular weight, amino acid composition, and tryptic map of coat protein: the sugarcane subgroup with three strains and Johnsongrass subgroup with one strain.[185] The same four SCMV isolates were placed in the same two subgroups on the basis of biological and serological properties. Thus amino acid composition, tryptic peptide map, and serology gave identical results. Similarly, serological typing and peptide mapping of tymovirus coat proteins also gave good correlation.[186] On the other hand, serologically indistinguishable strains of some viruses could be differentiated on the basis of symptomatology. This is true of six aphid-transmissible British isolates of potato leaf roll virus[187] and five strains of peanut stunt virus.[188]

Different PVX strains cannot be easily distinguishable by either direct ELISA or latex agglutination tests[189,190] and are generally classified on the basis of their interactions with host resistance genes.[191,192] Later, however, a range of monoclonal antibodies could successfully distinguish PVX strains[190a] but this serological classification does not correlate well with the classification of Cockerham.[191,192] The migration rates in SDS-PAGE and the peptide patterns of capsid proteins of three PVX isolates belonging to different strain groups on the basis of their reactions with host resistance genes were compared by Adams et al., who observed considerable differences both in the migration rates as well as peptide patterns of the three isolates.[193]

It is clear from the above that no single criterion is universally applicable for differentiating strains of a virus. However, antigenic properties and host ranges on the whole are the best criteria upon which to decide whether two viruses are distinct or strains of a virus.[9]

## REFERENCES

1. **McKinney, H. H.**, Virus mixtures that may not be detected in young tobacco plants, *Phytopathology,* 16, 893, 1926.
2. **Jensen, J. H.**, Isolation of yellow mosaic viruses from plants infected with tobacco mosaic, *Phytopathology,* 23, 964, 1933.
3. **Jensen, J. H.**, Studies on the origin of yellow mosaic viruses, *Phytopathology,* 26, 266, 1936.
4. **Kunkel, L. O.**, Genetics of viruses pathogenic to plants, *Am. Assoc. Adv. Sci.,* 12, 22, 1940.
5. **Price, W. C.**, Isolation and study of some yellow strains of cucumber mosaic, *Phytopathology,* 24, 743, 1934.
6. **Garcia-Arenal, F., Palukaitis, P., and Zaitlin, M.**, Strains and mutants of tobacco mosaic virus are both found in virus derived from single-lesion-passaged inoculum, *Virology,* 132, 131, 1984.
7. **Sehgal, O. P.**, Biological and physico-chemical properties of an atypical mutant of tobacco mosaic virus, *Mol. Gen. Genet.,* 121, 15, 1973.
8. **Hennig, B. and Wittman, H. G.**, Tobacco mosaic virus: mutants and strains, in *Principles and Techniques in Plant Virology,* Kado, C. I. and Agrawal, H. O., Eds., Van Nostrand-Reinhold, New York, 1972, 546.
9. **Hamilton, R. I., Edwardson, J. R., Francki, R. I. B., Hsu, H. T., Hull, R., Koenig, R., and Milne, R. G.**, Guidelines for the identification and characterization of plant viruses, *J. Gen. Virol.,* 54, 223, 1981.
10. **Bock, K. P.**, The identification and partial characterization of plant viruses in tropics, *Trop. Pest Manag.,* 28, 399, 1982.
11. **Bos, L.**, Applied plant virus research: An analysis and a scheme of organization, *Neth. J. Plant Pathol.,* 82, 24, 1976.

12. **Boswell, K. F., Dallwitz, M. J., Gibbs, A. J., and Watson, L.,** The VIDE (Virus Identification Data Exchange) Project: a data bank for plant viruses, *Rev. Plant Pathol.,* 65, 221, 1986.
13. **Raccah, B., Loebenstein, G., and Singer, S.,** Aphid-transmissibility variants of citrus tristeza virus in infected citrus trees, *Phytopathology,* 70, 89, 1980.
14. **Zitter, T. A. and Murakishi, H. H.,** Nature of increased virulence in tobacco mosaic virus after passage in resistant tomato plants, *Phytopathology,* 59, 1736, 1969.
15. **Reddy, D. V. R. and Black, L. M.,** Deletion mutations of the genome segments of wound tumor virus, *Virology,* 61, 458, 1974.
16. **Reddy, D. V. R. and Black, L. M.,** Isolation and replication of mutant populations of wound tumor virus virions lacking certain genome segments, *Virology,* 80, 336, 1977.
17. **Visvader, J. E. and Symons, R. H.,** Comparative sequence and structure of different isolates of citrus exocortis viroid, *Virology,* 130, 232, 1983.
18. **Schwinghamer, M. W. and Broadbent, P.,** Detection of viroids in dwarfed orange trees by transmission to chrysanthemum, *Phytopathology,* 77, 210, 1987.
19. **Hull, R. and Plaskitt, A.,** Electron microscopy on the behaviour of two strains of alfalfa mosaic virus in mixed infections, *Virology,* 42, 773, 1970.
20. **Bawden, F. C.,** *Plant Viruses and Virus Diseases,* 4th ed., Ronald Press, New York, 1964, 561.
21. **Siegel, A.,** Artificial production of mutants of tobacco mosaic virus, *Adv. Virus Res.,* 11, 25, 1965.
22. **Sehgal, O. P. and Krause, G. F.,** Efficiency of nitrous acid as an inactivating and mutagenic agent of intact tobacco mosaic virus and its isolated nucleic acid, *J. Virol.,* 2, 966, 1968.
23. **Fraser, L. and Matthews, R. E. F.,** Efficient mechanical inoculation of turnip yellow mosaic virus using small volumes of inoculum, *J. Gen. Virol.,* 44, 565, 1979.
24. **Milne, R. G.,** Alternatives to the species concept for virus taxonomy, *Intervirology,* 24, 94, 1985.
25. **Bos, L.,** The identification of three new viruses isolated from *Wisteria* and *Pisum* in the Netherlands and the problem of variation within potato virus Y group, *Neth. J. Plant Pathol.,* 76, 8, 1970.
26. **Hollings, M. and Brunt, A. A.,** Potyvirus group, CMI/AAB Descriptions of Plant Viruses, No. 245, 1981.
27. **Wetter, C. and Milne, R. G.,** Carlaviruses, in *Handbook of Plant Virus Infections and Comparative Diagnosis,* Kurstak, E., Ed., Elsevier/North-Holland, Amsterdam, 1981, 695.
28. **Uyemoto, J. K., Grogan, R. G., and Wakeman, J. R.,** Selective activation of satellite virus strains by strains of tobacco necrosis virus, *Virology,* 34, 410, 1968.
28a. **Rochow, W. F. and Duffus, J. E.,** Luteoviruses and yellow diseases, in *Handbook of Plant Virus Infections and Comparative Diagnosis,* Kurstak, E., Ed., Elsevier/North-Holland, Amsterdam, 1981, 147.
29. **Jones, R. T. and Diachun, S.,** Serologically and biologically distinct bean yellow mosaic virus strains, *Phytopathology,* 67, 831, 1977.
30. **Beczner, L., Maat, D. Z., and Bos, L.,** The relationship between pea necrosis virus and bean yellow mosaic virus, *Neth. J. Plant Pathol.,* 82, 41, 1976.
31. **Bos, L., Cz. Kowalska, and Maat, D. Z.,** The identification of bean mosaic, pea yellow mosaic and pea necrosis strains of bean yellow mosaic virus, *Neth. J. Plant Pathol.,* 80, 173, 1974.
32. **Reddick, B. R. and Barnett, O. W.,** A comparison of three potyviruses by direct hybridization analysis, *Phytopathology,* 73, 1506, 1983.
33. **Barnett, O. W., Randles, J. W., and Burrows, P. M.,** Relationships among Australian and North American isolates of the bean yellow mosaic potyvirus subgroup, *Phytopathology,* 77, 791, 1987.
34. **Koenig, R. and Lesemann, D. E.,** Tymoviruses, in *Handbook of Plant Virus Infections and Comparative Diagnosis,* Kurstak, E., Ed., Elseiver/North-Holland, Amsterdam, 1981, 33.
35. **Harrison, B. D.,** Usefulness and limitation of the species concept for plant viruses, *Intervirology,* 24, 71, 1985.
36. **Milne, R. G.,** The species problem in plant virology, *Microbiol. Sci.,* 1, 113, 1984.
37. **Gibbs, A. J.,** Plant virus classification, *Adv. Virus Res.,* 14, 263, 1969.
37a. **Reanney, D. C.,** The molecular evolution of viruses, in *The Microbe,* Part I: *Viruses,* Mahy, B. W. J., and Pattison, J. R., Eds., Cambridge University Press, Cambridge, 1984, 175.
38. **Tsugita, A.,** The proteins of mutants of TMV: Composition and structure of chemically evoked mutants of TMV RNA, *J. Mol. Biol.,* 5, 284, 1962.
39. **Wittmann, H. G. and Wittmann-Liebold, B.,** Tobacco mosaic virus mutants and the genetic coding problem, *Cold Spring Harbor Symp. Quant. Biol.,* 28, 589, 1963.
40. **Sarachu, A. N., Nassuth, A., Roosien, J., Van Vloten-Doting, L., and Bol, J. F.,** Replication of temperature-sensitive mutants of alfalfa mosaic virus in protoplasts, *Virology,* 125, 64, 1983.
41. **Jones, R. A. C.,** Further studies of resistance-breaking strains of potato virus X, *Plant Pathol.,* 34, 182, 1985.
42. **Roosien, J. and Van Vloten-Doting, L.,** A mutant of alfalfa mosaic virus with an unusual structure, *Virology,* 126, 155, 1983.

43. **de Jager, C. P. and Wesseling, J. B. M.,** Spontaneous mutations in cowpea mosaic virus overcoming resistance due to hypersensitivity in cowpea, *Physiol. Plant Pathol.,* 19, 347, 1981.

44. **Robinson, D. J.,** A variant of tobacco rattle virus: evidence for a second gene in RNA-2, *J. Gen. Virol.,* 35, 37, 1977.

45. **Valverde, R. A. and Fulton, J. P.,** Characterization and variability of strains of southern bean mosaic virus, *Phytopathology,* 72, 1265, 1982.

46. **Lamptey, P. N. L. and Hamilton, R. I.,** A new cowpea strain of southern bean mosaic virus from Ghana, *Phytopathology,* 64, 1100, 1974.

47. **Nuss, D. L. and Summers, D.,** Variant dsRNAs associated with transmission-defective isolates of wound tumor virus represent terminally conserved remnants of genome segments, *Virology,* 133, 276, 1984.

48. **Shirako, Y. and Brakke, M. K.,** Two purified RNAs of soil-borne wheat mosaic virus are needed for infection, *J. Gen. Virol.,* 65, 119, 1984.

49. **Shirako, Y. and Brakke, M. K.,** Spontaneous deletion mutation of soil-borne wheat mosaic virus RNA II., *J. Gen. Virol.,* 65, 855, 1984.

50. **Brakke, M. K.,** Sedimentation coefficients of the virions of soil-borne wheat mosaic virus, *Phytopathology,* 67, 1433, 1977.

51. **Shirako, Y. and Ehara, Y.,** Comparison of the in vitro translation products of wild-type and a deletion mutant of soil-borne wheat mosaic virus, *J. Gen. Virol.,* 67, 1237, 1986.

52. **Hsu, Y. H. and Brakke, M. K.,** Cell-free translation of soil-borne wheat mosaic virus RNAs, *Virology,* 143, 272, 1985.

53. **Hsu, Y. H. and Brakke, M. K.,** Sequence relationships among soil-borne wheat mosaic virus RNA species and terminal structures of RNA II, *J. Gen. Virol.,* 66, 915, 1985.

54. **Ziegler, V., Richards, K., Guilley, H., Jonard, G., and Putz, C.,** Cell-free translation of beet necrotic yellow vein virus: readthrough of the coat protein cistron, *J. Gen. Virol.,* 66, 2079, 1985.

55. **Bouzoubaa, S., Guilley, H., Richards, K., and Putz, C.,** Nucleotide sequence analysis of RNA 3 and RNA 4 of beet necrotic yellow vein virus, isolates F2 and G1, *J. Gen. Virol.,* 66, 1553, 1985.

56. **Koenig, R., Burgermeister, W., Weich, H., Sebald, W., and Kothe, C.,** Uniform RNA patterns of beet necrotic yellow vein virus in sugar beet roots, but not in leaves from several plant species, *J. Gen. Virol.,* 67, 2043, 1986.

57. **Verkleij, F. N. and Peters, D.,** Characterization of a defective form of tomato spotted wilt virus, *J. Gen. Virol.,* 64, 677, 1983.

58. **Robinson, D. J.,** RNA species of tobacco rattle virus strains and their nucleotide sequence relationships, *J. Gen. Virol.,* 64, 657, 1983.

59. **Hull, R.,** Particle differences related to aphid-transmissibility of a plant virus, *J. Gen. Virol.,* 34, 183, 1976.

60. **Adam, G., Sander, E., and Shepherd, R. J.,** Structural differences between pea enation mosaic virus strains affecting transmissibility by *Acyrthosiphon pisum* (Harris), *Virology,* 92, 1, 1979.

61. **Howarth, A. J., Gardner, R. C., Messing, J., and Shepherd, R. J.,** Nucleotide sequence of naturally occurring deletion mutants of cauliflower mosaic virus, *Virology,* 112, 678, 1981.

61a. **Armour, S. L., Melcher, U., Pirone, T. P., Lyttle, D. J., and Essenberg, R. C.,** Helper component for aphid transmission encoded by region II of cauliflower mosaic virus DNA, *Virology,* 129, 25, 1983.

62. **Woolston, C. J., Covey, S. N., Penswick, J. R., and Davies, J. W.,** Aphid transmission and a polypeptide are specified by a defined region of the cauliflower mosaic virus genome, *Gene,* 23, 15, 1983.

62a. **Woolston, C. J., Czaplewski, L. G., Markham, P. G., Goad, A. S., Hull, R., and Davies, J. W.,** Location and sequence of a region of cauliflower mosaic virus gene 2 responsible for aphid transmissibility, *Virology,* 160, 246, 1987.

62b. **Harker, C. L., Woolston, C. J., Markham, P. G., and Maule, A. J.,** Cauliflower mosaic virus aphid transmission factor is expressed in cells infected with some aphid non-transmissible isolates, *Virology,* 160, 252, 1987.

63. **Motjahedi, N., Volovitch, M., Mazzolini, L., and Yot, P.,** Comparison of the predicted secondary structure of aphid transmission factor for transmissible and nontransmissible cauliflower mosaic virus strains, *FEBS Lett.,* 181, 223, 1985.

63a. **Givord, L., Xiong, C., Giband, M., Koenig, I., Hohn, T., Lebeurier, G., and Hirth, L.,** A second cauliflower mosaic virus gene product influences the structure of the viral inclusion body, *EMBO J.,* 3, 1423, 1984.

63b. **Wang, X. F., Xie, D. Z., Xu, S., and Pei, M.,** Studies on the Xinjiang isolate of cauliflower mosaic virus, *Acta Microbiol. Sin.,* 20, 365, 1980.

64. **Ohno, T., Takamatsu, N., Meshi, T., Okada, Y., Nishiguchi, M., and Kiho, Y.,** Single amino acid substitution in 30K protein of TMV defective in virus transport function, *Virology,* 131, 255, 1983.

64a. **Nishiguchi, M., Kikuchi, S., Kiho, Y., Ohno, T., Meshi, T., and Okada, Y.,** Molecular basis of plant viral virulence; the complete nucleotide sequence of an attenuated strain of tobacco mosaic virus, *Nucleic Acids Res.,* 13, 5585, 1985.

65. **Dawson, J. R. O., Motoyoshi, F., Watts, J. W., and Bancroft, J. B.,** Production of RNA and coat protein of a wild-type isolate and a temperature-sensitive mutant of cowpea chlorotic mottle virus in cowpea leaves and tobacco protoplast, *J. Gen. Virol.,* 29, 99, 1975.

66. **Dawson, J. R. O. and Watts, J. W.,** Analysis of the products of mixed infection of tobacco protoplasts with two strains of cowpea chlorotic mottle virus, *J. Gen. Virol.,* 45, 133, 1979.

67. **Barker, H. and Harrison, B. D.,** Double infection, intereference and superinfection in protoplasts exposed to two strains of raspberry ringspot virus, *J. Gen. Virol.,* 40, 647, 1978.

68. **Otsuki, Y. and Takebe, I.,** Double infection of isolated tobacco leaf protoplasts by two strains of tobacco mosaic virus, in *Biochemistry and Cytology of Plant-Parasite Interaction,* Tomiyama, K., Daly, I. M., Uritani, I., Oku, H., and Ouchi, S., Eds., Kodanaha, Tokyo, 1976, 213.

69. **Best, R. J.,** Tomato spotted wilt virus, *Adv. Virus Res.,* 13, 65, 1968.

70. **Majorana, G. and Paul, H. L.,** The production of new types of symptoms by mixtures of different components of two strains of alfalfa mosaic virus, *Virology,* 38, 145, 1969.

71. **Piazolla, P., Diaz-Ruiz, J. R., and Kaper, J. M.,** Nucleic acid homologies of eighteen cucumber mosaic virus isolates determined by competition hybridization, *J. Gen. Virol.,* 45, 361, 1979.

72. **Gould, A. R. and Symons, R. H.,** Determination of the sequence homology between the four RNA species of cucumber mosaic virus by hybridization analysis with complementary DNA, *Nucleic Acids Res.,* 4, 3787, 1977.

73. **Robinson, D. J., Hamilton, W. D. O., Harrison, B. D., and Baulcombe, D. C.,** Two anomalous tobravirus isolates: evidence for RNA recombination in nature, *J. Gen. Virol.,* 68, 2551, 1987.

74. **Bujarski, J. J. and Kaesberg, P.,** Genetic recombination between RNA components of a multipartite plant virus, *Nature (London),* 321, 528, 1986.

75. **Dixon, L. K., Nyffenegger, T., Delley, G., Martinez-Izquierdo, J., and Hohn, T.,** Evidence for replicative recombination in cauliflower mosaic virus, *Virology,* 150, 463, 1986.

76. **Schoelz, J. E., Shepherd, R. J., and Richins, R. D.,** Properties of an unusual strain of cauliflower mosaic virus, *Phytopathology,* 76, 451, 1986.

77. **Carsner, E.,** Attenuation of the virus of sugar beet curly top, *Phytopathology,* 15, 745, 1925.

78. **Johnson, J.,** Virus attenuation and mutation, *Phytopathology,* 37, 12, 1947.

79. **Wilkins, P. W.,** Decreased severity of ryegrass mosaic virus after passage through resistant perennial ryegrass, *Ann. Appl. Biol.,* 89, 429, 1978.

80. **Yarwood, C. E.,** Host passage effects with plant viruses, *Adv. Virus Res.,* 25, 169, 1979.

81. **Atabekov, J. G.,** Host-specificity of plant virus, *Annu. Rev. Phytopathol.,* 13, 127, 1975.

82. **Yarwood, C. E.,** Reversible host adaptation in cucumber mosaic virus, *Phytopathology,* 60, 1117, 1970.

83. **McKinney, H. H. and Greeley, I. W.,** Biological characteristics of barley stripe mosaic virus strains and their evolution, *USDA Tech. Bull.,* No. 1324, 1965, 84.

84. **Chiko, A. W.,** Increased virulence of barley stripe mosaic virus for wild oats: evidence of strain selection by host passage, *Phytopathology,* 74, 595, 1984.

85. **Kuhn, C. W. and Wyatt, S. D.,** A variant of cowpea chlorotic mottle virus obtained by passage through beans, *Phytopathology,* 69, 621, 1979.

86. **Wyatt, S. D. and Shaw, J. G.,** Derivation of a new strain of cowpea chlorotic mottle virus from resistant cowpea, *J. Gen. Virol.,* 49, 289, 1980.

87. **Lakshman, D. K., Gonsalves, D., and Fulton, R. W.,** Role of *Vigna* species in the appearance of pathogenetic variants of cucumber mosaic virus, *Phytopathology,* 75, 751, 1985.

88. **Lakshman, D. K. and Gonsalves, D.,** Genetic analysis of two large-lesion isolates of cucumber mosaic virus, *Phytopathology,* 75, 758, 1985.

89. **Hurtt, S. S.,** Detection and comparison of electrophoretypes of hibiscus chlorotic ringspot virus, *Phytopathology,* 77, 845, 1987.

90. **Bawden, F. C.,** Reversible changes in strains of tobacco mosaic virus from leguminous plants, *J. Gen. Microbiol.,* 18, 751, 1958.

91. **Wetter, C.,** Antigenic relationships between isolates of mild dark-green tobacco mosaic virus, and the problem of host-induced mutation, *Phytopathology,* 74, 1308, 1984.

92. **Bald, J. G., Gumpf, D. J., and Heick, J.,** Transition from common tobacco mosaic virus to the *Nicotiana glauca* form, *Virology,* 59, 467, 1974.

93. **Sasaki, A., Tsuchizaki, T., and Saito, Y.,** Discrimination between mild and severe strains of citrus tristeza virus by fluorescent antibody technique, *Ann. Phytopathol. Soc. Japan,* 44, 205, 1978.

94. **Douine, L., Marchoux, G., Quoit, J. B., and Clement, M.,** Phénoménes d'interference entre souches du virus de la mosaique du concombre (CMV). II. Effect de la temperature d'incubation sur la multiplication de deux souches de sensibilites thermiques différentes, inoculées simultanément ou successivement á un hôte sensible *Nicotiana tabacum* var. *Xanthi* nc, *Ann. Phytopathol.,* 11, 421, 1979.

95. **Roosien, J., van Klaveren, P., and Van Vloten-Doting, L.,** Competition between the RNA 3 molecules of wild-type alfalfa mosaic virus and the temperature-sensitive mutant *Tbts* 7(uv), *Plant Mol. Biol.,* 2, 113, 1983.

96. **Kassanis, B.,** Effects of changing temperature in plant virus diseases, *Adv. Virus Res.,* 4, 221, 1957.

97. **Bar-Joseph, M.,** Cross-protection incompleteness: a possible cause for natural spread of citrus tristeza virus after a prolonged lag period in Israel, *Phytopathology,* 68, 1110, 1978.

98. **Raccah, B., Bar-Joseph, M., and Loebenstein, G.,** The role of aphid vectors and variation in virus isolates in the epidemiology of tristeza disease, in *Plant Disease Epidemiology,* Scott, P. R. and Bainbridge, A., Eds., Blackwell, Oxford, 1978, 221.

98a. **Kimura, I., Minobe, Y., and Omura, T.,** Changes in a nucleic acid and a protein component of rice dwarf virus particles associated with an increase in symptom severity, *J. Gen. Virol.,* 68, 3211, 1987.

99. **Abu-Samah, N. and Randles, J. W.,** A comparison of Australian bean yellow mosaic virus isolates using molecular hybridization analysis, *Ann. Appl. Biol.,* 103, 97, 1983.

100. **Plumb, R. T.,** Properties and isolates of barley yellow dwarf virus, *Ann. Appl. Biol.,* 77, 87, 1974.

101. **Koenig, R., Fribourg, C. E., and Jones, R. A.,** Symptomatological, serological, and electrophoretic diversity of isolates of Andean potato latent virus from different regions of the Andes, *Phytopathology,* 69, 748, 1979.

102. **Yeh, S.-D., Gonsalves, D., and Provvidenti, R.,** Comparative studies on host range and serology of papaya ringspot virus and watermelon mosaic virus, *Phytopathology,* 74, 1081, 1984.

103. **Bennett, C. W. and Tanrisever, A.,** Sugar beet curly top disease in Turkey, *Plant Dis. Rep.,* 41, 721, 1957.

104. **Pelham, J., Fletcher, J. T., and Hawkins, J. H.,** The establishment of a new strain of tobacco mosaic virus resulting from the use of resistant varieties of tomato, *Ann. Appl. Biol.,* 65, 293, 1970.

105. **Rochow, W. F.,** Field variants of barley yellow dwarf virus: detection and fluctuation during twenty years, *Phytopathology,* 69, 655, 1979.

106. **Paliwal, Y. C.,** Role of perennial grasses, winter wheat, and aphid vectors in the disease cycle and epidemiology of barley yellow dwarf virus, *Can. J. Plant Pathol.,* 4, 367, 1982.

107. **Torrance, L., Pead, M. T., Larkins, A. P., and Butcher, G. W.,** Characterization of monoclonal antibodies to a U.K. isolate of barley yellow dwarf virus, *J. Gen. Virol.,* 67, 549, 1986.

108. **Huth, W., Lesemann, D. E., and Paul, H. L.,** Barley yellow mosaic virus: Purification, electron microscopy, serology and other properties of two types of the virus, *Phytopathol. Z.,* 111, 37, 1984.

109. **Magyarosy, A. C. and Duffus, J. E.,** The occurrence of highly virulent strains of the beet curly top virus in California, *Plant Dis. Rep.,* 61, 248, 1977.

110. **Murant, A. F. and Mayo, M. A.,** Satellites of plant viruses, *Annu. Rev. Phytopathol.,* 20, 49, 1982.

111. **Francki, R. I. B.,** Plant virus satellites, *Annu. Rev. Microbiol.,* 39, 151, 1985.

112. **Fritsch, C. and Mayo, M. A.,** Satellites of plant viruses, in *Plant Viruses,* Vol. 1, *Structure and Replication,* Mandahar, C. L., Ed., CRC Press, Boca Raton, 1989, 289.

113. **Gibbs, A. J. and Harrison, B. D.,** *Plant Virology: The Principles,* Edward Arnold, London, 1976.

114. **Harrison, B. D., Murant, A. F., Mayo, M. A., and Roberts, I. M.,** Distribution of determinants for symptom production, host range, and nematode transmissibility between the two RNA components of raspberry ringspot virus, *J. Gen. Virol.,* 22, 233, 1974.

115. **Lister, R. M.,** Tobacco rattle, NETU, viruses in relation to functional heterogeneity in plant viruses, *Fed. Proc., Fed. Am. Soc. Expt. Biol.,* 28, 1875, 1969.

116. **Randles, J. W., Harrison, B. D., Murant, A. F., and Mayo, M. A.,** Packaging and biological activity of the two essential RNA species of tomato black ring virus, *J. Gen. Virol.,* 36, 187, 1977.

117. **Beier, H., Issinger, O. G., Deuschle, M., and Mundry, K. W.,** Translation of the RNA of cowpea severe mosaic virus in vitro and in cowpea protoplasts, *J. Gen. Virol.,* 54, 379, 1981.

118. **Habili, N. and Francki, R. I. B.,** Comparative studies on tomato aspermy and cucumber mosaic viruses. III. Further studies on relationship and construction of a virus from parts of the two viral genomes, *Virology,* 61, 443, 1974.

119. **Peterson, J. F. and Brakke, M. K.,** Genomic masking in mixed infections with brome mosaic and barley stripe mosaic viruses, *Virology,* 51, 174, 1973.

120. **Kassanis, B. and Bastow, C.,** In vivo phenotypic mixing between two strains of tobacco mosaic virus, *J. Gen. Virol.,* 10, 95, 1971.

121. **Kassanis, B. and Bastow, C.,** Phenotypic mixing between two strains of tobacco mosaic virus, *J. Gen. Virol.,* 11, 171, 1971.

122. **Atabekov, J. G., Schaskolskaya, N. D., Atabekova, T. I., and Sacharovskaya, G. A.,** Reproduction of temperature-sensitive strains of TMV under restrictive conditions in the presence of temperature-resistant helper strain, *Virology,* 41, 397, 1970.

123. **Sarkar, S.,** Evidence of phenotypic mixing between two strains of tobacco mosaic virus, *J. Mol. Gen. Genet.,* 105, 87, 1969.

124. **Hamilton, R. I. and Dodds, J. A.,** Infection of barley by tobacco mosaic virus in single and mixed infection, *Virology,* 42, 266, 1970.

125. **Dodds, J. A. and Hamilton, R. I.,** The influence of barley stripe mosaic virus on the replication of tobacco mosaic virus in *Hordeum vulgare* L., *Virology,* 50, 404, 1972.

126. **Dodds, J. A. and Hamilton, R. I.,** Masking of the RNA genome of tobacco mosaic virus by the protein of barley stripe mosaic virus in doubly infected barley, *Virology,* 59, 418, 1974.
127. **Rochow, W. F.,** Barley yellow dwarf virus: phenotypic mixing and vector specificity, *Science,* 167, 875, 1970.
128. **Elnagar, S. and Murant, A. F.,** Relations of carrot red leaf and carrot mottle virus with their aphid vector, *Cavariella aegopodii, Ann. Appl. Biol.,* 89, 237, 1978.
129. **Elnagar, S. and Murant, A. F.,** Aphid infection experiments with carrot mottle virus and its helper virus, carrot red leaf, *Ann. Appl. Biol.,* 89, 245, 1978.
130. **Waterhouse, P. M. and Murant, A. F.,** Further evidence on the nature of the dependence of carrot mottle virus on carrot red leaf virus for transmission by aphids, *Ann. Appl. Biol.,* 103, 455, 1983.
131. **Murant, A. F., Waterhouse, P. M., Raschke, J. H., and Robinson, D. J.,** Carrot red leaf and carrot mottle viruses: observations on the composition of the particles in single and mixed infections, *J. Gen. Virol.,* 66, 1575, 1985.
132. **Falk, B. W., Duffus, J. E., and Morris, T. J.,** Transmission, host range, and serological properties of the viruses that cause lettuce speckles disease, *Phytopathology,* 69, 612, 1979.
133. **Cockbain, A. J., Jones, P., and Woods, R. D.,** Transmission characteristics and some other properties of bean yellowing vein-banding virus, and its association with pea enation mosaic virus, *Ann. Appl. Biol.,* 108, 59, 1986.
134. **McWhorter, F. P. and Price, W. C.,** Evidence that two different plant viruses can multiply simultaneously in the same cell, *Science,* 109, 116, 1949.
135. **Fujisawa, I., Hayashi, T., and Matsui, C.,** Electron microscopy of mixed infections with two plant viruses. I. Intracellular interactions between tobacco mosaic virus and tobacco etch virus, *Virology,* 33, 70, 1967.
136. **Lee, Y.-S. and Ross, J. P.,** Top necrosis and cellular changes in soybean doubly infected by soybean mosaic and bean pod mottle viruses, *Phytopathology,* 62, 839, 1972.
137. **Hamilton, R. I. and Nichols, C.,** The influence of bromegrass mosaic virus on the replication of tobacco mosaic virus in *Hordeum vulgare, Phytopathology,* 67, 484, 1977.
138. **Mayer, C. D. and Sarkar, S.,** The ultrastructure of *Nicotiana tabacum* infected with potato virus X and potato virus Y, *J. Ultrastruct. Res.,* 81, 124, 1982.
139. **Barker, H. and Harrison, B. D.,** The interaction between raspberry ringspot and tobacco rattle viruses in doubly infected protoplasts, *J. Gen. Virol.,* 35, 135, 1977.
140. **Otsuki, Y. and Takebe, I.,** Double infection of isolated tobacco mesophyll protoplasts by unrelated viruses, *J. Gen. Virol.,* 30, 309, 1976.
141. **Jensen, S. G., Long-Davidson, B., and Seip, L.,** Size variation among proteins induced by sugarcane mosaic virus in plant tissue, *Phytopathology,* 76, 528, 1986.
142. **Brandes, J. and Wetter, C.,** Classification of elongated plant viruses on the basis of their particle morphology, *Virology,* 8, 99, 1959.
143. **Corbett, M. K.,** Some distinguishing characteristics of the orchid strain of tobacco mosaic virus, *Phytopathology,* 57, 164, 1967.
144. **Knight, C. A.,** Preparation and properties of plant virus proteins, in *Techniques in Experimental Virology,* Harris, R. J. C., Ed., Academic Press, London, 1964, 1.
144a. **Kado, C. I., van Regenmortel, M. H. V., and Knight, C. A.,** Studies on some strains of tobacco mosaic virus in orchids. I. Biological, chemical and serological studies, *Virology,* 34, 17, 1968.
145. **Gibbs, A. J.,** Tobamovirus group, CMI/AAB Descriptions of Plant Viruses, No. 184, 1977.
145a. **Van Regenmortel, M. H. V.,** Tobamoviruses, in *Handbook of Plant Virus Infections and Comparative Diagnosis,* Kurstak, E., Ed., Elsevier/North-Holland, Amsterdam, 1981, 541.
146. **Dayhoff, M. O., Ed.,** *Atlas of Protein Sequence and Structure,* Vol. V, National Biological Research Foundation, Washington, D.C., 1972.
146a. **Gibbs, A.,** How ancient are the tobamoviruses?, *Intervirology,* 14, 101, 1980.
147. **Meshi, T., Kiyama, R., Ohno, T., and Okada, Y.,** Nucleotide sequence of the coat protein cistron and the 3' noncoding region of cucumber green mottle mosaic virus (watermelon strain) RNA, *Virology,* 127, 54, 1983.
148. **Van Regenmortel, M. H. V.,** Serological studies on naturally occurring strains and chemically induced mutants of tobacco mosaic virus, *Virology,* 31, 467, 1967.
149. **von Sengbusch, P.,** Aminosaureaustausche und tertiar Struktur eines Proteins, Verleich von Mutanten des Tabakmoaisk Virus mit serologischen und physikochemischen Methoden, *Z. Vererbungsl.,* 96, 364, 1965.
150. **Van Regenmortel, M. H. V.,** Antigenic relationships between strains of tobacco mosaic virus, *Virology,* 64, 415, 1975.
151. **Koenig, R.,** A loop-structure in the serological classification system of tymoviruses, *Virology,* 72, 1, 1976.
151a. **Paul, H. L., Gibbs, A., and Wittmann-Liebold, B.,** The relationships of certain tymoviruses assessed from the amino acid composition of their coat proteins, *Intervirology,* 13, 99, 1980.

152. **Van Regenmortel, M. H. V. and Burckard, J.,** Detection of a wide spectrum of tobacco mosaic virus strains by indirect enzyme-linked immunosorbent assays (ELISA), *Virology*, 106, 327, 1980.

153. **Devergne, J. C., Cardin, L., Burckard, J., and Van Regenmortel, M. H. V.,** Comparison of direct and indirect ELISA for detecting antigenically related cucumoviruses, *J. Virol. Methods*, 3, 193, 1981.

154. **Gracia, C., Koenig, R., and Lesemann, D.-E.,** Properties and classification of a potexvirus isolated from three plant species in Argentina, *Phytopathology*, 73, 1488, 1983.

155. **Clement, D. L. and Converse, R. H.,** Serological relationships among four tobacco streak virus isolates, *Phytopathology*, 76, 842, 1986.

156. **Koenig, R. and Gibbs, A.,** Serological relationships among tombusviruses, *J. Gen. Virol.*, 67, 75, 1986.

157. **Symons, R. H., Rees, M. W., Short, M. N., and Markham, R.,** Relationships between the ribonucleic acid and protein of some plant viruses, *J. Mol. Biol.*, 6, 1, 1963.

158. **Van De Walle, M. J. and Siegel, A.,** A study of nucleotide sequence homology between strains of tobacco mosaic virus, *Virology*, 73, 413, 1976.

159. **Van De Walle, M. J. and Siegel, A.,** Relationships between strains of tobacco mosaic virus and other selected plant viruses, *Phytopathology*, 72, 390, 1982.

160. **Palukaitis, P. and Symons, R. H.,** Nucleotide sequence homology of thirteen tobamovirus RNAs as determined by hybridization analysis with complementary DNA, *Virology*, 107, 354, 1980.

161. **Bol, J. F., Brederode, F. Th., Janze, G. C., and Rauh, D. K.,** Studies on sequence homology between the RNAs of alfalfa mosaic virus, *Virology*, 65, 1, 1975.

162. **Palomar, M. K., Brakke, M. K., and Jackson, A. O.,** Base sequence homology in the RNAs of barley stripe mosaic virus, *Virology*, 77, 471, 1977.

163. **Gonda, T. J. and Symons, R. H.,** The use of hybridization analysis with complementary DNA to determine the RNA sequence homology between strains of plant viruses: its application to several strains of cucumoviruses, *Virology*, 88, 361, 1978.

163a. **Wilson, P. A. and Symons, R. H.,** The RNAs of cucumoviruses: 3′-terminal sequence analysis of two strains of tomato aspermy virus, *Virology*, 112, 342, 1981.

164. **Masseski, P. R. and Cooper, J. I.,** Comparison of the genomic RNA sequence homologies between isolates of cherry leaf roll virus by complementary DNA hybridization analysis, *J. Gen. Virol.*, 67, 1169, 1986.

165. **Atabekov, J. G., Novikov, V. K., Vishnichenko, V. K., and Javakhia, V. G.,** A study of the mechanisms controlling the host range of plant viruses. II. The host range of hybrid viruses reconstituted in vitro and of free viral RNA, *Virology*, 41, 108, 1970.

166. **Atabekov, J. G., Novikov, V. K., Vishnichenko, V. K., and Kaftanova, A. S.,** Some properties of hybrid viruses reassembled in vitro, *Virology*, 41, 519, 1970.

167. **Taliansky, M. E., Malyshenko, S. I., Pshennikova, E. S., and Atabekov, J. G.,** Plant virus-specific transport function. II. A factor controlling virus host range, *Virology*, 122, 327, 1982.

168. **Sulzinski, M. A. and Zaitlin, M.,** Tobacco mosaic virus replication in resistant and susceptible plants: in some resistant species virus is confined to a small number of initially infected cells, *Virology*, 121, 12, 1982.

169. **De Zoeten, G. A.,** Early events in plant virus infection, in *Plant Diseases and Vectors: Ecology and Epidemiology*, Maramorosch, K. and Harris, K. F., Eds., Academic Press, New York, 1981, 221.

170. **Marrou, J., Quiot, J. B., and Marchoux, G.,** Caractérisation de Douze souches du VMC par leurs aptitudes pathogénes: Tentative classification, *Meded. Fac. Landbouwwet. Rijksuniv. Gent.*, 40, 107, 1975.

171. **Devergne, J. C. and Cardin, L.,** Contribution a etude du virus de la mosaique concombre (CMV), *Ann. Phytopathol.*, 5, 409, 1973.

172. **Edwards, M. C. and Gonsalves, D.,** Grouping of seven biologically defined isolates of cucumber mosaic virus by peptide mapping, *Phytopathology*, 73, 1117, 1983.

173. **Lot, H. and Kaper, J. M.,** Physical and chemical differentiation of three strains of cucumber mosaic virus and peanut stunt virus, *Virology*, 74, 209, 1976.

174. **Siegel, A. and Wildman, S. G.,** Some natural relationships among strains of tobacco mosaic virus, *Phytopathology*, 44, 277, 1954.

175. **Ginoza, W. and Atkinson, D. E.,** Comparison of some physical and chemical properties of eight strains of tobacco mosaic virus, *Virology*, 1, 253, 1955.

176. **Zaitlin, M., Beachy, R. N., and Bruening, G.,** Lack of molecular hybridization between RNAs of two TMV strains: a reconsideration of the criteria for strain relationships, *Virology*, 82, 237, 1977.

177. **Edwardson, J. R., Zettler, F. W., Christie, R. G., and Evans, I. R.,** A cytological comparison of inclusions as a basis for distinguishing two filamentous legume viruses, *J. Gen. Virol.*, 15, 113, 1972.

178. **Zettler, F. W. and Evans, I. R.,** Blackeye cowpea mosaic virus in Florida: host range and incidence in certified cowpea seed, *Proc. Fla. State Hortic. Soc.*, 85, 99, 1972.

179. **Lima, J. A. A., Purcifull, D. E., and Hiebert, E.,** Purification, partial characterization, and serology of blackeye cowpea mosaic virus, *Phytopathology*, 69, 1252, 1979.

180. **Taiwo, M. A. and Gonsalves, D.,** Serological grouping of isolates of blackeye cowpea mosaic and cowpea aphid-borne mosaic viruses, *Phytopathology,* 72, 583, 1982.

181. **Taiwo, M. A., Gonsalves, D., Provvidenti, R., and Thurston, H. D.,** Partial characterization and grouping of isolates of blackeye cowpea mosaic and cowpea aphid-borne mosaic viruses, *Phytopathology,* 72, 590, 1982.

182. **Mang, K. G., Ghosh, A., and Kaesberg, P.,** A comparative study of the cowpea and bean strains of southern bean mosaic virus, *Virology,* 116, 264, 1982.

183. **Rosner, A. and Bar-Joseph, M.,** Diversity of citrus tristeza virus strains indicated by hybridization with cloned cDNA sequences, *Virology,* 139, 189, 1984.

184. **Gallitelli, D., Hull, R., and Koenig, R.,** Relationships among viruses in the tombusvirus group: nucleic acid hybridization studies, *J. Gen. Virol.,* 66, 1523, 1985.

185. **Gough, K. H. and Shukla, D. D.,** Coat protein of potyviruses. I. Comparison of the four Australian strains of sugarcane mosaic virus, *Virology,* 111, 455, 1981.

186. **Koenig, R., Francksen, H., and Stegemann, H.,** Comparison of tymovirus capsid proteins in SDS-polyacrylamide-porosity gradient gels after partial cleavage with different proteases, *Phytopathol. Z.,* 100, 347, 1981.

187. **Massalski, P. R. and Harrison, B. D.,** Properties of monoclonal antibodies to potato leaf roll luteovirus and their use to distinguish virus isolates differing in aphid transmissibility, *J. Gen. Virol.,* 68, 1813, 1987.

188. **Paguio, O. R. and Kuhn, C. W.,** Strains of peanut mottle virus, *Phytopathology,* 63, 976, 1973.

189. **Fribourg, C. E.,** Studies on potato virus X strains isolated from Peruvian potatoes, *Potato Res.,* 18, 216, 1975.

190. **Moreira, A., Jones, R. A. C., and Fribourg, C. E.,** Properties of a resistance-breaking strain of PVX, *Ann. Appl. Biol.,* 95, 93, 1980.

190a. **Torrance, L., Larkins, A. P., and Butcher, G. W.,** Characterization of monoclonal antibodies against potato virus X and comparison of serotypes with resistance groups, *J. Gen. Virol.,* 67, 57, 1986.

191. **Cockerham, G.,** Strains of potato virus X, *Proc. Second Conf. Potato Virus Diseases,* Lisse, Wageningen, 1954, 89.

192. **Cockerham, G.,** General studies on resistance to potato viruses X and Y, *Heredity,* 25, 309, 1970.

193. **Adams, S. E., Jones, R. A. C., and Coutts, R. H. A.,** A comparison between the capsid proteins and the products of in vivo translation of three strains of potato virus X, *J. Gen. Virol.,* 68, 3207, 1987.

Chapter 6

# GENETIC BASIS OF HOST SPECIFICITY OF PLANT VIRUSES

**J. E. Schoelz**

## TABLE OF CONTENTS

## I. INTRODUCTION

The ability of a plant virus to systemically infect its hosts is the result of an interaction between viral gene products and host gene products. Although it is not known how viruses and their hosts interact to produce an infection, the genetic basis for the interaction may be quite simple. In a few cases it has been shown that resistance of the host to infection is determined by a single dominant gene while the ability to overcome host resistance is also determined by a single viral gene. This chapter presents studies in which viral host determinants have been identified.

In one sense a great amount of progress has been made in identification of viral host determinants. It has been possible to map a determinant of host specificity to a single genome segment in many multicomponent plant virus groups. In several cases, determinants of host specificity have been mapped to single genes. However, it has been difficult to make any generalizations about viral host determinants because virtually every viral gene in one study or another has been implicated as a host determinant.

Host specificity is defined here as the ability to systemically infect a host. Consequently, only the papers that focus on the differences between viral strains that remain localized in the inoculated leaf and strains that spread systemically through the host have been chosen. A virus that remains localized in an inoculated leaf would be considered an avirulent virus, while a virus that systemically infects the host would be a virulent virus. In each study that has been chosen the basic questions has been, "What viral genes determine systemic spread in the host?" The subject of host specificity of plant viruses has been covered in at least three other reviews.[1-2a] However, this chapter will be useful because several new papers have appeared since the previous reports.

The first of the two sections of this chapter is a general review of the host specificity of plant viruses. In the second, the host specificity of a single plant virus, cauliflower mosaic virus, has been reviewed in detail. In each section the genome structure of the virus is presented, followed by the discussion of host specificity.

The first section is based on papers in which the mutation may have resulted in a conversion of a virulent virus into an avirulent virus, while papers in which the lack of systemic spread was due to a defective viral protein have been excluded. An example of each type of mutation can be found in tobacco mosaic virus (TMV). A mutation in the 30-kDa gene of TMV has been shown to inactivate the protein product, resulting in localization of the virus in the inoculated leaf.[3-7] The localization probably does not involve an active defense response in the host. In contrast, mutations in the coat protein gene of TMV can convert a virus that systemically infects *Nicotiana sylvestris* into a form that is localized within a necrotic lesion on the inoculated leaf.[8,9] In this case, it is thought that localization is due to an active defense response in the host. Eventually this distinction may seem artificial. However, in this chapter attention is focused on genetic differences in viruses that allow them to specifically infect different hosts.

## II. A REVIEW OF THE HOST SPECIFICITY OF PLANT VIRUSES

### A. Alfalfa Mosaic Virus

Studies with alfalfa mosaic virus (AMV) have shown that all three of its RNAs will affect host specificity. The genome of AMV consists of three RNAs that have approximate sizes of 3700, 2600, and 2000 nucleotides.[10-12] The 3.7-kb RNA encodes a 126-kDa protein, the 2.6-kb RNA encodes a 90-kDa protein, and the 2.0-kb RNA codes for two proteins, 32 and 24 kDa in size. The 126- and 90-kDa proteins are thought to be involved in replication.[13,14] The 90-kDa protein is believed to be a RNA-dependent RNA polymerase because its amino acid sequence has some homology with the polymerases of poliovirus and cowpea mosaic virus (CpMV).[15] The 24-kDa protein is the coat protein.[16] The 32-kDa protein may be a movement protein.[12]

Dingjan-Versteegh et al.[17] constructed pseudorecombinant ("hybrid") viruses between two serologically distinct strains of AMV in order to map symptomatology and host specificity. AMV strain 425 induces chlorotic symptoms in tobacco and pinpoint necrotic lesions on the inoculated leaves of *Phaseolus vulgaris* var 'Berna'. Strain 425 is unable to move systemically in bean. AMV strain YSMV causes chlorosis in *P. vulgaris* and is transported to the trifoliate leaves. The authors exchanged nucleoprotein components between the two strains and found that symptoms in tobacco were determined by the RNA3, whereas ability to infect systemically *P. vulgaris* was determined by RNA2.

Hartmann et al.[18] generated three mutants of AMV strain S and tested them for their ability to systemically infect *N. tabacum* cv. Xanthi-nc. AMV strain S induces a systemic mosaic in 'Xanthi-nc'. All three mutants induced different symptoms on inoculated leaves, and two of the three mutants were defective in their ability to systemically infect Xanthi-nc. The mutant $A_2$fi induced small, white necrotic lesions at both 22 and 28°C. It could not systemically infect 'Xanthi-nc'. The mutant AMV-246 was temperature sensitive. At 22°C, it induced well-defined necrotic lesions with white centers and could not systemically infect 'Xanthi-nc'. At 28°C, AMV-246 did not induce necrotic lesions and systemic spread in 'Xanthi-nc' was restored. The third mutant, $F_8$a, induced necrotic lesions and systemically infected 'Xanthi-nc' at both 22 and 28°C.

Hartmann et al.[18] constructed pseudorecombinant viruses by exchange of RNAs between the mutants and the wild-type virus and found that the mutations all mapped to RNA3. Because RNA3 encodes the AMV coat protein, the authors compared the tryptic peptides of the coat proteins of the mutant to the wild type. All three coat proteins of the mutant differed from the AMV strain S at at least two points. Hartmann et al.[18] suggested that the mutant phenotypes were due to changes in the coat protein. However, they also noted that RNA3 also encodes a 35-kDa protein and that their study could not exclude the possibility that the mutations could map to that protein.

## B. Cowpea Chlorotic Mottle Virus

Host specificity of cowpea chlorotic mottle virus (CCMV) has been mapped in one study to RNA1. CCMV belongs to the brome mosaic virus (BMV) group. Because BMV has been better characterized at the molecular level than CCMV, the genome structure of BMV will be summarized. The genome of BMV is composed of three RNAs, which are approximately 3200, 2900, and 2100 nucleotides in length.[19-21] In vitro translational studies and cDNA sequencing have shown that RNA1 encodes a 109-kDa protein, RNA2 encodes a 94-kDa protein, and RNA3 encodes a 32-kDa protein and the 20-kDa coat protein.[20-23] The 109- and 94-kDa proteins are believed to be involved in replication because it has been demonstrated that RNA1 plus RNA2 are sufficient for replication in protoplasts,[24] and RNA2 has some homology to the RNA polymerase of poliovirus.[15] The function of the 32-kDa protein is unknown, although it may be involved in cell-to-cell movement.[25]

Wyatt and Kuhn[26] have shown that the host specificity of one CCMV strain was determined by RNA1. They isolated a new strain (R) of CCMV from cowpeas that were resistant to the type T strain. The R strain of CCMV apparently existed as a variant within the T strain population, because it could be isolated from 1 to 3% of the resistant cowpea plants that were inoculated with the T strain. Strain R virions were serologically distinct from strain T. Strain T did not induce any local or systemic symptoms in the resistant cowpea plant introduction 186465, although a small amount of strain T could be detected in the inoculated leaves. In contrast, strain R induced a bright chlorosis on both inoculated and systemically infected leaves.

Pseudorecombinant viruses were made between strains T and R by exchange of RNAs.

The pseudorecombinant viruses demonstrated that RNA1 determined the ability to systemically infect the resistant cowpea variety, and RNA3 could influence the level of replication in the inoculated leaf. Pseudorecombinant viruses that had the genome composition of $R_1T_2T_3$ (i.e., RNA1 of R, RNA2 of T, and RNA3 of T), $R_1R_2T_3$, or $R_1T_2R_3$ could systemically infect the resistant cowpea variety, while pseudorecombinant viruses that had the genome composition $T_1R_2T_3$, $T_1R_2R_3$, or $T_1T_2R_3$ remained localized within the inoculated leaf. All six pseudorecombinant viruses were consistent with the hypothesis that RNA1 of CCMV determined systemic infection of the resistant cowpeas. However, some of the pseudorecombinant viruses that were incompatible for systemic infection had an enhanced ability to replicate and cause symptoms in the inoculated leaf. For instance, pseudorecombinant viruses with the genome composition $T_1R_2R_3$ and $T_1T_2R_3$ could replicate in the inoculated leaf of the resistant host to a concentration that was approximately half that of the R strain, and they could induce a few chlorotic spots on the inoculated leaves, but neither virus could move systemically. It seems that although RNA1 controlled systemic infection in resistant cowpeas, RNA3 could influence replication of the virus in the inoculated leaf.

### C. Cucumber Mosaic Virus

Host specificity in cucumber mosaic virus (CMV) can be determined by either RNA1 or RNA2. The genomic organization of CMV is similar to BMV and AMV. The virus is composed of three RNAs.[27] RNA1 of CMV is approximately 3400 nucleotides in length and it encodes a single 110-kDa protein.[28] The deduced amino acid sequence of the 110-kDa protein has significant homology to the 109-kDa protein of BMV. RNA2 of CMV is approximately 3000 nucleotides in length and it encodes a single 94-kDa protein.[29] The deduced amino acid sequence of the CMV 94-kDa protein also has striking homology to the proteins encoded by RNA2 of BMV and AMV. RNA3 of CMV is approximately 2200 nucleotides in length. RNA3 codes for a 36.7-kDa protein and the 26-kDa coat protein.[30]

Three studies have shown that RNA2 of CMV determines host specificity. Hanada and Tochihara[31] made pseudorecombinant viruses between three strains of CMV that caused similar symptoms on cucumber, but could be distinguished by their reaction on cowpea. The strain CMV-Y induced necrotic lesions on cowpea and did not move systemically. The strains CMV-E and CMV-L induced chlorotic lesions on inoculated leaves and a systemic mosaic. The three strains were serologically indistinguishable. By exchanging RNA components to make all combinations of pseudorecombinant viruses between CMV-L and CMV-Y and between CMV-E and CMV-Y, the authors demonstrated that RNA2 controlled host specificity. For instance, pseudorecombinant viruses with the composition $E_1Y_2E_3$ or $L_1Y_2L_3$ induced necrotic lesions on cowpea, while a virus composed of $Y_1L_2Y_3$ systemically infected cowpea.

Rao and Francki[32] demonstrated that RNA2 determined the ability of CMV to systemically infect *Zea mays*. The authors were working with three strains that were designated UCMV, KCMV, and MCMV. Although all three CMV strains are related serologically, UCMV and KCMV are more closely related to each other than to MCMV. Rao and Francki made all possible combinations of pseudorecombinant viruses by exchange of RNAs, a total of 18 pseudorecombinants, and tested them on 10 different hosts. In order to confirm the identity of each RNA component of the pseudorecombinant viruses, the authors regenerated the parental viruses from the pseudorecombinants and compared the symptoms of the regenerated parental strains to the original strains.

In most cases, Rao and Francki[32] mapped determinants of symptomatology rather than determinants of host specificity because all three viruses could systemically infect nine of ten hosts. However, the CMV strains could be distinguished by their ability to systemically infect *Z. mays*. Neither UCMV nor MCMV could infect *Z. mays*, while KCMV induced a systemic mosaic in that host. The authors found that RNA2 of CMV determined systemic infection of maize. Viruses with the composition $U_1K_2U_3$ and $M_1K_2M_3$ induced the same systemic symptoms of KCMV, while viruses composed of $K_1U_2K_3$ and $K_1M_2K_3$ did not infect maize.

One unique aspect of the Rao and Francki[32] paper is that the authors found pseudorecombinant viruses that could not infect hosts that were susceptible to both parental viruses. For example, viruses that had the compositions $M_1U_2M_3$, $K_1U_2K_3$, and $K_1M_2K_3$ could not infect *Datura stramonium* and *Solanum melongena*, even though all three parental viruses induced systemic mosaics in those hosts. The authors concluded that RNA2 must interact with either RNA1 or RNA3 or both in order to infect the two solanaceous hosts. Interestingly, the defect in systemic infection exhibited by the three pseudorecombinant viruses was specific to *D. stramonium* and *S. melongena*, because the three pseudorecombinant viruses moved systemically in six of the ten hosts.

A third study has shown that RNA2 determines the ability of CMV to systemically infect bean, pea, and cowpea.[33] The same study also showed that ability to systemically infect lettuce was dependent upon both RNA2 and RNA3. The two strains used in this study were CMV-B, a strain originally isolated from bean, and CMV-LsS, a strain that had been isolated from a lettuce breeding line that was resistant to other CMV strains. The CMV-LsS strain did not induce any response in bean leaves, but it could induce local responses in cowpea and pea leaves. In cowpea, CMV-LsS induced necrotic lesions on inoculated leaves, while pea leaves inoculated with CMV-LsS completely collapsed. Conversely, CMV-B did not induce any symptoms in lettuce, but a systemic mosaic and stunting in the three legumes. The pseudorecombinant viruses were made by exchanging RNAs between the two viral strains. Pseudorecombinant viruses with the genome composition $L_1B_2L_3$ and $L_1B_2B_3$ systemically infected pea, bean, and cowpea, demonstrating that RNA2 controlled systemic infection of those hosts. The only pseudorecombinant virus that could systemically infect lettuce consisted of $B_1L_2L_3$, demonstrating that both RNA2 and RNA3 of CMV-LsS are essential for systemic infection of lettuce.

One study has shown that RNA1 of CMV can determine host specificity. Lakshman et al.[34] isolated a naturally occurring mutant of CMV, CMV-NL#1, that induced necrotic lesions in *Cuburbita pepo* and could not systemically infect *C. pepo* or *N. tabacum*. The strain from which it was derived, CMV-C, induced chlorotic lesions in *C. pepo* and systemically infected both *C. pepo* and *N. tabacum*. By exchanging RNAs between the two strains to form pseudorecombinant viruses, Lakshman and Gonsalves[35] showed that RNA1 determined systemic infection.

## D. Dianthoviruses

Host specificity in the dianthovirus group can be controlled by either of its two RNAs. RNA1 in the dianthovirus group varies in size from $1.35 \times 10^6$ to $1.55 \times 10^6$, and RNA2 varies from $0.5 \times 10^6$ to $0.6 \times 10^6$.[36] RNA1 encodes proteins that have sizes of 36-, 50-, and 90-kDa, and a 39-kDa coat protein.[37,38] RNA2 of one dianthovirus, red clover necrotic mosaic virus (RCNMV), has a size of 1.4 kb and it encodes a 35-kDa protein.[37,38]

Osman et al.[39] made pseudorecombinant viruses between two serologically distinct strains of RCNMV to show that RNA2 determines host specificity. RCNMV strain TpM-34 can systemically infect cowpeas. It induces chlorotic local lesions approximately 3 to 4 d after inoculation and the lesions gradually expand until they become confluent. Systemically infected plants become stunted and chlorotic after 15 d. In contrast, RCNMV strain H induces reddish brown lesions 4 to 5 d after inoculation and these lesions do not expand. No systemic symptoms develop after 15 d. Host range tests of pseudorecombinant viruses made between strains TpM-34 and H demonstrated that RNA2 controlled host specificity. The authors noted that many plant viruses code for 30 to 35 kDa proteins and that these proteins may play a part in cell-to-cell movement. They suggested that the 34-kDa protein encoded by RNA2 of RCNMV might be a movement protein.

A second study with dianthoviruses has shown that RNA1 may also encode a determinant for host specificity. Okuno et al.[40] made pseudorecombinant viruses between RCNMV and sweet clover necrotic mosaic virus (SCNMV). The two viruses are serologically distinct. On sweet clover, SCNMV induces a severe systemic mosaic symptom at both 17 and 26°C. RCNMV induces local necrotic spots on sweet clover at 17°C and no response at 26°C. The authors found that RNA1 determined the ability of the viruses to systemically infect sweet clover.

### E. Nepoviruses

Studies with raspberry ringspot virus have shown that host specificity is determined by RNA1. Studies with cherry leaf roll virus (CLRV) have shown that host specificity in one host is determined by RNA1 and that both RNA1 and RNA2 determine systemic infection of a second host. The genomes of nepoviruses consist of two RNAs. The larger RNA (RNA1) has a molecular weight of $2.4 \times 10^6$, while the smaller RNA (RNA2) may vary from $1.4 \times 10^6$ to $2.2 \times 10^6$, depending on the strain of the virus.[41] It is probable that the translational strategy of the nepoviruses is similar to the translational strategy of CpMV. Both RNA1 and RNA2 encode polyproteins that are subsequently cleaved by proteolysis.[42] RNA1 of the nepoviruses encodes the genome-linked protein (VPg) and all of the functions necessary for replication.[43] RNA2 encodes the coat protein.[44]

Harrison et al.[45] have shown that systemic infection of 'Lloyd George' raspberry and *P. vulgaris* cv. "The Prince" is determined by RNA1 by making pseudorecombinant viruses between three strains of raspberry ringspot virus. The strain LG systemically infects 'Lloyd George' raspberry, but not *P. vulgaris*. Neither strain S nor E infects 'Lloyd George' raspberry, but both systemically infect *P. vulgaris*.

Harrison et al.[45] exchanged RNA components between strains LG and S to demonstrate that RNA1 determined systemic infections of 'Lloyd George' raspberry. Because it is very difficult to initiate infections in raspberry by manual inoculations, the authors inoculated *Chenopodium quinoa* with the pseudorecombinant viruses and then used the nematode vector, *Longidorus elongatus*, to transfer the virus to raspberry. The pseudorecombinant virus with the composition $LG_1S_2$ systemically infected 'Lloyd George' raspberry, while the pseudorecombinant $S_1LG_2$ did not infect that host.

The authors exchanged RNA components between strains LG and E to demonstrate that RNA1 determined systemic infection of *P. vulgaris*. In this case, the pseudorecombinants were tested by direct inoculation of *P. vulgaris* leaves. The pseudorecombinant virus $E_1LG_2$ induced a systemic chlorotic mottle, while the pseudorecombinant virus $LG_1E_2$ induced a symptomless infection in inoculated leaves and did not systemically infect *P. vulgaris*.

Jones and Duncan[46] have shown that the ability of CLRV to systemically infect *Gomphrena globosa* is determined by RNA1, while ability to systemically infect *Petunia hybrida* is determined by RNA1 and RNA2. The authors made exchanges of nucleoprotein components between the CLRV strains G and R. The CLRV strain G does not induce any symptoms in either *G. globosa* or *P. hybrida*. In contrast, CLRV-R induces necrotic local lesions and systemic necrosis in *G. globosa* and necrotic local lesions and a systemic mottle in *P. hybrida*. A pseudorecombinant virus with the composition $R_1G_2$ induced necrotic lesions and a symptomless systemic infection in *G. globosa*, while the pseudorecombinant virus $G_1R_2$ did not induce any response in *G. globosa*. These pseudorecombinants suggested that RNA1 determined systemic movement in *G. globosa*. In *P. hybrida*, neither the pseudorecombinant virus $R_1G_2$ nor $G_1R_2$ induced any local or systemic symptoms, an indication that both RNAs are involved in determining systemic movement in this host.

### F. Tobacco Mosaic Virus

Host determinants of TMV have been mapped to the coat protein and putative replicase genes. The genome of TMV is composed of positive-sense RNA approximately 6300 nucleotides in length.[47] The viral RNA encodes four proteins, which have sizes of 183, 126, 30, and 17.5 kDa. The 183-kDa protein is synthesized by read-through of an amber termination codon at the end of the 126-kDa protein.[48] The 126- and 183-kDa proteins are believed to be involved in replication.[49] The 30-kDa protein is involved in cell-to-cell movement[3-7] and the 17.5-kDa protein is the coat protein.[47] It has been possible to map host range determinants to viral genes because the TMV genome has been cloned in infectious form.[50,51]

Two studies have demonstrated that the coat protein gene TMV determines whether or not TMV strains can systemically infect *N. sylvestris*.[8,9] *N. sylvestris* is resistant to TMV infection because of the presence of the N' gene, a single, dominant gene. However, the N' resistance gene is not effective against all TMV strains. For instance, the strain TMV-OM can systemically infect *N. sylvestris*, while the TMV-L strain induces a resistant response. Saito et al.[8] made recombinant viruses between TMV-OM and TMV-L to demonstrate that a viral factor within the coat protein was responsible for eliciting the resistant response in the host. Knorr and Dawson[9] showed that a point mutation within the coat protein gene could convert a compatible virus into an incompatible virus.

A third study with TMV has demonstrated that the putative replicase genes may also serve as host range determinants.[52] The TMV-L strain is unable to replicate in tomatoes that contain the Tm-1 resistance gene. A naturally occurring TMV strain called TMV Lta 1 was recently isolated, and this strain could overcome the Tm-1 resistance gene. The complete nucleotide sequence of Lta 1 was determined and then compared to the sequence of the L strain. Only two substitutions were found that caused amino acid changes. Both substitutions were in the overlapping portion of the 126- and 183-kDa proteins. Oligonucleotide mutagenesis of an infectious cDNA clone demonstrated that both changes were involved in overcoming the Tm-1 resistance gene.

## G. Tobacco Rattle Virus

Host specificity of tobacco rattle virus (TRV), a two-component virus, has been mapped to RNA1. RNA1 of TRV, which has an approximate length of 6800 nucleotides,[53] can be translated in cell-free systems to yield a 134-kDa protein and a 196-kDa protein.[53,54] Because the molecular weights of the two proteins exceed the coding capacity of RNA1, the 196-kDa protein is considered to be a read-through product that results from a leaky stop codon at the end of the 134-kDa protein.[55] RNA1 also contains two other ORFs that code for 29- and 16-kDa proteins.[53,56,57] The 134- and 196-kDa proteins of TRV may be involved in viral RNA synthesis, based on similarities in size, translational strategy, and amino acid sequences to the 126- and 183-kDa proteins of TMV. The 29-kDa protein of TRV may be involved in cell-to-cell transport. Its sequence has some homology to the 30-kDa protein of TMV, a protein which has been demonstrated to be a movement protein.[3-7] A TRV protein involved in transport would have to be encoded by RNA1 because previous studies have shown that RNA1 can replicate and move systemically in the absence of RNA2.[58]

The other portion of the TRV genome, RNA2, may vary in size from 1800 to 4000 nucleotides. Sequencing studies of two TRV strains have indicated that RNA2 encodes only the coat protein.[58,59] However, the RNA2 of a third TRV strain, the TCM strain, carries a second gene which may code for an additional 29 kDa protein.[60] The 29-kDa protein present on RNA2 of strain TCM has no homology to the 29-kDa protein on RNA1.

A single study has shown that host specificity of TRV may be determined by RNA1, the larger RNA. Ghabrial and Lister[61] made pseudorecombinant viruses between two TRV strains that were serologically distinct by exchanging nucleoproteins between the two strains. One of the strains, a Brazilian strain designated TRV-Z, systemically infected *N. clevelandii*, but induced a hypersensitive response in *Petunia hybrida*. An Oregon strain, TRV-Y, systemically infected both hosts. The authors found that RNA1 determined whether the pseudorecombinant virus would systemically infect *P. hybrida* or induce a hypersensitive response. A pseudorecombinant virus with the composition $Z_1Y_2$ induced necrotic local lesions on *P. hybrida*. The reciprocal virus, $Y_1Z_2$, could not be constructed. It is probable that the long particle of TRV-Y is not compatible with the short particle of TRV-Z because they are too distantly related. However, the pseudorecombinant virus $Z_1Y_2$ provided evidence that systemic spread of *P. hybrida* was determined by RNA1.

## H. Conclusion

It is difficult to draw many conclusions that would encompass all of the research presented in this section. Virtually every viral gene has been implicated as a determinant of host specificity in one study or another. However, there are a few points that might be emphasized. In many of the studies presented in this section, the authors were looking at the difference between a viral strain that could systemically infect its host and another strain that induced a hypersensitive response. Kiraly[62] defines the hypersensitive response as early tissue necrosis of a host or nonhost plant that is more than normally sensitive to a pathogen. He notes a strong correlation exists between the hypersensitive response and disease resistance.

In four of the seven virus groups that have been presented in this section, a factor which induced a hypersensitive response could be mapped to a single gene product. Host range determinants have been mapped to the 90-kDa protein of AMV,[17] the 109-kDa protein of CCMV,[26] the 110-kDa protein of CMV,[35] the 94-kDa protein of CMV,[31-33] the TMV 17.5-kDa coat protein,[8,9] and the overlapping 126- and 183-kDa proteins of TMV.[52]

Attempts to understand how individual viral genes interact with the host to control compatibility have generally been unsuccessful. There is no adequate hypothesis to explain host specificity at the molecular level. However, a question which can be answered is, "What is the difference between virulent and avirulent strains of a virus?" As has been demonstrated, this question is very amenable to molecular biology techniques. It should be possible to identify sequences within proteins that determine host specificity. Such sequence information will contribute to an understanding of how viral gene products determine host range.

## III. HOST SPECIFICITY OF CAULIFLOWER MOSAIC VIRUS

## A. Introduction

Cauliflower mosaic virus (CaMV) is the type member of the caulimovirus group, a plant virus group composed of circular, double-stranded DNA.[63,64] Ever since the discovery that the nucleic acid of CaMV was composed of double-stranded DNA, CaMV has proven to be a very useful virus to study at the molecular level. The genome of CaMV was the first plant viral genome to be sequenced. Four strains of CaMV have been completely sequenced.[65-67a] The double-stranded DNA of CaMV was also the first plant viral genome to be cloned in infectious form.[68,69] The fact that full length clones of CaMV are infectious when excised from the vector has made it possible to exchange DNA segments between cloned strains in order to associate viral genes and sequences within genes with certain characteristics of the virus. For example, it has been possible to map aphid transmissibility to gene II by making exchanges of DNA segments between aphid-transmissible and aphid non-transmissible CaMV strains.[70,71] In one case, a single amino acid alteration within gene II has been demonstrated to be responsible for the lack of aphid transmission of the Campbell isolate of CaMV.[72] The ability to cause chlorosis in turnips, a characteristic symptom of many CaMV strains, has been shown to be associated with gene VI by making exchanges of genetic material between CaMV strains.[71]

CaMV may also contribute to our knowledge of host specificity. Although the host range of most strains of CaMV is limited to crucifers, at least two CaMV strains can systemically infect solanaceous hosts in addition to crucifers.[73,74] Recombinant viruses have been con-

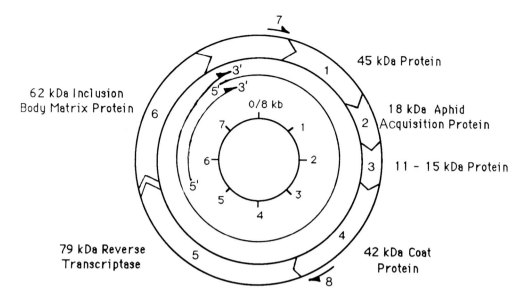

FIGURE 1.    The genome structure of cauliflower mosaic virus.

structed between strains that systemically infect solanaceous species and strains that are limited to crucifers in order to map the genes that determine host specificity. The following sections will describe research efforts directed towards an understanding of the molecular basis of host specificity of CaMV.

## B. Genome Structure of CaMV

The genome of CaMV, which is composed of circular, double-stranded DNA approximately 8000 bp in length, is organized into six major ORFs (Figure 1).[65-67] Gene products have been identified in vivo for five of the six ORFs, and functions have been determined for ORFs II, IV, and V. The product of ORF I has an apparent molecular weight of 41 to 45 kDa and is part of the viral inclusion body.[75,76] The 41-kDa protein may be involved in viral transport because it has some sequence homology to the 30-kDa transport protein of TMV.[77,78] The product of ORF II is an 18-kDa protein that is found in inclusion bodies and is required for aphid transmission.[70,71,79] The product of ORF III, an 11- to 15-kDa protein, has been reported to be found within the virions and to be a DNA-binding protein.[80-82] ORF IV encodes the 42-kDa capsid protein.[65,83] The product of ORF V is a reverse transcriptase.[84-86] ORF VI encodes a 62-kDa inclusion body matrix protein.[87,88] The 62-kDa protein is involved in controlling host specificity of CaMV strains[71,89] and in the production of characteristic symptoms.[71,90] Two smaller ORFs (VII and VIII) are also present on the CaMV genome, but no protein products have been associated with these ORFs.

There are two noncoding regions in the genome of CaMV. One noncoding region is between the end of gene V and the beginning of gene VI. This noncoding region contains the promoter for the CaMV 19S RNA, the mRNA for gene VI.[87,91,92] The other noncoding region is found between genes VI and I. This DNA segment is called the large intergenic region. It contains the promoter for the 35S-RNA, an RNA that serves as a replication intermediate during reverse transcription of the viral genome and may also serve as the mRNA for genes I to V.[93-96]

## C. Host Specificity of CaMV Strains

The host range of CaMV was originally reported to be limited to crucifers.[97,98] Since that time there have been four reports of the host range of CaMV being extended to solanaceous

## Table 1
### RESPONSE OF THREE SOLANACEOUS HOSTS TO FOUR CAULIFLOWER MOSAIC VIRUS STRAINS

| Virus | *Datura stramonium* | | *Nicotiana bigelovii* | | *Nicotiana edwardsonii* | |
|---|---|---|---|---|---|---|
| | Local response | Systemic response | Local response | Systemic response | Local response | Systemic response |
| CM1841 | N | — | — | — | — | — |
| Cabb-B | N | SN | C | — | — | — |
| D4 | C | CM | C | CM | C | CM |
| W260 | N | SN | C | CM | N | — |

*Note:*  C = Chlorotic local lesions; N = necrotic local lesions; CM = systemic chlorotic mottle; SN = systemic necrosis; — = no response

species. In 1968, a CaMV isolate was found in England that could systemically infect *N. clevelandii*.[99] In 1972 it was discovered that many isolates of CaMV would induce a hypersensitive response on *D. stramonium*.[100] CaMV strains such as CM1841 and NY8153 induced necrotic lesions on inoculated leaves and did not spread systemically throughout the plant. In 1985 an isolate was found in Argentina that could systemically infect *N. bigelovii* and induce local necrotic lesions on *D. stramonium* and *N. edwardsonii*.[73] In 1986 an isolate of CaMV from California was reported that could systemically infect *D. stramonium*, *N. clevelandii*, *N. bigelovii*, *N. glutinosa*, and *N. edwardsonii*, and it induced local lesions on *D. innoxia*.[74] The isolate, designated D4, did not infect *N. tabacum*, *N. glauca*, *N. debneyi*, *N. plumbaginifolia*, or *N. benthamiana*.

To map the CaMV genes that are responsible for determining host range, we chose four CaMV strains that differ in their ability to systemically infect the solanaceous hosts *D. stramonium*, *N. bigelovii*, and *N. edwardsonii* (Table 1).[89,101] All three of the viruses can infect turnips, but each of the viruses can be distinguished from the other two by its reaction on the three solanaceous hosts. The strain CM1841 induces necrotic lesions on *D. stramonium*, but does not spread systemically. This response is identical to what Lung and Pirone[100] reported for several CaMV strains, including CM1841. CM1841 does not induce any response in *N. bigelovii* or *N. edwardsonii*. The strain Cabb-B induces necrotic local lesions on *D. stramonium* and will occasionally induce systemic necrosis in that host. Cabb-B induces chlorotic local lesions on *N. bigelovii*, but does not systemically infect that host. When inoculated to *N. edwardsonii*, Cabb-B does not induce either a local or systemic response. The strain D4 induces chlorotic local lesions and a systemic chlorotic mottle in *D. stramonium*, *N. bigelovii*, and *N. edwardsonii*.

The CaMV strain W260 has a host range that is intermediate between D4 and Cabb-B. When W260 is inoculated to *D. stramonium*, it induces necrotic lesions and systemic necrosis, a response that is identical to Cabb-B. When W260 is inoculated to *N. bigelovii*, it induces chlorotic lesions and a systemic chlorotic mottle, a response that is similar to D4. The one difference between D4 and W260 infections of *N. bigelovii* concerns the appearance of the systemic symptoms. The systemic symptoms of W260 are delayed by approximately 5 d compared to the systemic symptoms of D4. At 7 weeks the concentration of W260 in *N. bigelovii* is approximately one half that of D4. W260 can also be distinguished from D4 by its reaction on *N. edwardsonii*. W260 induces necrotic lesions on *N. edwardsonii* and does not systemically infect this host, while D4 induces chlorotic lesions and a systemic chlorotic mottle.

The host specificity of W260 suggests that there must be at least two genetic sequences on CaMV that are important for host specificity. The genetic determinants for host specificity could be arranged in one of two ways. The first possibility is that D4 and W260 share a genetic determinant which allows them to systemically infect *N. bigelovii*. If this were the case, then it would be expected that D4 would have at least one other genetic determinant which would allow it to systemically infect *D. stramonium* and *N. edwardsonii*. The second possibility would be that D4 and W260 do not share any genetic determinant for host specificity. In this case, the genetic determinant of W260 would allow it to systemically infect *N. bigelovii* but not *D. stramonium* or *N. edwardsonii*. D4 would have a different set of genetic determinants, or possibly just one, that would allow it to systemically infect *N. bigelovii*, *N. edwardsonii*, and *D. stramonium*.

## D. Construction and Testing of Recombinant Viruses

Although CM1841, Cabb-B, D4, and W260 induce distinctly different symptoms in *N. bigelovii*, *D. stramonium*, and *N. edwardsonii*, they are all very closely related. The restriction enzyme maps are very similar, an indication of extensive homology at the nucleotide level. For instance, in a comparison of the restriction enzyme maps of D4 and CM1841, the two viruses had 14 out of 28 restriction enzyme sites in common.[74] In contrast, the restriction enzyme maps of other caulimoviruses such as figwort mosaic virus, carnation etched ring virus, dahlia mosaic virus, and mirabilis mosaic virus had no similarity to CM1841. Because of similarities in the restriction enzyme maps between the four CaMV strains, we have been able to construct recombinant viruses by exchanging DNA segments between cloned, infectious copies of the four viral genomes. After construction of the recombinant viruses, the viral DNA would be excised from the vector and the mixture inoculated to turnips to establish an infection.

The turnips infected with each of the recombinant viruses then would serve as the inoculum source for host range experiments. The ability of a recombinant virus to induce local and systemic symptoms in a particular solanaceous host was evaluated over a 7-week period. The number of plants that developed local lesions, the time at which the local lesions appeared and the type of local lesions that formed (necrotic vs. chlorotic) were scored. The number of plants that developed systemic symptoms and the type of systemic symptoms (systemic mottle vs. systemic necrosis) were also scored. At 7 weeks, we tested for the presence or absence of the viruses in systemically infected leaves by the use of a horseradish peroxidase enzyme linked immunosorbent assay test (ELISA). To confirm that certain recombinant viruses could systemically infect solanaceous hosts, the recombinant viral DNAs were purified from systemically infected leaves of solanaceous hosts, and key restriction enzyme sites were verified.

## E. Identification of a Determinant of Host Specificity of CaMV Strain D4

In order to identify sequences of CaMV strain D4 that determine its ability to systemically infect *D. stramonium* and *N. bigelovii*, nine recombinant viruses were constructed between D4, CM1841, and Cabb-B (Figure 2).[71,89] The results obtained with all nine viruses were consistent with the hypothesis that the host specificity determinant of D4 mapped within the first half of gene VI (Tables 2 and 3). If the 5' half of gene VI was derived from D4, the recombinant virus could systemically infect *D. stramonium* and *N. bigelovii*. A chlorotic mottle developed on noninoculated leaves and the virus could be detected immunologically in those plants. If the 5' half of gene VI was derived from Cabb-B or CM1841, test plants were resistant to systemic infection. In most cases, no systemic symptoms developed on plants inoculated with H9, H10, H13, and H16. When symptoms did appear, they consisted of chlorotic and necrotic spots scattered on the older leaves. Even though noninoculated leaves of some plants infected with H9, H10, H13, and H16 had symptoms, the concentration of virus in those plants was very low. When plants were sampled randomly and tested by ELISA for the presence of the virus, the absorbance was not different from that of the healthy plant extract.

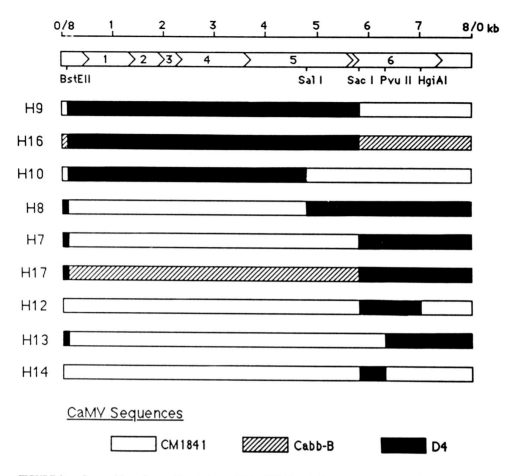

FIGURE 2.    Composition of recombinant viruses H7 to H17. The viral genomes are presented in a linear fashion to facilitate a comparison between them. The zero position of the circular map (Figure 1) corresponds to each end of these linear diagrams. (From Schoelz, J. E., Shepherd, R. J., and Daubert, S. D., *Mol. Cell. Biol.*, 6, 2632, 1986. With permission.)

The main conclusion from this work is that determinants of host specificity can be mapped to specific regions of the virus genome. However, in studying the results in Tables 2 and 3, three other interesting observations can be made. The first observation concerns the nature of the systemic necrosis symptoms of Cabb-B. Although plants inoculated with Cabb-B occasionally developed a systemic symptom, the appearance of the systemic necrosis symptom was variable and the amount of virus in leaves that developed systemic necrosis was very low (Table 2). For instance, in test 1, four plants developed systemic necrosis, while in test 2 no Cabb-B inoculated plants developed this symptom. Even when plants developed the systemic necrosis symptom, very little virus could be detected in the systemically infected leaves. In contrast, *D. stramonium* or *N. bigelovii* inoculated with D4 consistently developed a systemic mottle, and the amount of virus detected by ELISA was considerably higher than the healthy reaction. A conclusion that can be drawn from these results is that systemic movement of Cabb-B in *D. stramonium* is inhibited when compared to that of D4. Although both viruses can induce systemic symptoms in *D. stramonium*, there is a qualitative difference in their abilities. Therefore, *D. stramonium* can be considered to be resistant to infection by Cabb-B.

A second observation concerns the nature of resistance in *N. bigelovii*. *N. bigelovii* is resistant to many of the same CaMV strains and recombinants as *D. stramonium*, and in

**Table 2**
## RESPONSE OF *DATURA STRAMONIUM* TO INFECTION BY WILD-TYPE AND RECOMBINANT STRAINS OF CAULIFLOWER MOSAIC VIRUS

| | Test 1 | | | Test 2 | | |
|---|---|---|---|---|---|---|
| Virus | Local response | Systemic response | $A_{450}$[a] | Local response | Systemic response | $A_{450}$[a] |
| D4 | 10C[b] | 10[c] | 0.492 | 10C | 10 | 1.024 |
| Cabb-B | 10N | 4[d] | 0.032 | 10N | 0 | 0.032 |
| CM1841 | 10N | 0 | 0.017 | 6N | 0 | 0.016 |
| H9 | 10N | 0 | 0.007 | 10N | 0 | 0.028 |
| H16 | 10N | 0 | 0.009 | 10N | 0 | 0.013 |
| H10 | 10N | 0 | 0.012 | 10N | 0 | 0.003 |
| H8 | 10C | 10 | 0.805 | 10C | 10 | 2.326 |
| H7 | 10C | 10 | 0.787 | 10C | 10 | 2.262 |
| H17 | 10C | 10 | 0.269 | 10C | 10 | 0.167 |
| H12 | 10C | 10 | 0.321 | 10C | 10 | 2.281 |
| H13 | 10N | 9[d] | 0.006 | 10N | 0 | 0.013 |
| H14 | 10C | 10 | 0.421 | 10C | 10 | 0.859 |
| Healthy | 0 | 0 | 0.007 | 0 | 0 | 0.057 |

[a] The abosrbence at 450 nm is the spectrophotometric value obtained in the enzyme-linked immunosorbent tests. The $A_{450}$ value given is the average of all plants in the test.

[b] Number of plants of ten inoculated that reacted with local lesions 2 to 3 weeks after inoculation with virus. (C = chlorotic local lesions; N = necrotic lesions.)

[c] Number of plants of ten inoculated that reacted with systemic symptoms 4 to 5 weeks after inoculation.

[d] Symptoms consisted of necrotic or chlorotic spots scattered on the older leaves (see text).

From Schoelz, J. E., Shepherd, R. J., and Daubert, S. D., *Mol. Cell. Biol.*, 6, 2632, 1986. With permission.

both cases avirulence is determined by a sequence within the first half of gene VI. One difference between the resistance of *N. bigelovii* and that of *D. stramonium* is that avirulent viruses such as Cabb-B, H9, H13, and H16 induce necrotic local lesions on *D. stramonium*, while on *N. bigelovii* the same viruses induce chlorotic local lesions. Although the primary response is different, the same viruses are unable to spread systemically in both hosts. These results suggest that necrosis may not have a role in inhibition of systemic spread in this system.

The third observation is that viral sequences that map to regions beyond the 5' half of gene VI can enhance or diminish the ability of a recombinant virus to systemically infect *D. stramonium* or *N. bigelovii*. Sequence variations that enhance virus concentration are revealed in a comparison of the concentrations of D4, H7, and H14 in systemically infected leaves of *D. stramonium* and *N. bigelovii* (Tables 2 and 3). The concentration of H7 is nearly twice that of D4 and H14 in both hosts. The conclusion that can be made from these three viruses is that the concentration of virus in a plant can be enhanced by interactions between viral genes. In this case, a D4 sequence within the 5' half of gene VI determines whether or not a virus will systemically infect *D. stramonium* and *N. bigelovii*. However, one D4 sequence within the 3' half of gene VI and large intergenic region and a second CM1841 sequence within genes I to V enhance the concentration of virus in those solanaceous hosts. The D4 sequence in the 3' half of gene VI and large intergenic region must differ from the CM1841 counterpart in order for the enhancement in concentration to occur. Likewise, a CM1841 sequence in genes I to V must differ from its D4 counterpart.

**Table 3**
### RESPONSE OF *NICOTIANA BIGELOVII* TO INFECTION BY WILD-TYPE AND RECOMBINANT STRAINS OF CAULIFLOWER MOSAIC VIRUS

| Virus | Local response | Systemic response | $A_{450}^{a}$ |
|-------|----------------|-------------------|---------|
| D4 | 10C[b] | 10[c] | 0.156 |
| Cabb-B | 6C | 0 | 0.029 |
| CM1841 | 0 | 0 | 0.023 |
| H9 | 5C | 0 | 0.001 |
| H16 | 9C | 0 | 0.005 |
| H7 | 2C | 7 | 0.297 |
| H17 | 9C | 9 | 0.108 |
| H12 | 9C | 9 | 0.298 |
| H13 | 4C | 0 | 0.004 |
| H14 | 8C | 8 | 0.179 |
| Healthy | 0 | 0 | 0.028 |

[a]  The absorbence at 450 nm is the spectrophotometric value obtained in the enzyme-linked immunosorbent tests. The $A_{450}$ value given is the average of all plants in the test.
[b]  Number of plants of ten inoculated that reacted with local lesions 2 to 3 weeks after inoculation with virus. (C = chlorotic local lesions.)
[c]  Number of plants of ten inoculated that reacted with systemic symptoms 4 to 5 weeks after inoculation.

From Schoelz, J. E., Shepherd, R. J., and Daubert, S. D., *Mol. Cell. Biol.*, 6, 2632, 1986. With permission.)

Sequence variations that diminish the ability of a recombinant virus to systemically infect *D. stramonium* and *N. bigelovii* are revealed in a comparison of Cabb-B, H17, H16, and D4. Recombinant virus H17 induced chlorotic local lesions and systemically infected both *D. stramonium* and *N. bigelovii*. However, the systemic symptoms were milder and the concentration of virus in systemically infected leaves was lower than in leaves systemically infected with D4. Similarly, the recombinant virus H16 may have a diminished ability to induce systemic necrosis in *D. stramonium*. In one test, ten out of ten plants inoculated with Cabb-B developed the systemic necrosis symptom, while only one H16-inoculated plant developed systemic necrosis. The results obtained with H16 and H17 again suggest that we have detected interactions between gene VI and genes I to V. However, in this case the viral sequences are not as compatible as the parental sequences, resulting in a diminished ability to induce systemic symptoms.

The interactions described in the two preceding paragraphs might at first seem contradictory to the premise that host specificity of D4 is determined by the 5' half of gene VI. The contradiction can be resolved by recognizing that control of host specificity is not always a yes or no phenomenon. For instance, a virus might have the proper virulence gene required to systemically infect a particular host, but might not be able to replicate or move due to adverse interactions between viral genes. Conversely, a virus that has an avirulence gene might also have a combination of viral genes that are optimal for replication and movement, leading to a systemic necrosis symptom like that of Cabb-B. Our results with our recombinant viruses are consistent with the hypothesis that the basic interaction is determined by sequences within the 5' half of gene VI of D4. However, it is obvious that other viral genes can influence this basic host-pathogen interaction.

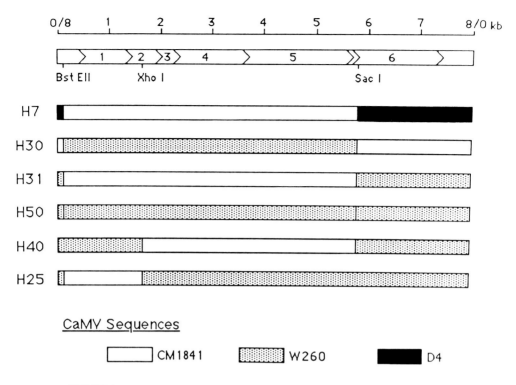

FIGURE 3.    Composition of recombinant viruses H7, H30, H31, H25, H40, and H50.

## F. Identification of Determinants of Host Specificity of CaMV Strain W260

In order to identify sequences of CaMV strain W260 which determine its ability to systemically infect *N. bigelovii*, five recombinant viruses have been constructed between W260 and CM1841 (Figure 3).[101] In the case of W260, it is apparent that more than one viral gene determines systemic infection of *N. bigelovii*. This is illustrated by a comparison of the reactions of H7, H30, and H31 (Table 4). The recombinant virus H7 was included as a control in this test. Recombinant virus H7 systemically infected nine of ten plants in this test, and the concentration of H7 was actually higher than that of D4, demonstrating that gene VI of D4 controls the ability of D4 to systemically infect *N. bigelovii*. In contrast, H30 and H31 induced chlorotic primary lesions in some plants, but there was no evidence that they could systemically infect *N. bigelovii*. No systemic symptoms appeared on plants inoculated with H30 and H31 and no virus was detected in noninoculated leaves. These results show that the ability of W260 to systemically infect *N. bigelovii* does not map to gene VI alone or to genes I to V.

Because H7 could systemically infect *N. bigelovii*, we had expected that the comparable W260/CM1841 recombinant, H31, would also systemically infect this solanaceous host. It was quite surprising to find that H31 could not move systemically in *N. bigelovii*. In order to eliminate the possibility that the lack of systemic spread of H30 or H31 might be due to an artifact of the cloning procedure, the parent W260 clone was reconstructed. Recombinant virus H50 was made by exchanging SacI — BstEII DNA segments between H30 and H31. Recombinant virus H50 was then tested for its ability to move systemically in *N. bigelovii*. Out of ten plants inoculated with H50, all ten developed a systemic mottle, and these systemic symptoms appeared at the same time as the parent W260-inoculated controls. This test demonstrated that no mutation had been introduced into H30 or H31. Recombinant virus H50 validated results obtained with H30 and H31 which suggested that at least two regions of the virus are necessary for systemic infection of *N. bigelovii*.

## Table 4
### RESPONSE OF *NICOTIANA BIGELOVII* TO INFECTION BY WILD-TYPE AND RECOMBINANT STRAINS OF CAULIFLOWER MOSAIC VIRUS

| Virus | Test 1 | | | Test 2 | | |
|---|---|---|---|---|---|---|
| | Local response | Systemic response | $A_{450}^{a}$ | Local response | Systemic response | $A_{450}$ |
| D4 | 10C[b] | 9/9[c] | 0.245 | 10C | 10 | 0.847 |
| W260 | 10C | 8/9 | 0.141 | 10C | 10 | 0.326 |
| CM1841 | 0 | 0/7 | 0.004 | 0 | 0 | 0.053 |
| H7 | 9C | 9 | 0.993 | nt | nt | nt |
| H30 | 0 | 0 | 0.025 | nt | nt | nt |
| H31 | 3C | 0/9 | 0.009 | nt | nt | nt |
| H50 | nt | nt | nt | 10C | 10 | 0.233 |
| H25 | nt | nt | nt | 3C | 0 | 0.108 |
| H40 | nt | nt | nt | 10C | 0 | 0.068 |
| Healthy | 0 | 0/9 | 0.016 | 0 | 0 | 0.057 |

[a] The absorbence at 450 nm is the spectrophotometric value obtained in the enzyme-linked immunosorbent tests. The $A_{450}$ value given is the average of all plants in the test.

[b] Number of plants of ten inoculated that reacted with local lesions 2 to 3 weeks after inoculation with virus. (C = chlorotic local lesions.)

[c] Number of plants of ten inoculated that reacted with systemic symptoms 4 to 5 weeks after inoculation. The numerator is the number of plants that developed systemic symptoms.

The two recombinants H25 and H40 have been constructed recently in order to test whether or not gene I of W260 was involved in systemic infection of *N. bigelovii*. Gene I might be a likely candidate for involvement because it has been shown to have homology to the 30 kDa movement protein of TMV.[77,78] It has been demonstrated that a defective movement protein of TMV can prevent systemic spread.[3-5] In the case of CaMV, the gene I product of CM1841 might not function properly in combination with the gene VI product of W260, resulting in an inhibition of systemic movement. The recombinant virus H40 was made to test whether or not a combination of gene VI and I were required for systemic infection, while H25 directly tested for involvement of gene I. The two recombinant viral DNAs were initially inoculated to turnips to establish an infection, and then infected turnips served as the inoculum source for the host range test. Neither H25 nor H40 induced systemic symptoms in *N. bigelovii*. H40 could not be detected in systemically infected leaves by ELISA tests. In the case of H25, it appears that there might be a low level of replication and movement into systemically infected leaves. The results suggest that both genes VI and I are necessary, but not sufficient for systemic spread of W260 in *N. bigelovii*. Results obtained with H25 and H40 also suggest that at least three W260 genes are involved in determining systemic infection of *N. bigelovii*. We are now in the process of trying to determine which of W260 genes II to V might also have a role in systemic infection of *N. bigelovii*.

### G. Evidence for Interactions between CaMV Genes
It has been demonstrated that the first half of gene VI plays an important role in determining whether or not a CaMV isolate can systemically infect solanaceous hosts. However, as mentioned previously, it is obvious that at least two other sequences can either enhance, diminish, or abolish the ability of a particular isolate to systemically infect solanaceous hosts. One sequence maps within the 3′ half of gene VI and large intergenic region, while at least one other sequence maps within genes I to V. The evidence suggests that these sequences interact to produce a particular viral trait. These sequences are responsible for at

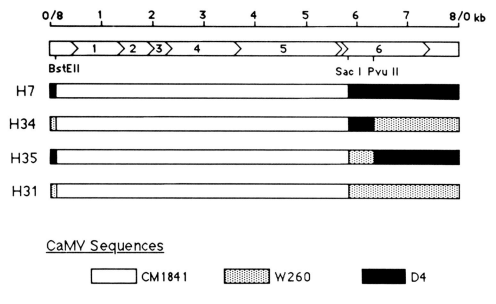

FIGURE 4.    Composition of recombinant viruses H7, H31, H34, and H35.

least four distinct viral traits. The viral traits affected by these sequences are (1) virus concentration in *D. stramonium* and *N. bigelovii*, a subject covered in an earlier section; (2) necrotic lesion size in *N. edwardsonii*; (3) ability to systemically infect *N. edwardsonii*; and (4) ability of W260 to systemically infect *N. bigelovii*.[101]

Sequence variations that affect necrotic lesion size can be seen in a comparison of the lesion sizes of W260, H31, and H35 on *N. edwardsonii* (Figure 4, Table 5). W260 and H35 both induce large necrotic lesions on *N. edwardsonii*, while H31 induces very small necrotic lesions on that host. Results obtained with recombinant virus H31 indicate that lesion size is determined by at least two factors. One is located within gene VI and the large intergenic region, while the other is found within genes I to V. Results obtained with recombinant virus H35 indicate that one of the factors maps to the back half of gene VI and large intergenic region, because the D4 Pvu II-Bst EII DNA segment within H35 restores the lesion size to that of W260. The results suggest that the 5′ half of gene VI may determine the basic response, in this case whether the lesion will be chlorotic or necrotic. At least two other factors can modify that response. One factor is located within the 3′ half of gene VI and large intergenic region, and a second factor within genes I to V.

Sequence variations that affect the ability of a virus to systemically spread in *N. edwardsonii* can be seen in a comparison of systemic infections of D4, H7, and H34 in *N. edwardsonii* (Figure 4, Table 5). In this case, systemic spread of *N. edwardsonii* is dependent on a sequence within the 5′ half of gene VI and an additional sequence within the 3′ half of gene VI or in the neighboring large intergenic region. H7 clearly was able to systemically infect *N. edwardsonii*. The concentration of H7 in systemically infected leaves was considerably higher than D4 in this test. In contrast, the ability of H34 to spread systemically in *N. edwardsonii* was limited. Of ten H34-inoculated plants, six developed a chlorotic mottle on noninoculated leaves, but very little virus could be detected in those plants. Results obtained with *N. edwardsonii* and D4, H7, and H34 mirror the results obtained with W260, H31, H33, and H35. Again, the basic host response, whether the lesion will be chlorotic or necrotic, is determined by the 5′ half of gene VI. A second sequence that affects the ability of a virus to spread systemically in *N. edwardsonii* maps to the 3′ half of gene VI.

The requirements for systemic infection of *N. edwardsonii* by D4 must be slightly different

**Table 5**
**RESPONSE OF *NICOTIANA EDWARDSONII* TO**
**INFECTION BY WILD-TYPE AND**
**RECOMBINANT STRAINS OF CAULIFLOWER**
**MOSAIC VIRUS**

| Virus | Local response | Systemic response | $A_{450}^{a}$ |
|---|---|---|---|
| D4 | 10C[b] | 10[c] | 0.171 |
| W260 | 10N | 0 | 0.058 |
| H7 | 2C | 7 | 0.653 |
| H34 | 10C | 6 | 0.083 |
| H35 | 10N | 0 | 0.049 |
| H31 | 8N | 0 | 0.062 |
| Healthy | 0 | 0 | 0.053 |

[a]   The absorbence at 450 nm is the spectrophotometric value obtained in the enzyme-linked immunosorbent tests. The $A_{450}$ value given is the average of six plants.

[b]   Number of plants of ten inoculated that reacted with local lesions 2 to 3 weeks after inoculation with virus. (C = chlorotic local lesions; N = necrotic local lesions.)

[c]   Number of plants of ten inoculated that reacted with systemic symptoms 4 to 5 weeks after inoculation.

From Schoelz, J. E. and Shepherd, R. J., *Virology*, 162, 30, 1988. With permission.

than the requirements for systemic infection of either *D. stramonium* or *N. bigelovii*. With either *D. stramonium* or *N. bigelovii*, determinants of host specificity of D4 map within the first half of gene VI. This has been demonstrated by both H14 and H34. When H34 was inoculated to *D. stramonium,* it spread systemically and attained a concentration that was slightly lower than D4. In contrast, the amount of H34 detected in systemically infected *N. edwardsonii* was only slightly above the healthy reaction. Systemic infection of *N. edwardsonii* must be dependent on the whole of gene VI and the large intergenic region of D4.

Sequence variations within genes I to V and the 3′ half of gene VI also affect the ability of W260 to systemically infect *N. bigelovii*. This can be seen by comparing the abilities of W260, H30, H31, H33, H35, and H13 to systemically infect *N. bigelovii* (Figure 5, Table 6). Although W260 is able to systemically infect *N. bigelovii,* neither H30 nor H31 can systemically infect that host. These results suggest that at least two sequences affect the ability of W260 to systemically infect *N. bigelovii.* One sequence maps within genes I to V and a second maps within gene VI and the large intergenic region. The recombinant viruses H33, H13, and H35 indicate that sequences within the 3′ half of gene VI can also have an influence on ability of a virus to move systemically in solanaceous hosts. Neither H13 nor H33 can systemically infect *N. bigelovii.* What is surprising is that H35 is able to move systemically in this host. This is surprising because neither the DNA segment in H33 that is derived from W260 nor the DNA segment in H35 that is derived from D4 determine systemic spread in *N. bigelovii.* However, when the 5′ half of gene VI of W260 is linked to the back half of gene VI and large intergenic region of D4, a virus is created that can systemically infect *N. bigelovii.* Therefore, each DNA segment must contribute an essential function in order for systemic infection to occur.

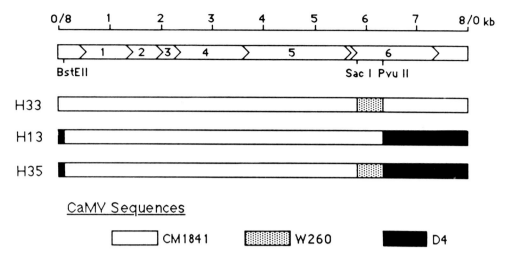

FIGURE 5.    Composition of recombinant viruses H13, H33, and H35.

**Table 6**
**RESPONSE OF *NICOTIANA BIGELOVII* TO**
**INFECTION BY WILD-TYPE AND**
**RECOMBINANT STRAINS OF CAULIFLOWER**
**MOSAIC VIRUS**

| Virus | Local response | Systemic response | $A_{450}^a$ |
|---|---|---|---|
| D4 | 10C[b] | 10[c] | 0.170 |
| W260 | 10C | 10 | 0.127 |
| CM1841 | 0 | 0 | 0.020 |
| H33 | 10C | 0 | 0.010 |
| H13 | 10C | 0 | 0.020 |
| H35 | 10C | 10 | 0.196 |
| Healthy | 0 | 0 | 0.020 |

[a]    The absorbence at 450 nm is the spectrophotometric value obtained in the enzyme-linked immunosorbent tests. The $A_{450}$ value is the average of all plants in the test.

[b]    Number of plants of ten inoculated that reacted with local lesions 2 to 3 weeks after inoculation with virus. The numerator is the number of plants that developed symptoms. (C = chlorotic local lesions; N = necrotic local lesions.)

[c]    Number of plants of ten inoculated that reacted with systemic symptoms 4 to 5 weeks after inoculation. The numerator is the number of plants that developed symptoms.

From Schoelz, J. E. and Shepherd, R. J., *Virology*, 162, 30, 1988. With permission.

## H. Conclusion

The results obtained from 14 recombinant CaMV viruses suggest that D4 and W260 do not have a common genetic determinant for systemic infection of *N. bigelovii*. The sequence of D4 that determines systemic infection of *N. bigelovii* maps within the 5′ half of gene VI. In contrast, sequence changes have occurred in at least three genes of W260 relative to CM1841 in order for W260 to spread systemically in *N. bigelovii*. Therefore, it is probable that W260 and D4 have evolved in different ways in order to infect a common host.

An explanation for the host specificity of W260 may be somewhat more complicated than for D4. At this point, there are two different ways that W260 sequences can contribute to the ability of a recombinant virus to systemically infect *N. bigelovii*. The first option is revealed in the construction of recombinants between W260 and CM1841. In this case, at least three genes of W260 determine systemic infection of *N. bigelovii*; genes VI, I, and at least one of genes II to V. The other option is revealed in the construction of H35, a three-way construct made between D4, CM1841, and W260. In this case, two sequences within a single gene, gene VI, may determine systemic infection of *N. bigelovii*. One of the sequences maps within the first half of gene VI and is derived from W260. The second sequence, derived from D4, is found within the back half of gene VI and large intergenic region. Neither sequence is sufficient for systemic infection of *N. bigelovii*. However, when the two sequences have been joined together to form H35, the result is a virus that looks similar to W260 on *N. bigelovii*. The alternative ways that W260 sequences contribute to systemic infection of *N. bigelovii* may provide some insight into how CaMV genes interact to establish an infection in different hosts.

# REFERENCES

1. **Matthews, R. E. F.,** *Plant Virology,* Academic Press, New York, 1981, 474.
2. **Fraser, R. S. S. and Gerwitz, A.,** The genetics of resistance and virulence in plant virus disease, in *Genetics and Plant Pathogenesis,* Day, P. R. and Jellis, G. J., Eds., Blackwell Scientific, Oxford, 1987, chap. 4.
2a. **Atabakov, J. G.,** Host-specificity of plant viruses, *Annu. Rev. Phytopathol.,* 13, 127, 1975.
3. **Nishiguchi, M., Motoyoshi, F., and Oshima, N.,** Behaviour of a temperature sensitive strain of tobacco mosaic virus in tomato leaves and protoplasts, *J. Gen. Virol.,* 39, 53, 1978.
4. **Leonard, D. A. and Zaitlin, M.,** A temperature sensitive strain of tobacco mosaic virus defective in cell-to-cell movement generates an altered viral-coded protein, *Virology,* 117, 416, 1982.
5. **Ohno, T., Takamatsu, N., Meshi, T., Okada, Y., Nishiguchi, M., and Kiho, Y.,** Single amino acid substitution in 30K protein of TMV defective in virus transport function, *Virology,* 131, 255, 1983.
6. **Deom, C. M., Oliver, M. J., and Beachy, R. N.,** The 30-kilodalton gene product of tobacco mosaic virus potentiates virus movement, *Science,* 237, 389, 1987.
7. **Meshi, T., Watanabe, Y., Saito, T., Sugimoto, A., Maeda, T., and Okada, Y.,** Function of the 30 kDa protein of tobacco mosaic virus: involvement in cell-to-cell movement and dispensability for replication, *EMBO J.,* 6, 2557, 1987.
8. **Saito, T., Meshi, T., Takamatsu, N., and Okada, Y.,** Coat protein gene sequence of tobacco mosaic virus encodes a host response determinant, *Proc. Natl. Acad. Sci. U.S.A.,* 84, 6074, 1987.
9. **Knorr, D. A. and Dawson, W. O.,** A point mutation in the tobacco mosaic virus capsid protein gene induces hypersensitivity in *Nicotiana sylvestris, Proc. Natl. Acad. Sci. U.S.A.,* 85, 170, 1988.
10. **Cornelissen, B. J. C., Brederode, F. Th., Moorman, R. J. M., and Bol, J. F.,** Complete nucleotide sequence of alfalfa mosaic virus RNA 1, *Nucleic Acids Res.,* 11, 1253, 1983.
11. **Cornelissen, B. J. C., Brederode, F. Th., Veeneman, G. H., van Boom, J. H., and Bol, J. F.,** Complete nucleotide sequence of AMV RNA 2, *Nucleic Acids Res.,* 11, 3019, 1983.
12. **Barker, R. F., Jarvis, N. P., Thompson, D. V., Loesch-Fries, L. S., and Hall, T. C.,** Complete nucleotide sequence of alfalfa mosaic virus RNA 3, *Nucleic Acids Res.,* 11, 2881, 1983.
13. **Nassuth, A., Alblas, F., and Bol, J. F.,** Localization of genetic information involved in the replication of alfalfa mosaic virus, *J. Gen. Virol.,* 53, 207, 1981.
14. **Nassuth, A. and Bol, J. F.,** Altered balance of the synthesis of plus- and minus-strand RNAs induced by RNAs 1 and 2 of alfalfa mosaic virus in the absence of RNA 3, *Virology,* 124, 75, 1983.
15. **Kamer, G. and Argos, P.,** Primary structural comparison of RNA-dependent polymerases from plant, animal and bacterial viruses, *Nucleic Acids Res.,* 12, 7269, 1984.
16. **Bol, J. F., van Vloten-Doting, L., and Jaspars, E. M. J.,** A functional equivalence of top-component a RNA and coat protein in the initiation of infection by alfalfa mosaic virus, *Virology,* 46, 73, 1971.
17. **Dingjan-Versteegh, A., van Vloten-Doting, L., and Jaspars, E. M. J.,** Alfalfa mosaic virus hybrids constructed by exchanging nucleoprotein components, *Virology,* 49, 716, 1972.

18. **Hartmann, D., Mohler, E., Leroy, C., and Hirth, L.,** Genetic analysis of alfalfa mosaic virus mutants, *Virology,* 74, 470, 1976.
19. **Lane, L. C. and Kaesberg, P.,** Multiple genetic components in bromegrass mosaic virus, *Nature (London), New Biol.,* 232, 40, 1971.
20. **Ahlquist, P., Luckow, V., and Kaesberg, P.,** Complete nucleotide sequence of brome mosaic virus RNA 3, *J. Mol. Biol.,* 153, 23, 1981.
21. **Ahlquist, P., Dasgupta, R., and Kaesberg, P.,** Nucleotide sequence of the brome mosaic virus genome and its implications for viral replication, *J. Mol. Biol.,* 172, 369, 1984.
22. **Shih, D. S. and Kaesberg, P.,** Translation of the RNAs of brome mosaic virus ribonucleic acid in a cell-free system derived from wheat embryo, *Proc. Natl. Acad. Sci. U.S.A.,* 70, 1799, 1973.
23. **Shih, D. S. and Kaesberg, P.,** Translation of the RNAs of brome mosaic virus: the monocistronic nature of RNA 1 and RNA 2, *J. Mol. Biol.,* 103, 77, 1976.
24. **Kiberstis, P. A., Loesch-Fries, L. S., and Hall, T. C.,** Viral protein synthesis in barley protoplasts inoculated with native and fractionated brome mosaic virus RNA, *Virology,* 112, 804, 1981.
25. **Zaitlin, M. and Hull, R.,** Plant virus-host interactions, *Annu. Rev. Plant Physiol.,* 38, 291, 1987.
26. **Wyatt, S. D. and Kuhn, C. W.,** Derivation of a new strain of cowpea chlorotic mottle virus from resistant cowpeas, *J. Gen. Virol.,* 49, 289, 1980.
27. **Peden, K. W. C. and Symons, R. H.,** Cucumber mosaic virus contains a functionally divided genome, *Virology,* 53, 487, 1973.
28. **Rezaian, M. A., Williams, R. H. V., and Symons, R. H.,** Nucleotide sequence of cucumber mosaic virus RNA1: presence of a sequence complementary to part of the viral satellite RNA and homologies with other viral RNAs, *Eur. J. Biochem.,* 150, 331, 1985.
29. **Rezaian, M. A., Williams, R. H. V., Gorden, K. H. J., Gould, A. R., and Symons, R. H.,** Nucleotide sequence of cucumber-mosaic-virus RNA 2 reveals a translation product significantly homologous to corresponding proteins of other viruses, *Eur. J. Biochem.,* 143, 277, 1984.
30. **Gould, A. R. and Symons, R. H.,** Cucumber mosaic virus RNA 3: determination of the nucleotide sequence provides the amino acid sequences of protein 3a and viral coat protein, *Eur. J. Biochem.,* 126, 217, 1982.
31. **Hanada, K. and Tochihara, H.,** Genetic analysis of cucumber mosaic peanut stunt, and chrysanthemum mild mottle viruses, *Ann. Phytopathol. Soc. Jpn.,* 46, 159, 1980.
32. **Rao, A. L. N. and Francki, R. I. B.,** Distribution of determinants for symptom production and host range on the three RNA components of cucumber mosaic virus, *J. Gen. Virol.,* 61, 197, 1982.
33. **Edwards, M. C., Gonsalves, D., and Providenti, R.,** Genetic analysis of cucumber mosaic virus in relation to host resistance: location of determinants for pathogenicity to certain legumes and *Lactuca saligna, Phytopathology,* 73, 269, 1983.
34. **Lakshman, D. K., Gonsalves, D., and Fulton, R. W.,** Role of *Vigna* species in the appearance of pathogenic variants of cucumber mosaic virus, *Phytopathology,* 75, 749, 1984.
35. **Lakshman, D. K. and Gonsalves, D.,** Genetic analysis of two large-lesion isolates of cucumber mosaic virus, *Phytopathology,* 75, 758, 1985.
36. **Hollings, M. and Stone, O. M.,** Red clover necrotic mosaic virus, *CMI/AAB Descriptions of Plant Viruses,* No. 181, 1977.
37. **Morris-Krsinich, B. A. M., Forster, R. L. S., and Mossop, D. W.,** Translation of red clover necrotic mosaic virus RNA in rabbit reticulocyte lysate: identification of the virus coat protein cistron on the larger RNA strand of the bipartite genome, *Virology,* 124, 349, 1983.
38. **Lommel, S. A., Weston-Fina, M., Xiong, Z., and Lomonossoff, G. P.,** The nucleotide sequence and gene organization of red clover necrotic mosaic virus RNA-2, *Nucleic Acids Res.,* 16, 8587, 1988.
39. **Osman, T. A. M., Dodd, S. M., and Buck, K. W.,** RNA 2 of red clover necrotic mosaic virus determines lesion morphology and systemic invasion in cowpea, *J. Gen. Virol.,* 67, 203, 1986.
40. **Okuno, T., Hiruki, C., Rao, D. V., and Figueiredo, G. C.,** Genetic determinants distributed in two genomic RNAs of sweet clover necrotic mosaic, red clover necrotic mosaic, and clover primary leaf necrosis viruses, *J. Gen. Virol.,* 64, 1907, 1983.
41. **Harrison, B. D. and Murant, A. F.,** Nepovirus group, *CMI/AAB Descriptions of Plant Viruses,* No. 185, 1977.
42. **Fritsch, C., Mayo, M. A., and Murant, A. F.,** Translation product of genome and satellite RNAs of tomato black ring virus, *J. Gen. Virol.,* 46, 381, 1980.
43. **Robinson, D. J., Barker, H., Harrison, B. D., and Mayo, M. A.,** Replication of RNA-1 of tomato black ring virus independently of RNA-2, *J. Gen. Virol.,* 51, 317, 1980.
44. **Harrison, B. D., Murant, A. F., and Mayo, M. A.,** Two properties of raspberry ringspot virus determined by its smaller RNA, *J. Gen. Virol.,* 17, 137, 1972.
45. **Harrison, B. D., Murant, A. F., Mayo, M. A., and Roberts, I. M.,** Distribution of determinants for symptom production, host range, and nematode transmissibility between the two RNA components of raspberry ringspot virus, *J. Gen. Virol.,* 22, 233, 1974.

46. **Jones, A. T. and Duncan, G. H.,** The distribution of some genetic determinants in the two nucleoprotein particles of cherry leaf roll virus, *J. Gen. Virol.,* 50, 269, 1980.

47. **Goelet, P., Lomonossoff, G. P., Butler, P. J. G., Akam, M. E., Gait, M. J., and Karn, J.,** Nucleotide sequence of tobacco mosaic virus RNA, *Proc. Natl. Acad. Sci. U.S.A.,* 79, 5818, 1982.

48. **Pelham, H. R. B.,** Leaky UAG termination codon in tobacco mosaic virus RNA, *Nature,* 272, 469, 1978.

49. **Ishikawa, M. T., Meshi, T., Motoyoshi, F., Takamatsu, N., and Okada, Y.,** *In vitro* mutagenesis of the putative replicase genes of tobacco mosaic virus, *Nucleic Acids Res.,* 14, 8291, 1986.

50. **Dawson, W. O., Beck, D. L., Knorr, D. A., and Grantham, G. L.,** cDNA cloning of the complete genome of tobacco mosaic virus and production of infectious transcripts, *Proc. Natl. Acad. Sci. U.S.A.,* 83, 1832, 1986.

51. **Meshi, T., Ishikawa, M., Motoyoshi, F., Semba, K., and Okada, Y.,** *In vitro* transcription of infectious RNAs from full-length cDNAs of tobacco mosaic virus, *Proc. Natl. Acad. Sci. U.S.A.,* 83, 5043, 1986.

52. **Meshi, T., Motoyoshi, F., Adachi, A., Watanabe, Y., Takamatsu, N., and Okada, Y.,** Two concomitant base substitutions in the putative replicase genes of tobacco mosaic virus confer the ability to overcome the effects of a tomato resistance gene, Tm-1, *EMBO J.,* 7, 1575, 1988.

53. **Hamilton, W. D. O., Boccara, M., Robinson, D. J., and Baulcombe, D. C.,** The complete nucleotide sequence of tobacco rattle virus RNA-1, *J. Gen. Virol.,* 68, 2563, 1987.

54. **Fritsch, C., Mayo, M. A., and Hirth, L.,** Further studies on the translation products of tobacco rattle virus RNA *in vitro, Virology,* 77, 722, 1977.

55. **Pelham, H. R. B.,** Translation of tobacco rattle virus RNAs in vitro: four proteins from three RNAs, *Virology,* 97, 256, 1979.

56. **Cornelissen, B. J. C., Linthorst, H. J. M., Brederode, F. Th., and Bol, J. F.,** Analysis of the genome structure of tobacco rattle virus strain PSG, *Nucleic Acids Res.,* 14, 2157, 1986.

57. **Boccara, M., Hamilton, W. D. O., and Baulcombe, D. C.,** The organization and interviral homologies of genes at the 3′ end of tobacco rattle virus RNA 1, *EMBO J.,* 5, 223, 1986.

58. **Harrison, B. D. and Robinson, D. J.,** The tobraviruses, *Adv. Virus Res.,* 23, 25, 1978.

59. **Bergh, S. T., Koziel, M. G., Huang, S., Thomas, R. A., Gilley, D. P., and Siegel, A.,** The nucleotide sequence of tobacco rattle virus RNA-2 (CAM strain), *Nucleic Acids Res.,* 13, 8507, 1985.

60. **Angenent, G. C., Linthorst, H. J. M., van Belkum, A. F., Cornelissen, B. J. C., and Bol, J. F.,** RNA 2 of tobacco rattle virus strain TCM encodes an unexpected gene, *Nucleic Acids Res.,* 14, 4673, 1986.

61. **Ghabrial, S. A. and Lister, R. M.,** Coat protein and symptom specification in tobacco rattle virus, *Virology,* 52, 1, 1973.

62. **Kiraly, Z.,** Defenses triggered by the invader: hypersensitivity, in *Plant Disease: An Advanced Treatise,* Vol. V, Horsfall, J. and Cowling, E. B., Eds., Academic Press, New York, 1980, 201.

63. **Shepherd, R. J.,** DNA plant viruses, *Annu. Rev. Plant Physiol.,* 30, 405, 1979.

64. **Hull, R.,** Caulimovirus group, *CMI/AAB Descriptions of Plant Viruses,* No. 295, 1984.

65. **Frank, A., Guilley, H., Jonard, G., Richards, K., and Hirth, L.,** Nucleotide sequences of cauliflower mosaic virus DNA, *Cell,* 21, 285, 1980.

66. **Gardner, R. C., Howarth, A., Hahn, P., Brown-Leudi, M., Shepherd, R. J., and Messing, J.,** The complete nucleotide sequence of an infectious clone of cauliflower mosaic virus M13mp7 shotgun sequencing, *Nucleic Acids Res.,* 9, 2871, 1981.

67. **Balazs, E., Guilley, H., Jonard, G., and Richards, K.,** Nucleotide sequence of DNA from an altered-virulence isolate D/H of cauliflower mosaic virus, *Gene,* 19, 239, 1982.

67a. **Fong, R., Wu, X., Bu, M., Tian, Y., Cai, F., and Mang, K.,** Complete nucleotide sequence of camv (Xinjing isolate) genomic DNA, *Chin. J. Virol.,* 1, 247, 1985.

68. **Howell, S. H., Walker, L. L., Dudley, R. K.,** Cloned cauliflower mosaic virus DNA infects turnips, *(Brassica rapa), Science,* 208, 1265, 1980.

69. **Lebeurier, G., Hirth, L., Hohn, T., and Hohn, B.,** Infectivities of native and cloned DNA of cauliflower mosaic virus, *Gene,* 12, 139, 1980.

70. **Woolsten, C., Covey, S., Penswick, J., and Davies, J.,** Aphid transmission and a polypeptide are specified by a defined region of the cauliflower mosaic virus genome, *Gene,* 23, 15, 1983.

71. **Daubert, S., Schoelz, J., Debao, L., and Shepherd, R.,** Expression of disease symptoms in cauliflower mosaic virus genomic hybrids, *J. Mol. Appl. Genet.,* 2, 537, 1984.

72. **Woolsten, C. J., Czaplewski, L. G., Markham, P. G., Goad, A. S., Hull, R., and Davies, J. W.,** Location and sequence of a region of cauliflower mosaic virus gene 2 responsible for aphid transmissibility, *Virology,* 160, 246, 1987.

73. **Gracia, O. and Shepherd, R. J.,** Cauliflower mosaic virus in the nucleus of *Nicotiana, Virology,* 146, 141, 1985.

74. **Schoelz, J. E., Shepherd, R. J., and Richins, R. D.,** Properties of an unusual strain of cauliflower mosaic virus, *Phytopathology,* 76, 451, 1986.

75. **Martinez-Izquierdo, J. A., Futterer, J., and Hohn, T.,** Protein encoded by ORF I of cauliflower mosaic virus is part of the viral inclusion body, *Virology,* 160, 527, 1987.

76. **Young, M. J., Daubert, S. D., and Shepherd, R. J.,** Gene I products of cauliflower mosaic virus detected in extracts of infected tissue, *Virology,* 158, 444, 1987.

77. **Hull, R. and Covey, S. N.,** Cauliflower mosaic virus: pathways of infection, *BioEssays,* 3, 160, 1985.

78. **Hull, R., Sadler, J., and Longstaff, M.,** The sequence of carnation etched ring virus DNA: comparison with cauliflower mosaic virus and retroviruses, *EMBO J.,* 5, 3083, 1986.

79. **Armour, S. L., Melcher, U., Pirone, T. P., Lyttle, D. J., and Essenberg, R. C.,** Helper component for aphid transmission encoded by region II of cauliflower mosaic virus DNA, *Virology,* 129, 25, 1983.

80. **Xiong, C., Lebeuier, G., and Hirth, L.,** Detection *in vivo* of a new gene product (gene III) of cauliflower mosaic virus, *Proc. Natl. Acad. Sci. U.S.A.,* 81, 6608, 1984.

81. **Mesnard, J. M., Geldreich, A., Xiong, C., Lebeurier, G., and Hirth, L.,** Expression of a putative plant viral gene in *Escherichia coli, Gene,* 31, 39, 1984.

82. **Giband, M., Mesnard, J. M., and Lebeurier, G.,** The gene III product (P15) of cauliflower mosaic virus is a DNA-binding protein while an immunologically related P11 polypeptide is associated with virions, *EMBO J.,* 5, 2433, 1986.

83. **Daubert, S., Richins, R., Shepherd, R., and Gardner, R.,** Mapping of the coat protein gene of CaMV by its expression in a prokaryotic system, *Virology,* 122, 444, 1982.

84. **Toh, H., Hayashida, H., Miyata, T.,** Sequence homology between retroviral reverse transcriptase and putative polymerases of hepatitis B virus and cauliflower mosaic virus, *Nature,* 305, 827, 1983.

85. **Laquel, P., Ziegler, V., and Hirth, L.,** The 80K polypeptide associated with the replication complexes of cauliflower mosaic virus is recognized by antibodies to gene V translation product, *J. Gen. Virol.,* 67, 197, 1986.

86. **Takatsuji, H., Hirochika, H., Fukushi, T., and Ikeda, J. E.,** Expression of cauliflower mosaic virus reverse transcriptase in yeast, *Nature,* 319, 240, 1986.

87. **Covey, S. and Hull, R.,** Transcription of cauliflower mosaic virus DNA. Detection of transcripts, properties, and location of the gene encoding the virus inclusion body protein, *Virology,* 111, 463, 1981.

88. **Xiong, C., Muller, S., Lebeurier, G., and Hirth, L.,** Identification by immunoprecipitation of cauliflower mosaic virus *in vitro* major translation product with a specific serum against viroplasm protein, *EMBO J.,* 1, 971, 1982.

89. **Schoelz, J. E., Shepherd, R. J., and Daubert, S. D.,** Region VI of cauliflower mosaic virus encodes a host range determinant, *Mol. Cell. Biol.,* 6, 2632, 1986.

90. **Baughman, G. A., Jacobs, J. D., and Howell, S. H.,** Cauliflower mosaic virus gene VI produces a symptomatic phenotype in transgenic tobacco plants, *Proc. Natl. Acad. Sci. U.S.A.,* 85, 733, 1988.

91. **Odell, J. T. and Howell, S. H.,** The identification, mapping, and characterization of mRNA for P66, a cauliflower mosaic virus-coded protein, *Virology,* 102, 349, 1980.

92. **Guilley, H., Dudley, R. K., Jonard, G., Balazs, E., and Richards, K.,** Transcription of cauliflower mosaic virus DNA: detection of promoter sequences, and characterization of transcripts, *Cell,* 30, 763, 1982.

93. **Pfeiffer, P. and Hohn, T.,** Involvement of reverse transcription in the replication of cauliflower mosaic virus: a detailed model and test of some aspects, *Cell,* 33, 781, 1983.

94. **Hull, R. and Covey, S. N.,** Does cauliflower mosaic virus replicate by reverse transcription?, *Trends Biochem. Sci.,* 8, 119, 1983.

95. **Hull, R.,** A model for the expression of CaMV nucleic acid, *Plant Mol Biol.,* 3, 121, 1984.

96. **Dixon, L. K. and Hohn, T.,** Initiation of translation of the cauliflower mosaic virus genome form a polycistronic mRNA: evidence from deletion mutagenesis, *EMBO J.,* 3, 2731, 1984.

97. **Tompkins, C. M.,** A transmissible mosaic disease of cauliflower, *J. Agric. Res.,* 55, 33, 1937.

98. **Walker, J. C., Lebeau, F. J., and Pound, G. S.,** Viruses associated with cabbage mosaic, *J. Agric. Res.,* 70, 379, 1945.

99. **Hills, G. J. and Campbell, R. N.,** Morphology of broccoli necrotic yellows virus, *J. Ultrastruct. Res.,* 24, 134, 1968.

100. **Lung, M. C. Y. and Pirone, T. P.,** *Datura stramonium,* a local lesion host for certain isolates of cauliflower mosaic virus, *Phytopathology,* 62, 1473, 1972.

101. **Schoelz, J. E. and Shepherd, R. J.,** Host range control of cauliflower mosaic virus, *Virology,* 162, 30, 1988.

Chapter 7

# APHID TRANSMISSION OF PLANT VIRUSES

## K. F. Harris

### TABLE OF CONTENTS

# I. INTRODUCTION

There are about 383 known species of animal vectors of plant viruses.[1] About 94% of these vectors are arthropods, and the remainder are nematodes. Of the 358 known arthropod vectors, 356 are insects and 2 are mites. About 273 (75.4%) of the insect vectors belong to the order Homoptera: 214 species in the suborder Sternorrhyncha and 59 in the suborder Auchenorrhyncha. The transmission systems discussed here include only viruses and primarily aphid vectors in the suborder Sternorrhyncha.

Aphid transmission of plant viruses can be influenced by a number of factors: geographical distribution and availability of viruses, vectors, and virus host plants, and their proximity to each other; species, variety, age, phytochemistry, and condition of the virus-source, aphid-host, or test plant; host plant susceptibility to inoculation and infection; distribution of virus in the source plant and the area of the plant probed or fed upon by the vector; vector species, biotype, morph, seasonal form, age, and clonal origin; weather conditions; vector interactions with other flora and fauna; conditions favoring or hindering the production of alates in a given population; agricultural practices; vector behavior; virus species, strain, or variant; possible dependency of transmission on helper agents or mixed infections; and, finally, the most complex biotic factor of all, the human element, the grower (or vector researcher), and the extent to which he or she participates in, manipulates, and directs conditions. The study of how environmental factors such as these affect virus-vector-plant compatibility, as measured by virus spread or vector transmission efficiency, might be referred to as transmission ecology.[2,3] Many of the aforementioned ecological considerations are discussed in detail in chapters of a recent series of books on vector transmission of plant disease agents.[4-8]

The scope of transmission ecology discussed here is mainly limited to the times when virus, aphid, and plant come together during the acquisition, carryover, and inoculation phases of transmission. Special emphasis is placed on how virus-aphid-plant interactions determine the mechanisms by which viruses are transmitted.

# II. CATEGORIZING TRANSMISSIONS

Virus transmissions, or the viruses transmitted, may be classified as *circulative*, including nonpropagative and propagative subcategories, and *noncirculative,* including nonpersistent and semipersistent subcategories (Table 1). In circulative transmission, virus enters (acquisition phase) and leaves (inoculation phase) the vector by different routes. Virus is acquired via the maxillary food canal, absorbed, translocated, and, following a latent (nonpropagative) or incubation (propagative) period in the vector, inoculated to plants in virus-laden salivary secretions ejected from the maxillary salivary canal during probing and feeding: an *ingestion-salivation mechanism* of transmission. Circulative viruses may be subcategorized as either propagative or nonpropagative, depending on demonstrability of virus multiplication or nonmultiplication, respectively, in the vector. In noncirculative transmission (some believe stylet-borne), virus is acquired and inoculated by the same route, the maxillary food canal: an *ingestion-egestion mechanism.* The noncirculative mode of transmission is characterized by the absence of a detectable latent period, loss of vector inoculativity through molting (nontransstadiality), and the lack of evidence for transmissible virus entering the hemocoel *and* exiting via the salivary system of the vector. Assumedly, all aphid-borne viruses, or their transmissions, referred to in the literature as nonpersistent or semipersistent, meet at least the first two of these criteria, but relatively few reports have been made on the basis of the third characteristic. Similarly, many persistent viruses have been classified as circulative solely on the basis of transstadial passage, the presence of a latent period, and analogy with known circulative viruses. Thus far, this assumed synonymity of terminologies

**Table 1**
**CATEGORIZING AND COMPARING PLANT VIRUS TRANSMISSIONS BY**
**APHIDS**

| Characteristic | Noncirculative | | Circulative (persistent)[a] |
|---|---|---|---|
| | **Nonpersistent** | **Semipersistent** | |
| Virus entry | Maxillary food canal[b] | | Maxillary food canal |
| Virus retention | Foregut Lumen | | Hemocoel |
| Virus release | Maxillary food canal | | Maxillary salivary canal |
| Virus traverses aphid membrane systems | No | | Yes |
| Virus detectable in hemolymph | No | | Yes |
| Aphid rendered inoculative by viral injection | No | | Yes |
| Transstadial inoculativity | No | | Yes |
| Transmission mechanism | Ingestion-egestion | | Ingestion-salivation |
| Acquisition threshold | Seconds | Minutes | Minutes to hours |
| Effect of preacquisition starvation on probability of transmission | Positive | None | None |
| Duration of acquisition-access feeding period leading to maximal probability of transmission | Seconds to minutes | Hours | Hours |
| Correlation between the duration of the acquisition-access feeding period and the probability of transmission | Negative | Positive | Positive |
| Latent period | None | None | Hours to days |
| Inoculation threshold | Seconds | Minutes | Minutes to hours |
| Transmission threshold | Seconds to minutes | Minutes to hours | Hours to days |
| Effect of prolonged feeding probe on probability of inoculativity during subsequent probe | Negative | None | None |
| Rate of decrease in inoculativity of feeding aphids measurable in | Minutes | Hours | Days |
| Retention of inoculativity by probing and feeding aphids | Minutes | Hours to days | Days to life |
| Effect of postacquisition starvation on duration of retention | Positive | None | None |
| Number of plants in a series inoculated by a single aphid | One to few | Few to many | Many |
| Acquisition tissues | Epidermis | Phloem[c] | Phloem[c] |
| Inoculation tissues | Epidermis | Phloem[c] | Phloem[c] |
| Sap inoculable | Yes | Yes[d] | No |
| Level of virus-vector specificity | Low | Medium | High |
| Typical host-plant symptomatology | Mosaic | Mosaic-yellowing | Yellowing-leaf rolling |

*Note:* There are obvious exceptions to the generalized characteristics listed here. For example, the circulative pea enation mosaic virus (PEMV) is unusual in that it infects both superficial and deep tissues of host plants, has acquisition, inoculation, and transmission thresholds of 5 min or less, 5 to 7 s, and 4 to 6 h, respectively, and is readily sap inoculable.

[a] Circulative transmission can be subcategorized as nonpropagative or propagative depending on demonstrability of virus nonmultiplication or multiplication in the vector, respectively.

[b] Whether or not the external surfaces of the stylet, paired mandibles, and maxillae can serve as the site as a supplementary site for nonpersistent or semipersistent noncirculative virus uptake, retention, and release (i.e., stylet-borne transmission) has yet to be determined.

[c] In semipersistent noncirculative and circulative transmissions, the intervening mesophyll might serve as an acquisition or inoculation site for those viruses that are not phloem-restricted. Bimodal noncirculative transmission (having both nonpersistent and semipersistent transmission parameters) presumably results when an appropriate aphid vector is paired with a noncirculative virus that infects both superficial and deep tissues of

**Table 1 (continued)**
**CATEGORIZING AND COMPARING PLANT VIRUS TRANSMISSIONS BY**
**APHIDS**

the host plant. Similarly, the unusually brief, nonpersistent-like inoculation threshold (<15 s) for PEMV indicates that this unique circulative virus can be inoculated by aphids to both the epidermis and mesophyll. And, compared to the intervening mesophyll, the phloem appears an inefficient, if not nonsusceptible, site for PEMV inoculation.[150]

d    Some are sap inoculable only with difficulty, if at all.

(i.e., nonpersistent and semipersistent with noncirculative, and persistent with circulative) appears to be a prescient conclusion.

Numerous observable phenomena serve to separate noncirculative transmissions into the aforementioned nonpersistent and semipersistent subcategories (Table 1).[1] Obviously, by definition, they can be differentiated on the basis of the duration of retention of inoculativity by viruliferous aphids that are allowed access to uninfected plants, but disallowed additional access to a virus source (a few minutes to an hour for nonpersistent vs. several hours to 3 to 4 d for semipersistent). Furthermore, (1) starving aphids prior to allowing them acquisition probes increases the level of nonpersistent transmission severalfold over that obtained using nonstarved controls. This "preacquisition starvation effect" does not occur in semipersistent transmission. (2) The acquisition and inoculation thresholds in nonpersistent transmission are generally measurable in seconds, vs. several minutes for semipersistent. (3) In nonpersistent transmission, full vector inoculative capacity is generally realized, and optimally so, following single, naturally terminated acquisition probes of 15 s to 1 min. As the duration of acquisition probes exceeds 1 min, corresponding marked decreases in the level of transmission occur. In semipersistent transmission, vector inoculative capacity generally increases with increases (for periods up to several hours) in the duration of the acquisition-access feeding period (AAFP), suggesting that virus can accumulate in the vector up to a point of saturation or maximal carrying capacity. (4) Although aphids may inoculate nonpersistent viruses to plants during single, prolonged, phloem-seeking or feeding probes, they are rarely inoculative afterward, whereas such probes are optimal for semipersistent transmission and do not have this drastic effect on continued vector inoculativity. (5) In nonpersistent transmission, groups of starved, viruliferous aphids remain inoculative longer than feeding ones, whereas in semipersistent transmission, the rate of loss of inoculativity is independent of whether the insects are starved or allowed to feed.

Whether a specific noncirculative virus is transmitted nonpersistently, semipersistently, or "bimodally" by a specific aphid vector depends on which tissues and cells of the host plant contain transmissible virus and are susceptible to virus inoculation by aphids, on whether or not the virus is able to survive, accumulate, and persist in a transmissible state in the fore alimentary canal of the vector, and on whether or not the probing and feeding behavior of the vector is conducive to plant inoculation with virus and infection-site development.

## III. TRANSMISSION PARTICIPANTS

### A. Noncirculative Viruses

There are approximately 321 known, animal-borne plant viruses, of which 298 (93%) are vectored by arthropods and 23 (7%) by nematodes. About 55% of the arthropod-borne viruses are transmitted by aphids.[1] Of the 164 known aphid-borne viruses, 109 are noncirculative (101 nonpersistent and 8 semipersistent), 38 are circulative, and the status of 17 is uncertain.[9,10]

The Plant Virus Subcommittee of the International Committee for Taxonomy of Viruses (ICTV) has endorsed eight groups of plant viruses, each of which contains one or more noncirculative, aphid-transmitted members: potyviruses, carlaviruses, caulimoviruses, cucumoviruses, alfalfa mosaic virus (AMV), broad bean wilt virus, closteroviruses, and parsnip yellow fleck virus (PYFV).[11,12] The first six groups, up to and including broad bean wilt virus, contain at least one nonpersistently transmitted member, and some contain members that appear to be transmitted both nonpersistently and semipersistently (bimodally). Most members of these groups are readily sap-inoculable. Closteroviruses and PYFV are transmitted semipersistently and are sap-inoculable with difficulty, if at all.

Several potyviruses are known to require a helper agent for transmission.[13] Two potyviruses, bean yellow mosaic virus (BYMV) and potato virus Y (PVY), have been the most commonly used viruses in studies on the mechanism of noncirculative transmission. Most of the potyviruses are nonpersistently transmitted, i.e., turnip mosaic (TuMV), celery mosaic, cocksfoot streak, plum pox, tobacco etch (TEV), beet mosaic, tulip breaking, bean common mosaic, carnation vein mottle, papaya ringspot, parsnip mosaic, soybean mosaic, henbane mosaic, pokeweed mosaic, passion fruit woodiness, and clover yellow vein viruses.

Nonpersistently transmitted members of the carlavirus group include carnation latent, potato virus M, chrysanthemum virus B, and red clover vein mosaic viruses. Some potato virus S isolates are known to be aphid transmissible; vectors have not yet been found for cowpea mild mottle and poplar mosaic viruses. Pea streak virus is a candidate for bimodal transmission.[14]

Cauliflower mosaic (CaMV) caulimovirus is transmitted nonpersistently by *Myzus persicae* Sulz, and bimodally by *Brevicoryne brassicae* L.[14] Some isolates of this virus appear to require a helper agent.[15] Dahlia mosaic caulimovirus possibly is transmitted bimodally.[16] Cucumber mosaic virus (CMV) and other cucumoviruses, peanut stunt and tomato aspermy viruses, are nonpersistently transmitted. Aphids transmit AMV and broad bean wilt virus nonpersistently. PYFV is semipersistently transmitted by its aphid vector.[17,18] Among the closteroviruses, beet yellows virus and carnation necrotic fleck virus are semipersistently transmitted members, whereas citrus tristeza virus may be transmitted in a bimodal manner.

## B. Helpers

Nonpersistent, noncirculative aphid transmission of purified preparations of potyviruses is dependent on the presence of a protein, helper component (HC), which is present in extracts of potyvirus-infected, but not healthy plants.[13,19-27] Several types of evidence have clearly established that HC is a virus-coded protein product rather than a product of the host plant produced in response to virus infection.[22,28,29] Evidence for the latter is based on in vitro translation studies of viral RNA. Posttranslational cleavage of a 78K polyprotein generates the HCs of TuMV and pepper mottle virus,[30] whereas the HC of the watermelon mosaic virus (WMV) 1 strain of papaya ringspot virus results from cleavage of a 110K polyprotein.[31] The coding region of HC in all three of these viruses is near the 5′ terminus of the RNA genome.[30,31] The molecular weights of TuMV HC and PVY HC are 53K and 58K, respectively.[27,29] It has been suggested that HC of both these viruses exists as a dimer with a native molecular weight of 106K for TuMV and 116K for PVY.[29]

Serological studies which show that serologically distinct HC proteins are produced in response to specific potyvirus infections also suggest that HC is virus coded.[23,25,27,32] Antiserum to PVY HC precipitated a unique PVY RNA translation product,[28] and TuMV HC antiserum also reacted with TuMV RNA-induced product.[22] And, lastly, serological tests indicate that the biologically functional HC of TuMV is related to a component of amorphous inclusion bodies[31] which appears to be of viral origin.[32]

Specificity of HC for preferential and efficient transmission of its virus by aphids (*Myzus persicae*) was tested by studying the transmission of various virus (WMV 1 strain of papaya ringspot virus, WMV 2, and zucchini yellow mosaic virus)-HC mixtures in homologous and heterologous combinations.[33] Transmission of homologous virus (from a mixture of HC with both homologous and heterologous virus) was nearly the same as when it was alone with its helper, whereas transmission of heterologous virus from the mixture was reduced by four to six times.[33] Similarly, TuMV transmission was consistently more efficient and more common than PVY transmission from a TuMV-PVY mixture in the presence of TuMV HC.[23]

Lack of strict HC-virus specificity has also been noted. The HCs of TuMV and PVY differ in size and serological specificity, but are nevertheless functionally interchangeable.[23] Similarly, purified TEV is transmitted equally efficiently in the presence of PVY HC and TuMV HC.[28] BYMV HC is effective in the transmission of purified TuMV or PVY, but neither TuMV HC nor PVY HC are effective in the transmission of purified BYMV.[23] Such one-sided effectiveness of HC also occurs in the case of TuMV with either WMV HC or PVY HC.[24,25] Addition of PVY HC resulted in aphid transmission of two other potyviruses, TEV and henbane mosaic virus.[19] Sako and Ogata[25] observed that WMV HC from infected pumpkin leaves or in the soluble fraction enabled *M. persicae* to transmit TuMV, but failed to effect the transmission of purified PVY. In contrast, TuMV HC from infected turnip leaves or in the soluble fraction failed to effect transmission of purified WMV and PVY by *M. persicae*; PVY HC helped *M. persicae* to transmit purified TuMV, but not WMV. And no heterologous HC could effect the transmission of PVY. Thus, some specificity exists in the activity of the HCs associated with TuMV, WMV, and PVY.

All pairs of dependent and helper viruses known are potyviruses except for the viruses of the rice tungro virus complex. Green leafhoppers, *Nephotettix virescens*, transmit both rice tungro bacilliform virus (RTBV) and rice tungro spherical virus (RTSV) in a semipersistent manner.[34-36] RTSV can be transmitted independently, whereas RTBV depends on RTSV for its transmission by *N. virescens*.[34-37] RTBV from RTBV-infected plants can be transmitted only by leafhoppers that have earlier fed on RTSV-infected plants, suggesting that some HC is acquired by leafhoppers from the RTSV-infected plants.[34-37] Leafhoppers could transmit RTSV for 3 to 4 d, whereas their ability to acquire and transmit RTBV lasted for 7 d.[35,37] RTBV and RTSV are not serologically related.

Based on the need for its presence at the time of virus acquisition, a number of different roles have been proposed for HC in aphid transmission: regulation of virus uptake, binding of virions to receptor sites in the foregut from which virus is later eluted,[20] protection of virus from adverse conditions within the foregut, or effecting the inoculation process.[20,38-42] Indirect support for its proposed role in binding virions to receptor sites in aphids, from which egestion could occur, comes from work with the potyvirus-HC system. Radiolabeled potyvirus (PVY or TEV) fed to aphids in the presence of HC (and hence transmissible) accumulated in all regions of the foregut, whereas labeled virus in the absence of HC (and hence not transmissible) was not retained at these sites and passed beyond the esophageal or cardiac valve into the midgut.[42] It was further noted that in preparations where the mandibles of the stylet bundle separated from the inner interlocked maxillae, label was specifically associated with the maxillae which form the food canal portion of the foregut. These data support the hypothesis that HC may function to bind virus to retention sites in the vector, although they do not rule out alternate or additional functions for HC.

Pirone and Thornbury[40] identified three isolates of TEV as either highly aphid-transmissible (HAT), poorly aphid-transmissible (PAT), or nonaphid-transmissible (NAT) from infected plants. However, purified virus of each of these isolates was transmitted with high, intermediate, and low efficiency, respectively, when acquired through membranes in the presence of HC. When the activities of HC isolated from each of the three isolates were compared, HC from PAT-infected plants proved the most active. The latter suggests that differences in properties of the virions, probably differences in the coat proteins, are responsible for the observed differences in isolate transmissibility. However, since aphids given access first to HC and then to infected plants were able to transmit the NAT isolate and transmitted the

PAT isolate with increased efficiency, a role for HC in regulating the efficiency of aphid transmission from plants cannot be ruled out.[40] Similar data have been reported with respect to the aphid transmissibilities of TuMV isolates from TuMV-infected mustard plants (Pirone, unpublished data, quoted in Hiebert et al.[28]), suggesting that aphid-transmissibility differences among potyvirus isolates (TEV or TuMV) cannot be fully explained on the basis of HC absence or deficiency, but are due, at least in part, to varying properties of the capsid protein.

Helper dependency has been noted for some isolates of cauliflower mosaic virus (CaMV), a noncirculatively transmitted ds-DNA virus.[15] Most of the natural CaMV isolates are nonpersistently or bimodally aphid transmitted (AT⁺), but some are not aphid-transmissible (AT⁻).[15,43] However, AT⁻ isolates can be readily transmitted if the aphids first probe a plant infected with an AT⁺ isolate, or another transmissible caulimovirus such as carnation etched ring virus or figwort mosaic virus,[44] which presumably supplies the necessary HC.[41] Aphid transmissibility of CaMV is controlled by open reading frame (ORF) II of the CaMV genome, through its translation product protein (p18) of molecular weight 18,000.[41,45-48] The ORF II product copurifies with viroplasm preparations. Moreover, Rodriguez et al.[49] found that CaMV HC activity was associated only with the viroplasm preparations of the AT⁺ isolate. Thus, functional HC is present in the purified viroplasms of AT⁺ isolates and the p18 protein is either the active CaMV helper or part of it.

## C. Circulative Viruses

Circulative aphid-borne viruses can be separated into at least four distinct taxonomic categories or groups: luteoviruses, plant rhabdoviruses, a monotypic group based on pea enation mosaic virus (PEMV), and a probable group based on carrot mottle virus.

Members of the luteovirus group are confined to the phloem tissues of infected plants, circulatively (persistently) transmitted by aphids, and noninoculable by sap. Some of the luteoviruses reported to have a circulative relationship with their aphid vectors include barley yellow dwarf (BYDV), beet western yellows, carrot red leaf, potato leaf roll (PLRV), soybean dwarf, subterranean clover red leaf, and groundnut rosette assistor viruses.

Many plant rhabdoviruses are known to be transmitted by aphid, leafhopper, or planthopper vectors.[50] The circulative aphid-borne plant rhabdoviruses include broccoli necrotic yellows, lettuce necrotic yellows, potato yellow dwarf, raspberry vein chlorosis, sonchus yellow net, sowthistle yellow vein, and strawberry crinkle viruses.[50,51] All these viruses are known to multiply in their aphid vectors and are therefore transmitted in a propagative, circulative manner.

## D. Virus Capsids

The primary amino acid sequence of HAT and NAT isolates of TEV have been determined by nucleotide sequencing of the coat protein cistron and its flanking region.[52,53] Three of the six amino acid differences between the capsid proteins of the HAT and NAT isolates occur within the 29 N-terminal amino acid residues of the TEV capsid protein. And the 29 N-terminal amino acid residues are located at or near the surface of virus particles.[52] Thus, capsid protein modifications at or near the N-terminus, leading to some change(s) in the surface properties of the virion, appear to determine aphid transmissibility of the isolates.

Massalski and Harrison[54] found that all of ten monoclonal antibodies (MAbs) reacted strongly with 28 HAT British isolates of PLRV, indicating considerable antigenic uniformity among the isolates. Particle capsid proteins play some crucial role during aphid transmission of luteoviruses, and transmission of PLRV isolates by *M. persicae* could be the selection pressure responsible for antigenic conservation.[54] The latter is further confirmed by the fact that the two antigenically variant isolates (PLRV-15 and PLRV-V) were also only poorly aphid transmissible. Serological studies with MAbs showed that two epitopes, which occur on the particles of the 28 aphid-transmissible isolates, were missing from the PAT isolates. These two epitopes are part of the quaternary structure of the capsid protein and are therefore of the conformational dependent type.[54]

The nucleotide sequences of two aphid-transmissible CaMV isolates (Cabb S and D/H) and one NAT isolate (CM 1841) are known.[55-57] Comparisons of the derived amino acid sequences of the isolates indicate that those of the aphid-transmissible Cabb S and D/H are very similar, whereas the amino acid sequence of the NAT CM 1841 isolate differs from both Cabb S and D/H isolates at five amino acid positions, and two of these five amino acid changes, glycine to arginine at position 94 and lysine to asparagine at position 21, appear critical with respect to aphid transmissibility.[45]

Adam et al.[58] demonstrated that NAT isolates of PEMV lack a second coat protein typical of aphid-transmissible ones and have an RNA1 with a molecular weight at $1.2 \times 10^5$ less than that of aphid-transmissible isolates. An analysis of the available data suggests that loss of, or a deletion in, the larger RNA1 (and thus of the second protein) during successive mechanical inoculations of aphid-transmissible isolates to plant hosts (exclusion of aphid transmission) is responsible for loss of aphid transmissibility.[50] The second coat protein presumably also accounts for the observed differential permeability of accessory salivary glands of the aphid vector to virions of aphid-transmissible vs. NAT isolates.[50,59]

## E. Vectors

Relationships among plants, viruses, and insects have evolved during the last 200 million years. The biology, feeding behavior, and worldwide distribution of aphids make them ideally suited for transmitting plant viruses.[4] The total number of described species of Aphididae is 3801.[60] Only about 300 species have been tested as vectors of any of 300 different viruses in about the same number of plant species. About 193 of the 300 species tested have been reported as vectors of at least one of 164 viruses. Thus, these insects alone account for about 71% of all homopterous vectors and transmit about 55% of the 298 or so known arthropod-borne viruses.[1]

All aphid vectors, except for one species in Adelgidae, belong to the family Aphididae, which accounts for about 97% of the 3917 described species of Aphidoidea (Aphididae, Adelgidae, and Phylloxeridae). Furthermore, more than half of all aphid species and most economically important virus vectors are in the Aphidinae. Although this subfamily contains a few polyphagous aphids, most of its members are host specific. Many genera of Aphidinae have Rosaceae as overwintering primary hosts, and migrate to herbaceous secondary hosts in the summer.

The amount of virus associated with individual viruliferous aphid vectors has been measured in some instances of circulative transmission. The average content of subterranean clover red leaf luteovirus was greater than 170 pg per aphid as measured in individual and groups of aphid vectors by [3]H-labeled cDNA transcribed from viral RNA.[61] ELISA detected about 0.01 ng of PLRV per aphid.[62] Using radioimmunosorbent assay, it was possible to detect as little as 0.15 ng of blueberry shoestring virus (an unassigned virus) per aphid vector, *Illinoia pepperi*.[63]

Isotopically labeled potyvirus (PVY or TEV) has been detected in association with the maxillary stylets and foregut (maxillary food canal, precibarium, cibarial pump, and esophagus) of aphids fed through membranes on HC-treated purified virus preparations.[42] Taylor and Robertson[64] sectioned stylets of *M. persicae* for electron microscopy after 10-s acquisition probes of TEV-infected leaves and found viruslike particles lining the distal 20 μm of the maxillae. Latex spheres coated with antipea seed-borne mosaic virus (PSbMV) immunoglobulins adhere to the stylet surfaces of aphids that have previously probed PSbMV-infected plant material.[65,66] All of the above instances of noncirculative virus association with aphid vectors suggest that during the probes that result in acquisition of these viruses, the feeding apparatus of the vector comes in contact with virus-laden cytoplasm of the host plant. How this intimate contact between vector and host is established during probing and its significance with respect to noncirculative virus acquisition and transmission will be discussed in detail shortly.

## F. Plants

Eastop[10] has collated the geographical, climatological, and host-plant distributions of aphids and plant viruses. The known plant viruses are distributed fairly equally through the major plant groups, considering the number of plant species in each group. The proportion of viruses transmitted by aphids does not appear to be a function of the number of aphids specific to a particular plant group. As might be expected, however, plant families containing species of economic importance, such as Chenopodiaceae, Rosaceae, Leguminosae, Solanaceae, and Gramineae, have been recorded as hosts of most viruses.

Most researchers have used very few plant species, primarily tobaccos and legumes, to study the mechanism of noncirculative virus transmission. This bias undoubtedly results from the preferential use of PVY, CMV, and BYMV in such studies.[67] There are about 100 species of aphids specific to Leguminosae, and these aphids are responsible for the transmission of many of the viruses of Leguminosae. In contrast to aphids of Rosaceae, few Leguminosae-specific aphids have host-plant alternation. Thus, the source of winged aphids colonizing Leguminosae is often another legume of the same species, a situation which presumably facilitates the development of virus transmission.

Complete or partial loss of aphid transmissibility occurs in some noncirculative viruses, such as BYMV, CMV, PSbMV, and tomato aspermy virus. The loss is generally correlated with repeated successive mechanical inoculation of virus from plant host to plant host (exclusion of vector transmission), which probably leads to the selection of a NAT isolate from the virus population or the selection of an induced mutant species.[68,69] Maintenance of TuMV in turnip by repeated mechanical inoculations with crude sap also caused the loss of its aphid transmissibility.[70] An isolate of sugarcane mosaic virus strain H, originally from a field-grown sugarcane plant, lost its transmissibility by the aphid *Dactynotus ambrosiae* after having been maintained in sugarcane in the glasshouse by mechanical inoculation or vegetative propagation since 1963.[71,72] And, as mentioned earlier, NAT isolates of the circulative PEMV can be produced by subjecting aphid-transmissible isolates to a regime of constant plant-to-plant transfer by sap inoculation only, thus selecting for a deletion mutant with respect to the RNA1 of the faster sedimenting bottom component.[50,58]

## G. Man

Man obviously has influenced the distribution of aphid vectors and aphid-borne viruses by his agricultural practices and commerce. For instance, over a third of the 62 known viruses of Solanaceae are aphid transmitted, and most are transmitted by vectors that did not encounter potatoes until about 400 years ago. With the possible exception of *Macrosiphum euphorbiae* Thos., most virus transmission by aphids in North America is by introduced vector species on introduced plants.[10] The number of present potato viruses that originated in potato is not known.

In the laboratory, experimenters can directly influence the "performance" of transmission systems both by their choice of virus-vector-plant combinations and environmental conditions, and by their manner of formulating and carrying out particular experimental designs. Individual experimenters conducting similar experiments at the same time in the same laboratory and presumably under the same conditions often obtain strikingly different results, especially in terms of the level of transmission. The causes of such discrepancies are rarely evident; however, they might be explained, at least in part, by subtle differences in the handling of test aphids and the interpretation of what the insects are doing during the acquisition and inoculation phases of transmission. The latter point can be especially crucial when selected, timed, source, or inoculation "probes" are used. As pointed out by Pirone and Harris,[67] "One person's 'feed' may well be another's 'probe'."

In most retention-time experiments involving aphid vectors and nonpersistent noncirculative viruses, active viruliferous insects are removed from virus-source plants and confined for various periods of time in a glass container before being placed on assay plants to test for inoculativity. During confinement, the aphids can be seen to repeatedly probe and salivate against the glass walls or, in the case of cellophane-covered petri dishes, through the tops of the container. Aphid saliva has been shown to have virus-inactivating properties[73,74] and therefore could be expected to inactivate virus carried on the stylet tips or, if ingested, in the anterior portion (maxillary food canal, precibarium, and cibarial pump) of the foregut of the vector. Aphids have been observed to ingest previously secreted watery saliva while probing and feeding, through membranes, in feeding solutions.[75] Also, the presence of virus in the salivary deposits left on the surfaces of glass containers indicates at least some contact between virus and saliva.[76] Since nonprobing aphids retain inoculativity for much longer periods than do ones that are allowed to probe and salivate during confinement,[76,77] retention times for nonprobing, air-borne vectors in nature might be considerably longer than those suggested by traditional methods for measuring retention time.[77,78]

## IV. VIRUS-VECTOR-PLANT INTERACTIONS

### A. Noncirculative Transmission
#### 1. Nonpersistent, Noncirculative Transmission

The transmission event appears epidermal and intracellular in nature. Sap-sampling or host-selection behavior, especially sap sampling in the epidermis, appears to play an important, if not crucial, role in the transmission process. Sap-sampling behavior is stimulated by subjecting aphids to preacquisition starvation. Sap sampling on a virus-infected plant serves to contaminate the foregut with virus-laden plant material (cell sap or protoplasm). The transmission cycle is completed when an infective dose of virus, possibly retained at and released from specific retention sites within the foregut of the vector, is egested during subsequent sap-sampling probes of healthy plants: an *ingestion-egestion mechanism* of transmission. This sap-sampling behavior serves to bring plant material in contact with three sets of gustatory chemosensilla in the vicinity of the precibarial valve, thus permitting a chemical analysis of the suitability of the plant as a host.[79,80]

The amounts of plant material and virus egested during sap sampling in the epidermis can be expected to vary greatly among probes involving either the same or different virus-vector-plant combinations. Initiation of an infection site presumably requires a living cell and, in the case of unipartite viruses, an absolute minimal viral inoculum of one infective particle. The volumes of fluid involved when transmission occurs are presumably small in comparison to the total volume of protoplasm of an epidermal cell. Involvement of the maxillary food canal portion of the foregut is obvious, since all material must enter or exit the aphid by this route during the acquisition (ingestion) and inoculation (egestion) phases of transmission, respectively. Virus-laden material moved into or through the maxillary food canal from the precibarium (epipharynx and hypopharynx) and cibarial pump regions of the foregut during egestion is likewise suspect. Whether or not virus can be egested from farther back in the foregut (i.e., the pharynx and esophagus) remains to be seen. However, even if infrequent or abnormal, such egestion might at least serve to recontaminate more anterior regions of the foregut with virus which might be transmitted during a subsequent sap-sampling probe of a host plant.

The volumes of plant sap transmitted during sap-sampling probes can be many times greater than what could be physically accommodated on the mandibular surfaces and tips and external surfaces of the maxillae or, for that matter, solely in the maxillary food canal portion of the foregut which has a maximum carrying capacity of about 60 $\mu m^3$. For transmission of CMV to occur, aphids must transmit from 480 to 841 $\mu m^3$ of virus-laden cell sap.[81] When aphids were allowed 1-, 3-, 4-, or 5-min source probes on CMV-infected or $P^{32}$-labeled plants, the percentages of them that transmitted virus or 570 $\mu m^3$ or more of plant sap were 22.5, 52.5, 65.0, and 70.0%, and 20.0, 50.0, 55.0, and 82.5%, respectively. Preacquisition starvation significantly increases both the number of aphids that transmit 640 $\mu m^3$ or more of plant sap and the number that transmit CMV (Garrett, personal communication quoted in Harris).[82]

This unique sap-sampling behavior of aphids, coupled with their finely tipped stylets and their habit of inserting only the maxillary tips into cells from which they ingest sap (and then closing these feeding sites with a salivary plug during stylet withdrawal), makes them ideally suited as vectors of nonpersistent, noncirculative plant viruses. Aphids apparently function more like flying syringes than "flying needles". However, sap-sampling behavior provides for a continuum of transmission types ranging from simple external stylet contamination (although such a mechanism has yet to be demonstrated) to any number of simple to complex variations on the basic ingestion-egestion theme. Minimal infective inocula, as well as acquisition and inoculation thresholds, can be expected to vary from one virus-vector-plant combination to another. Reported acquisition and inoculation thresholds for the same transmission system might vary considerably from one laboratory to another depending on the methods of timing the acquisition and inoculation phases that experimenters use: e.g., electronic recording, duration of contact between the labium of an aphid and the surface of a virus-source or test plant, or simply the total time an aphid spends on a plant. Vector behavior also might have a marked effect on the duration of the transmission cycle. For example, an aphid such as *M. persicae* is prone to making one or more initial brief probes of several seconds to 1 to 2 min in duration, whereas this behavior is unusual for *B. brassicae* and *Schizaphis graminum* Rondani, even following preacquisition starvation. For the latter vectors, the subepidermal mesophyll, not the epidermis, might serve as the preferred site for sap sampling and hence nonpersistent virus acquisition and inoculation. Vector behavior and virus-vector compatibility presumably could affect not only the volume and titer of the virus inoculum, but also the duration of its retention by the vector.

Several lines of evidence can be cited in support of foregut involvement in nonpersistent noncirculative virus transmission. A recent scanning electron microscopy study of the morphology of the feeding apparatus of the pea aphid confirms earlier reports of aphid ingestion-egestion behavior and sap sampling in the epidermis.[80] Presumably, ingested fluids that fail to stimulate or negatively stimulate the precibarial chemosensilla elicit a motor response to prevent closure of the precibarial valve during contraction of the cibarial pump, resulting in egestion of fluids from the maxillary food canal.[80]

Harris and Bath[75] reported direct light microscopy observations of aphid ingestion-egestion behavior during probing and feeding through parafilm membranes in sucrose solutions, thus confirming that egestion was not only possible, but apparently represented a normal aspect of aphid probing and feeding behavior. The concept of ingestion-egestion behavior and sap sampling during brief probes of plants was further strengthened by simultaneous independent research by Garrett[81] who, as mentioned earlier, was able to relate certain minimal uptakes and releases of isotopically labeled sap during probing with the percentages of aphids known to acquire and inoculate nonpersistent virus (CMV). And the relative amounts of plant sap transmitted were far greater than what could be accommodated as stylet-surface contaminant or, for that matter, solely in the maxillary food canal portion of the foregut.

At about the same time as the foregoing reports of ingestion-egestion behavior, Taylor and Robertson[64] reported observing tobacco severe etch virus (TSEV)-like particles lining the distal 20 $\mu m$ of the food canals of aphids (*M. persicae*) that were allowed 10 s acquisition probes of TSEV-infected source plants. It was originally postulated that virus contaminated the stylet when the maxillae penetrated plant tissue in a staggered state. However, it is now known that the stylet bundle is always preceded and surrounded by salivary sheath saliva during plant penetration, and 10-s probes allow adequate time for epidermal cell entry and ingestion of TSEV-laden sap.[82,83]

As discussed earlier, many noncirculative viruses, whether aphid or leafhopper-borne, appear to require virus-coded HC for vector transmission.[84] Recent research of potyvirus-HC systems has provided indirect support for nonpersistent virus retention at sites in the foregut.[42] Radiolabeled potyvirus fed to aphids in the presence of HC (and hence aphid transmissible) accumulated in all regions of the foregut, whereas labeled virus in the absence of HC (and hence nonaphid transmissible) was not retained at these sites and passed beyond the cardiac or esophageal valve into the "stomach" portion of the midgut. It was further noted that virus was specifically associated with the interlocked maxillae of the stylet which form the food canal portion of the foregut. These data support the hypothesis that virus is retained in the foregut and transmitted by an ingestion-egestion mechanism, and that one possible function of HC is to bind virus to retention sites in the vector.

Finally, the results of a light and electron microscopical study of the stylet paths made by aphids during brief virus-acquisition probes make possible the following statements regarding aphid probing behavior and nonpersistent virus transmission:[83]

1.  Aphids nearly always initiate probes intercellularly in anticlinal grooves formed between adjacent epidermal cells.
2.  Initiation of probing and stylet penetration is always accompanied by salivation and sheath formation and, except during feeding, the salivary sheath is closed distally and completely surrounds the stylet bundle.
3.  The length of stylet bundle extension beyond the tip of the labium after a probe is not necessarily an accurate measure of the depth of penetration of that probe.
4.  The presence of smooth salivary sheath saliva is not necessarily indicative of intercellular penetration, since smooth sheath saliva also occurs during intracellular stylet penetration in the cytoplasm.
5.  Fasted aphids generally make one or more brief (15 to 59 s) probes within the epidermis, whereas nonfasted aphids generally make long probes (>59 s) that penetrate past the epidermis.
6.  Only aphids that make intracellular stylet paths, and apparently a very high percentage of them, acquire and subsequently transmit virus.
7.  The rate of intracellular stylet paths (and, hence, transmission) by fasted aphids is far greater (preacquisition starvation effect) than that for nonfasted aphids, regardless of acquisition probe duration and depth.

These observations further strengthen the hypothesis that nonpersistent virus transmission is an intracellular epidermal event, and that aphids and leafhoppers carry noncirculative viruses in their foreguts and transmit them by an ingestion-egestion mechanism.[50,82,85] A stylet-borne mechanism of nonpersistent virus transmission, *sensu stricto*, seems less likely, but as yet cannot be completely ruled out.[67,82] However, a recent study of the rate of loss of inoculativity of maize dwarf mosaic virus (nonpersistent noncirculative) by the aphid *S. graminum* after different AAFPs suggests that the mode of transmission is mono- rather than duomechanistic.[86] Cell entry and sap ingestion provide for a possible continuum ranging from simply stylet-tip contamination to any number of simple to complex variations on the ingestion-egestion theme involving differences among virus-vector-plant combinations in the number, type, quality, location, and functioning of internal virus retention sites.

## 2. *Semipersistent, Noncirculative Transmission*

This type of transmission is also compatible with an ingestion-egestion mechanism of transmission.[2,50,82,85,87] Semipersistence and increases both in the probability of transmission and the duration of inoculativity with increases in the duration of the AAFP suggest that semipersistent viruses can accumulate in the foregut of the vector and resist being quickly dissociated from the vector by egestion or flushing through with virus-free sap ingested from healthy plants. Noncirculative viruses that do not occur in an infectious state in host-plant superficial tissues and are able to persist in the vector can be expected to be transmitted solely in a semipersistent manner. For example, the noncirculative aphid-borne beet yellows virus (BYV) is restricted to phloic tissue and surrounding vascular parenchymatous cells; and, rather predictably, it is semipersistently transmitted.[88,89] Vectors require about 5 min to reach the phloem of a BYV-infected plant. And, once having reached a sieve element, aphids require a minimal, additional 5 min and 22 s of *fluid ingestion* to become viruliferous.[90] Similarly, the inoculation threshold for BYV transmission approximates the time required for an inoculative insect to reach phloic tissues. AAFPs of more than 12 h and inoculation feeding periods of at least 6 h are necessary for optimal transmission.[91,92] Feeding in the phloem would allow for maximal virus passage through, and accumulation in, the foregut of the vector: the longer the feed, the greater the virus accumulation (up to the saturation point for a particular virus-vector-plant combination) and the longer the persistence of vector inoculativity. Beet yellows virus is retained by aphids for up to 3 d, with a half-life of about 8 h.

Semipersistent and possibly some nonpersistent viruses presumably persist in their aphid vectors by selectively adsorbing (either directly or indirectly by a helper agent) to surfaces lining the foregut. The foregut of an aphid is hypodermal in origin and therefore lined with cuticula. Like the stylets, the cuticular lining or intima of the foregut is shed as part of the exuviae during ecdysis, thus explaining the nontransstadiality of noncirculative transmission. There are precedents for the kinds of aphid feeding behaviors and virus retention sites alluded to here and in preceding paragraphs.[50,75,81,82,93] Moreover, similar ingestion-egestion behavior and virus-vector relationships have been demonstrated for semipersistent virus transmissions by leafhoppers and semipersistent-like ones by nematodes.[50,85,94]

Involvement of the foregut and ingestion-egestion behavior in semipersistent noncirculative transmission is supported by electron microscopical observations of presumed virus retention sites in both aphid and leafhopper vectors. Murant and associates[93] observed virus-like particles (VLP) believed to be the semipersistent anthriscus yellows virus (AYV) in a matrix material (M-material) attached to a 15- to 20-μm portion of the ridged intima lining the ventral wall of the posterior region of the pharynx where it passes over the tentorial bar of the aphid vector *Cavariella aegopodii*. Particles in M-material were seen in aphids transmitting AYV or AYV-PYFV complex and in aphids allowed adequate AAFPs on AYV-infected plants. Similar VLP were not seen in insects allowed minimal or subminimal AAFPs, or in ones fed on healthy plants or on a source of PYFV alone. The forces binding the M-material-virion complex to the intima were obviously strong because the VLP were still present in aphids that had fed for 2 h on sucrose solutions after leaving an AYV source. In molting insects, the M-material with associated VLP remained attached to the shed intima of the exuviae.

Childress and Harris[95] observed VLP believed to be maize chlorotic dwarf virus (MCDV) in viruliferous and MCDV-inoculative leafhoppers. The leafhopper *Graminella nigrifrons* (Forbes) (Homoptera: Cicadellidae) transmits MCDV in a semipersistent manner and loses its inoculativity following ecdysis. VLP resembling MCDV in purified preparations and in MCDV-infected plants and measuring 23 to 31 nm in diameter were observed by electron microscopy adhering to the cuticula lining the precibarium, cibarial pump, pharynx, and fore esophagus in viruliferous leafhoppers carrying MCDV.[95] No VLP were observed in the

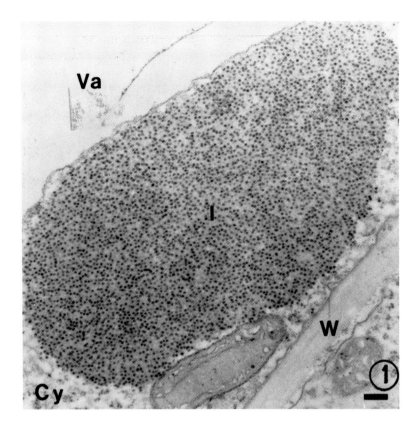

FIGURE 1.    Photograph of a transmission electron micrograph of maize chlorotic dwarf virus (MCDV) in infected maize, *Zea mays* L. A dense accumulation of MCDV virions fills the electron-dense matrix (helper component?) of a viroplasm-like inclusion (I) between the vacuole (Va) and cell wall (W) in the cytoplasm (Cy) of a sieve cell. Magnification bar represents 0.25 μm.

alimentary canal beyond the cardiac valve, in any organ or tissue in the hemocoel, or in the maxillary salivary duct. No VLP were observed associated with the stylet tips or anywhere on the cuticular surfaces of either the mandibular or maxillary stylets. The VLP were not found associated with the cuticular lining of the foregut in a similarly treated nonvector species, *Dalbulus maidis* (Delong and Wolcott), or in vector or nonvector species given either a 48-h AAFP on MCDV-infected source plants followed by a 48-h postacquisition feeding on healthy plants (renders inoculative leafhoppers noninoculative) or solely a 48-h feeding on healthy plants. No differences were observed between MCDV-inoculative and noninoculative *Graminella* given similar access to virus. The VLP occurred in thin layers or small to large clusters embedded in a lightly stained M-material or a densely stained substrate (DSS) which, in turn, apparently were attached by M-material to the cuticula. It is thought that the VLP are those of MCDV and that the M-material, which resembles materials previously reported in association with virus binding sites in the fore alimentary canals of aphid and nematode vectors, is actually MCDV HC and functions in binding virus to retention sites on the cuticula of the vector.[95]

The origin (plant or vector) and composition of the M-material and DSS are not known. The latter resembles the dense staining granular matrix of viral inclusions in host phloem parenchyma cells, whereas the less dense and smoother M-material more closely resembles the matrix (HC?) of viroplasm-like inclusions seen in phloem sieve cells (Figure 1).[96] HC activity has been associated with viroplasms of an aphid-transmissible isolate of CaMV.[49] M-material was so named because of its resemblance to the ''M-material'' believed to bind

AYV to cuticular retention sites in the pharynx of its aphid vector.[93] Virus particles at cuticular retention sites on the feeding apparatus and foregut of vector nematodes are reportedly embedded in a similar "mucus-like layer".[97,99] Recent experiments indicate that HC is required for leafhopper transmission of MCDV;[100] and one possible function of HC is to bind noncirculative viruses to retention sites in the foregut of the vector.[101]

The aforementioned data confirm the hypothesis that leafhoppers carry semipersistent viruses such as MCDV and rice tungro virus in their foreguts.[1,3,50,85,87,95] The foregut anatomy of aphids and leafhoppers is similar;[80,102-104] however, they differ with respect to the mode of action of the precibarial valve. In aphids, the valve is normally closed unless opened by dilator muscles, whereas in leafhoppers it is normally open unless closed by extensor muscles. As in aphids, one possible function of the valve in leafhoppers is to prevent egestion of fluid to the food canal while it is being pumped to the esophagus.[102,103] To what extent the valve might preclude egestion of virus retained in various parts of the foregut remains to be determined. However, observations of the membrane feeding behavior of leafhoppers indicated that the valve and pump can be coordinated to accommodate movement of fluid either into or out of the maxillary food canal.[85]

Overall, the characteristics of semipersistent, noncirculative virus transmission seem compatible with an internal retention site.[1,3,50,82] Loss of vector inoculativity through molting (nontransstadial) is characteristic of noncirculative transmission; shedding of the intima during molting would result in loss of MCDV held by HC to retention sites in the foregut and, hence, vector inoculativity. The semipersistence of MCDV and the fact that both the probability of transmission and the duration of vector inoculativity increase with increases in the duration of the AAFP suggest that MCDV can accumulate in the foregut and resist being quickly dissociated from the vector by egestion or flushing through with virus-free sap ingested from healthy plants.[1,3,50,82,95] Indeed, the basic difference between nonpersistent and semipersistent noncirculative transmission may prove to be the tenacity with which virus can be carried in a transmissible state and titer at retention sites in the foregut of aphid and leafhopper vectors.[83,95]

On the basis of an ingestion-egestion mechanism of noncirculative transmission, one might suspect or predict the existence of "bimodal" viruses: viruses that can be transmitted both nonpersistently and semipersistently by some aphids.[2,50,65] Apparently, such viruses can be acquired from or inoculated to either superficial or deeper tissues of the host plant. In addition, they would be able to accumulate and persist to varying degrees in the foreguts of certain vector species, but not others.

## B. Circulative Transmission

Circulative, aphid-borne viruses encompass at least four taxonomic groups, three of which have been formally described and endorsed by the ICTV. The reader is referred to a recent review for detailed information on the characteristics and virus-vector relationships of nonpropagative and propagative viruses in these groups.[50]

Studies on the fate of PEMV in its pea aphid vector, *Acyrthosiphon pisum* Harris, elucidated the mechanism of circulative transmission of small isometric viruses by aphids, and provided the first direct evidence in support of the hypothetical role of the aphid salivary system in determining virus-vector specificity phenomena.[50,59,105] Virions of PEMV were observed *in situ* in the gut epithelium, connective tissue, and fat body of viruliferous aphids. In these tissues, virions appear to be engulfed in the vacuolar apparatus or phagosome-lysosome system of individual cells. Virions of an aphid-transmissible isolate (but not of its NAT variant) were observed in the salivary gland systems of aphids that were rendered viruliferous either per os by feeding on PEMV-infected source plants or by injecting suspensions of purified virus directly into the hemocoel: a first for an isometric aphid-borne virus.

Except for the occasional observance of a few isolated particles in the basal lamina only of the primary glands, virions were not seen in these paired, bilobed organs. In contrast, virions were highly concentrated in the basal laminae of the paired accessory glands and in the labyrinth of cisternae formed by the extensive infolding and anastomosing of the plasmalemma of accessory gland cells. Virions were not observed in the basal laminae of any other organs in the aphid, indicating a selective role for the basal lamina of the accessory glands in allowing passage of aphid-transmissible, but not NAT, PEMV from the hemocoel. Moreover, when a comparative study was made of the fate of aphid-transmissible PEMV in a highly efficient vs. an inefficient vector biotype, it was found that the accessory glands of the efficient biotype accommodated the movement of far greater numbers of virions through their basal laminae.[151] Aphid transmissibility of PEMV appears to be linked to the presence of a second coat protein which is not present in NAT variants or isolates.[50,58,106]

The foregoing data suggest that reciprocity between recognition sites on virus coat protein and accessory gland membranes is required for passage of virions from the hemocoel through the salivary system to plants. Also, even slight variations in the capsid protein of viruses, virus strains, or virus variants, or in the accessory gland membrane systems of vector species or biotypes, can apparently affect the aphid transmissibility of virus, as well as vector specificity and efficiency, by altering the permeability of the accessory glands to virions.[50]

Additional support for the selective role of aphid salivary glands in virus transmission comes from electron microscopical studies of two isolates of the luteovirus BYDV that are differentially transmitted by certain aphid vector species.[107-111] These studies made it possible to propose hypothetical models for the mechanism by which luteoviruses enter and leave their aphid vectors. Luteovirus ingestion occurs when the aphid feeds on infected phloem cells. Ingested virus moves through the forgut in its passage to the midgut; however, unlike the noncirculatively transmitted viruses, attachment of virus to sites in the foregut are not thought to play a role in circulative transmission. Virus moves from the foregut through the cardiac or esophageal valve into the enlarged stomach region of the anterior midgut and, from there, into the narrower intestinal region of the posterior midgut. From the midgut, virus moves to the hindgut. Once in the hindgut, virus may either continue through the alimentary canal to the rectum and be eliminated with honeydew, or move from the lumen through the hindgut epithelium to the hemocoel. If the latter occurs, the aphid can be said to have acquired potentially transmissible virus.

Acquired virus is presumably carried by hemolymph which moves through the open circulatory system of the aphid, bathing all of the organs and tissues within the hemocoel. Thus, the aphid becomes a reservoir for the virus, aiding virus dispersal and survival in the absence of suitable plant host species. However, to be transmitted by an aphid to another plant, acquired virus must pass through the accessory glands of its potential vector into the salivary duct. Once within the salivary duct, the virus is external to the tissues of the vector and in a position to be inoculated to plants in virus-laden salivary gland secretory products during feeding. Gildow[109] has proposed models for luteovirus acquisition and transmission based on morphological information provided by static ultrastructural images of virus-vector interactions. The models are intended as reference points to build upon, alter, or discard as more information becomes available.[109]

In the hypothetical model of virus acquisition, the first step leading to luteovirus transport through the hindgut is attachment of virus particles to the apical plasmalemma of a hindgut epithelial cell by luteovirus-recognizing receptor molecules in the membrane.[108] Luteovirus-receptor recognition stimulates endocytosis of the virion into a coated pit[112] which either invaginates deep into the cytoplasm or buds off to become a coated vesicle.[113] Many virus-containing coated vesicles could eventually fuse together, forming a larger, noncoated spherical vesicle (receptosome) containing many virions. Larger, virion containing, noncoated, spherical vesicles or receptosomes might be formed by the fusion of many virion-containing coated vesicles. Elongated tubular vesicles might form directly by repeated end-to-end fusion of coated vesicles, or indirectly by a budding off process from the receptosome.[114] Virions in receptosomes that eventually fuse with lysosomes would presumably be degraded. However, particles in tubular vesicles directed toward the basal end of the hindgut epithelial cell would be released to the hemocoel when the vesicle eventually contacted and fused with the basal plasmalemma. Virions thus released would be free to diffuse through the extracellular basal lamina and, suspended in hemolymph, be carried throughout the open circulatory system of the vector. According to this model, the virus itself never comes into direct contact with hindgut cell cytoplasm, but is instead transported through the hindgut continuously packaged by a membrane system.

The hypothetical model describing luteovirus movement through the aphid accessory salivary gland system[107,109] suggests that luteoviruses suspended in hemolymph diffuse through the fibrous accessory gland basal lamina and bind with the basal plasmalemma by virus-specific receptors embedded in the membrane. The process by which virus is then internalized is not clear. Endocytosis by coated pits does not appear to be involved, as evidenced by the lack of virion-containing coated vesicles in the basal cytoplasm of the accessory gland cells. However, frequent localization of virions singly and in linear arrays deep within basal plasmalemma invaginations suggests that virus is endocytosed when these virion-containing invaginations bud off the plasmalemma to form smooth-membraned tubular vesicles which might then fuse together to form tubular vesicles analogous to receptosomes. Virions in tubular vesicles are apparently packaged in a specific manner in coated vesicles, transported to the apical plasmalemma, and exocytosed into the apical canal lumen. As indicated by the lack of luteovirus particle accumulation in lysosomes and secretory vesicles, virion transport in the coated vesicles is apparently very site-specific. Having migrated to the apical canal membrane, coated vesicles fuse with it, becoming coated pits, and release their lumen contents into the canal. At this point, the virions are external to the membrane systems of the vector and in a position to be transmitted (by the chitin-lined salivary duct) in salivary gland secretion products to a plant host during feeding.[109]

## V. APHID-PLANT INTERACTIONS

### A. Aphid Feeding Apparatus and Sensory Transduction

The mouthparts of aphids consist of an upper lip, or labrum, a stylet bundle (paired mandibles and maxillae), and a lower lip, or labium (Figure 2). A longitudinal groove on the inner surface of the labrum serves to position and guide the stylet bundle as it exits the head (Figure 3). The stylet bundle is contained within a deep longitudinal groove along the anterior surface of the segmented, telescopic labium (Figure 2). At rest, the labium is held in a posteroventral direction with its tip extending along the thoracic sternites between the coxae. The labium is protracted during probing and feeding. The tip of the labium is pressed firmly against the plant surface and the labium telescopes to allow the stylet bundle to emerge from the tip to penetrate the plant tissue.

The paired mandibular and maxillary stylets are adapted for penetrating plant tissue and extracting juices. The paired maxillae are grooved and interlocked to form salivary and food canals. Associated with the salivary canal is a complex and well-developed salivary system. Probing is accompanied by salivation, from the formation of a surface salivary flange when the labial tip first contacts the plant surface (Figure 4) to the continuous secretion of a stylet sheath as the stylet bundle penetrates between cells to the phloem.

The sensory transduction system of the feeding apparatus equips aphids to detect and analyze both mechanical and chemical stimuli from potential host plants. Aphids have no maxillary or labial palpi. The sensory function of such palpi apparently has been taken over by other receptors on the labium. Tactile hairs are present on all segments of the labium.

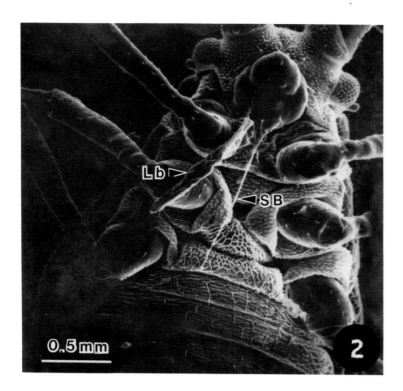

FIGURE 2.    Photograph of a scanning electron micrograph of the underside of the pea aphid, *Acyrthosiphon pisum* Harris. The stylet bundle (SB) has been removed from its normal position within the longitudinal groove on the anterior surface of the quadruply segmented, telescoping lower lip or labium (Lb). (From Harris, K. F., *Adv. Virus Res.*, 28, 113, 1983. With permission.)

Additionally, the distal tip of the fourth segment bears short sensory pegs around the opening through which the stylets emerge (Figure 5). The number and arrangement of the pegs (bilaterally symmetrical but eccentric groups of eight on either side of the labial groove) seem similar for a number of different aphid species.[115] On the basis of their arrangement and fine structure, it is possible to predict that the receptors detect both surface contact (pressure) and surface profile.[116] Additionally, in *A. pisum* a pore is present at the base of each peg (Figure 6). These pores superficially resemble ecdysial pores and are located above the peg socket on the side opposite the point membrane.[117] This positioning would result in maximal stimulation when a peg is deflected toward its pore side,[3] thus confirming the theory that these pegs are mechanoreceptors.[116] Aphids are known to tap or otherwise explore leaf surfaces with the tip of the labium before probing, presumably to locate vein contours and intercellular anticlinal grooves.

   Sensory innervation of the mandibular stylets consists of two groups of sensory neurons, each with a short dendrite extending into and ending in the base of the stylet and another dendrite with a long microtubular process extending to the distal tip of the mandible. On the basis of preliminary scanning electron microscopic examination and in an effort to explain phloem-seeking behavior, I earlier postulated that the dendrites serving the mandibular tips were chemoreceptors. However, more detailed studies using both scanning and transmission electron microscopy have confirmed Wensler's[118] description of the dendrites as mechano-receptors that monitor movement and positioning of the stylets.[3]

   Chemoreception via the feeding apparatus appears to be limited to gustatory chemosensilla in the vicinity of the precibarial valve.[79,80,119] Each sensillum is innervated by neurons with

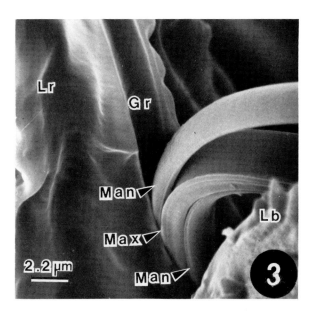

FIGURE 3.    View of the stylet bundle, paired mandibles (Man), and interlocked maxillae (Max) of a green peach aphid, *Myzus persicae* Sulz. The longitudinal groove (Gr) on the posterior surface of the labrum (Lr) serves to guide the stylets as they exit the head capsule between the labrum and the labium (Lb). (From Harris, K. F., *Adv. Virus Res.*, 28, 113, 1983. With permission.)

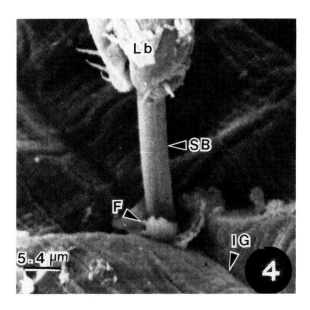

FIGURE 4.    View of the mouthparts of a green peach aphid, *M. persicae*, as it feeds in a turnip leaf, *Brassica rapa* L. Separation of the labium (Lb) from the intercellular groove (IG) on the leaf surface and the attendant exposure of a portion of the stylet bundle (SB) are artifacts introduced during preparation of the specimen for scanning electron microscropy. Note that a portion of the salivary flange (F) remains attached to the labium. (From Harris, K. F., *Adv. Virus Res.*, 28, 113, 1983. With permission.)

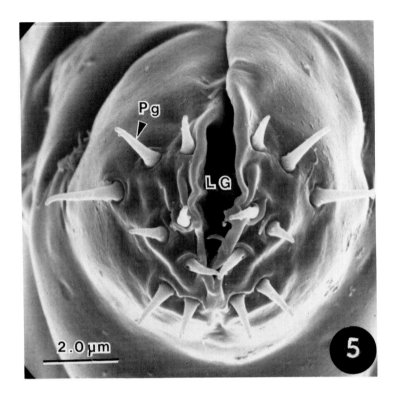

FIGURE 5.    View of the mechanoreceptor pegs (Pg) on the tip of the labium of *A. pisum*. The pegs are arranged in bilaterally symmetrical, but eccentric, groups of eight on either side of the labial groove (LG). (From Harris, K. F., *Adv. Virus Res.*, 28, 113, 1983. With permission.)

dendrites that end at a definite pore to the lumen of the alimentary canal. Consequently the sensory cells of the chemosensilla make direct contact with ingested fluids passing through the canal and equip the aphid to detect chemical stimuli in fluids ingested during cell entry and sap sampling in the epidermis or mesophyll, or during feeding in a phloem sieve element.

## B. Probing and Feeding Behavior and Control-Oriented Research

The sensory transduction system of the feeding apparatus equips aphids to detect and analyze mechanical and chemical stimuli from potential host plants. It is this system and its interaction with the plant that ultimately defines probing and feeding behavior. The role of labial mechanoreceptors in detecting surface contact (pressure) and surface profile (e.g., vein contours and intercellular grooves) has already been mentioned. Each stylet is equipped with protractor and retractor muscles. Following secretion of the salivary flange, stylet penetration of plant tissues is accomplished by a series of alternate protractions of the mandibular stylets followed by protraction of the paired maxillary stylets as a unit. Thus it is the mandibles that lead the way, usually intercellularly, to the goal tissue, usually the phloem. When not relying on host-cell configuration for guidance, aphids may actively control the direction of the stylet bundle by unequal protractions of the mandibular stylets or by unequal compression of the maxillary stylets.

The nature of the guidance system that aphids use in penetrating plant tissues has long been a controversial issue. Most aphid species seem to produce characteristic paths and, in most cases, penetration and phloem localization seem more than fortuitous. Given our present knowledge of aphid sensory systems, however, one can predict that the processes of superficial probing, stylet penetration, epidermal cell entry, and sap sampling, as well as deep

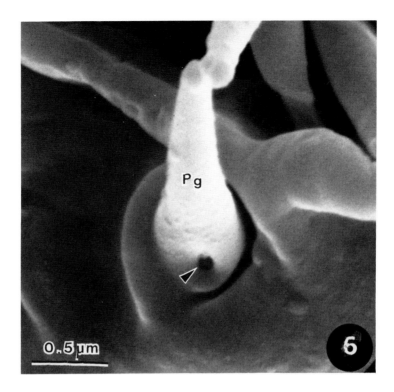

FIGURE 6.    High-magnification view of a mechanoreceptor peg (Pg) on the distal
tip of the labium of *A. pisum*. Note the presence of basal pore (arrow). The positioning
of these pores is such that presumably maximal stimulation results when a peg is
deflected toward its pore side. (From Harris, K. F., *Adv. Virus Res.*, 28, 113, 1983.
With permission.)

probing, cell and tissue localization, feeding site selection, sieve element entry, and prolonged ingestion, are determined by motor responses to various combinations of tactile and
chemical stimuli received by the labial and mandibular mechanoreceptors and the precibarial
gustatory sensilla, respectively.[3,80,116,118] It follows too that such a sensory transduction
system would enable aphids to differentiate tissue and cell types based on mechanical and
chemical stimuli. The primary movement of a mandibular tip as it is protracted past the
stylet bundle is in the lateromedial plan. And the structure of the long sensory dendrites of
the mandibles is suited to the detection of lateromedial movement and its direction.[118]
Resistance to lateromedial movement at the stylet tip would presumably be far greater within
the middle lamella of a compacted thick-walled tissue such as epidermal or vascular tissue
than in loosely arranged, thin-walled mesophyll. Similarly, differing resistances would permit
differentiation between intercellular and intracellular stylet penetration. Chemical guidance
would require occasional cell entry and ingestion of sap to the precibarial chemosensilla
during penetration, such as during sap sampling in the epidermis or in the mesophyll during
deeper, phloem-seeking probes.[3] Final feeding site selection would result from favorable
stimulation of the gustatory chemosensilla during ingestion, as from a sieve tube.

With respect to the plant, it is essential for the plasma membrane to remain intact, if the
sieve tube is to maintain its function. Mechanical injury of the tube, resulting in a sudden
loss of turgor pressure, would elicit a surge of liquid and P-protein fibrils to block the sieve
plate, followed by massive callous deposition. Aphids have bypassed this potential problem
by an ingenious adaptation. A small hole is made in the sieve tube wall, probably by the
mandibles, which taper to a sharp point about 0.04 μm in diameter at the tip.[120-122] The

paired tapered maxillae are then forced in, constricted by the perimeter of the hole, causing them to gape and exposing the food canal which terminates about 1 μm from the stylet tip.[120,122] Stylet penetration into the cell is not accompanied by salivation, but, once feeding is terminated, the aphid seals the stylet entry hole with a saliva plug while withdrawing.[122,123]

Given the foregoing description of deep probing and feeding behavior, it is not difficult to relate aphid feeding behavior with the process of virus acquisition by ingestion of virus-laden sap, with respect to both semipersistent noncirculative and circulative virus transmission.[3,50] Until recently, however, the same association was difficult, if not impossible, to make with respect to nonpersistent noncirculative transmission because of our limited conceptualization of brief probing behavior. Most studies of probing behavior have been "in depth" to determine how aphids penetrate to vascular tissues, the time required to reach these tissues, whether the stylets passed through or between cells on the way, rate of feeding and volumes ingested, nature and composition of salivary secretions, mechanism of cell entry into sieve tubes, etc. However, brief superficial probes in the epidermis, not feeding probes, are conducive to nonpersistent virus transmission.[124-128] We now know that aphids do enter cells and do ingest sap during superficial probing in the epidermis, and cell entry and ingestion seem essential to virus acquisition and transmission.[81,83] Superficial probing is now viewed as a normal preliminary part of aphid host-selection behavior. Favorable stimulation of the precibarial chemosensilla during sap sampling in the epidermis presumably leads to deeper, more time-consuming, phloem-seeking probes and test feeding in a sieve element.

Virus transmission is a by-product of aphid probing and feeding behavior. It therefore would seem worthwhile to research ways of altering that behavior in favor of nontransmission.[129] One approach might be to inhibit the normal functioning of the sensory transduction system of the feeding apparatus. It is known, for example, that oil can be effective in preventing the spread of nonpersistent, semipersistent, and possible even circulative (persistent) viruses by aphids, but how it does so is not known.[130-136] Those adhering to a "stylet-borne" view of noncirculative transmission propose a surface adherence hypothesis in which oil modifies the surface charge of the virion or stylets, or both, thus impeding virus absorption to, or its elution from, the stylets. If this is the mode of action, it seems equally applicable to ingested virus and virus adsorption sites in the foregut of a vector. Aphids are known to ingest oil from oil-treated leaves.[137] Oil also might act on the sensory transduction system, and thus modify the probing and feeding behavior responsible for transmission.[82] It might interfere with normal stylet operation by moving up and between the stylets or by inhibiting or altering mechanoreception via the labial pegs or mandibular stylets. Ingested oil might insulate the precibarial chemosensilla, inhibiting such activities as sap sampling (ingestion-egestion behavior), feeding-site localization, and prolonged feeding, thus preventing acquisition of transmissible titers of virus. The latter effect might also be important in semipersistent noncirculative and circulative transmissions.[134,135]

There have been reports that oil does not affect aphid behavior.[131,137,138] However, more recent data indicate that the mechanism of oil inhibition of transmission relates to aphid-plant interactions (vector sensory transduction) rather than oil-virus or oil-plant interactions. Using an electrical measurement system for more critically monitoring aphid behavior,[139,140] it was possible to detect subtle (but presumably crucial with respect to transmission) behavioral changes in aphids on oil-treated leaves.[141] Aphids allowed brief probes on oil-treated leaves showed a significant increase (compared to control aphids on nontreated leaves) in preprobing times and, more significantly, a drastic reduction in sap sampling activity.[141] Also, cell entry and sap ingestion seem as critical to nonpersistent virus transmission as they are to both semipersistent noncirculative and circulative transmission.[83]

It is well established that aphid probing behavior, transmission efficiency, and host-plant specificity can be a function of the phytochemistry of the aphid-host, virus-source, or test plant used in a given transmission system. Aphids can discern a wide range of chemicals presented to them in artifical diets, and a number of naturally occurring substances have been identified as influencing aphid probing and feeding behavior.[142-147] These and other substances could be screened for their effects on probing and feeding behavior using membrane feeding systems, electronic monitoring systems, or both.[75,85,139,140,148] Those showing behavior-altering properties could then be bioassayed for their effect, if any, on virus transmission. Gibson et al.[149] recently demonstrated that certain aphid alarm pheromone derivatives can inhibit the acquisition or inoculation, or both, of some nonpersistent and semipersistent noncirculative viruses.

# REFERENCES

1. **Harris, K. F.,** Arthropod and nematode vectors of plant viruses, *Annu. Rev. Phytopathol.,* 19, 391, 1981.
2. **Harris, K. F.,** Aphid-borne viruses: ecological and environmental aspects, in *Viruses and Environment,* Kurstak, E. and Maramorosch, K., Eds., Academic Press, New York, 1978, 311.
3. **Harris, K. F.,** Sternorrhynchous vectors of plant viruses: virus-vector interactions and transmission mechanisms, *Adv. Virus Res.,* 28, 113, 1983.
4. **Harris, K. F. and Maramorosch, K.,** *Aphids as Virus Vectors,* Academic Press, New York, 1977, 559 pp.
5. **Harris, K. F. and Maramorosch, K.,** *Vectors of Plant Pathogens,* Academic Press, New York, 1980, 467 pp.
6. **Harris, K. F. and Maramorosch, K.,** *Pathogens, Vectors, and Plant Diseases. Approaches to Control,* Academic Press, New York, 1982, 310 pp.
7. **Maramorosch, K. and Harris, K. F.,** *Leafhopper Vectors and Plant Disease Agents,* Academic Press, New York, 1979, 654 pp.
8. **Maramorosch, K. and Harris, K. F.,** *Plant Diseases and Vectors, Ecology and Epidemiology,* Academic Press, New York, 1981, 368 pp.
9. **Kennedy, J. S., Day, M. F., and Eastop, V. F.,** *A Conspectus of Aphids as Vectors of Plant Viruses,* Commonwealth Institute of Entomology, London, 1971, 114 pp.
10. **Eastop, V. F.,** Worldwide importance of aphids as virus vectors, in *Aphids as Virus Vectors,* Harris, K. F. and Maramorosch, K., Eds., Academic Press, New York, 1977, 3.
11. **Shepherd, R. J.,** Intrinsic properties and taxonomy of aphid-borne viruses, in *Aphids as Virus Vectors,* Harris, K. F. and Maramorosch, K., Eds, Academic Press, New York, 1977, 121.
12. **Matthews, R. E. F.,** *A Critical Appraisal of Viral Taxonomy,* CRC Press, Boca Raton, FL, 1983, 264 pp.
13. **Pirone, T. P.,** Accessory factors in nonpersistent virus transmission, in *Aphids as Virus Vectors,* Harris, K. F. and Maramorosch, K., Eds., Academic Press, New York, 1977, 221.
14. **Lim, W. L. and Hagedorn, D. J.,** Bimodal transmission of plant viruses, in *Aphids as Virus Vectors,* Harris, K. F. and Maramorosch, K., Eds., Academic Press, New York, 1977, 237.
15. **Lung, M. C. Y. and Pirone, T. P.,** Studies on the reason for differential transmissibility of cauliflower mosaic virus isolates by aphids, *Phytopathology,* 63, 910, 1973.
16. **Brierley, P. and Smith, E. F.,** Some vectors, hosts, and properties of dahlia mosaic virus, *Plant Dis. Rep.,* 34, 363, 1950.
17. **Elnager, S. and Murant, A. F.,** Relations of the semi-persistent viruses, parsnip yellow fleck and anthriscus yellows, with their vector, *Cavariella aegopodii, Ann. Appl. Biol.,* 84, 153, 1976.
18. **Elnager, S. and Murant, A. F.,** The role of the helper virus, anthriscus yellows in the transmission of parsnip yellow fleck virus by the aphid *Cavariella aegopodii, Ann. Appl. Biol.,* 84, 169, 1976.
19. **Govier, D. A. and Kassanis, B.,** Evidence that a component other than the virus particle is needed for aphid transmission of potato virus Y, *Virology,* 57, 285, 1974.
20. **Govier, D. A. and Kassanis, B.,** A virus-induced component of plant sap needed when aphids acquire potato virus Y from purified preparations, *Virology,* 61, 420, 1974.
21. **Govier, D. A., Kassanis, B., and Pirone, T. P.,** Partial purification and characterization of the potato virus Y helper component, *Virology,* 78, 306, 1977.
22. **Hellman, G., Thornbury, D., Hiebert, E., Shaw, J., Pirone, T., and Rhoads, R.,** Cell-free translation of tobacco vein mottling virus RNA. II. Immunoprecipitation of products to antisera to cylindrical inclusion, nuclear inclusion, and helper component proteins, *Virology,* 124, 434, 1983.

23. **Pirone, T. P.,** Efficiency and selectivity of the helper-component-mediated aphid transmission of purified potyviruses, *Phytopathology,* 71, 922, 1981.
24. **Sako, N. M. and Ogata, K.,** Different helper factors associated with aphid transmission of some potyviruses, *Virology,* 112, 762, 1981.
25. **Sako, N. and Ogata, K.,** A helper factor essential for aphid transmissibility of turnip mosaic virus, *Ann. Phytopathol. Soc. Jpn.,* 47, 68, 1981.
26. **Simons, J. M.,** Aphid transmission of a nonaphid-transmissible strain of tobacco etch virus, *Phytopathology,* 66, 652, 1976.
27. **Thornbury, D. W. and Pirone, T. P.,** Helper components of two potyviruses are serologically distinct, *Virology,* 125, 487, 1983.
28. **Hiebert, E., Thornbury, D. W., and Pirone, T. P.,** Immunoprecipitation analysis of potyviral in vitro translation products using antisera to helper component of tobacco vein mottling virus and potato virus Y, *Virology,* 135, 1, 1984.
29. **Thornbury, D. W., Hellmann, G. M., Rhoads, R. E., and Pirone, T. P.,** Purification and characterization of potyvirus helper component, *Virology,* 144, 260, 1985.
30. **Hellmann, G. M., Shaw, J. G., and Rhoads, R. E.,** On the origin of the helper component of tobacco vein mottling virus: translational initiation near the 5' terminus of the viral RNA and termination by UAG codons, *Virology,* 143, 23, 1985.
31. **De Mejia, M. V. G., Hiebert, E., Purcifull, D. E., Thornbury, D. W., and Pirone, T. P.,** Identification of potyviral amorphous inclusion protein as a non-structural virus-specific protein related to helper component, *Virology,* 142, 34, 1985.
32. **De Mejia, M. V. G., Hiebert, E., and Purcifull, D. E.,** Isolation and partial characterization of the amorphous cytoplasmic inclusions associated with infections caused by two potyviruses, *Virology,* 142, 24, 1985.
33. **Lecoq, H. and Pitrat, M.,** Specificity of the helper-component mediated aphid transmission of three potyviruses infecting muskmelon, *Phytopathology,* 75, 890, 1985.
34. **Hibino, H.,** Transmission of two rice-tungro-associated viruses and rice waika virus from doubly and singly infected source plants by leafhopper vectors, *Plant Dis.,* 67, 774, 1983.
35. **Hibino, H.,** Relations of rice tungro bacilliform and rice tungro spherical viruses with their vector *Nephotettix virescens, Ann. Phytopathol. Soc. Jpn.,* 49, 545, 1983.
36. **Hibino, H., Saleh, N., and Roechan, M.,** Transmission of two kinds of rice tungro-associated viruses by insect vectors, *Phytopathology,* 69, 1266, 1979.
37. **Cabauatan, P. Q. and Hibino, H.,** Transmission of rice tungro bacilliform and spherical viruses by *Nephotettix virescens* Distant, *Philipp. Phytopathol.,* 21, 103, 1985.
38. **Lopez-Abella, D., Pirone, T. P., Mernaugh, R. E., and Johnson, M. C.,** Effect of fixation and helper component on the detection of potato virus Y in alimentary tract extracts of *Myzus persicae, Phytopathology,* 71, 807, 1981.
39. **Raccah, B. and Pirone, T. P.,** Characteristics of and factors affecting helper-component-mediated aphid transmission of a potyvirus, *Phytopathology,* 74, 305, 1984.
40. **Pirone, T. P. and Thornbury, D. W.,** Role of virion and helper component in regulating aphid transmission of tobacco etch virus, *Phytopathology,* 73, 872, 1983.
41. **Pirone, T. P. and Thornbury, D. W.,** The involvement of a helper component in the nonpersistent transmission of plant viruses by aphids, *Microbiol. Sci.,* 1, 191, 1984.
42. **Berger, P. H. and Pirone, T. P.,** The effect of helper component on the uptake and localization of potyviruses in *Myzus persicae, Virology,* 153, 256, 1986.
43. **Lung, M. C. Y. and Pirone, T. P.,** Acquisition factor required for aphid transmission of purified cauliflower mosaic virus, *Virology,* 60, 260, 1974.
44. **Markham, P. G. and Hull, R.,** Cauliflower mosaic virus aphid transmission facilitated by transmission factors from other caulimoviruses, *J. Gen. Virol.,* 66, 921, 1985.
45. **Armour, S. L., Melcher, U., Pirone, T. P., Lyttle, D. J., and Essenberg, R. C.,** Helper component for aphid transmission encoded by region II of cauliflower mosaic virus DNA, *Virology,* 129, 25, 1983.
46. **Givord, L., Xiong, C., Giband, M., Koenig, I., Hohn, T., Lebeurier, G., and Hirth, L.,** A second cauliflower mosaic virus gene product influences the structure of the viral inclusions body, *EMBO J.,* 3, 1423, 1984.
47. **Woolston, C. J., Covey, S. N., Penswick, J. R., and Davies, J. W.,** Aphid transmission and a polypeptide are specified by a defined region of the cauliflower mosaic virus genome, *Gene,* 23, 15, 1983.
48. **Woolston, C. J., Czaplewski, L. G., Markham, P. G., Goad, A. S., Hull, R., and Davies, J. W.,** Location and sequence of a region of cauliflower mosaic virus gene 2 responsible for aphid transmissibility, *Virology,* 160, 246, 1987.
49. **Rodriguez, D., Lopez-Abella, D., and Diaz-Ruiz, J. R.,** Viroplasms of an aphid-transmissible isolate of cauliflower mosaic virus contain helper component activity, *J. Gen. Virol.,* 68, 2063, 1987.

50. **Harris, K. F.,** Leafhoppers and aphids as biological vectors: vector-virus relationships, in *Leafhopper Vectors and Plant Disease Agents,* Maramorosch, K. and Harris, K. F., Eds., Academic Press, New York, 1979, 217.

51. **Martelli, G. P. and Russo, M.,** Rhabdoviruses of plants, in *The Atlas of Insect and Plant Viruses,* Maramorosch, K., Ed., Academic Press, New York, 1977, 181.

52. **Allison, R. R., Sorenson, J. C., Kelley, M., Armstrong, F. B., and Dougherty, W. G.,** Sequence determination of the capsid protein gene and flanking regions of tobacco etch virus: evidence for the synthesis and processing of a polyprotein in potyvirus genome expression, *Proc. Natl. Acad. Sci. U.S.A.,* 82, 3969, 1985.

53. **Allison, R. F., Dougherty, W. G., Parks, T. W., Willis, L., Johnston, R. E., Kelley, M., and Armstrong, F. B.,** Biochemical analysis of the capsid protein gene and capsid protein of tobacco etch virus: N-terminal amino acids are located on the virion's surface, *Virology,* 147, 309, 1985.

54. **Massalski, P. R. and Harrison, B. D.,** Properties of monoclonal antibodies to potato leafroll luteovirus and their use to distinguish virus isolates differing in aphid transmissibility, *J. Gen. Virol.,* 68, 1813, 1987.

55. **Balazs, E., Guilley, H., Jonard, G., and Richards, K.,** Nucleotide sequence of DNA from an altered-virulence isolate D/H of the cauliflower mosaic virus, *Gene,* 19, 239, 1982.

56. **Frank, A., Guilley, H., Jonard, G., Richards, K., and Hirth, L.,** Nucleotide sequence of cauliflower mosaic virus DNA, *Cell,* 21, 285, 1980.

57. **Gardner, R. C., Howarth, A. J., Hahn, P., Brown-Luedi, M., Shepherd, R. J., and Messing, J.,** The complete nucleotide sequence of an infectious clone of cauliflower mosaic virus by M 13 and mp[7] shot-gun cloning, *Nucleic Acids Res.,* 9, 2871, 1981.

58. **Adam, G., Sander, E., and Shepherd, R. J.,** Structural differences between pea enation mosaic virus strains affecting transmissibility by *Acyrthosiphon pisum* (Harris), *Virology,* 92, 1, 1979.

59. **Harris, K. F., Bath, J. E., Thottappilly, G., and Hooper, G. R.,** Fate of pea enation mosaic virus in PEMV-injected pea aphids, *Virology,* 65, 148, 1975.

60. **Eastop, V.,** Sternorrhyncha as angiosperm toxonomists, *Symb. Bot. Ups.* XXII, 4, 120, 1979.

61. **Jayasena, K. W., Randles, J. W., and Barnett, O. W.,** Synthesis of a complementary DNA probe specific for detecting subterranean clover red leaf virus in plants and aphids, *J. Gen. Virol.,* 65, 109, 1984.

62. **Tamada, T. and Harrison, B. D.,** Quantitative studies on the uptake and retention of potato leafroll virus by aphids in laboratory and field conditions, *Ann. Appl. Biol.,* 98, 261, 1981.

63. **Morimoto, K. M., Ramsdell, D. C., Gillett, J. M., and Chaney, W. G.,** Acquisition and transmission of blueberry shoestring virus by its aphid vector *Illinoia pepperi, Phytopathology,* 75, 709, 1985.

64. **Taylor, C. E. and Robertson, W. M.,** Electron microscopy evidence for the association of tobacco severe etch virus with the maxillae in *Myzus persicae* (Sulz.), *Phytopathol. Z.,* 80, 257, 1974.

65. **Lim, W. L. and Hagedorn, D. J.,** Bimodal transnmission of plant viruses, in *Aphids as Virus Vectors,* Harris, K. F. and Maramorosch, K., Eds., Academic Press, New York, 1977, 237.

66. **Lim, W. L., de Zoeten, G. A., and Hagedorn, D. J.,** Scanning electron-microscopic evidence for attachment of a nonpersistently transmitted virus to its vector's stylets, *Virology,* 79, 121, 1977.

67. **Pirone, T. P. and Harris, K. F.,** Nonpersistent transmission of plant viruses by aphids, *Annu. Rev. Phytopathol.,* 15, 55, 1977.

68. **Evans, I. R. and Zettler, F. W.,** Aphid and mechanical transmission properties of bean yellow mosaic virus isolates, *Phytopathology,* 60, 1170, 1970.

69. **Gonzalez, L. C. and Hagedorn, D. J.,** The transmission of pea seed-borne mosaic virus by three aphid species, *Phytopathology,* 61, 825, 1971.

70. **Sako, N.,** Loss of aphid-transmissibility of turnip mosaic virus, *Phytopathology,* 70, 647, 1980.

71. **Koike, H.,** Loss of aphid transmissibility in an isolate of sugarcane mosaic virus strain H, *Plant Dis. Rep.,* 63, 373, 1979.

72. **Koike, H.,** Loss of aphid transmissibility in an isolate of sugarcane mosaic virus strain H, *Sugarcane Pathol. Newsl.,* 22, 19, 1979.

73. **Hashiba, T. and Misawa, T.,** Studies on the mechanism of aphid transmission of stylet-borne virus. VI. Effect of the saliva of the aphid, *Tohoku J. Agric. Res.,* 21, 73, 1970.

74. **Nishi, Y.,** Inhibition of viruses by vector saliva, in *Viruses, Vectors, and Vegetation,* Maramorosch, K., Ed., Wiley-Interscience, New York, 1969, 579.

75. **Harris, K. F. and Bath, J. E.,** Regurgitation by *Myzus persicae* during membrane feeding: its likely function in transmission of nonpersistent plant virus, *Ann. Entomol. Soc. Am.,* 66, 793, 1973.

76. **Hashiba, T. and Misawa, T.,** Studies on the mechanism of aphid transmission of stylet-borne virus. III. On the adherence of the virus to the stylet, *Tohoku J. Agric. Res.,* 20, 159, 1969.

77. **Berger, P. H., Zeyen, R. J., and Groth, J. V.,** Aphid retention of maize dwarf mosaic virus (potyvirus): epidemiological implications, *Ann. Appl. Biol.,* 111, 337, 1987.

78. **Zeyen, R. J., Stromberg, E. L., and Kuehnast, E. L.,** Long-range aphid transport hypothesis for maize dwarf mosaic virus: history and distribution in Minnesota, USA, *Ann. Appl. Biol.,* 111, 325, 1987.

79. **Wensler, R. J. and Filshie, B. K.,** Gustatory sense organs in the food canal of aphids, *J. Morphol.,* 129, 473, 1969.

80. **McLean, D. L. and Kinsey, M. G.,** The precibarial valve and its role in the feeding behavior of the pea aphid, *Acyrthosiphon pisum, Bull. Entomol. Soc. Am.,* 30, 26, 1984.

81. **Garrett, R. G.,** Non-persistent aphid-borne viruses, in *Viruses and Invertebrates,* Gibbs, A. J., Ed., North-Holland, Amsterdam, 1973, 476.

82. **Harris, K. F.,** An ingestion-egestion hypothesis of noncirculative virus transmission, in *Aphids as Virus Vectors,* Harris, K. F. and Maramorosch, K., Eds., Academic Press, New York, 1977, 165.

83. **Lopez-Abella, D., Bradley, R. H. E., and Harris, K. F.,** Correlation between stylet paths made during superficial probing and the ability of aphids to transmit nonpersistent viruses, in *Advances in Disease Vector Research,* Vol. 5, Harris, K. F., Ed., Springer-Verlag, New York, 1988, 251.

84. **Harrison, B. D. and Murant, A. F.,** Involvement of virus-coded proteins in transmission of plant viruses of vectors, in *Vectors in Virus Biology,* Mayo, M. A. and Harrap, K. A., Eds., Academic Press, New York, 1984, 1.

85. **Harris, K. F., Treur, B., Tsai, J., and Toler, R.,** Observations of leafhopper ingestion-egestion behavior: its likely role in the transmission of noncirculative viruses and other plant pathogens, *J. Econ. Entomol.,* 74, 446, 1981.

86. **Berger, P. H., Harris, K. F., and Toler, R. W.,** Rate of loss of infectivity of maize dwarf mosaic virus by *Schizaphis graminum* Rondani after different acquisition access periods, *J. Phytopathology,* 125, 336, 1989.

87. **Harris, K. F.,** Auchenorrhynchous vectors of plant viruses: virus-vector interactions and transmission mechanisms, in *Proc. 1st Int. Workshop on Leafhoppers and Planthoppers of Economic Importance,* Knight, W. J., Pant, N. C., Robertson, T. S., and Wilson, M. R., Eds., Commonwealth Institute of Entomology, London, 1983, 405.

88. **Esau, K. and Hoefert, L. L.,** Cytology of beet yellows virus infection in *Tetragonia.* II. Vascular elements in infected leaf, *Protoplasma,* 72, 459, 1971.

89. **Esau, K. and Hoefert, L. L.,** Cytology of beet yellows virus infection in *Tetragonia.* III. Conformation of virus in cells, *Protoplasma,* 73, 51, 1971.

90. **Chang, V. V.-S.,** Intraspecific Variation in the Ability of the Green Peach Aphid, *Myzus persicae* (Sulz.) to Transmit Sugar Beet Yellows Virus, Ph.D. thesis, University of California, Davis, 1968.

91. **Watson, M. A.,** The transmission of beet mosaic and beet yellows viruses by aphids: a comparative study of a non-persistent virus and a persistent virus having host plants and vectors in common, in *Proc. R. Soc. London. Ser. B.* 133, 200, 1946.

92. **Russell, G. E.,** Beet yellow virus, in *C. M. I/A. A. B. Descriptions of Plant Viruses,* Gibbs, A. J., Harrison, B. D., and Murant, A. F., Eds., Commonwealth Mycological Institute and Association of Applied Biology, Surrey, England, 1970, No. 13, 3 pp.

93. **Murant, A. F., Roberts, I. M., and Elnager, S.,** Association of virus-like particles with the foregut of the aphid *Cavariella aegopodii* transmitting the semi-persistent viruses anthriscus yellows and parsnip yellow fleck, *J. Gen. Virol.,* 31, 47, 1976.

94. **Taylor, C. E.,** Nematodes, in *Vectors of Plant Pathogens,* Harris, K. F. and Maramorosch, K., Eds., Academic Press, New York, 1980, 375.

95. **Childress, S. A. and Harris, K. F.,** Localization of viruslike particles in the foreguts of viruliferous *Graminella nigrifrons* leafhoppers carrying the semipersistent maize chlorotic dwarf virus, *J. Gen. Virol.,* 70, 247, 1989.

96. **Harris, K. F. and Childress, S. A.,** Cytology of maize chlorotic dwarf virus infection in corn, *Int. J. Trop. Plant Dis.,* 1, 35, 140, 1983.

97. **Taylor, C. E. and Robertson, W. M.,** The localization of raspberry ringspot and tomato blackring viruses in the nematode vector, *Longidorus elongatus* (de Man), *Ann. Appl. Biol.,* 64, 233, 1969.

98. **Taylor, C. E. and Robertson, W. M.,** Sites of virus retention in the alimentary tract of the nematode vectors *Xiphinema diversicaudatum* (Micol.) and *X. index* (Thorne and Allen), *Ann. Appl. Biol.,* 66, 375, 1970.

99. **Taylor, C. E. and Robertson, W. M.,** The location of tobacco rattle virus in the nematode vector, *Trichodorus pachydermus* Seinhorst, *J. Gen. Virol.,* 6, 179, 1970.

100. **Hunt, R. E., Nault, L. R., and Gingery, R. E.,** Evidence for infectivity of maize chlorotic dwarf virus and a helper component for its leafhopper transmission, *Phytopathology,* 78, 499, 1988.

101. **Berger, P. H. and Pirone, T. P.,** The effect of helper component on the uptake and localization of potyviruses in *Myzus persicae, Virology,* 153, 256, 1986.

102. **Backus, E. A. and McLean, D. L.,** The sensory systems and feeding behavior of leafhoppers. I. The aster leafhopper, *Macrosteles fascifrons* Stal (Homoptera: Cicadellidae), *J. Morphol.,* 127, 369, 1982.

103. **Backus, E. A. and McLean, D. L.,** The sensory systems and feeding behavior of leafhoppers. II. A comparison of the sensillar morphologies of several species (Homoptera: Cicadellidae), *J. Morphol.,* 176, 3, 1983.

104. **Backus, E. A. and McLean, D. L.,** Behavioral evidence that the precibarial sensilla of leafhoppers are chemosensory and function in host discrimination, *Entomol. Exp. Appl.,* 37, 219, 1985.

105. **Harris, K. F. and Bath, J. E.**, The fate of pea enation mosaic virus in its pea aphid vector, *Acyrthosiphon pisum* (Harris), *Virology*, 50, 778, 1972.
106. **Hull, R.**, Particle differences related to aphid-transmissibility of a plant virus, *J. Gen. Virol.*, 34, 183, 1977.
107. **Gildow, F. E.**, Coated-vesicle transport of luteoviruses through salivary glands of *Myzus persicae*, *Phytopathology*, 72, 1289, 1982.
108. **Gildow, F. E.** Transcellular transport of barley yellow dwarf virus into the hemocoel of the aphid vector, *Rhopalosiphum padi*, *Phytopathology*, 75, 292, 1985.
109. **Gildow, F. E.**, Virus-membrane interactions involved in circulative transmission of luteoviruses by aphids, in *Current Topics in Vector Research*, Vol. 4, Harris, K. F., Ed., Springer-Verlag, New York, 1987, 93.
110. **Gildow, F. E. and Rochow, W. F.**, Importance of capsid integrity for interference between two isolates of barley yellow dwarf virus in an aphid, *Phytopathology*, 70, 1013, 1980.
111. **Gildow, F. E. and Rochow, W. F.**, Role of accessory salivary glands in aphid transmission of barley yellow dwarf virus, *Virology*, 104, 97, 1980.
112. **Roth, T. F. and Porter, K. R.**, Yolk protein uptake in the oocyte of the mosquito *Aldes aegypti* L., *J. Cell Biol.*, 20, 313, 1964.
113. **Bowers, B.**, Coated vesicles in the pericardial cells of the aphid (*Myzus persicae* Sulz.)., *Protoplasma*, 59, 351, 1981.
114. **Pastan, I. and Willingham, M. C.**, Journey to the center of the cell: role of the receptosome, *Science*, 214, 504, 1981.
115. **Harris, K. F. and Childress, S. A.**, Preliminary observations on the morphology of apical sensory pegs on aphid labia, *Phytopathology*, 71 (Abstr.), 879, 1981.
116. **Wensler, R. J.**, The fine structure of distal receptors on the labium of the aphid, *Brevicoryne brassicae* L. (Homoptera), *Cell Tissue Res.*, 181, 409, 1977.
117. **French, A. S. and Sanders, E. J.**, The mechanism of sensory transduction in the sensilla of the trochanteral hair plate of the cockroach, *Periplaneta americana*, *Cell Tissue Res.*, 198, 159, 1979.
118. **Wensler, R. J. D.**, Sensory innervation monitoring movement and position in the mandibular stylets of the aphid, *Brevicoryne brassicae*, *J. Morphol.*, 143, 349, 1974.
119. **Ponsen, M. B.**, The site of potato leafroll virus multiplication in its vector, *Myzus persicae*. An anatomical study, *Meded. Landbouwhogesch. Wageningen*, 72, 1, 1972.
120. **Forbes, A. R.**, The mouthparts and feeding mechanism of aphids, in *Aphids as Virus Vectors*, Harris, K. F. and Maramorosch, K., Eds., Academic Press, New York, 1977, 83.
121. **Kennedy, J. S. and Fosbrooke, I. H. M.**, The plant in the life of an aphid, in *Insect/Plant Relationships*, van Emden, H. F., Ed., John Wiley & Sons, New York, 1973, 129.
122. **Pollard, D. G.**, Aphid penetration of plant tissues, in *Aphids as Virus Vectors*, Harris, K. F. and Maramorosch, K., Eds., Academic Press, New York, 1977, 105.
123. **Kunkel, H.**, Membrane feeding systems in aphid research, in *Aphids as Virus Vectors*, Harris, K. F. and Maramorosch, K., Eds., Academic Press, New York, 1977, 311.
124. **Bradley, R. H. E.**, Studies on the mechanism of transmission of potato virus Y by the green peach aphid, *Myzus persicae* (Sulz.) (Homoptera: Aphididae), *Can. J. Zool.* 32, 64, 1954.
125. **Bradley, R. H. E.**, Aphid transmission of stylet-borne viruses, in *Plant Virology*, Corbett, M. K. and Sisler, H. D., Eds., University of Florida Press, Gainesville, 1964, 148.
126. **Bradley, R. H. E. and Rideout, D. W.**, Comparative transmission of potato virus Y by four aphid species that infest potato, *Can. J. Zool.*, 31, 333, 1953.
127. **Misawa, T. and Hashiba, T.**, Studies on the mechanism of aphid transmission of stylet-borne viruses (1), *Tohoku J. Agric. Res.*, 18, 87, 1967.
128. **Sylvester, E. S.**, Aphid transmission of nonpersistant plant viruses with special reference to the *Brassica nigra* virus, *Hilgardia*, 23, 53, 1954.
129. **Harris, K. F.**, Aphids, leafhoppers, and planthoppers, in *Vectors of Plant Pathogens*, Harris, K. F. and Maramorosch, K., Eds., Academic Press, New York, 1980, 1.
130. **Bradley, R. H. E., Wade, C. V., and Wood, F. A.**, Aphid transmission of potato virus Y inhibited by oils, *Virology*, 18, 327, 1962.
131. **Bradley, R. H. E.**, Some ways in which a paraffin oil impedes aphid transmission of potato virus Y, *Can. J. Microbiol.*, 9, 369, 1963.
132. **Loebenstein, G., Alper, M., and Deutsch, M.**, Preventing aphid-spread cucumber mosaic virus with oils, *Phytopathology*, 54, 960, 1964.
133. **Vanderveken, J.**, Inhibition de la transmission du virus de la mosaique de la luzerne par *Myzus persicae* à l'aide d'huile, *Parasitica*, 28, 39, 1972.
134. **Vanderveken, J. J.**, Oils and other inhibitors of nonpersistent virus transmission, in *Aphids as Virus Vectors*, Harris, K. F. and Maramorosch, K., Eds., Academic Press, New York, 1977, 435.
135. **Simons, J. N. and Zitter, T. A.**, Use of oils to control aphid-borne viruses, *Plant Dis.*, 64, 542, 1980.

136. **Simons, J. N.**, Use of oil sprays and reflective surfaces for control of insect-transmitted viruses, in *Pathogens, Vectors and Plant Diseases. Approaches to Control*, Harris, K. F. and Maramorosch, K., Eds., Academic Press, New York, 1982, 71.

137. **Vanderveken, J.**, Recherche du mécanisme de l'inhibition de la transmission aphidienne des phytovirus par des substances huileuses, *Parasitica*, 29, 1, 1973.

138. **Hein, A.**, Untersuchungen zur Wirkung von Ölen bei der Virusübertragung durch Blattläuse. II. Wirkung von Öl auf *Myzus persicae* Sulz., *Phytophatol. Z.*, 75, 241, 1972.

139. **McLean, D. L. and Weigt, W. R., Jr.**, An electronic measuring system to record aphid salivation and ingestion, *Ann. Entomol. Soc. Am.*, 61, 180, 1968.

140. **McLean, D. L.**, An electrical measurement system for studying aphid probing behavior, in *Aphids as Virus Vectors*, Harris, K. F. and Maramorosch, K., Eds., Academic Press, New York, 1977, 277.

141. **Simons, J. N., McLean, D. L., and Kinsey, M. G.**, Effects of mineral oil on probing behavior and transmission of stylet-borne viruses by *Myzus persicae, J. Econ. Entomol.*, 70, 309, 1977.

142. **Wensler, R. J. D.**, Mode of host selection by an aphid, *Nature*, 195, 830, 1962.

143. **Smith, B. D.**, Effects of the plant alkaloid sparteine on the distribution of the aphid *Acyrthosiphon spartii* (Koch), *Nature (London)*, 212, 213, 1966.

144. **Klingauf, F.**, Die wirkung des Glucosids Phlorizin auf das Wirtswahlverhalten von *Rhopalosiphum insertum* (Walk.) und *Aphis pomi* DeGeer (Homoptera: Aphididae), *Z. Angew. Entomol.*, 68, 41, 1971.

145. **Klingauf, F., Sengona, C., and Bennewita, H.**, Einfluss von Sinigrin auf die Nahrungsaufnahme polyphager und oligophager Blattlausarten (Aphididae), *Oecologia (Berlin)*, 9, 53, 1972.

146. **Edwards, L. J., Siddall, J. B., Dunham, L. L., Uden, P., and Kislow, C. J.**, Trans-β-farnesene, alarm pheromone of the green peach aphid, *Myzus persicae* (Sulzer), *Nature (London)*, 241, 126, 1973.

147. **Montgomery, M. E. and Arn, H.**, Feeding response of *Aphis pomi, Myzus persicae* and *Amphorphora agathonica* to phlorizin, *J. Insect Physiol*, 20, 413, 1974.

148. **Tarn, T. R. and Adams, J. B.**, Aphid probing and feeding, electronic monitoring, and plant breeding, in *Pathogens, Vectors, and Plant Diseases. Approaches to Control*, Harris, K. F. and Maramorosch, K., Eds., Academic Press, New York, 1982, 221.

149. **Gibson, R. W., Pickett, J. A., Dawson, G. W., Rice, A. D., and Stribley, M. F.**, Effects of aphid alarm pheromone derivatives and related compounds on non- and semi-persistent plant virus transmission by *Myzus persicae, Ann. Appl. Biol.*, 104, 203, 1984.

150. **Toros, S., Schotman, C. Y. L., and Peters, D.**, A new approach to measure the $LP_{50}$ of pea enation mosaic virus in its vector *Acyrthosiphon pisum, Virology*, 90, 235, 1978.

151. **Harris, K. F.**, Unpublished data.

Chapter 8

# VIRUS TRANSMISSION

## C. L. Mandahar

## TABLE OF CONTENTS

# I. INTRODUCTION

Plant viruses are transmitted mechanically and through a wide range of agents like vegetative propagative organs, grafting, seeds and pollen, fungi, and through insects and nematodes. The last two agents are very effective transmitting media. They feed on plants and then move from one plant to another, during which viruses are effectively transmitted to the plants visited. Many viruses depend exclusively or largely on insects and nematodes for their transmission.

Any part of systemically infected plants used for vegetative propagation will inevitably give rise to an infected plant. Systemic virus diseases in this way attain considerable significance in vegetatively propagated plants like fruit trees, potato, rose, raspberry, and sugarcane. Lily symptomless, narcissus mosaic, and tulip breaking viruses are spread through infected bulbs; abutilon mosaic, cassava common mosaic, and sugarcane Fiji viruses spread through infected cuttings; and dahlia mosaic and potato leaf roll viruses are spread through infected tubers.[1]

Grafting plays a significant role in propagation of fruit and ornamental trees, as also the spread of their viruses. A notorious example is the tristeza disease of sweet orange plants. It is still known to ravage plantings throughout the world. A more recent example of natural spread of a virus through grafting is cherry leaf roll virus,[2] which causes in the U.S. the blackline disease of English walnut when propagated on 'Northern California Black' and 'Paradox' walnut rootstocks.[3] Natural root grafts possibly lead to the spread of carnation mosaic virus from diseased to healthy carnations.

## II. MECHANICAL (CONTACT) TRANSMISSION

Many plant viruses are experimentally mechanically transmissible, but only a few of these exploit this mechanism for natural spread in the field. Andean potato latent, cucumber green mottle mosaic,[4] cucumber mosaic, potato viruses S, X, and F, potato spindle tuber viroid, tobacco mosaic (TMV), and tomato mosaic viruses (ToMV), are transmitted in nature by mechanical means. All these viruses are stable and reach high concentrations in infected plants. Rubbing of leaves against each other by foliage movement by wind and/or cultural practices such as cutting, pruning, shearing, etc., and even contaminated hands and clothes spread these viruses in the field.

TMV/ToMV is the most infectious plant virus, so much so that rubbing healthy tomato plants with infected tomato sap diluted with water to 0.2 ppm can still cause infection.[5]

Leaves of tomato plants grown in ToMV-infested soil become contaminated with the virus by splash dissemination during watering by deposition of virus-adsorbed soil.[6] Rubbing of these contaminated leaves causes infection. This virus is also readily transmitted in the field and in the greenhouse by workmen through their clothes, hands, body parts, implements, and machinery. Man, in fact, is the most important vector of this disease. This virus may remain infective for several months on clothes kept in light and for several years on clothes kept in darkness.[5] Potato virus S, in plots containing 10% infected plants initially, spread through contact to about 10% more plants in one season.[7] It is also spread by the cutting knife. Moreover, this virus remains in the infective state under greenhouse conditions for up to 180 h on or in materials that come in contact with infected plants or tubers during cultivation, storage, or processing.[8] The normal means of spread of Andean potato latent virus in potato fields in the Andean highlands is probably through plant-to-plant contact.[9] Brome mosaic virus isolates infecting Gramineae and Commelinaceae in towns in northeastern Arkansas appear to spread by mechanical transmission by mowing.[10] Odontoglossum ringspot and cymbidium mosaic are the two most prevalent viruses infecting cultivated orchids. Both are readily transmitted to healthy plants by cutting tools used during propagation and flower harvesting. Carnation ringspot virus in carnation spreads rapidly by careless handling and by leaf and root contact when management of carnation stocks is neglected. Barley stripe mosaic virus (BSMV) is not contact transmitted from healthy to wild oats, but one of its strains was transmitted by contact from wild oats to Herta and Conquest barley cultivars.[11] It is possible that such a situation may prevail in nature and that passage of this virus through wild oats may evolve a mechanically transmissible isolate.[12]

## A. Mechanical Transmission through Soil (Soilborne Viruses)

Viruses which occur in soil and are spread from one plant to another through soil are the soilborne viruses. They can be spread by mechanical means, by transmission through nematodes, and/or by transmission through fungi. Only the soilborne, mechanically transmissible viruses are discussed in this section. Unequivocal evidence has established that several of these viruses are released into soil from the roots of systemically infected plants. This is true of tomato bushy stunt virus (TBSV) from roots of *Celosia argentata* and *Chenopodium quinoa*,[13] of carnation ringspot virus from the roots of its hosts,[14] and of red clover necrotic mosaic virus from the roots of *Nicotiana clevelandii* and *Trifolium pratense*.[15] Various types of circumstantial evidence suggest that several other viruses are released by plant roots into soil. Sweet clover necrotic mosaic virus is present in the soil around the roots of inoculated or infected plants.[16] The fruit necrosis strain of ToMV was recovered from debris and soil in a field where infected tomato plants had been grown the previous year.[17] The chenopodium necrosis strain of tobacco necrosis virus (TNV) was isolated from leachates of soil in which systemically infected *C. quinoa* plants were grown.[18] Musk melon necrotic spot virus was also isolated from leachates of contaminated soil.[19] TBSV was isolated from leachate of soil in which systemically infected tomato or *C. quinoa* plants were grown.[20] Cucumber soilborne virus occurs in the soil of Beirut.[21]

The healthy plants/seedlings transplanted in virus-infested soil become infected. This presumably occurs by mechanical transmission through wounds and abrasions inflicted on roots of transplanted plants as happens during ToMV infection of tomato, cucumber fruit streak virus infection of cucumber, and red clover necrotic mosaic virus infection of *N. clevelandii* and red clover plants.[15,22] The incidence of ToMV infection increased 10 to 15 times in transplanted vs. nontransplanted tomato plants. All *N. clevelandii* plants grown in sterilized soil became infected with red clover necrotic mosaic virus when the roots were artifically damaged prior to pouring of infective sap into the soil.[22] In contrast, undisturbed or nontransplanted red clover and tomato seedlings grown in red clover necrotic mosaic virus and ToMV-infested soils, respectively, show low, erratic, or no infection at all.[5,22] Healthy *C. quinoa*, tomato, and melon plants became infected with the chenopodium necrosis strain of TNV, TBSV, and melon necrotic spot virus, respectively, when grown in contaminated soil or in soil watered with virus suspension.[18-20] Galinsoga mosaic virus infects roots of *Galinsoga parviflora* possibly through mechanical transmission in soil between roots.[23] TBSV was taken up by *Celosia argentata* plants through their roots from nutrient solutions, agar, and quartz sand.[13] Carnation ringspot virus can also be absorbed by roots of healthy plants from sterile or nonsterile soil.[14] Petunia asteroid mosaic virus, which is serologically related to TBSV, infected petunia plants grown in soil in which virus-infected plants had been grown earlier.[24] It appears in all the above mentioned cases that virus particles released from the roots of infected plants are transported in drainage water (see next section) and subsequently infect host roots which may have been abraded during their growth in soil, or possibly damaged by nematodes or fungi. Cucumber fruit streak virus may also be soilborne and acquired by cucumber roots from soil, but this may or may not be the natural means of its spread.[25]

Successful infection of healthy plants by a virus released earlier into soil from the roots of an infected plant presupposes that such a virus is stable and can persist in soil in the infectious state for short-to-long periods. This has been found to be the situation in all cases investigated. Carnation ringspot virus and TBSV in soil can resist extreme temperatures of 121°C and remain infectious for at least 7 months without access to plants.[13] Most heat-pasteurization procedures fail to decontaminate TMV/ToMV in soils used for potting mixes or ground beds.[6] Plant debris appears to protect viruses against inactivation. ToMV survives for the longest period, as compared to other viruses, in dead and dry tissues or tissue debris. Powdered leaf debris in dry soil maintained its infectivity for 2 years, but lost infectivity within a month in moist soil. The tomato root debris located at a depth of at least 120 cm in fallow soil continued to contain infectious virus for at least 22 months. In wet soil, however, root debris remains infectious for 6 months only.[5] This virus is much more resistant to heat inactivation in dead tissue than in sap or when "free". The efficiency of soil transmission of ToMV[17] and cucumber green mottle mosaic virus[4] is greatly increased when present in decaying plant material rather than in a "free" state. Reduction of infectivity of "free" TMV, amended into soils, correlated positively with the extent of soil dehydration.[6] "Free" TNV was inactivated within 1 d by drying the soil, while it could be detected for at least 130 d in infected roots in drying soil.[26] The soilborne viruses were considered earlier to survive in soil solely by adsorption on colloidal particles. Recent evidence shows that this may not be the case since chenopodium necrosis strains of TNV and TBSV were not found to be adsorbed to soil.[18,20]

## B. Mechanical Transmission through Water

Yarwood[27] was the first to demonstrate the release of TNV and some TMV strains into drainage water from roots of infected plants grown in pots. Later, there were several other such reports (Table 1). It was suggested that these viruses were probably released from injured, decaying, or dead roots.[27] This was confirmed later when it was found that ToMV and cucumber green mottle mosaic virus were released from debris of infected tomato[17,29,30] and infected cucumber[4] plants, respectively. However, some viruses (Table 1) are also released by roots into soil-free nutrient solutions, suggesting that they have presumably been released from undamaged and undecaying, i.e., living, roots.

Release of viruses in drainage water indicated that they may also be present in bigger water bodies such as rivers, canals, and lakes. It was, however, only recently that viruses were shown to be present in considerable amounts in environmental waters[18,20,28] (Table 2). Koenig[28] has reviewed the origin, state of occurrence, characteristics, and epidemiological importance of plant viruses occurring in rivers and lakes.

Tomlinson et al.[31] and Tomlinson and Faithfull[20] established for the first time that sewage is also an important source of virus (TBSV) in rivers. Consumption of TBSV-infected tomato

## Table 1
## VIRUSES RELEASED BY ROOTS IN DRAINAGE WATER AND/OR SOIL-FREE NUTRIENT SOLUTIONS

| Virus | Drainage water | Soil-free nutrient solution[a] |
|---|:---:|:---:|
| Carnation ringspot | + | + |
| Cucumber green mottle mosaic | + | + |
| Cucumber necrosis | + | + |
| Cymbidium ringspot | + | |
| Galinsoga mosaic | + | |
| Petunia asteroid mosaic | + | |
| Red clover necrotic mosaic | + | |
| Southern bean mosaic | + | + |
| Sweet clover necrotic mosaic | + | |
| Tobacco mosaic | + | + |
| Tobacco necrosis | + | |
| Tobacco rattle | + | |
| Tomato bushy stunt | + | + |
| Tomato mosaic | + | + |

[a] Investigations concerning the presence of only some viruses in soil-free nutrient solution (shown by + ) have been conducted; the presence or absence of the remaining viruses in soil-free nutrient solutions has not been investigated.

Based on information reviewed by Koenig.[28]

fruits by man results in passage of intact infectious TBSV virions through the alimentary canal into human feces, then into sewage, and ultimately into rivers by sewage dispersal.[20,31] Even "free" TBSV passes through the human alimentary tract in an infective state.

The presence of viruses in drainage water, rivers, and lakes can have some epidemiological importance only if those viruses can infect plants through their roots. It is recognized now that several plant viruses can infect plants through their roots, when watered with virus suspension, without the aid of any vector. Roberts[32] was probably the first to show this with potato virus X, TMV, and TBSV. Since then several other viruses have been shown to cause infection in this way: carnation mottle, carnation ringspot, cymbidium ringspot, galinsoga mosaic, red clover necrotic mosaic, southern bean mosaic, sowbane mosaic (SMV), ToMV, and TBSV.[28] Roots also take up ToMV and cucumber green mottle mosaic virus from nutrient solutions in nutrient film technique.[33] These viruses are presumably taken up by healthy roots. However, the transmission of TBSV and carnation ringspot virus to roots was greater in sterile sand than in sterile nutrient solution, possibly because of the damage or wounds caused in the former case.[14] Similarly, transmission of red clover necrotic mosaic virus to roots increased in the presence of zoospores of *Olpidium brassicae*.[28]

Consequently, Koenig[28] suggested that long-distance spread (in the absence of any known efficient aerial vectors) of these viruses is ensured by dissemination in drainage water, lakes, streams, and rivers. These viruses may have remained restricted to limited areas in the absence of their capacity to be water transmitted. Man acts as a vector, and directly or indirectly helps in rapid as well as long-distance spread of these viruses through feces by consuming raw vegetables and fruits, through agricultural practices (shearing, pruning, application of solid or liquid manure, use of machinery, presence of plant debris in or on soil, etc.), and through disposal of agricultural and horticultural waste products on dumps and in sewage.

The majority of the viruses occurring in drainage water, streams, lakes, and rivers share the following common characteristics; they are very stable, reach high concentrations in infected plants, are released into soil from infected roots, infect plants through roots without

**Table 2**
**VIRUSES FOUND IN CANALS, RIVERS, AND LAKES**

| Virus | Name of river | Country |
|---|---|---|
| Carnation mottle | Oker | West Germany |
| Cucumber green mottle mosaic | Canals | |
| Cucumber mosaic | Bradano | Italy |
| Satellite tobacco necrosis | Loire | France |
| Tobacco (tomato) mosaic | Danube | Yugoslavia |
| | Sava | Yugoslavia |
| | Lato | Italy |
| | Various rivers | West Germany |
| Tobacco necrosis (*Chenopodium* necrosis strain) | Several rivers and lakes | England |
| | Loire | France |
| Tomato bushy stunt | Several rivers and lakes | England |
| | Amazon | Brazil |
| | Mississippi | U.S. |
| | Rhine | West Germany |
| | Po | Italy |
| | Manjira | Hyderabad (India) |
| | Le Gbangbo | Ivory Coast |
| | Several rivers and lakes | West Germany |
| Unidentified tobamovirus(es) | Rhine | West Germany |
| | Po | Italy |
| | Nile | Egypt |
| | Amazon | Brazil |
| | Bramfontein-Spruit | South Africa |
| Unidentified potexvirus | Sieg | West Germany |
| Unidentified tombusvirus | Neckar | West Germany |
| Unidentified isometric viruses | Several rivers | West Germany |
| | Several rivers | Italy |

Based on information reviewed by Koenig.[28]

the help of a vector, are mechanically transmissible to other plants through contact of aerial parts (leaves), are also transmissible to other plants through soil via roots which may not, but most likely may, have been damaged or injured earlier, generally have a wide host range, and generally have no known efficient aerial vectors.

TBSV was demonstrated to be adsorbed on clay particles.[13] Petunia asteroid mosaic virus was also postulated to be adsorbed on clay particles.[24] Adsorption of viruses on clay or viruses present in roots and other plant debris in soil and in water probably protects them against inactivation by physical or chemical factors. Thus, TBSV adsorbed on clay-containing soils was more resistant to extreme temperatures than nonadsorbed ("free") virus or "free virus" in sand lacking colloids.[34] Petunia asteroid mosaic virus survived in soil containing clay more than twice as long as it survived in sap.[24] This also explains the presence of labile cucumber mosaic virus in rivers since it was sediment protected.[35] In contrast, Tomlinson and Faithfull[20] failed to find any evidence of TBSV adsorption to soils containing clay and organic carbon content. This may have been due to the presence of different adsorbing materials or the prevalence of different environmental conditions like pH, salts, and occurrence of some other materials.[28]

## III. TRANSMISSION BY SEEDS

Transmission of viruses by seeds was earlier considered to be an insignificant factor in spread of virus diseases. The picture over the years has changed. Bennett[36] mentioned 53 viruses as being transmitted by seeds of some 124 plant species. A total of 138 viruses were reported by Mandahar[37] to be transmitted through the seeds of about 170 plants. Seed transmission of plant viruses has been reviewed several times.[36-43] Certain generalizations can now be made about them. Some of the more important of them are as follows:

1. Only those viruses present *in situ* at the time of differentiation of male and female sporogenous cells and before the callose walls are laid down will be able to infect microspores, megaspores, and embryos, and will be seed transmitted. In other words, seed transmissibility of a virus depends upon its capacity to attack the floral meristem during the earliest stage of its differentiation.

2. Seed-transmitted viruses are always carried within the embryo. Viruses may be present in seed coats, nucellus, endosperm, and other seed parts, but are not transmissible if not present within the embryo. TMV is the only exception to this general behavior. However, not all virus-containing embryos would produce virus-infected seeds/plants. Thus, although 17% of the seeds from cucumber green mottle mosaic virus-infected plants contain the virus in embryo, only 3% of the seeds give rise to infected seedlings.

3. A few seed-borne viruses are present in immature seeds (or seeds nearing maturity), but are eliminated from mature, ripened seeds. Such viruses are transmitted to a high degree by immature seeds and are either not transmitted, or transmitted to a very low level by mature seeds.

4. The extent and frequency of seed transmission of viruses vary over the whole range from 0 to 100%. Several plant viruses show 100% seed transmission, but seed transmission of the great majority of seed-borne viruses rarely exceeds 50%.

5. The success and extent of seed transmission depends on both the host species and the virus strain. Thus seed transmissibility of a particular virus (strain) varies greatly in different species or even varieties and in cultivars of a particular host species. On the other hand, seed transmissibility of different strains or isolates of a virus can differ even in the same cultivar.

6. Time of infection of a host by a virus influences its rate of seed transmissibility. The earlier the infection of host plants, the greater the number of seeds transmitting a virus. In general, systemic infection of a host before or shortly after flowering is necessary for seed transmissibility.

7. Virus-infected seeds and/or seeds derived from severely infected plants are small, wrinkled, and have abnormal testa. Such seeds show decreased rate of germination as compared to the healthy normal seeds.

8. Seed-transmitted viruses are mechanically transmissible and attack the parenchymatous tissue extensively. They are therefore largely aphid transmitted in a nonpersistent manner. Phloem-limited viruses are not seed transmitted.

9. Nematode-transmitted viruses are generally transmitted by seeds. No virus persistently transmitted by leafhoppers and aphids is seed transmitted.

10. Longevity of viruses within infected seeds varies widely from a few months to more than 10 years.

11. Seed transmission of viruses is epidemiologically important because: it helps the virus to tide over and be perpetuated during unfavorable environmental conditions; it provides early and randomized infection foci within a crop from which the virus spreads to other plants in a field; it helps in the dispersal, introduction, and establishment of a virus into new and distant localities and countries; and it may be present inobtrusively in germplasm banks with all the attendant complications in plant-breeding programs.

12. The best way to control seed-transmitted viruses is to plant certified virus-free seed stocks.

13. Stace-Smith and Hamilton[43a] arrived at the following conclusions in their review on inoculum threshold of viruses: (1) Seed transmission is of economic importance in

viroids and in only the following ten virus groups: alfalfa mosaic virus, bromovirus, comovirus, cucumovirus, hordeivirus, ilarvirus, nepovirus, potyvirus, tobamovirus, and tobravirus. (2) The level of seed transmission (inoculum threshold) of viruses is exceedingly low, but still is critically important in the epidemiology of the disease in a crop, provided the crop is an annual crop, the vector is an aphid which transmits the virus in a nonpersistent .pa manner, and the virus has a narrow natural host range. Such a seed-transmitted virus is likely to be a potyvirus or a cucumovirus and will have an inoculum threshold of zero or close to zero. (3) Seed transmission plays an insignificant role in epidemiology of those seed-borne viruses which have a broad natural host range consisting of both annual and perennial plants. A high inoculum threshold is tolerated for these viruses.

## IV. TRANSMISSION BY POLLEN

Several viruses are transmitted by the pollen of infected plants. These viruses enter healthy plants through any one or more of several avenues. First, a pollen tube may become mechanically inoculated by the virus present on the external surface of the exine as the germ tube elongates through and disrupts the exine. Second, the stigma may become infected on contact with virus-contaminated pollen. Third, internally pollen-borne viruses may be carried by sperms to eggs, resulting in infection of developing zygote. Fourth, virus present in the cytoplasm of vegetative cells may be directly transmitted to the embryo. Thus, internally pollen-borne viruses may infect seed by either maternal transmission through the vegetative cell or through sperm during fertilization. The last method of pollen transmission of viruses occurs when externally pollen-borne viruses may back-infect the plants. The efficiency of virus spread by pollen may be increased by open pollination rather than self-pollination, high virus titer in pollen, and pollination through an insect rather than through wind.[44]

As already mentioned, some viruses are carried internally by pollen. Tobacco ringspot virus is present in the intine and cytoplasm and the wall of generative cells of pollen,[45] while BSMV is situated as single particles or small aggregates in the cytoplasm and/or nucleus of sperm and vegetative cells of pollen.[46] Cherry leaf roll virus occurs in sperm cell cytoplasm of immature pollen grains of birch as well as within tubules in anther cells and pollen grains of walnut.[47] Blueberry leaf mottle virus (BBLMV) and prune dwarf virus are present within pollen grains of infected highbush blueberry bushes and infected sweet cherry plants, respectively.[48,49] Both long and short particles of Brazilian tobacco rattle virus were present in anther cells during microsporogenesis, in microspores, in mature pollen grains, in cytoplasm of vegetative and generative cells, but not in their nuclei.[50] BSMV particles were also seen in pollen grains, pollen tube, and pollen tube discharge within the zygote and degenerating synergids of barley plants.[51] The internally pollen-borne viruses reach the developing zygote through sperms or directly reach the embryo through vegetative cells. Both avenues result in the formation of infected seeds which, on germination, produce infected seedlings. This virus transmission from parent to progeny is the vertical transmission of viruses through pollen.

Several viruses are borne on external surfaces (exine) of pollen grains of infected plants. These viruses fall into two categories. Viruses of category A in Table 3 are not transmissible by pollen to other plants. Hence, these externally pollen-borne viruses are of no epidemiological importance. Externally pollen-borne viruses of the second category (category D; Table 3) are transmissible by pollen to other plants which become infected. These externally pollen-borne viruses are of great epidemiological importance. All externally pollen-borne viruses, of whatever category, can be easily removed from the pollen surface by washing several times with phosphate-buffered saline.[48,49,53]

Pollen-transmissible plant viruses borne on the exine of pollen obtained from infected plants are prunus necrotic ringspot virus (PNRSV) on pollen of infected sweet cherry,[54]

**Table 3**
**CATEGORIZATION OF POLLEN-TRANSMITTED VIRUSES INTO VARIOUS
GROUPS/CATEGORIES BASED ON THE EPIDEMIOLOGICAL IMPORTANCE
OF POLLEN TRANSMISSION OF THOSE VIRUSES IN DISEASE SPREAD IN
NATURE/GREENHOUSES**

| Virus | Host | Percent transmission through pollen, if known[a] |
|---|---|---|
| **Category A** | | |
| a. Externally pollen-borne viruses | | |
| Apple chlorotic leafspot | *Chenopodium quinoa* and *C. amaranticolor* | |
| Potato virus X | *Petunia hybrida* | |
| Sowbane mosaic | *Atriplex coulteri* | |
| Brome mosaic, southern bean mosaic, and tobacco mosaic | *Phaseolus vulgaris* and *Vigna unguiculata* | |
| An unidentified virus | Decline-infected pear | |
| EAMV | *Vicia faba* | |
| Potato virus T | *Datura stramonium* | |
| | *Nicotiana physaloides* | |
| Squash mosaic | Squash | |
| b. Viruses causing pollen sterility | | |
| Beet yellows | *Beta vulgaris* | |
| *Datura* quercina virus (a strain of tobacco streak virus) | *Datura stramonium* | |
| Onion mosaic | *Allium cepa* | |
| Tobacco ringspot | *Glycine max* | |
| Tomato ringspot | *Pelargonium hortorum* | |
| **Category B** | | |
| Bean southern mosaic | *Phaseolus vulgaris* | |
| Bean yellow mosaic | *Phaseolus vulgaris* | |
| Cherry raspleaf | Cherry | |
| Cucumber mosaic | *Stellaria media* | |
| Onion yellow dwarf | *Allium cepa* | |
| Pelargonium zonate spot | *Nicotiana glutinosa* | |
| **Category C** | | |
| Arracacha virus B | *Solanum tuberosum* | 2 |
| Beet cryptic | *Beta vulgaris* | |
| Broad bean strain | *Vicia faba* | 0.52 |
| Cowpea aphid-borne mosaic | *Vigna sinensis* | |
| Cowpea banding mosaic | *Vigna sinensis* | 16.3 |
| Elm mosaic | *Sambucus racemosa* | 30.5 |
| | *Ulmus americana* | |
| Elm mottle | *Syringa vulgaris* | |
| Lettuce mosaic | *Lactuca sativa* | 0.48 |
| Lychnis ringspot | *Lychnis divaricata* | 18.6 |
| | *Silene noctiflora* | 18.6 |
| Potato virus T | *Solanum tuberosum* | 8.0 |
| Raspberry ringspot | *Fragaria virginiana* | 5.5 |
| | *Rubus* sp. | 5.2 |
| Tomato black ring | *Rubus* sp. | 12.9 |
| Tomato ringspot | *Pelargonium hortorum* | 1.0 |

**Table 3 (continued)**
## CATEGORIZATION OF POLLEN-TRANSMITTED VIRUSES INTO VARIOUS GROUPS/CATEGORIES BASED ON THE EPIDEMIOLOGICAL IMPORTANCE OF POLLEN TRANSMISSION OF THOSE VIRUSES IN DISEASE SPREAD IN NATURE/GREENHOUSES

| Virus | Host | Percent transmission through pollen, if known[a] |
|---|---|---|
| | **Category C₁** | |
| Alfalfa mosaic | *Medicago sativa* | 0.5—26.5, 1.0—14.0 |
| Barley stripe mosaic | *Hordeum vulgare* | 10, 45—59, 68.4 |
| Bean common mosaic | *Phaseolus vulgaris* | 25, 18—76 |
| | **Category D₁** | |
| Blueberry leaf mottle | *Vaccinium corymbosum* | |
| Cherry yellows | *Prunus cerasus* | |
| Prune dwarf | *P. cerasus* | 19.3, 25, 33 |
| Prunus necrotic ringspot | *Cucurbita maxima* | |
| Tobacco streak (black raspberry latent strain) | *Rubus occidentalis* | |
| | **Category D₂** | |
| Cherry leaf roll | *Juglans regia* | |
| Prunus necrotic ringspot | *Prunus cerasus* | 21, 27.8 |
| Raspberry bushy dwarf | *Rubus idaeus* | |
| | *R. occidentalis* | |
| | *R. ursinus* var. *loganobaccus* | |

[a]   Multiple values or a range of values, where given, indicate the percent virus transmission through pollen as reported by different workers.

Based on Mandahar[37,38] and Mandahar and Gill.[52]

SMV on pollen of infected *C. quinoa* and spinach,[55] cherry leaf roll virus on pollen from infected birch and English walnut,[47] prune dwarf virus on pollen from infected sweet cherry,[49] BBLMV on pollen from infected highbush blueberry,[48] and Brazilian tobacco rattle virus on pollen from infected tomato plants.[50] As already mentioned, some of these viruses are also carried within the pollen grains as well and are vertically transmitted by pollen to seeds.

All the above-mentioned pollen-transmissible, externally pollen-borne viruses have evolved the capacity to invade and systemically infect the ovule-bearing mother plants — called back-infection. This in turn leads to the formation and release of more and more infected/infested pollen and hence more and more infections of the contemporary adult, mature, healthy, ovule-bearing plants. This is the horizontal spread of viruses through pollen. Cherry yellows, tobacco streak (black raspberry latent strain), and raspberry bushy dwarf viruses are the other viruses, apart from the ones already mentioned above, that are horizontally transmitted.

But how do the adult mature trees become back-infected by viruses? The mystery now appears to have been solved on the basis of two types of evidence. First, movement of virus-contaminated pollen by foraging honey bees was suggested to be responsible for the spread of PNRSV[56,57] and prune dwarf virus[58,59] in Montmorency cherry orchards. Two, infectivity of PNRSV and prune dwarf virus particles borne at the surface of pollen of infected almond and sweet cherry plants was retained for several days.[60] Cole et al.[60] speculated on these bases that foraging bees might mechanically spread and infect the flower parts with virus through abrasions or wounds caused by them. The virus can then push backwards and systemically back-infect the main plant through plasmodesmata that connect the various flower parts (except the male and female sporogenous cells) with the sporophyte. This speculation is confirmed by experimental evidence. Particles of PNRSV[54,61] and BBLMV[62] were associated with pollen collected by bees from infected sweet cherry and highbush blueberry plants, and rub-inoculation of pollen obtained from BBLMV-infected highbush blueberry bushes readily transmitted the virus to healthy *C. quinoa* plants.[48] Similarly, spinach and *C. quinoa* plants became infected when SMV-infested pollen blown onto their surfaces was rubbed onto the leaves.[55] Thus, all horizontally pollen-transmitted viruses may be spread to healthy plants by bees carrying infested pollen.

This speculative mechanical transmission of the virus by bees may be important in the epidemiology of PNRSV causing cherry rugose mosaic disease in some western states of the U.S.[61] About 50,000 beehives are shifted each year from Washington to California to pollinate various stone fruit orchards, and then back to Washington to pollinate sweet cherry.[61] Bees collect and store huge quantities of PNRSV-infected/infested pollen in beehives in California and, on reentry into Washington, carry the contaminated pollen. Virus-infested pollen was attached to the bodies of bees leaving commercial hives in cherry orchards, and this surface-borne virus remained infective for several days.[61] Pollen-borne virus may thus be spread for several days by bees. This can be important epidemiologically in various ways: even nonviable pollen can act as a virus carrier and transmit the virus to healthy plants through abrasions on flowers caused by bees; pollen from even noncompatible species may be equally effective; and the disease can spread rapidly over extensive areas.[61] However, practically all virus disappeared from the pollen during prolonged storage in hives.[63]

## A. Effects of Virus Infection on Pollen

Virus infection of pollen and/or host plants affects the pollen in diverse ways.[64-66] These effects can be grouped under three broad categories: morphological, physiological, and cytological.

### 1. Morphological Changes

The morphological abnormalities reported in the virus-infected pollen and/or pollen obtained from virus-infected plants generally are distorted, crescent-shaped, and shrivelled pollen, and grooved pollen with sunken opercula. Haight and Gibbs[66] examined the morphology of pollen from the following virus-infected and comparable healthy plants by light microscope and scanning electron microscope: BSMV-infected *Hordeum vulgare, H. spontaneum,* and *Triticum aestivum;* broad bean wilt virus- or ribgrass mosaic virus-infected *Plantago lanceolata;* SMV-infected *C. quinoa;* and turnip yellow mosaic virus-infected *Cardamine* species. Of these, only the BSMV is pollen transmitted in barley; the rest of the above-mentioned viruses are not transmitted through the pollen of their respective infected plants.

Haight and Gibbs[66] found that pollen from the early (first) flowers of SMV-infected *Chenopodium quinoa* were collapsed, grooved, and had sunken opercula compared to the smooth, rounded pollen grains with protuberant opercula from identical flowers of healthy plants, while, surprisingly, pollen from late flowers of uninfected plants were similar in morphology and appearance to those from the early flowers of virus-infected plants.[66] Moreover, pollen from all other virus-infected plants either did not show any or exhibited only little morphological abnormality. This was also true of the BSMV-*Hordeum* complex.

### 2. Physiological Changes

Pollen germination and pollen germ tube growth of virus-infected pollen and/or pollen obtained from virus-infected trees are adversely affected, leading to decreased competitive ability of such pollen.[45,48,67,68] The germination rate of pollen from tobacco ringspot virus-infected soybean plants was 47% compared to 77% for pollen from healthy plants.[45] In addition, pollen from these infected plants produced shorter germ tubes which also elongated more slowly than the normal germ tube of healthy pollen.

Poor pollen germination, poor pollen tube growth, and the overall shortened length of pollen tube were also observed in the healthy pollen (i.e., pollen which do not carry virus particle) derived from maize dwarf mosaic virus (MDMV)-infected sweet corn plants.[68] Pollen obtained from such plants did not carry any virus, but pollen vigor and pollen receptivity of silk (stigma and style) were adversely affected. Moreover, in vitro germination of pollen from diseased and healthy sweet corn plants was almost the same. Nevertheless, the length of the germ tube of pollen from infected plants was significantly shorter than that of the germ tube of pollen grains from healthy plants in vitro as well as in vivo. Thus growth of pollen germ tubes in vitro and in vivo was significantly reduced in pollen obtained from diseased plants. Moreover, when pollen from healthy plants was employed to pollinate silks of diseased plants, the pollen germ tube length was significantly reduced, indicating that virus infection adversely affects the growth and development of pollen germ tube inside the silks of infected plants.

Pollen from BBLMV-infected highbush blueberry plants showed poor and delayed germination in nutrient solution as compared to that of the healthy pollen. Only 6% of pollen grains from diseased plants germinated in 12.5% sucrose solution, compared to 44% of the healthy pollen germinated.[48] Moreover, germ tube formation was also delayed, and considerably fewer germ tubes were produced by pollen from infected plants. Thus pollen from infected highbush blueberry bushes possesses reduced viability and undergoes pollen abortion so that contaminated pollen cannot compete with pollen from healthy plants. This may limit virus spread during fertilization, and many of the seeds may not carry the virus. Only 27% of blueberry seeds collected from infected plants carried the virus, but only 1.5% of these seeds produced infected seedlings.[48] Pollen from Brazilian tobacco rattle virus-infected plants germinated poorly compared to that from healthy plants.[50] Pollen obtained from potato spindle tuber viroid-infected potato plants was less viable than the pollen from healthy plants.[69]

### 3. Amount of Pollen Produced

The number of pollen grains produced per flower or anther is considerably reduced in virus-infected plants as compared to healthy plants. The number of pollen grains per flower produced by tobacco ringspot virus-infected soybean plants ranged from 0 to 2000, compared to 4000 to 6500 produced by flowers on healthy plants.[45] Similarly, the average number of pollen grains in a microscope field was 61.3 (range 50 to 75) in anthers of BSMV-infected barley plants, compared to 84.7 (range 57 to 111) in anthers of healthy plants.[46] The pollen sterility was 10% in infected and 6% in healthy barley plants.

### 4. Cytological Changes

Virus infection may cause abnormal and incomplete microsporogenesis in infected plants. Abnormal microsporogenesis has been observed in TMV-infected tobacco and tomato, mosaic-infected *Datura quercifolia*, and mosaic-infected brinjal. In all these cases, chromosomes in pollen mother cells at metaphase I largely fail to pair so that they generally occur as a large number of univalents. This leads to an unequal and abnromal chromosomal movement associated with laggards and bridges at anaphase I. In addition, decreased or complete degeneration of cytoplasm, collapse of pollen grain walls, and collapse of pollen grains have also been observed.[46,66,70] Depending upon the degree and extent of microsporogenesis, there may be complete or incomplete pollen sterility (i.e., pollen abortion). Examples of complete pollen sterility or abortion are listed in Table 3.

## B. Techniques

The most common method employed to detect pollen-transmitted viruses is to prepare a pollen triturate in buffer and then to inoculate the triturate to some indicator plant or subject it to conventional serology to test for the presence of virus. A positive reaction has always been taken to indicate the presence of a pollen-transmitted virus. However, this technique is faulty and the conclusions drawn on this basis (about the pollen transmissibility of the virus) are suspect as is clearly borne out by the following two examples.

Rader et al.[71] had detected squash mosaic virus in triturates of presumably unwashed pollen, while Alvarez and Campbell[53] could not detect the same virus in triturates of washed pollen. This indicates that the virus must have been externally pollen borne and that all reports on triturates of unwashed pollen are suspect. Moreover, the presence of viruses in triturates of washed or unwashed pollen has always been interpreted as indicating pollen transmission of these viruses without, in many of the cases, actually performing experiments to confirm whether this pollen does lead to the production of virus-infected seed. Evidence indicates that this is not always so. Thus, both broad bean stain virus (BBSV) and Echtes Ackerbohnenmosaik virus (EAMV) were present in triturates of pollen obtained from infected *Vicia faba* plants, but greenhouse experiments showed that EAMV was not transmitted through pollen to seeds, while BBSV was transmitted.[72] Potato virus T was also detected in the pollen of *D. stramonium*, *Nicandra physaloides*, and *Solanum demissum*, but was transmitted from pollen to seed only in *S. demissum*.

It is apparent from the above that only BBSV and potato virus T were transmitted by pollen of *V. faba* and *S. demissum*, respectively; and that squash mosaic virus in cantaloup, EAMV in *V. faba*, and potato virus T in *D. stramonium* and *N. physaloides* are not pollen transmitted, but must be only externally pollen borne. Consequently, all reports about the pollen transmissibility of viruses based exclusively on infectivity of or presence of virus particles in pollen triturates are suspect unless supported by more reliable evidence. Such viruses have unfairly been regarded or taken to be pollen transmitted. On critical examination of published information, the total number of viruses truly pollen-transmitted in this way was found to be only 19[52] (Table 3).

## C. Epidemiological Role of Pollen-Transmitted Viruses

Two contrasting opinions are available concerning the epidemiological role of pollen-transmitted viruses: virus transmission through pollen is of little consequence in nature since the infected pollen cannot compete with healthy pollen during fertilization,[45,48,67,68] while, according to others,[2,55,56,73] spread of viruses through pollen results in rapid and extensive spread of some diseases. The precise epidemiological role of pollen transmission of viruses therefore needs to be clearly spelled out. Pollen-transmitted viruses can be categorized into different groups on the basis of epidemiological role of pollen transmission in disease spread in nature (Table 3).

Some externally pollen-borne viruses and viruses causing pollen sterility belong to category A in Table 3. None of these viruses can be pollen transmitted. Thus, the fact of these viruses being externally pollen borne is of no epidemiological importance, while in the case of viruses causing pollen sterility, the seeds are not produced so that pollen infection has become a liability to these viruses. Some investigators have reported the detection of virus particles in pollen homogenates, but the viruses could not be pollen-transmitted to seeds, nor was any abnormality in morphology and development of anthers and pollen grains recorded. These viruses include EAMV in *V. faba*, potato virus T in *D. stramonium* and *N. physaloides*, and squash mosaic virus in squash. It appears that these viruses also must be externally pollen borne on the exine and should be regarded as such. None of these viruses is pollen transmitted in its respective host.

Viruses in category B of Table 3 were detected in triturates of pollen from infected plants, but no experiment was conducted in greenhouse/field to find out if cross-pollination of emasculated flowers on healthy plants leads to the formation of infected seeds. That is, it is not yet known whether these viruses are vertically pollen transmitted. Thus, these viruses should not at present be considered as vertically pollen transmitted unless more definite evidence is forthcoming. The vertical pollen transmission of these viruses has no epidemiological importance in the spread of disease in nature.

Viruses in category C were detected in pollen triturates and/or shown in greenhouse/field experiments on cross pollination to be vertically transmitted through pollen and lead to the production of infected seed on emasculated healthy female plants. However, their vertical pollen transmission to produce infected seeds has not been detected in nature and the mother plant does not become back-infected. These are the vertically pollen-transmitted viruses. Their pollen transmissibility may or may not be of potential danger, but at present it does not appear to have any epidemiological threat.

Viruses in category $C_1$ are the same as in C, but their pollen transmissibility has been shown or appears to be of some limited epidemiological value in disease spread. These viruses, however, cannot back-infect the female parent. The limited epidemiological value of vertical pollen transmission of these viruses is evidenced by the fact that pollen usually vertically transmits virus to more seeds than ovules, or is somewhat less efficient than ovule transmission, but still transmits virus to a considerable number of seeds.

Viruses in category D were shown by deblooming and/or cross-pollination tests in greenhouse/field experiments to be transmitted through pollen, leading to the production of infected seed on emasculated healthy female plants. All these viruses are vertically transmitted and all are also horizontally transmitted through pollen to systematically back-infect the pollinated (either by hand or in nature) mother plants. Vertical pollen transmission of these viruses is of little epidemiological value, particularly since the specific hosts in question are almost all perennial plants. However, from the point of view of epidemiological value of horizontal pollen transmission, these viruses are further divisible into two subcategories.

Horizontal pollen transmission of viruses in category $D_1$ has been shown or appears to be of some limited epidemiological importance.[58,59]

Horizontal pollen transmission of viruses in category $D_2$ is of great epidemiological importance in the spread of these diseases in nature. Many of these viruses are externally pollen-borne, and honeybees take part in their transmission.

Cherry leaf roll virus, causing the blackline disease of English walnut in the U.S., is spread horizontally by infected pollen and is a serious threat to the walnut industry in California.[2,3] Raspberry bushy dwarf virus horizontally spreads naturally in red raspberry (*Rubus idaeus*) orchards in Canada, England, New Zealand, Scotland, and the U.S. only through infected pollen.[73-75] Infected pollen is the major and, most probably, the only means of natural spread of prunus necrotic ringspot virus in sour cherry orchards in Canada and the U.S.[56,76,77] The spread by infected pollen is always horizontal. This was the first virus in which pollen transmission was found to back-infect some of the cross-pollinated mother plants.[56]

Viruses horizontally spread through pollen invade and systemically back-infect the ovule-bearing mother plants, which in turn leads to the formation and release of more and more infected pollen. This cycle can be repeated during subsequent flowering seasons, leading to rapid epidemiological buildup of infected pollen and so to disastrous consequences in a couple of years. This is the reason only horizontal transmission through pollen is epidemiologically important (viruses of category $D_2$) or appears to have at least some limited epidemiological importance (viruses of category $D_1$).

## D. Conclusions

(1) Virus infection may cause complete pollen abortion, resulting in male sterility. (2) Abnormal or incomplete microsporogenesis is the cause of complete or incomplete pollen abortion. (3) Decreased amount of pollen production, decreased pollen germination, de-

## Table 4
## VIRUSES AND DISEASE AGENTS TRANSMITTED BY FUNGI

| Virus/disease agent | Virus particle | Vector | Vector/virus relationship | Ref. |
|---|---|---|---|---|
| Viruses | | | | |
| Barley yellow mosaic | Filamentous | *Polymyxa graminis* | Persistent | 81 |
| Beet necrotic yellow vein | Straight tubular | *P. betae* | Persistent | 82 |
| Brome mosaic | Isometric | *Puccinia graminis tritici* | | 83,84 |
| | | *Puccinia recondita tritici* | | |
| Cucumber necrosis | Isometric | *Olpidium radicale* (syn. *O. cucurbitacearum*) | Nonpersistent | |
| India peanut clump | Straight tubular | *Polymyxa graminis* | | 85 |
| Oat golden mosaic | Straight tubular | ? *P. graminis* | ? | |
| Oat mosaic | Filamentous | *P. graminis* | Persistent | |
| Peanut (groundnut) clump | Straight tubular | ? *P. graminis* | ? | 86 |
| Pea stem necrosis | Isometric | *Olpidium* sp. | ? | |
| Potato mop top | Straight tubular | *Spongospora subterranea* | Persistent | |
| Potato X | Straight tubular | *Synchytrium endobioticum* | Persistent | |
| Rice necrosis mosaic | Filamentous | *Polymyxa graminis* | Persistent | |
| Satellite tobacco necrosis | Isometric | *O. brassicae* | Nonpersistent | |
| Tobacco necrosis | Isometric | *O. brassicae* | Nonpersistent | |
| Wheat mosaic (soilborne) | Straight tubular | *Polymyxa graminis* | Persistent | |
| Wheat spindle streak mosaic | Filamentous | *P. graminis* | Persistent | |
| Agents | | | | |
| Freesia leaf necrosis | Unknown | ? *O. brassicae* | ? Persistent | |
| Lettuce big vein | Unknown | *O. brassicae* | Persistent | |
| Tobacco stunt | Unknown | *O. brassicae* | Persistent | |

Based on Teakle,[78,79] Campbell,[80] and subsequent literature.

creased growth and development of pollen germ tubes, and reduced style receptivity are responsible for decreased competitiveness of pollen from infected plants. (4) Taking the results and suggestions both of Haight and Gibbs[66] and of Mikel et al.[68] together, it is almost certain that actual virus infection of the pollen grains is not necessary for the morphological and physiological abnormalities and that the general adverse effects of virus infection on the vigor and health of the infected plants, per se, probably lead to all those debilities in the pollen. It is known that reduced health of the plants does bring about all the above-mentioned morphological and physiological changes. The same may also hold true in connection with the decreased pollen production by infected plants. However, cytological abnormalities may be the direct effect of virus infection rather than due to the decreased vigor of the infected plants. (5) Horizontally pollen-transmitted viruses, whether epidemiologically important (category $D_2$ viruses) or not (category $D_1$ viruses), generally belong to ilarvirus and nepovirus groups, except cherry yellows virus, of which the taxonomic grouping is not yet known. (6) Only 20 plant viruses (of categories C and D) are truly pollen transmitted.

## V. TRANSMISSION BY FUNGI

Fungal-transmitted viruses are listed in Table 4.

### A. Vectors
Fungal vectors of plant viruses belong to four different genera, two (*Olpidium* and *Synchytrium*) to the order Chytridiales and the other two (*Polymyxa* and *Spongospora*) to the order Plasmodiophorales. The morphology and life cycles of these two orders of obligate parasites possess the following common features: absence of mycelium; thallus endophytic and permanently or temporarily in the form of plasmodium; asexual reproduction by zoos-

pores formed in zoosporangia, but released outside the host cells; sexual reproduction by fusion of motile gametes; zygotes (resting spores) can survive in soil for long periods; and infection of host occurs by reentry of zoospores and zygotes into the host cells. The endophytic nature and plasmodial stage are important parameters which occur only in these fungi and are the main reasons the fungi of only these two orders act as vectors.

Of the two chytridiaceous fungi, *Olpidium* is a much more important vector than *Synchytrium*. *O. brassicae* is the more important species and transmits four isometric viruses and three disease agents, while *O. radicale* (syn. *O. cucurbitacearum*) transmits only one isometric virus (Table 4). Of the two plasmodiophores, *Spongospora subterranea* transmits only one straight tubular virus and *Polymyxa graminis* transmits eight elongated viruses, while *P. betae* transmits only one elongated virus. It is a remarkable coincidence that the chytrids transmit only the isometric viruses and plasmodiophores transmit only the elongated viruses. Three of the above viruses and one disease agent are suspected to be transmitted by fungi, but definite proof of their fungal transmission has yet to be produced.

### B. Virus-Vector Relationships

Protoplasm of an infected host cell contains both the virus and the naked protoplast thallus of a vector. The two come into contact, and every fungal structure, zoospores or resting spores, formed subsequently will contain the virus within its interior. These are the thallus-acquired viruses and can be acquired only in vivo. The virus uptake has been suggested to be by endocytosis.[87] The majority of the fungal-transmitted viruses and other disease agents are in this way acquired by the fungus thallus from inside the host cell and are present within the body of zoospores and resting spores. These viruses, because of their presence inside the resting spores of the vector, survive in fungus-infested soils for long periods. Potato mop top virus was detected in soil in which potatoes had not been grown for 3 to 6 years.[88] Tobacco stunt agent can survive in air-dried soil for many years. Lettuce big vein agent can survive in soil for more than 8 years. The longevity endpoint of *O. brassicae* stored in air-dried silty clay loam soil at about 2% moisture content was 20.8 to 22.5 years, while the persistence endpoint of the lettuce big vein agent was 18.7 to 20.5 years.[89] Lettuce big vein-prone soils have slow water drainage, resulting in prolonged periods of low moisture tension which is necessary for germination of resting spores, release of zoospores from sporangia, and movement of zoospores in soil.[90] It is clear from the above that thallus-acquired viruses, being internally borne in zoospores and resting spores, have a persistent relationship with their vectors. These viruses again become active whenever resting spores germinate and cause infection of the host cells. Infected resting spores present in soil may therefore be a major source of virus inoculum.

The situation in two other fungal-transmitted viruses, TNV and cucumber necrosis virus, is different from the one outlined above. Protoplasm of an infected host cell contains TNV and the naked protoplast thallus of the vector *O. brassicae*. But the two are released outside the cell independently of each other with the result that TNV is not present within the body of zoospores. It is, on the other hand, acquired by zoospores from outside the host and is carried externally on zoospore surfaces as a surface contaminant. It is tightly adsorbed to surface membranes, plasmalemma, and axonemal sheath of zoospores by some forces which are not yet known. Electrostatic changes are at least not responsible for this surface adsorption. The adsorbed virus possibly enters the protoplast on withdrawl of flagellum and is later transmitted into an epidermal cell of a root when the zoospore cyst discharges its contaminated protoplast into that cell. The zoospore-acquired virus, acquired in vitro, is carried externally on zoospore surface and has a nonpersistent relationship with its vector. Satellite tobacco necrosis virus is also presumably carried by zoospores of *O. brassicae* in identical manner. Cucumber necrosis virus is also released outside the roots and zoospores of *O. radicale* (= *O. cucurbitacearum*) when initially discharged from resting spores and zoosporangia formed in roots infected by both the pathogens are free of virus. Zoospores subsequently acquire the virus on their external surfaces from outside the cells. This virus, like TNV, is also specifically adsorbed to plasmalemma and axonemal sheath of the zoospores, is never carried internally in zoospores, and apparently enters the zoospore cyst by flagellar retraction and endocytosis. The specific adsorption of cucumber necrosis virus on external surface of *O. radicale* zoospores is associated with viral coat protein, since viral RNA was not transmitted by zoospores to cucumber roots.

To sum up: two distinctly different virus-vector relationships occur: persistent relationship in which viruses survive inside the resting spores of the fungi and are internally borne by zoospores; and a nonpersistent relationship in which viruses do not survive in resting sporangia and are externally borne by zoospores.

Brome mosaic virus was transmitted through uredospores of *Puccinia graminis tritici* and *P. recondita tritici* to healthy barley and wheat plants.[83,84] These uredospores were obtained from plants doubly infected with the virus and the fungus and carried virus particles as contaminants on their external surface. The virus may also have been present within the uredospores. TMV-infected plants are also often infected by rusts and powdery mildews. It was suggested earlier on the basis of some indirect evidence that these fungi or their propagules may act as vectors.[91]

## C. Virus-Vector Specificity

Virus-vector specificity varies in different cases. Lettuce big vein agent is not specifically transmitted by any vector strains of *O. brassicae;* all vector isolates capable of multiplication in lettuce can acquire and transmit this agent. Specificity of TNV transmission by *O. brassicae* on the hand exists. All *O. brassicae* strains are not TNV vectors. Lettuce strains constantly transmit TNV while several crucifer strains are nonvectors and do not transmit it. Specificity of TNV transmission by different strains of the vector is associated with the ability or inability of zoospores to adsorb TNV on their surfaces.[92] Good vector strains adsorb TNV on their surfaces, poor vector isolates (oat) adsorb fewer particles of the specific TNV isolate, while zoospores of nonvector oat and mustard strains do not adsorb the relevant TNV isolate.

## VI. TRANSMISSION BY ARTHROPODS AND NEMATODES

About 94% of the 381 animal species transmitting plant viruses are arthropods and 6% are nematodes, while 99% or 356 of the arthropods are insect vectors.[93]

## A. Leafhopper Vectors (Cicadellidae, Homoptera, Insecta)
### 1. Viruses

Leafhoppers, after aphids, are the most important insect vectors in terms of the number of diseases transmitted (Table 5). There are 130 known leafhopper (Cicadellidae) vector species which belong to 58 genera and 10 subfamilies.[93b] Deltocephalinae contains the maximum number of 75 vector species, followed by Cicadellinae with 28 vector species, Agallinae with 13 vector species, Macropsinae and Typhlocybinae with 3 vector species each, and three subfamilies (Gyponinae, Coelidinae, and Aphrodinae) with 2 vector species each.[93b] However, the number of leafhopper vectors transmitting viruses, mycoplasmalike organisms (MLOs), and spiroplasmas is now around 142 (Tables 5 and 7). Leafhopper vectors transmit about 82 plant diseases, out of which about 40 are caused by viruses and the rest by MLOs and spiroplasmas (Tables 5 and 7).

Transmission of viruses by leafhopper vectors is under genetic control of the vector. Transmission rates of maize rayado fino virus (MRFV) by *Dalbulus maidis* can be tripled or increased to a still greater or lesser extent by selection, with or without subsequent matings of the selected individuals. Random mating of the selected populations bring down the MRFV transmission rate of a normal level after several generations. Some of the viruses

## Table 5
## KNOWN AND PRESUMED PLANT VIRUSES TRANSMITTED BY
## LEAFHOPPER VECTORS

| Virus/Presumed virus | Vector |
|---|---|
| Bean summer death | *Orosius argentatus* |
| Beet curly top | |
|   North American strain | *Circulifer tenellus, C. opacipennis* |
|   Argentina strain | *Agalliana ensigera* |
|   Brazilian strain | *Agallia albidula, Agalliana ensigera, A. stricticollis* |
| Beet yellow vein | *Aceratagallia calcaris* |
| Bermuda grass etched line | *Aconurella prolixa*[96] |
| Cereal chlorotic mottle | *Cicadulina bimaculata,*[97] *Nesoclutha pallida*[97] |
| Chloris striate | *Nesoclutha pallida* |
| Cotton yellow vein | *Scaphytopius albifrons* |
| Eastern wheat mosaic | *Cicadulina mbila* |
| Elm phloem necrosis | *Scaphoideus luteolus* |
| Maize chlorotic dwarf | *Graminella nigrifrons, G. sonora* (syn. *Deltocephalus sonorus*) |
| Maize leaf gall | *Cicadulina bipunctella bimaculata* |
| Maize mottle | *Cicadulina mbila, C. storeyi, C. bipunctella zeae* |
| Maize rayado fino | *Baldulus tripsaci, Dalbulus maidis, D. elimatus, Graminella nigrifrons, Stirellus bicolor* |
| Maize streak | *Cicadulina bipunctella zeae, C. bimaculata, C. latens, C. mbila, C. parazeae, C. storeyi* |
| Maize wallaby ear disease | *Cicadulina bipunctella bimaculata, Nesoclutha pallida* |
| Oat blue dwarf | |
|   North American strain | *Macrosteles fascifrons* |
|   Swedish strain | *M. laevis* |
| Oat striate mosaic | *Graminella nigrifrons* |
| Paspalum striate | *Nesoclutha pallida* |
| Potato yellow dwarf | |
|   New York strain | *Aceratagallia curvata, A. longula, A. obscura, A. sanguinolenta, Agallia quadripunctata, A. constricta, Agalliopsis novella* |
|   New Jersey strain | *Agallia constricta, A. quadripunctata, Agalliopsis novella* |
| Ragi streak | *Cicadulina chinai*[99] |
| Rice (bunchy) stunt | *Nephotettix cincticeps, N. nigropictus, N. virescens, Recilia dorsalis* |
| Rice dwarf | *Nephotettix cincticeps, N. nigropictus, N. virescens,*[100] *Recilia dorsalis* |
| Rice gall dwarf | *Nephotettix cincticeps, N. malayanus, N. virescens, N. nigropictus, Recilia dorsalis* |
| Rice leaf gall | *Cicadulina bipunctella bimaculata* |
| Rice leaf yellowing | *Nephotettix virescens* |
| Rice streak | *Cicadulina chinai* |
| Rice transitory yellowing | *Nephotettix cincticeps, N. nigropictus, N. virescens* |
| Rice tungro | *N. nigropictus, N. virescens* |
| Rice waika | *N. cincticeps, N. malayanus, N. nigropictus, N. virescens* |
| Rice yellow orange leaf | *N. nigropictus, N. virescens, Recilia dorsalis* |
| Russian winter wheat mosaic | *Psammotettix alienus, P. striatus* |
| Sowbane mosaic | *Circulifer tenellus* |
| Striate mosaic of grasses and cerals | *Nesoclutha pallida* |
| Sugarcane chlorotic streak | *Draeculacephala portola* |
| Tobacco yellow dwarf | *Orosius argentatus* |
| Tomato leaf crinkle | *Agallia venosa, Anaceratagallia venosa* |
| Wheat dwarf | *Psammotettix striatus, P. alienus* |
| Wheat striate mosaic (North American strain) | *Endria inimica, Elymana sulphurella, Nesoclutha pallida* |
| Wound tumor | *Agallia constricta, A. quadripunctata, Agalliopsis novella* |

**Table 5 (continued)**
## KNOWN AND PRESUMED PLANT VIRUSES TRANSMITTED BY LEAFHOPPER VECTORS

*Note:*  Some synonyms include: *Elymana sulphurella*, syn. *E. virescens; Nephotettix nigropictus*, syn. *N apicalis; Nephotettix virescens*, syn. *N. impicticeps; Nesoclutha pallida*, syn. *N. obscura;* and *Recilia dorsalis*, syn. *Inazuma dorsalis*

After Ishihara,[93a] Harris,[93] Carter,[94] and Nielson,[95] with additions.

are transmitted by two or more leafhopper vectors, while the majority of viruses are transmitted by only one leafhopper each. Similarly, a few leafhopper vectors can transmit more than one virus, while the rest can transmit only one virus each (Table 5).

Most of the leafhopper-transmitted disease agents (viruses, MLOs, and spiroplasmas) have a circulative-propagative relationship with their vectors. These viruses occur systemically in almost all body parts of a vector. They follow a definite route and sequence for spreading systemically within the body of a vector after ingestion. The virus is sucked into the gut (which is generally the primary site of virus multiplication), passes into the hemolymph through some part of the alimentary tract, reaches the blood-stream, is carried to the salivary glands by circulating blood, and is finally injected into healthy plants with salivary liquid during feeding. Blood carries the virus to other parts also, even if they do not lie on the above-mentioned route. The virus infects and multiplies in all these parts. Work on wound tumor virus (WTV) is particularly extensive and important in this connection. The primary site of infection, multiplication, and accumulation of this virus in *Agallia constricta* is the filter chamber of the intestine. The virus later penetrates into the gut wall and primarily or solely infects the spherule cells and plasmocytes of hemolymph. Hemocytes then act as virus carriers and carry WTV to different sites in the body of the vector. Spread and multiplication of WTV need suitable and susceptible hemocytes, a permeable gut, salivary glands, and an adequate level of sustained multiplication. A vector will not become viruliferous in the absence of any one of these factors.

Ability to penetrate the gut wall and enter hemolymph appears to be the most important single property of all circulative-propagative viruses, enabling them to spread in the body of the vector. The permeability of tissues to a particular virus cannot be a chance event, but must depend on the virus acting specifically on the tissues to make them permeable. Such detailed work on circulative-propagative viruses and aphid vectors has already been discussed in Chapter 7.

The above route of circulative-propagative viruses results in a long incubation period only after which the leafhoppers are able to transmit the virus. The minimum acquisition feeding period in all these cases is in days and may range, depending upon the virus-vector combination and various other factors, from 5 to 10 d to 50 to 60 d. The incubation period of potato yellow dwarf virus New York strain in *Acertagallia sanguinolenta* is 6 to 9 d, of WTV in *Agallia quadripunctata* and *Agalliopsis novella* is 14 d, of maize streak virus in *Cicadulina mbila* is 6 to 12 d, of rice gall dwarf virus in *Recilia dorsalis* is 5 to 11 d (average 8 d) and in *Nephotettix nigropictus* is 8 to 18 d, of rice transitory yellowing virus in *N. cincticeps* and *N. nigropictus* is 48 to 66 d, and of MRFV in *D. maidis* is 16 d.

Several factors influence the duration of the minimum incubation period of viruses in their vectors. Injection of a partially purified virus preparation into the leafhopper vector increases the transmission rate and decreases the mean latent period in vector as of MRFV in *D. maidis* and of oat blue dwarf virus in *Macrosteles fascifrons*. In the former case, it was reduced to 6.9 d from 16 d. Different strains of a virus may have different incubation periods in the same vector or in the different specific vectors.

MRFV multiplies in its vector *D. maidis* and achieves its peak concentration after 25 d of acquisition from maize and then declines.[101] Its transmission to test corn plants by *D.*

*maidis, D. elimatus, Stirellus bicolor,* and *Graminella nigrifrons* was 70.0, 25.0, 11.5, and 9.7%, respectively.[102] The transmission rate by single *D. maidis* was 15% when the virus was acquired by the vector from infected plants, but was 77% when virus was injected into the vector.[102] The mean latent period of virus in vector was 16 d in the former case and 6.9 d in the latter case.

Maize chlorotic dwarf virus (MCDV) was more efficiently transmitted by adult females (46%) than by adult males (29%) of *G. nigrifrons*.[103] In contrast, transmission efficiency of male (10.5%) and female (9.2%) leafhoppers was nearly the same.[102] Table 6 lists some transmission data of some viruses by their vectors.

A few of the leafhopper-transmitted viruses are nonpropagative and are transmitted in a noncirculative semipersistent manner. These viruses are beet curly top, MCDV, and rice tungro (RTV). The RTV and MCDV do not have a latent period, cannot enter the hemocoele, and cannot exit via the salivary system of their vectors. Moreover, inoculativity of viruliferous vectors, without further access to a virus source, gradually declines with time. The ingestion-egestion transmission mechanism is compatible with these transmission characteristics of RTV and MCDV.[93,112] The sucking pump of leafhoppers can function normally in both directions: ingestion of material from the plants followed by periods of intermittent egestion, which may last for as long as 10 min. Material is usually egested from the foregut one or more times during feeding, and the material steadily flows out of the maxillary food canal during egestion. MCDV particles in single- and multilayered aggregates as well as their dense aggregates in a matrix material are adsorbed to the intima lining, the cibarial pump, the pharyngeal, and particularly the esophageal parts of the gut of viruliferous leafhopper vectors. Virus particles are not present on stylets, in the gut region beyond the esophageal valve, or in other tissues or regions of the vector. The frequency of MCDV transmission increases with extension of the acquisition access period.[103] MCDV is lost after molting.[103] In vivo neutralization of infectivity of the two RTVs and their loss after molting confirm indirectly that both viruses are adsorbed in the mouth or fore alimentary canal of *N. virescens*.[113] Semipersistent transmission of MCDV and RTV by leafhopper vectors is easily explained by the accumulation and persistence of virus particles at retention sites in the foregut.[93,113]

Leafhopper vector-virus specificity exists at various levels. It occurs at the subfamily level whereby all vectors of potato yellow dwarf virus and WTV are limited to members of Agallinae. The specificity occurs at the genus level as well: only *Nephotettix* spp. transmit rice dwarf, rice transitory yellowing, and rice waika viruses, while maize streak virus transmission is limited to *Cicadulina*. Specificity at the species level is most common and several examples are known (Table 5). Beet curly top virus is transmitted by different leafhopper vectors in different countries: by *Circulifer tenellus* in North America and Turkey; by *Agalliana ensigera*, but not by *C. tenellus*, in Argentina; and of the two strains in Brazil, one is transmitted by *Agallia albidula*, but not by *Agalliana ensigera*, while the second strain is transmitted by *A. ensigera* and *A. stricticollis*, but not by *Agallia albidula*.

Vector specificity of circulative-propagative viruses is due to their intimate biological contact with their vectors. Any factor which inhibits multiplication of such a virus in its vector will control the specificity of the latter. Other factors can be the inability of a virus to pass through gut wall, or inability of a virus to enter, survive, multiply, or accumulate in salivary glands. Any one of these factors can prevent an insect from acting as a vector and can thus account for vector specificity.

The differences in capsid protein have been involved in vector specificity. Adam and Hsu[114] determined the differences in structural proteins of two potato yellow dwarf virus isolates, the *sanguinolenta* variety (SYDV) transmitted by *Acertagallia sanguinolenta*, and the *constricta* variety (CYDV) transmitted by *Agallia constricta*, in order to determine the cause of vector specificity. They discovered that of the four structural proteins, two ($M_1$ and N) did not differ much between the two isolates, while the remaining two proteins ($M_2$

## Table 6
## DATA ON TRANSMISSION OF SOME PLANT VIRUSES BY LEAFHOPPER, PLANTHOPPER, MITES, AND PIESMID VECTORS

| Virus | Vector | Minimum | | Incubation period | Maximum retention period | Ref. |
|---|---|---|---|---|---|---|
| | | Acquisition feeding period | Inoculation feeding period | | | |
| **Leafhopper Vectors** | | | | | | |
| Maize chlorotic dwarf | Graminella nigrifrons | 15 min | 15 min | Absent | 37—40 h (about 2 d) | 103 |
| Rice gall dwarf | Recilia dorsalis | 8 h | 1 h | 5—11 d (av 8 d) | — | 104 |
| Rice tungro | Nephotettix virescens | 5—30 min | 7—30 min | Absent | 2—5 d | 93—95 |
| Rice waika | Nephotetix cincticeps | 30 min | 30 min | Absent | 36 h in N. nigropictus | 105 |
| Ragi streak | Cicadulina chinae | 6 h | 30 min | 1 h | Lifelong | 99 |
| **Planthopper Vectors** | | | | | | |
| Barley yellow striate mosaic | Laodelphax striatella | 1—5 h | 15 min | 15—16 d (mean) | 36 d | 106 |
| Rice grassy stunt | Nilaparvata lugens | 30 min | 5 min | — | — | 107 |
| Rice ragged stunt | Nilaparvata lugens | 2—3 h | 1 h | 4—17 d | 9—41 d or more | 108, 109 |
| **Eriophyid Mite Vectors** | | | | | | |
| Fig mosaic | Eriophyes ficus | 5 min | 5 min | few hours | 6—10 d | 93, 110 |
| Ryegrass mosaic | Abacarus hystrix | 2 h | — | — | 24 h | 93, 110 |
| Wheat streak mosaic | Eriophyes tulipae | 15 min or longer | — | — | 9 d | 93, 110 |
| **Piesmid Vector** | | | | | | |
| Beetle leaf curl | Piesma quadratum | 30 min | 40 min | 14 d | Lifetime | 111 |

and G) were distinctly different in the two isolates. However, differences between the G proteins seem to be correlated with selective transmission of the two virus isolates. Accordingly, it has been suggested that the G protein governs virus-vector specificity by regulating the entrance of infectious virus into the cytoplasm.[114] Coat protein also controls the virus-aphid vector specificity. Vector salivary gland-virus coat protein reciprocity and interactions control virus-vector specificity so that specific viruses can pass through the salivary system of specific vectors and become transmissible.

### 2. Mycoplasmalike Organisms (MLOs)

Only phloem-feeding insects are able to transmit MLOs. Leafhoppers do so and in the process transmit a great majority of MLOs from one plant to another (Table 7). A leafhopper once infected remains viruliferous for the rest of its life and can transmit an MLO after relevant incubation period to all plants on which it feeds for as long as it lives. This implies multiplication of MLO in insect vectors. Two types of evidence show that MLOs do actually multiply in their leafhopper vectors. First, serial passage of aster yellows MLO was made through ten generations of leafhoppers.[121] The MLO was calculated to have been much diluted at the tenth passage. Instead its concentration at the tenth passage was the same as at the first passage. The MLO must therefore have multiplied in vectors. Second, MLOs sequentially spread to different organs of vectors after increasing numbers of days after inoculation or after test feeding on diseased plants. They therefore occur systemically in almost all parts of the vectors and have been found in intestine, ventral ganglion, nervous system, brain, fat bodies, dorsal vessel, Malpighian tubules, salivary glands, and connective tissue of various organs.

*Circulifer tenellus* can transmit *Spiroplasma citri* acquired either by injection (into abdomen), membrane feeding, or feeding directly on infected host plants. The minimum length of the incubation period varied with the three methods and was 10, 16, and 24 d for injection, feeding on infected plants, and membrane feeding, respectively.[122] There are two possible explanations for this difference in latent period. First, *S. citri* injected directly into the hemocoel multiplies and is then transported to the salivary glands from which the leafhopper can then transmit it. This takes place within 10 d. *S. citri* therefore does not have to pass through the midgut and through the gut wall into hemocoel as in the case of natural feeding on infected plants.[122] Second, both the physical and chemical environment of the culture medium differ from those of hemolymph. It is reasonable to expect that *S. citri* would require a longer latent period to adapt to the insect hemolymph after membrane feeding before it could multiply and spread into salivary glands.

Electron microscope studies of sequential sections of the natural vector *C. tenellus* following membrane feeding of *S. citri* can be summarized as follows: *S. citri* enters the gut lumen, passes through the gut wall into the epithelial cells where it multiplies, continues to move towards and through the basement membrane of the intestine into the hemocoel where further multiplication occurs, is transported by the hemolymph to the salivary glands, and from the salivary glands, into plants via salivary secretion during feeding.[123] *S. citri* is first observed in salivary glands 10 d after its injection into insects; large *S. citri* concentrations are observed adjacent to the acini of salivary glands 15 d after inoculation.

*S. citri* also multiplies in *Macrosteles fascifrons, Dalbulus elimatus, Draeculacephala* sp., *D. pseudoobscura*, and *Euscelis plebeja*, none of which are natural vectors of *S. citri*. The multiplication pattern of *S. citri* in *C. tenellus* is very similar to that in *M. fascifrons*. Purcell et al.[124] tested the relative titer of MLO of X-disease of stone fruits in the nonvector *M. fascifrons*. The MLO multiplies in this leafhopper and was present intercellularly as well as appressed to various organs in hemocoele. This suggests that the MLO, despite its multiplication in vivo, cannot be transmitted by *M. fascifrons* because barriers in the gut and salivary glands prevent it from doing so.[124]

Several leafhoppers can transmit more than one type of plant pathogen. *Dalbulus maidis*

**Table 7**
**LEAFHOPPER VECTORS OF PLANT DISEASES INCITED BY MLOs**

| MLO Diseases | Vector |
|---|---|
| Alfalfa witches' broom | *Scaphytopius acutus cirrus* (10) |
| Aster yellows | |
|    North American strain | *Acinopterus angulatus* (11—26), *Chlorotettix similis*, *Colladonus commissus*, *C. flavocapitatus*, *C. geminatus* (18—36), *C. holmesi*, *C. intricatus* (12), *C. kirkaldyi* (66), *C. montanus* (8—40), *C. rupinatus*, *Elymana sulphurella*, *Endria inimica* (18—81), *Euscelidius variegatus*, *Excultanus incurvatus*, *Fieberiella flori*, *Gyponana angulata*, *G. hasta* (19—35), *Idiodonus heidmanni*, *Macrosteles fascifrons*, *M. cristatus*, *M. laevis*, *M. quadripunctulatus*, *M. sexnotatus*, *Paraphlepsius apertinus*, *Scaphytopius acutus acutus*, *S. irroratus*, *Texananus lathropi* (7—8), *T. latipex* (8—33), *T. oregonus*, *T. pergradus*, *T. spatulatus* |
|    European strain | *Aphrodes bicincta*, *Macrosteles laevis*, *M. quadripunctulatus*, *M. sexnotatus* |
|    Japanese strain | *M. masatonis*, *Scleroracus flavopictus* |
| Blueberry stunt | *Scaphytopius magdalensis* |
| Brinjal (eggplant) little leaf | *Amrasca biguttula*, *biguttula*, *Empoasca devastans*, *Hishimonus phycitis* (1—7) |
| *Bupleurum falcatum* yellows | *Macrosteles orientalis* |
| Citrus stubborn | *Circulifer tenellus*, *Scaphytopius nitridus* |
| Clover phyllody | *Aphrodes albifrons*, *A. bicincta* (20—49), *Euscelis lineolatus* (23—45), *E. galiberti*, *E. plebeja* (20—24), *Euscelidius variegatus*, *Macrosteles fascifrons* (20—43), *M. viridigriseus*, *M. cristatus*, *Scaphytopius acutus acutus* (32—35), *Scleroracus dasidus*, *S. balli*, *Speudotettix subfusculus* |
| Clover stolbur | *Euscelis plebeja* (30—40) |
| Clover stunt | *Euscelis plebeja* |
| Clover witches' broom | *Euscelis lineolatus* (23—45), *E. plebeja*, *Macrosteles viridigriseus* |
| Corn stunt | *Aphrodes bicincta*, *Baldulus tripsaci*,[116] *Dalbulus elimatus*, *D. maidis* (16), *D. gelbus*, *D. guevari*, *D. quinquenotatus*, *D. tripsacoides*, *Graminella nigrifrons*, *G. sonora* |
| Cranberry false blossom | *Scleroracus vaccinii* |
| Crotolaria witches' broom | *Orosius orientalis* |
| Flavescence doree of grapevine | *Scaphoideus littoralis* |
| Groundnut witches' broom | *Orosius argentatus* |
| Legume little leaf | *Orosius argentatus*[117] |
| Legume witches' broom | *Nesophrosyne orientalis* |
| Horse radish brittle root | *Circulifer tenellus*[118] |
| Little cherry | *Colladonus geminatus*, *Macropsis trimaculata* |
| Mulberry dwarf | *Hishimonoides sellatiformis*, *Hishimonus sellatus* |
| Papaya bunchy top | *Empoasca dilitare*, *E. papayae* |
| Parastolbur (potato) | *Euscelis plebeja* |
| Peach Eastern X | *Colladonus clitellarius*, *Fieberiella flori*, *Gyponana lamina* (35), *Paraphlepsius irroratus*, *Scaphytopius acutus cirrus* (5—25) |
| Peach Western X | *Colladonus germinatus* (22—40), *C. montanus* (45—70), *Euscelidius variegatus*, *Graphocephala confluens*, *Fieberiella flori*, *Keonolla confluens*, *Osbornellus borealis*, *Scaphytopius acutus cirrus* |
| Peach yellows | *Macropsis trimaculata* (8) |
| Peach yellow leaf roll | *Scaphytopius nitridus* and *Acinopterus angulatus* |
| Potato stolbur | *Aphrodes bicinctus*, *Hyalesthes obsoletus*, *Loepotettix dilutior*, *Macrosteles cristatus*, *M. laevis* |
| Potato purple top roll | *Alberoides deravidonus*, *Orosius albicinctus*, *O. argentatus*, *Seriana equata* |

**Table 7 (continued)**
**LEAFHOPPER VECTORS OF PLANT DISEASES INCITED BY MLOs**

| MLO Diseases | Vector |
| --- | --- |
| Potato witches' broom | *Hyalesthes obsoletus, Peragallia sinuata, Scleroracus balli, S. dasidus, S. flavopictus* |
| Rice yellow dwarf | *Nephotettix cincticeps, N. impicticeps, N. nigropictus* |
| Rhynchosia little leaf | *Ollarianus balli*[119] |
| Rubus stunt | *Macropsis fuscula, M. scotti* |
| Safflower phyllody | *Circulifer fenestratus, Neoaliturus fenestratus* (20—25) |
| Sandalwood spike | *Coelidia indica, Moonia albomaculata* |
| Sesamum phyllody | *Orosius albicinctus* |
| Soybean witches' broom | *Orosius sp.*[120] |
| *Spiroplasma citri* | *Circulifer tenellus* (10—24) |
| Sweet potato (little leaf) witches' broom | *Nesophrosyne ryukyuensis, N. lotophagorum* |
| Tobacco yellow dwarf | *Nesophrosyne lotophagorum, Orosius argentatus* |
| Tomato big bud | *Empoasca devastans, Hishimonus phycitis, Hyalesthes obsoletus, Nesophrosyne lotophagorum* |
| Yellow wilt of sugar beet | *Paratanus exitiosus* |

*Note:* Figure given in parenthesis after the name of some vectors is the incubation period in days of that MLO in that vector.

Based on Ishihara,[93a] Whitcomb and Davis,[115] Nielson,[95] Harris,[93] and subsequent literature.

is the principal field vector of corn stunt spiroplasma (CSS), maize bushy stunt spiroplasma (MBBS), and MRFV. All *Dalbulus* species tested are vectors of both CSS and MBBS and these mollicutes are pathogenic to them.[116] Moreover, a leafhopper vector may show considerable diversity in its ability to transmit different types of pathogens. Thus *Nephotettix nigropictus* can tansmit an MLO (rice yellow dwarf MLO), a large polyhedral virus (rice dwarf virus), a small polyhedral virus (RTVs), and a rhabdovirus (rice transitory yellowing virus) in a persistent (rice dwarf virus and rice yellowing virus), semipersistent (RTVs), and transovarially transmitted virus (rice dwarf virus). Other examples of transmission of different etiologic agents by the same leafhopper vector species are potato yellow dwarf bacilliform virus, WTV, and clover club leaf rickettsia-like organism (RLO) by *Agalliopsis novella*; and oat blue dwarf virus and aster yellows MLO by *M. fascifrons*. Several other examples can be found from Tables 5 and 7.

Several other types of parameters exist between MLO-leafhopper vector systems. Males of *M. fascifrons* (59% transmission) were less efficient vectors of aster yellows MLO than females (68% transmission). More leafhopper vectors generally become viruliferous after feeding on leaves showing severe symptoms or on susceptible cultivars than on leaves of recovered plants or on resistant cultivars. *S. citri* isolate MV 101 lost pathogenicity and/or transmissibility after the fourth transfer in a culture medium,[122] so that the *S. citri* isolates (C 189, C3B, and Marco) that are nonpathogenic or nontransmissible to citrus or periwinkle may have originated by some such phenomenon. The pathogen may exert a deleterious or beneficial effect on the leafhopper vector.

A viruliferous leafhopper in certain cases cannot acquire and transmit another related strain of the same MLO, since the first strain protects or prevents the transmission of the second related strain. This type of cross-protection exists in transmission of strains of aster yellows MLO and corn stunt MLO by leafhoppers. Cross-protection may be reciprocal or unilateral. In reciprocal cross-protection, the strain taken up first, whichever of the two strains it may be, does not permit the vector to acquire and transmit the second strain. Such reciprocal cross-protection occurs between two western X strains and between severe and dwarf strains of aster yellows MLO. In unilateral cross-protection, only one particular strain inhibits the uptake of the second strain. Thus while dwarf strain of aster yellows MLO

## Table 8
## PLANT VIRUSES AND MYCOPLASMALIKE ORGANISMS (MLOs) TRANSMITTED BY PLANTHOPPER VECTORS

| Virus/MLO | Vector |
|---|---|
| Barley yellow striate mosaic | *Laodelphax striatella* |
| Bobone disease | *Tarophagus proserpina* |
| Cereal tillering | *Dicranotropis hamata, L. striatella* |
| Digitaria striate mosaic | *Sogatella kolophon* |
| European wheat striate mosaic | *Javesella dubia, J. obscurella, J. pellucida* |
| Maize mosaic | *Peregrinus maidis, Unkanodes albifascia*[98] |
| Maize rough dwarf | *J. pellucida, L. striatella, P. maidis, S. vibix, Delphacodes propinqua* (syn. *J. propinqua*) |
| Maize sterile stunt | *Sogatella longifurcifera* |
| Maize stripe | *P. maidis* |
| Northern cereal mosaic | *L. striatella, Muellerianella fairmairei, Terthron albovittatus, Unkanodes albifascia, U. sapporonus* |
| Oat pseudorosette | *L. striatella* |
| Oat sterile dwarf (MLO) (= Arrhenatherum blue dwarf and Iolium enation) | *Dicranotropis hamata, J. discolor, J. dubia, J. obscurella, J. pellucida* |
| Pangola stunt | *S. furcifera* |
| Phleum green stripe | *Megadelphax sordidulus* |
| Rice black streaked dwarf | *L. striatella, U. albifascia, U. sapporonus* |
| Rice grassy stunt (MLO) | *Nilaparvata lugens* |
| Rice hoja blanca | *Sogatodes cubanus, S. oryzicola* |
| Rice ragged stunt | *J. pellucida, N. bakeri,*[126] *N. lugens* |
| Rice stripe (MLO) | *L. striatella, T. albovittatus, U. albifascia, U. sapporonus* |
| Sugarcane Fiji | *Perkinsiella saccharicida, P. vastatrix, P. vitiensis* |
| Wheat chlorotic streak | *L. striatella* |

Based on Harris,[93] Lindsten,[125] and subsequent literature.

protects the vector against uptake of the Tulelake strain, the reverse is not true. Similarly, Rio Grande strain of corn stunt MLO unilaterally protects the vector leafhopper against transmission of the Mesa Central strain. Any one or more of the several possibilities may be responsible for reciprocal or unilateral cross-protection in vectors; multiplication of the first strain may exclude the second strain; the second strain may be unable to multiply or may undergo only limited multiplication for lack of multiplication sites or the necessary intermediates; or only one strain out of the two survives because of an antagonistic reaction between them.

### B. Planthopper Vectors (Delphacidae, Homoptera, Insecta)

About 26 planthopper species of 14 genera (Tables 6 and 8) transmit 21 plant diseases caused by plant viruses and MLOs in a persistent-circulative manner. Some of these agents have a circulative-propagative relationship and are transovarially transmitted.

Transmission of rice ragged stunt virus occurred in 28% of vector planthoppers, *Nilaparvata lugens*.[108] The incubation period of several viruses in planthopper vectors is 10 to 20 d. A disease agent may be transmitted by only one or more vector species and a single planthopper vector can transmit more than one virus. In the latter cases, these viruses may variously interact in planthopper vectors. Rice grassy stunt MLO and rice ragged stunt virus are acquired by *N. lugens* both from singly as well as doubly infected plants, and each agent is transmitted independently of the other. The number of delphacid planthopper *Peregrinus maidis* capable of transmitting maize stripe virus decreased significantly if they had access to maize mosaic virus-infected maize plants 0 to 14 d before or after being fed on maize stripe virus-infected plants.[127] Moreover, insects exposed to both these viruses transmitted maize stripe virus after much delay, possibly because maize mosaic virus interferes with replication of maize stripe virus in the vector. This was suggested to be so by the longer latent period and lower titer of maize stripe virus in insects exposed to both viruses.[127] In contrast, access to maize stripe virus-infected plants generally did not influence the acquisition and transmission of maize mosaic virus so that the interference was unilateral.[127] Similarly, European wheat striate mosaic virus (probably of the rice stripe virus group) was suppressed unilaterally by oat sterile dwarf MLO in the delphacid planthopper *Javesella pellucida*.[128]

*Perkinsiella saccharicida* is an inefficient vector of Fiji disease virus of sugarcane.[129] There were several bases for this conclusion. Only a low proportion (less than one in four) of planthoppers could acquire the virus, even after their rearing on infected sugarcane plants for at least three generations which spanned 4 months. Even then less than half of the viruliferous planthoppers could transmit the virus despite their having access to test plants for a week.

## C. Transovarial Transmission

Passage of a pathogen from eggs to progeny insects is transovarial (vertical) transmission. Most of these pathogens are viruses, some are mycoplasma-like organisms and RLOs, while their vectors are aphids, leafhoppers, and planthoppers (Table 9). All these pathogens have a circulative-propagative relationship with their vectors, are sucked up from the infected plants, reach and multiply in the intestine, are passed through the gut and released into the hemolymph, and invade and multiply in various internal organs including ovaries, salivary glands, as well as in the brain and other organs of the nervous system. These pathogens spread in two directions: they are passed on to progeny insects through eggs (vertical or transovarial transmission) as well as to new plants via salivary secretions during the feeding process (horizontal transmission). However, potato leaf roll virus may be an exception to the above generalization.

Obviously, transovarial transmission can occur only through the infected female leafhoppers or planthoppers; infected males fail to transmit the pathogen to the progeny. In the case of aphids, vertical transmission of viruses occurs in infective virginoparae.

Fukushi[137,138] was the first to demonstrate convincingly the transovarial transmission of a plant virus (rice dwarf plant reovirus) in the leafhopper vector *Nephotettix cincticeps*. The virus could be serially passed through the egg to six succeeding generations over a span of more than a year without any noticeable progressive decrease in the infectivity of progeny leafhoppers. About 85% of the progeny of viruliferous female leafhoppers were viruliferous. Later this virus was found to be congenitally transmitted in two other leafhoppers, *N. nigropictus* and *R. dorsalis*. However, the percent transmission in both these vectors was less than in *N. cincticeps*.

Clover club leaf "virus" (now known to be an RLO) was the second "virus" shown to be transovarially transmitted.[139] It was transovarially transmitted serially to 21 generations of *A. novella* over a period of 5 years, and 75% of the progeny insects of viruliferous female leafhoppers were infective.[139] The original agent would be diluted to $2.8 \times 10^{-26}$ dilution at the end of the 21st generation if the agent did not multiply. This dilution, completely out of range of that known for any mechanically transmissible agent, does not occur. The third example of transovarial transmission was that of the aster yellows MLO (then known as a "virus").[121] It was serially passaged through ten generations of *M. fascifrons* leafhoppers, and the MLO was calculated to have been diluted by approximately $10^{-40}$ by the tenth passage. Instead, its concentration at the tenth passage was the same as at the first passage. WTV was the next agent found to be transovarially transmitted.[140] The original dose of this virus would be diluted to about $10^{-18}$ at the seventh passage from insect to insect if no multiplication occurred; instead, the concentration of virus in the vector was approximately the same at each passage. Potato yellow dwarf plant rhabdovirus is only occasionally transovarially transmitted to progeny insects of *Agallia constricta*.[141] Rice stripe MLO in a single viruliferous planthopper *Laodelphax striatella* could be transovarially transmitted to progeny

**Table 9**
**TRANSOVARIAL TRANSMISSION OF VIRUSES, MYCOPLASMALIKE**
**ORGANISMS (MLOs) AND RICKETTSIALIKE ORGANISMS (RLOs) IN**
**VARIOUS INSECT VECTORS**

| Pathogen | Vector | Transmission to progeny |
|---|---|---|
| **Aphids** | | |
| Lettuce necrotic yellows virus | *Hyperomyzus lactucae* | 1% |
| Potato leaf roll virus | *Myzus persicae* | 3% |
| Sowthistle yellow vein virus | *H. lactucae* | 1% |
| **Leafhoppers** | | |
| Aster yellows MLO | *Macrosteles fascifrons* | — |
| Clover club leaf RLO | *Agalliopsis novella* | Up to 75% |
| Clover phyllody MLO | *Euscelis plebeja* | <1% |
| Clover rugose leaf curl RLO | *Austroagallia torrida* | Up to 42% |
| Clover witches' broom MLO | *Euscelis plebeja* | <1% |
| Maize wallaby ear agent | *Cicadulina bimaculata* | 21% |
| Potato yellow dwarf virus | *Agallia constricta* | <1% |
| Rice dwarf virus | *Nephotettix cincticeps* | 85% |
| | *N. nigropictus* | — |
| | *Recilia dorsalis* | — |
| Rice gall dwarf virus[132] | *Nephotettix nigropictus* | 87% |
| | *N. cincticeps* | — |
| | *N. malayanus* | — |
| Russian winter wheat mosaic virus | *Psammotettix striatus* | — |
| Wound tumor virus | *Agallia constricta* | Up to 80% |
| | *Agalliopsis novella* | 2—10% |
| **Planthoppers** | | |
| Barley yellow striate mosaic virus[106] | *Laodelphax striatella* | — |
| European wheat striate mosaic virus | *Javesella pellucida* | Up to 50% |
| Maize mosaic virus[133] | *Peregrinus maidis* | — |
| Maize rough dwarf virus | *L. striatella* | 4% |
| Maize stripe virus[134-136] | *Peregrinus maidis* | — |
| Oat sterile dwarf MLO | *J. pellucida* | 0.2% |
| Rice hoja balanca virus | *Sogatodes oryzicola* | Up to 94% |
| Rice stripe MLO | *L. striatella* | 95% |

Based on Nielson,[95,130] Carter,[94] Sinha,[131] and subsequent information.

insects up to 40 following generations over a period of about 6 years without any decline in virus concentration.[142,143] As many as 95% of the progeny insects of the last (40th) generation were viruliferous without any decline in titer. Vertical transmission of virus by aphids is of a very low order (1 to 3%).

Transovarial transmission clearly suggests that the disease agent can infect the ovaries and penetrate the egg — a legitimate presumption that has been confirmed by infectivity, serologic tests, and electron microscopy. Infectivity bioassays demonstrated the presence of rice dwarf virus in the eggs of viruliferous leafhoppers,[144] while electron microscopy showed them to be present in the ovariole.[145] Fluorescent antibody technique demonstrated the presence of WTV antigen in the ovaries and eggs of leafhopper vectors.[146] Electron microscopy also detected WTV particles in ovaries of viruliferous vectors. Infectivity bioassays showed a low level of potato yellow dwarf virus infection in the ovaries of viruliferous leafhoppers.[147] However, how the transovarially transmitted pathogens actually enter the egg is not precisely known in the majority of cases, although mycetocytes of the ovarioles have been suggested to play an important role in vertical transmission of rice dwarf virus.[148,149] Two spheroids, L and H symbiotes, occur free in hemolymph. Only the L symbiotes are infected with the virus, which is carried on or between their surface membrane structures; H symbiotes are not infected. The L symbiotes later migrate and pass selectively into the mycetocytes, which then soon migrate into oocytes of the yolk-forming stage. The virus that entered the oocytes multiplies in the cytoplasm of the mycetome during embryonic development so that the embryo developed from the infected oocyte contains the virus from its inception.

No information has yet been adduced to suggest that transovarial transmission is a genetic character of the viruses concerned. It will not be surprising if it actually is so; but at present it can also be rationally explained on the basis of circulative-propagative relationship of the virus-vector concerned, leading to systemic infection of internal organs including the ovaries and eggs of a vector by the relevant virus. However, experimental evidence is available which suggests that vertical transmission of a virus may be genetically controlled by the vector. Congenital transmission of WTV was as high as 80% in efficient races of the *A. constricta* leafhopper, but was only 0 to 28% in inefficient races of the same vector.[150] The efficient races of *A. constricta* also transmitted the virus to plants efficiently, while the inefficient transovarial transmittters also transmitted the virus to plants inefficiently. Such a correlation also exists between the European wheat striate mosaic virus and races of the planthopper vector *J. pellucida*. Some planthopper vector races transovarially transmitted the virus to about 50% of the progeny insects (efficient races), while transovarial passage of the virus did not occur in other races of the planthopper vector (inefficient races).

## D. Whitefly Vectors (Aleyroididae, Homoptera, Insecta)
### 1. Viruses
Many plant diseases/pathogens are naturally transmitted by whiteflies (Table 10). Some of them cause important diseases on crops of chilli, cotton, legume, and tomato. The majority of the pathogens belong to the geminivirus group so that geminiviruses are typically transmitted by whitefly vectors. A few other isometric viruses and one straight filamentous virus are also whitefly transmitted. However, the nature of the causal agent of several of these diseases is not yet known. Whitefly-transmitted diseases were grouped by Costa[152] into mosaic, leaf curl, and yellowing diseases, but have been grouped by Muniyappa[154] into yellow mosaic, yellow vein mosaic, leaf curl, mosaic, and other types of diseases. Such a grouping of diseases has obvious disadvantages, but for the time being, no better system is available. Several of these diseases are caused by the same virus or its different strains. Mung bean yellow mosaic virus and its strains cause yellow mosaic on *Cajanus cajan*, *Centrosema* sp., *Dolichos biflorus*, *Glycine max*, *Phaseolus mungo*, and six other *Phaseolus* species. Most of the whitefly-transmitted diseases are located in tropical areas and a few in temperate areas. Only a few of the whitefly-transmitted viruses are mechanically transmissible.

### 2. Vectors
Only two whitefly genera, *Bemisia* and *Trialeurodes*, with three species in each genus, are involved in virus transmission. *B. tabaci* is the major and most important vector, while the remaining whitefly vector species transmit only one or a few viruses each. Whiteflies are piercing and sucking insects. They penetrate plant tissues intercellularly, feed in phloem, and suck plant juices through their stylets.

Three different viruses, namely, yellow vein mosaic of pumpkin and bhendi and yellow mosaic of lima beans were simultaneously carried by *B. tabaci* for 6 d. Similarly, individual insects of *B. tabaci* could simultaneously carry euphorbia mosaic and abutilon mosaic viruses. Several other such cases are known.

**Table 10**
**WHITEFLY-TRANSMITTED PLANT VIRUSES**

| Vector | Virus |
|---|---|
| *Bemisia tabaci* | Yellow mosaic diseases of acalypha, cowpea, dolichos, euphorbia, French bean (bean golden yellow mosaic), hollyhock, horse-gram, jacquemontia, jatropha, jute, lima bean, merremia, mung bean, pigeon pea, rhynchosia, rose, sida, and of soybean; tomato yellow leaf curl, tomato golden yellow mosaic, and tomato yellow top. |
| | Yellow vein mosaic diseases of ageratum, bhendi, blumea, *Casmos sulphureus*,[155] croton, cucumber vein yellowing, cucurbit, eclipta, legendra, leucas, malvastrum, mulberry yellow net vein, pumpkin, rose, salvia, tobacco yellow net, zinnia yellow net. |
| | Leaf curl diseases of balsam, cape-gooseberry, chilli, cotton leaf crumple, cotton leaf curl, geranium, hibiscus, jatropha, lupin, malvaviscus, papaya, potato, sesamum, soapwort, soybean crinkle mosaic, squash, tobacco, tomato enation leaf curl (strain of tobacco leaf curl virus), tomato leaf curl, tomato yellow leaf curl, zinnia. |
| | Mosaic and other types of diseases: abutilon mosaic (syn. abutilon infectious chlorosis, abutilon infectious variegation, and infectious chlorosis of malvaceae), anthurium disease, bean dwarf mosaic (strain of abutilon mosaic virus), cassava mosaic, cotton leaf crinkle,[156] cotton mosaic, cowpea golden mosaic, cowpea mild mottle, eupatorium pseudomosaic, jasmine chlorotic ringspot, lettuce infectious yellow,[157] potato necrosis (mild strain of tobacco leaf curl virus), soybean leaf crinkle,[158] sweet potato mosaic (sweet potato virus B), sweet potato vein clearing (chlorotic streak, medium chlorosis, and angular spot strains), tomato pale chlorosis,[159] wissadula mosaic. |
| *Bemisia gossypiperda* | Cassava mosaic, cotton leaf curl, tobacco leaf curl |
| *B. tuberculata* | Potato leaf curl |
| *Trialeurodes abutilonea* | Sweet potato feathery mottle (sweet potato yellow dwarf) |
| *T. lubia* | Cotton mosaic |
| *T. vaporariorum* | Beet pseudo yellows |

Based on Costa,[151,152] Carter,[94] Bird and Maramorosch,[153] Muniyappa,[154] Harris,[93] and subsequent literature.

Females of *B. tabaci* transmitted abutilon mosaic virus more efficiently than the males; they also transmitted yellow mosaic of euphorbia (*Euphorbia prunifolia*) nearly twice as efficiently as the males. Female *B. tabaci* is six times more efficient a vector of tomato leaf curl virus than males. Other diseases transmitted more efficiently by female whiteflies are abutilon infectious variegation, bhendi yellow vein mosaic, tomato yellow leaf curl, and mung bean yellow mosaic. Puerto Rican *Rhynchosia* and bean golden yellow mosaic viruses are transmitted nearly equally efficiently by the male and female whiteflies. Single adult whiteflies can generally acquire and transmit the infectious agent.

The ability and efficiency of whitefly vectors to acquire and transmit viruses depends upon the host used, the feeding site, the fasting period prior to acquisition and inoculation feeding, and the sex of vectors. Fasting periods of up to 4 h increased the efficiency of *B. tabaci* in acquisition of bhendi yellow vein mosaic virus, and its maximum acquisition feeding period could be reduced from 1 h to 30 min by fasting it for 1 to 6 h prior to acquisition. Similarly, the inoculation feeding period of this virus by *B. tabaci* decreased from 30 to 10 min if viruliferous whiteflies were earlier fasted for 3 to 4 h.

Whiteflies generally acquire viruses more readily from young leaves and recently infected plants than from old leaves or long-infected plants. Virus transmission and disease development are also easier when viruliferous insects feed on young leaves, while, if fed on old leaves, the disease may not develop at all. This is true of cassava mosaic, euphorbia mosaic, and yellow mosaic of jatropha (*Jatropha gossypifolia*).

The following conclusions are based on the data given in Table 11: these viruses are not

## Table 11
## TRANSMISSION DATA OF SOME PLANT VIRUSES TRANSMITTED BY WHITEFLIES

| Virus | Vector | Minimum acquisition feeding period | Minimum inoculation feeding period | Incubation period | Retention period |
|---|---|---|---|---|---|
| Ageratum yellow vein | *Bemisia tabaci* | — | — | — | 8 h |
| Beet pseudo yellows | *Trialeurodes vaporariorum* | 1 h | 6 h | 6 h | 4—6 d |
| Bhendi yellow vein mosaic | *B. tabaci* | 30 min | 10 min | 6—8 h | 24 d (lifetime) |
| Cotton leaf crumple | *B. tabaci* | 4—8 h | 1—2 h | 20 h | 5 d |
| Cotton leaf curl | *B. gossypiperda* | 3 h | 30 min | 8—24 h | 7 d |
| Cassava mosaic | *B. tabaci* | 4 h | 15 min | 8 h | 2+ d |
| Cowpea mild mottle | *B. tabaci* | 10 min | 5 min | — | — |
| Cowpea golden mosaic[160] | *B. tabaci* | 7 min | 2 min | 8 h | 21 d |
| *Euphorbia prunifolia* yellow mosaic | *B. tabaci* | 30—60 min | 30—60 min | 4—5 h | 20 d |
| French bean yellow mosaic | | | | | |
| Costa Rica isolate | *B. tabaci* | 3 h | 3 h | — | 21 d |
| Puerto Rico isolate | *B. tabaci* | 5 min | 5 min | — | 6 d |
| Horse-gram yellow mosaic | *B. tabaci* | 30 min | 10 min | 6 h | 12 d |
| *Jatropha gossypifolia* yellow mosaic | *B. tabaci* | 2 h | 10 min | — | 4 d |
| Lima bean yellow mosaic | *B. tabaci* | 1 h | 15 min | 8 h | 15—20 d |
| Malvaceae infectious chlorosis | *B. tabaci* | — | — | — | 22 d |
| *Malvastrum* yellow vein mosaic | *B. tabaci* | 30 min | 30—60 min | 4 h | 16 d |
| Mung bean yellow vein mosaic | *B. tabaci* | 15 min | 10 min | 4 h | 10 d |
| *Phaseolus lunatus* mosaic | *B. tabaci* | 1 h | 15 min | 8 h | 15—20 d |
| Sweet potato mosaic | *B. tabaci* | 5 min | — | — | — |
| *Sida* yellow mosaic | *B. tabaci* | 15 min | 20 min | — | 7 d |
| Soybean leaf crinkle[158] | *B. tabaci* | 30—60 min | 10—30 min | 8—10 h | 9 d |
| Tobacco leaf curl | | | | | |
| Indian isolate | *B. tabaci* | — | — | 4 h | Lifetime |
| Philippines isolate | *B. tabaci* | 1 h | 3 min | — | — |
| Tomato leaf curl[161] | *B. tabaci* | 30 min | 30 min | 6 h | Lifetime |
| Tomato yellow leaf curl | *B. tabaci* | 15—30 min | 15—30 min | 21 h | 20 d |
| Tomato yellow mosaic | *B. tabaci* | 30 min | 15 min | 21—24 h | Lifetime (25 + d) |
| Tomato yellow top | *B. tabaci* | Several hours | 1 h | 24 h | 10 d |
| Zinnia yellow net | *B. tabaci* | 1 h | 1 h | 15—18 h | Lifetime |

Based on data given in reviews by Costa,[151,152] Bird and Maramarosch,[153] Muniyappa,[154] and subsequent literature.

acquired as rapidly as mechanically transmitted viruses; they are persistent viruses (and are not stylet-borne), the transmission efficiency of vectors increases with longer feeding periods of up to several hours on virus sources; the acquisition threshold is 5 to 7 h, the acquisition feeding period is generally longer than the inoculation feeding period; transmission by individual vectors generally decreases gradually with increasing time subsequent to acquisition; inoculation thresholds usually range from 10 to 60 min, but longer thresholds of up to 6 h are also known; a definite but short incubation period generally of 4 to 10 h (range 4 to 24 h) occurs in most cases; viruses are retained in vectors from a few to up to 25 d; and serial transmission is generally intermittent and inefficient.

The relationship of whitefly-transmitted viruses with their vectors is of the circulative-nonpropagative type. However, a circulative-propagative relationship cannot be ruled out.[151] Multiplication of squash leaf curl geminivirus in *B. tabaci* is suggested by a relatively long incubation period in the vector as compared to other geminiviruses, by a circulative relationship, by a high frequency of transmission rates in serial transfers established by individual whiteflies, and by apparent harmful effects of the virus on vector.[162]

## E. Beetle Vectors (Coleoptera, Insecta)
### 1. Viruses
So far, 39 viruses and 1 viroid have been reported to be beetle transmitted (Table 12). All these viruses are icosahedral and about 25 to 30 nm in diameter, contain single-stranded RNA of positive polarity, are relatively stable, reach high concentrations in plant tissues, and are easily mechanically transmissible. They belong to the bromovirus, comovirus, sobemovirus, and tymovirus groups, besides some unclassified viruses or one or two viruses of a few other groups and a viriod.

### 2. Vectors
About 80 beetle vector species are known. Genera of beetles containing virus vector species belong to different families: one genus (*Epilachna*) to the family Coccinellidae (the leaf-feeding beetles); one genus (*Epicauta*) to the family Meloidae, three genera (*Apion, Cionus,* and *Sitona*) to the family Cuculionidae; and all the rest (about 28 genera) to the family Chrysomelidae (leaf beetles and flea beetles). The leaf beetles and flea beetles are widespread and are the most important beetle vectors. Some beetle vectors transmit only one virus each, while others can transmit more. *Ceratoma trifurcata* transmits about ten viruses.

Only a few studies have been published concerning transmission of viruses by larvae. Larvae of *Phaedon cochleariae* were better transmitters of turnip yellow mosaic virus (TYMV) than adults. The same is true of maize chlorotic mottle virus (MCMV) transmission by larvae and adults of *Oulema melanopus*. Possibly viruliferous larvae of western corn root worm (*Diabrotica virgifera*) transmit MCMV in soil in the U.S. In contrast, larval stages of *O. melanopus* transmitted phleum mottle virus and cocksfoot mottle virus at much lower levels than adults. However, viruses acquired by a larva cannot be passed on to the adult.

Bean pod mottle, cowpea mosaic (CPMV), squash mosaic, and southern bean mosaic (SBMV) viruses have been found in the hemolymph of beetle vectors. They appear in hemolymph soon after acquisition from infected plants. Moreover, beetles can be made viruliferous by injecting a virus into the hemocoel. Regurgitants contain virus particles when beetles feed on infected plants as well as when purified virus is injected into the hemocoel. It is assumed that viruses move from plant to the gut, from gut (midgut) to hemocoel, from hemocoel to regurgitant, and from regurgitant back to the plant which becomes infected. Some stable nontransmissible viruses are also taken in by beetles, are found in the hemolymph and regurgitant, and are later deposited on leaf surfaces during feeding, but cause no infection.[172] Beetles are able to transmit viruses only during feeding and by no other means.

**Table 12**
**VIRUSES TRANSMITTED BY BEETLE VECTORS**

| Virus | Vector |
|---|---|
| **Comovirus Group** | |

| Virus | Vector |
|---|---|
| Bean pod mottle | *Ceratoma trifurcata* (7), *C. ruficornis, Colapsis flavida, C. lata, Diabrotica balteata, D. undecimpunctata, Epicauta vittata, Epilachna varivestis* |
| Bean rugose mosaic | *Ceratoma ruficornis, Diabrotica balteata, D. adelpha* |
| Broad bean stain | *Apion vorax, A. aestivum, A. aethiops, Sitona lineatus, S. hispidula* |
| Cowpea mosaic | |
|    Arkansas strain | *Acalymma vittatum, Ceratoma arcuata, C. atrofaciata, C. ruficornis, C. trifurcata* (8), *C. variegata, Diabrotica adelpha, D. undecimpunctata, D. virgifera, D. laeta, Diphaulaca meridae, Epilachna varivestis, Epilachna* sp., *Gynandrobrotica variabilis, Systena* sp. |
|    Cuban strain | *Ceratoma ruficornis* |
|    Nigerian strain | *Ootheca mutabilis* (2) |
|    Severe strain | *Acalymma vittatum, Ceratoma arcuata, C. ruficornis, C. trifurcata, Diabrotica balteata, D. speciosa, D. undecimpunctata, Epilachna varivestis* |
|    Trinidad strain | *Ceratoma ruficornis* (6), *C. trifurcata* |
|    Yellow strain | *Acalymma vittatum, Ceratoma trifurcata, Diabrotica balteata, D. undecimpunctata, D. virgifera, Epilachna varivestis, Medythia quaterna, Nematocerus acerbus, Ootheca mutabilis, Paraluperodes quaternus* (syn. *Medythia quaterna*) |
| Echtes Ackerbohenen mosaic (broad bean true mosaic) | *Apion vorax, A. aethiops, A. aestivum, Sitona lineatus* |
| Quail pea mosaic | *Ceratoma ruficornis, C. trifurcata, Diabrotica undecimpunctata, Epilachna varivestis* |
| Radish mosaic | *Diabrotica undecimpunctata, Epitrix hirtipennis, Phyllotreta striolata, P. atra, P. cruciferae.* |
| Red clover mottle | *Apion apricans, A. varipes* |
| Squash mosaic | *Acalymma thiemei, A. trivittatum* (17), *A. vittatum, Diabrotica balteata, D. longicornis, D. soror, D. undecimpunctata* (20), *D. virgifera, Epilachna chrysomelina* |

| Virus | Vector |
|---|---|
| **Bromovirus Group** | |

| Virus | Vector |
|---|---|
| Broad bean mottle | *Acalymma trivittata* (Syn. *A. trivittatum*), *Colapsis brunnea, C. flavida, Diabrotica undecimpunctata* |
| Brome mosaic | *Diabrotica longicornis, Lema sexpunctata, Oulema melanopus* |
| Cowpea chlorotic mottle | *Ceratoma ruficornis, C. trifurcata* (3), *Diabrotica adelpha, D. balteata, D. undecimpunctata, Epilachna varivestis* |

| Virus | Vector |
|---|---|
| **Sobemovirus Group** | |

| Virus | Vector |
|---|---|
| Cocksfoot mottle | *Oulema lichenis, O. melanopus* (15) |
| Rice yellow mottle | *Chaetocnema abyssinica, C. kenyensis, C. pulla, Cryptocephalus chalybeipennis, C. nigrum, Dactylispa bayoni, Dicladispa paucispina, D. viridicyanea, Monolepta flaveola, M. haematura, Sesselia pusilla, Trichispa sericea, Oulema dunbrodiensis* |
| Southern bean mosaic | *Acalymma vittatum, Atrachya menetriesi, Ceratoma trifurcata* (5,14), *Epilachna varivestis, Madurasia obscurella*[166] |
| Turnip rosette | *Phyllotreta striolata* |

| Virus | Vector |
|---|---|
| **Tymovirus Group** | |

| Virus | Vector |
|---|---|
| Andean potato latent | *Epitrix* sp. |
| Belladona mottle | *Epitrix atropae* |
| Dulcamara mottle | *Psylliodes affinis* |
| Eggplant mosaic | *Epitrix* spp. particularly *E. fuscula* |

**Table 12 (continued)**
**VIRUSES TRANSMITTED BY BEETLE VECTORS**

| Virus | Vector |
|---|---|
| **Tymorvirus Group** | |
| Erysimum latent | *Phyllotreta atra, P. nigripes, P. undulata* |
| Okra mosaic | *Podagrica sjostedti, P. uniforma, Syagrus calcaratus* |
| Passion fruit yellow mosaic | *Diabrotica speciosa*[167] |
| Scrophularia mottle | *Cionus alauda, C. hortulanus, C. scrophulariae, C. tuberculosis* |
| Turnip yellow mosaic | *Phaedon cochleariae* (14), *Phyllotreta* spp., *Psylliodes* spp., *Pedilophorus* sp.[168] |
| Wild cucumber mosaic | *Acalymma trivattatum* |
| **Miscellaneous Viruses** | |
| Bean curly dwarf mosaic | *Ceratoma ruficornis, Diabrotica balteata, D. undecimpunctata, Epilachna varivestis, Gynandrobrotica variabilis* |
| Bean mild mosaic | *Diabrotica balteata, D. undecimpunctata, Epilachna varivestis, Gynandrobrotica variabilis* |
| Bean yellow stipple | *Ceratoma ruficornis, Diabrotica adelpha, D. balteata, D. undecimpunctata, Epilachna varivestis* |
| Blackgram mottle | *Ceratoma trifurcata, Epilachna varivestis, Monolepta signata*[169] |
| Cowpea mottle | *Ootheca mutabilis* |
| Cucumber green mottle mosaic | *Raphidopalpa fevicollis*[4] |
| Maize chlorotic mottle | *Chaetocnema pulicaria, Diabrotica longicornis, D. undecimpunctata, D. virgifera, Oulema melanopus, Systena frontalis* |
| Phleum mottle | *Oulema lichenis, O. melanopus* |
| Potato spindle tuber viroid | *Disonycha triangularis, Epitrix cucumeris, Leptinotarsa decemlineata, Systena taeniata* |
| Radish enation mosaic | *Diabrotica undecimpunctata, Phyllotreta striolata* |
| *Solanum nodiflorum* mottle | *Henosepilachna australica, H. guttatopustulate, H. sparsa*[170] |
| Tobacco ringspot | *Epitrix cucumeris, E. parvula* |
| Turnip crinkle | *Phyllotreta* spp. (1), *Psylloides cuprea, P. chrysocephala* (1). |
| Urdbean leaf crinkle | *Henosepilachna dodecastigma*[171] |

*Note:* Figures in parentheses after the names of some beetle vectors indicate the reported maximum retention period in days.

Based on Carter,[94] Walters,[163] Fulton et al,[164,165] Harris,[93] and subsequent literature.

## 3. Virus-Vector Relationship

Beetles have an acquisition access feeding period of 24 h or less. However, in several cases an acquisition threshold of a few minutes is sufficient, while in some cases vectors become viruliferous even after the first bite. Increased transmission percentages are achieved with increased acquisition feeding. Some of the viruses (cowpea [severe] mosaic, radish enation mosaic, radish mosaic, turnip crinkle, and turnip rosette) are retained by vectors for short periods usually ranging from 24 to 48 h, while others (Arkansas and Trinidad cowpea mosaic, bean pod mottle, cocksfoot mottle, cowpea strain of SBMV, squash mosaic, and TYMV) are retained by vectors for prolonged periods ranging from 7 to 20 d or even more.[163,173] *Pedilophorus* sp. transmits TYMV for 48 h.[168]

Beetle transmission of viruses like CPMV, SBMV, squash mosaic, etc., that are transmitted over several days exhibits one characteristic feature. The number of beetles transmitting any one of these viruses is highest on the first day after acquisition, but their transmission rates drop rapidly during subsequent days and are later transmitted for about 20 d or longer by only occasional beetles. Moreover, virus transmission may be erratic since a beetle may transmit a virus after several days of nontransmission. Retention and transmission of a virus varies greatly, depending on any one of several variables like virus strain, vector species, host plant, and environmental conditions. Virus-vector relationships of some beetle-transmitted viruses have been described as nonpersistent (for viruses retained for 1 to 2 d only) and of others as persistent (for viruses retained for 7 to 20 d, depending upon the virus retention period of the vector. However, Harris[93] suggests that these terms may better be avoided, primarily because the retention period of persistent viruses depends upon the duration of the acquisition access feeding period and also because retention times vary greatly, depending upon virus-vector-plant combination and environmental conditions. Grouping of viruses into two groups (those retained for short durations of up to 1 to 2 d and those retained for longer periods of up to 7 to 20 d) on the basis of their retention period by the vectors should also be avoided.[164] Viruses have no incubation period in beetles and are always transmitted immediately after acquisition.

A virus may be retained for different periods by different beetle vectors. Bean rugose mosaic virus is retained by *C. ruficornis* for longer periods (9 d) than by *D. balteata* (3 d). CPMV (severe strain) is retained by the Mexican bean beetle for only short periods of 1 to 2 d, but is retained by the bean leaf beetle for longer periods of several days and even up to 8 d.

Beetles have biting mouthparts. They bite the plant and push the plant material back into the gut. This can result in accumulation of infected material, and thereby of viruses, inside the beetles during acquisition feeding. Thus virus titer of bean pod mottle virus in *C. trifurcata* was higher than that in the infected plants on which they had fed,[174] and the concentration of the cowpea strain of SBMV in the regurgitant of beetles was higher than that in sap from infected tissues.[164] Moreover, the amount of SBMV in the regurgitants of *Epilachna varivestis* (Mexican bean beetle) and *C. trifurcata* and serial transmission levels of this virus by both these beetles was related to the virus concentration present in material from which acquisition occurred.[175]

The amount of infected tissue consumed during acquisition feeding determines the length of the period for which the beetles remain viruliferous. This is particularly valid in cases where viruses are retained by vectors for long durations. It obviously has something to do with the amount of virus taken in by the vector. Increased feeding time increases the percentage of vectors becoming viruliferous as well as the retention period of a virus by a beetle. Thus, the retention period of these viruses by vectors depends upon the quantity of infective material ingested or, in other words, on acquisition feeding periods. Retention of squash mosaic virus by beetle vectors increased with increasing length of acquisition feeding: 2 d acquisition feeding caused 2 d retention, 10 d feeding resulted in 8 d retention, and 13 d feeding in 17 d retention.[176] Each beetle has its own acquisition feeding period to achieve its maximum virus-retaining capacity, which, if once reached, will not increase further with longer acquisition feedings.

A population of beetles transmitting a virus after an acquisition feeding determines the vector efficiency. It can vary greatly in different beetle species. Transmission efficiency of EAMV was 43% with *Apion vorax*, but only 2% with *A. aethiops*, *A. aestivum*, or *Sitona lineatus*; that of bean rugose mosaic virus was 80% with *Ceratoma ruficornis*, 20% with *Diabrotica balteata*, and 10% with *D. adelpha*; and that of cowpea chlorotic mottle virus (CPMV) was 12% with *C. trifurcata* (bean leaf beetle), but 16.2% with *D. undecimpunctata* (spotted cucumber beetle.)[177] The same species of beetle vectors may transmit various viruses with greatly varying efficiency: *C. trifurcata* transmits CCMV at 20% or less efficiency, but transmits CPMV or SBMV at about 100% efficiency. According to another report, only 2% of *C. trifurcata* transmitted CCMV, but 12% of them transmitted CPMV.[177]

Transmission efficiency of viruses by beetles has also been expressed in other terms. *Raphidopalpa fevicollis* (cucumber green beetle) transmitted cucumber green mottle mosaic virus to 10% of the tested plants.[4] *Pedilophorus* species transmitted TYMV to 2.5% of *Cardamine lilacina* seedlings and to 10% of Chinese cabbage seedlings.[168] *Ceratoma arcuata* and *D. speciosa* transmitted cowpea (severe) mosaic virus from bean to bean at a rate of 50% and 43%, respectively.[178]

### 4. Transmission Mechanism

The presence of beetle-transmitted ciruses in hemolymph and the long retention periods of these viruses by beetles suggests virus replication in them. There is, however, no convincing evidence about such a presumption. Instead, a decrease of virus concentration in hemolymph with time suggests the absence of virus multiplication in vectors. This is firmly believed and has also been shown to be so. SBMV concentration in hemolymph of *C. trifurcata*, as well as the ability of this vector to transmit the virus, decreased with time. Beetles have biting mouthparts and transmit viruses mechanically, since presumably the latter are carried as contaminants on mouthparts of beetles. The noncirculative-nonpersistent nature of these viruses confirms this. However, long retention periods of viruses in vectors as well as their intermittent transmission needed proper explanation. Regurgitation of infected material from the foregut was related to the ability of beetles to transmit these viruses mechanically to all plants they bite during the period of mastication.[179,180] Such viruses, being stored away in infected material in gut and regurgitated later, will be retained and (intermittently) transmitted for prolonged periods. A relationship seemed to exist between the virus in regurgitant and transmission.

However, later studies indicated that virus transmission by beetles is not entirely a mechanical process due to contamination of the mouthparts of the beetle; a more complex biological phenomenon appears to be involved.[181] Both a transmitted virus (cowpea strain of SBMV) and a nontransmitted virus (cowpea strain of TMV) appear in the regurgitant after infection of purified virus preparations into the hemocoel and are later deposited on leaf surfaces during feeding.[172] However, only the transmitted virus is able to cause infection (and hence is transmitted), while the nontransmitted virus also is deposited on the leaf surface, but does not infect the plants and hence is not transmitted.[172] This is despite the fact that the host plant is equally susceptible to both viruses upon mechanical inoculation and that infectivity of both viruses is at the same level as that in the plant. Still, beetles transmit only SBMV. An identical situation prevails with regard to the beetle nontransmitted tobacco ringspot virus.[164] Similarly, beetles which earlier had been acquisition fed on a cowpea plant infected with both cowpea (severe) mosaic and cowpea aphid-borne mosaic viruses, could transmit only cowpea (severe) mosaic virus to cowpea plants. Both viruses were deposited on leaf surfaces. TMV was also liberally deposited on leaf surfaces by beetles during feeding without causing infection of the plants.[165] Clearly, the transmission mechanism seems more complicated since it fails to explain the failure of beetles to transmit some viruses in conditions in which they are able to transmit other viruses.

The regurgitant of leaf-feeding beetles was found to contain a factor that prevented infection by beetle-nontransmissible viruses, but has no effect on beetle-transmissible viruses.[182] Monis et al.[183] confirmed that regurgitant from *Epilachna varivestis* (Mexican bean beetle) and *Ceratoma trifurcata*, mixed with purified virus preparations and inoculated to systemic hosts, inhibited the transmission of beetle-nontransmissible viruses (zucchini yellow mosaic, tobacco ringspot, and TMV), but had little or no effect on beetle-transmissible viruses (cowpea severe mosaic and squash mosaic viruses). The regurgitant of *D. undecimpunctata*, *C. trifurcata*, and *E. varivestis* showed high levels of RNase activity and virus inoculation in the presence of pancreatic RNase, at an activity equivalent to that in beetle regurgitant, and simulated the selective virus transmission by beetles, indicating that RNases are responsible for this selectivity.[184] However, beetle-nontransmissible viruses (TMV, zucchini yellow mosaic) are not inactivated in vitro when mixed with beetle regurgitant.[182,183] This suggests that RNase may either be degrading viral RNA soon after its uncoating or may be acting on plant cells to make them somehow inhospitable for viral RNA replication. The selective transmission of viruses by beetles has now been ascribed to some other reason. It is suggested

## Table 13
## KNOWN OR PRESUMED PLANT VIRUSES OR DISEASE AGENTS TRANSMITTED BY ERIOPHYID MITES

| Virus/Agent | Vector |
|---|---|
| Disease agents/viruses infecting grasses | |
| Agropyron mosaic virus | *Abacarus hystrix* (<1) |
| Onion mosaic virus | *Eriophyes (Aceria) tulipae* |
| Ryegrass mosaic virus | *A. hystrix* (30) |
| Wheat spot mosaic agent | *E. tulipae* (65) |
| Wheat streak mosaic virus | *E. tulipae* (39) |
| Disease agents/viruses infecting dicots | |
| Black currant reversion agent | *Cecidophyopsis ribis* (syn. *Phytoptus* and *Eriophyes*) (<1) |
| Cherry leaf mottle agent | *Phytoptus inaequalis* |
| Fig mosaic agent | *E. ficus* (70) |
| Peach mosaic agent | *Phytoptus insidiosus* (2.5) |
| Diseases possibly transmitted by mites | |
| Cadang cadang viroid | Mite vector unknown |
| Hordeum mosaic virus | Mite vector unknown |
| Little cherry (in British Columbia) | Mite vector unknown |
| Doubtful or erroneous reports of mite transmission | |
| Mango malformation agent | *E. mangiferae* |
| Prunus latent virus | *Aculus* (= *Vastes*) *fockeui* |
| Vine Panachure agent | *Colomerus vitis* |
| Diseases possibly caused by mites without a virus/agent | |
| Citrus concentric ring blotch | *Calacarus citrifolii* |
| Pigeon pea sterility mosaic | *E. cajani* |
| Rose rosette | *Phyllocoptes fructiphilus* |
| Sugarcane chlorotic streak | *E. sacchari* |

*Note:* Figures given in parentheses represent the percent transmission by the vector of the relevant disease agent.

Based on Slykhuis[110,186,187] and Oldfield.[188]

that beetle-transmissible viruses can be translocated from the inoculation point to distant cells and infect them, while beetle-nontransmissible viruses cannot do so.[165,185] The translocation of beetle-transmissible viruses from inoculation sites presumably removes them from the sites where RNases of regurgitant are active. They are therefore able to cause infection.

### F. Mite Vectors (Eriophyidae, Acarina, Arachnida)
#### 1. Disease Agents
About 19 plant diseases have at one time or other been reported to be transmitted by mites. Their present disposition is as shown in Table 13. Some disease agents (three to four only) are viruses, while the nature of others is not known. It has not been conclusively established so far that some of the diseases listed in Table 13 are caused by an infectious agent; instead, these diseases may be caused by mites themselves.

#### 2. Vectors
All mite vectors are placed in the family Eriophyidae, order Acarina, class Arachnida. Eriophyid mites are small, usually less than 0.25 mm long, wormlike insects having only two pairs of legs, few setae, and mouthparts that are specifically adapted for the piercing and sucking processes. Generally, eriophyid mites complete the life cycle from egg to egg (egg, first nymph, second nymph, adult, egg) within 2 to 3 weeks, while some can do so in 6 to 8 d under ideal conditions in summer. They are all phytophagous, and several are

highly host specific. A pharyngeal pump and a rostrum provide the piercing and sucking action during feeding of eriophyid mites. Mouth stylets are five: two anterior chelicerae are small and penetrate slowly, probably only up to a few cell layers of the tissue; two auxiliary stylets; and one oral stylet.

Both nymphs and adults can transmit viruses with nearly equal efficiency, as, for example, wheat streak mosaic virus (WSMV) by *Eriophyes tulipae*, wheat spot mosaic agent by *E. tulipae*, and ryegrass mosaic virus by *Abacarus hystrix*. Large amounts of virus appear to be ingested and accumulated in the gut of mite nymphs. The virus passes through molt, remains in an infective state, and is transmitted later for almost the entire life span of mites (which is only 7 to 10 d).

*3. Transmission*

The relationship of WSMV with its vector *E. tulipae* is best known. The vector possesses a long acquisition feeding period (15 min), a long inoculation feeding period, and a long retention period (6 to 7 or even up to 9 d), and is retained by the vector through molting. All this shows that this virus has a persistent-circulative relationship with the vector.

Densely packed accumulations of WSMV particles occur in the alimentary canal (mainly in the hindgut) of *E. tulipae* during extended feeding periods on infected plants.[189] Later, however, Paliwal[190] found that the virus particles accumulate mainly as dense packs in the lumen of the sac-like midgut, which was earlier wrongly identified as the hindgut, that the posterior midgut contained large amounts of intact virus particles for at least up to 5 d after the removal of mites from infection sources, and that virus particles occur in salivary glands of mite vectors.

It has been speculated on the basis of the above considerations that ingested and accumulated virus particles are presumably released from the midgut, are circulated in the body cavity of the vector, pass into salivary glands, and are finally transmitted to plants via salivary secretion during feeding. Plants could also possibly become infected by the virus by its backflow from gut to mouthparts during feeding or by introduction of defecated virus into plant cells through feeding punctures or abrasions caused by anal setae or anal suckers.

Agropyron mosaic virus particles, in contrast to WSMV particles, do not occur in the alimentary canal or anywhere else in the body of its mite vector *A. hystrix*. Obviously, this virus does not have a circulative relationship with its vector, but is conceivably transmitted in some other manner. BSMV and brome mosaic virus, like WSMV, are also efficiently ingested and accumulated in an infective state in the midgut as well as found in the body cavity of the mite. But these viruses are not transmitted by mites because of their inability to enter the salivary glands.

Various evidence shows that virus-vector specificity exists. *E. tulipae* transmits WSMV in all countries of North America and Europe except Yugoslavia, where it is transmitted by *E. tosichella*. Mite-nontransmitted brome mosaic virus accumulated in densely packed masses in the midgut of *E. tulipae*, when fed on infected plants, but no transmission took place.[191] This is despite the fact that infective brome mosaic virus persisted in these mites for at least 6 d. The nonvector mites *Aculodes mackenzie* and *Abacarus hystrix* contained some WSMV particles in the lumen of the midgut after being fed on infected plants without being able to transmit this virus. Of the three mites (*A. hystrix, Vastes dubius*, and *Eriophyes* sp.) found on ryegrass mosaic virus-infected plants, only *A. hystrix* could transmit the virus. The nonvector mite *E. tulipae* efficiently ingested and accumulated large amounts of BSMV in the lumen of its midgut when fed on infected plants. The harbored virus was in an infectious state and was retained for at least 4 d after removal of mites from a virus source. However, these mites failed to transmit the virus.[190] All the above evidence shows that a transmissible virus has a highly specific biological interaction with its vector.

## Table 14
## DISEASE AGENTS TRANSMITTED BY VARIOUS VECTORS

| Disease agent | Vector |
|---|---|

**Mealybugs (Pseudococcidae) (based on Roivainen,[192] Harris,[93] and other literature)**

| | |
|---|---|
| Cocoa swollen shoot virus | *Delococcus tafoensis* (syn. *Formicoceus tafoensis*), *Dysmicoccus brevipes, Ferrisia virgata, Maconellicoccus ugandae* (syn. *Phenacoccus* sp.), *Paracoccus* sp. near *proteae* (syn. *Pseudococcus* sp. near *proteae*), *Paraputo anomalus* (syn. *P. ritchiei*), *Planococcoides njalensis, Planococcus* sp. near *celtis, P. citri, P. kenyae, Pseudococcus concavocerarii, Pseudococcus* sp. near *fragilis, P. hargreavesi* (syn. *P. bukobensis*), *P. longispinus* (syn. *P. adonidum*), *Pseudococcus* sp. near *masakensis, Tylococcus westwoodi* |
| Cocoa mottle leaf virus | *Dysmicoccus brevipes, Planococcoides njalensis, Planococcus* sp. near *celtis, P. citri, P. kenyae, Pseudococcus concavocerarii, Pseudococcus* sp. near *fragilis, P. hargreavesi* (syn. *P. bukobensis*) |
| Cocoa Trinidad virus | *Dysmicoccus brevipes, Dysmicoccus* sp. near *brevipes, Ferrisia virgata, Planococcus citri, Pseudococcus comstocki* |
| Ceylon cocoa virus | *Planococcus citri, P. lilacinus* |
| Dasheen mosaic virus | *Planococcus citri, Pseudococcus longispinus* (syn. *P. adonidum*) |
| Grapevine virus A | *Pseudococcus longispinus, Planococcus ficus*, and *P. citri*[193] |
| Rice chlorotic streak virus | *Heterococcus rehi*[194] |

**Psyllids (Psyllidae) (based on Kaloostian[195])**

| | |
|---|---|
| Carrot proliferation MLO | *Trioza nigricornis* |
| Citrus greening MLO | *T. erytreae, Diaphorina citri* |
| Pear decline MLO | *Psylla pyricola, P. pyri, P. pyrisuga* |

**Piesmids (Piesmidae) (based on Proeseler[111] and Harris[93])**

| | |
|---|---|
| Beet savoy virus | *Piesma cinereum* |
| Beet leaf curl virus | *P. quadratum* |
| Beet latent rosette RLO | *P. quadratum* |

**Thrips (Thripidae) (based on Ananthakrishnan,[196] Harris, [93] and other literature)**

| | |
|---|---|
| Tomato spotted wilt virus | *Frankliniella fusca, F. occidentalis, F. schultzei, Thrips tabaci, Scirtothrips dorsalis*[197] |
| Tobacco streak virus | *Frankliniella occidentalis* and/or *Thrips tabaci*,[198] *Frankliniella* sp. |

**Mirids (Miridae) (based on Harris[93] and subsequent literature)**

| | |
|---|---|
| Spinach blight virus | *Lygus pratensis* |
| Velvet tobacco mottle virus | *Cyrtopeltis nicotianae*[199] |

## G. Miscellaneous Vectors

### 1. Mealy Bugs (Pseudococcidae, Homoptera, Insecta)

Some vectors are known to transmit only a few plant disease agents (Table 14). Of these, mealy bugs transmit one of the most serious and devastating plant virus diseases — the cocoa swollen shoot (CSSV). The status of the other three cocoa viruses is not yet fully worked out; they may be independent viruses or isolates of CSSV, but are treated here as separate viruses. Some valid and some incompletely known mealy bug species transmit these viruses. The virus-vector relationship is generally considered to be of the persistent-circulative type. Infection rate increases with increased acquisition access period. The minimum acquisition access period has been reported to be 30 min in the case of cocoa Trinidad virus-*Planococcus citri* mealy bug complex, and 90 min in the case of an isolate of the

CSSV-*Planococcoides najalensis* combination. The optimum acquisition access period for maximum transmission of CSSV and cocoa mottle leaf virus by starved *Ferrisia virgata* and *P. najalensis,* respectively, was 48 to 72 h. The retention period of CSSV in various vectors varies from about 20 to 40 h. The minimum inoculation access period of cocoa viruses is about 15 min, but the probability of infection increases with an increase of the inoculation access period up to about 3 h. Virus-vector specificity exists, but transovarial transmission is not reported.

### 2. Piesmids (Piesmidae, Hemiptera, Insecta)
Only two species of a genus of lace bugs act as virus vectors. They are *Piesma cinereum* in North America and *P. quadaratum* in Europe, and they transmit two viruses and one RLO in all (Table 14). The beet leaf curl rhabdovirus multiplies in its vector and has a circulative-propagative relationship (Table 6). The average incubation period of this virus in its vector is reported to vary from 3 to 4 weeks, but the minimum incubation period is 14 d. The little available information suggests that transmission of beet savoy virus by *P. cinereum* is of the nonpropagative-circulative type.

### 3. Thrips (Thripidae, Thysanoptera)
Only two viruses are known to be transmitted by thrips (Table 14). Tomato spotted wilt virus has a definite latent period, a long retention period, a persistent, possibly circulative relationship with its vectors, and is acquired and transmitted only by nymphal or "larval" instars. The reported acquisition and inoculation thresholds are 30 and 5 min, respectively, but increased acquisition feeding periods (15 min, 1 h, 1 d, and 4 d) resulted in increased transmission (4, 33, 50, and 77%, respectively) of the virus. The incubation period in various vectors ranges from 4 to 16 d. The maximum retention period of the virus in *Frankliniella schultzei, F. occidentalis, Thrips tabaci,* and *F. fusca* are 22 to 24, 30, and 43 d, respectively. The virus has as yet neither been detected to multiply nor has it shown transovarial passage in vectors.

## H. Nematode Vectors (Nematoda)
Nettle-head disease of hops was claimed by Percival[200] to be spread by the nematode *Heterodera humuli.* It was confirmed much later when it was shown that this disease is caused by a strain of arabis mosaic virus and is transmitted by *Xiphinema diversicaudatum.* However, Hewitt et al.[201] were the first to prove experimentally that a virus, grapevine fanleaf virus, was transmitted by a nematode, *X. index.* Since then, much information has been published and reviewed.[202-205]

### 1. Viruses
All nematode-transmitted viruses (Table 15) are relatively stable in vitro, are sap transmissible, possess a wide host range of woody and herbaceous plants, and several are also transmitted through seed and pollen. They are divided into two groups on the basis of particle shape: nepoviruses and tobraviruses. Nepoviruses are icosahedral viruses of about 30 nm diameter, possess similar physicochemical properties, are transmitted by species of *Xiphinema* or *Longidorus,* and are also transmitted through seeds and pollen of infected plants. Tobraviruses (pea early browning and tobacco rattle viruses) have tubular particles of different sizes, have a bipartite genome, and are transmitted by species of *Trichodorus* and *Paratrichodorus.*

### 2. Vectors
The nematode genera involved in virus transmissions are cosmopolitan in distribution and are believed to be obligatory but migratory ectoparasites, getting their food mainly from root tips. The *Longidorus* and *Xiphinema* vectors of nepoviruses are placed in the family

**Table 15**
**VIRUSES TRANSMITTED BY NEMATODES**

| Virus | Vector |
| --- | --- |
| Arabis mosaic | |
|   Type strain | *Xiphinema diversicaudatum, X. coxi, X. bakeri* |
|   Hop strain | *X. diversicaudatum* |
| Artichoke Italian latent | *Longidorus apulus, L. attenuatus* |
| Ash ringspot (Type strain) | *X. americanum* |
| Brome mosaic | *L. macrosoma, X. diversicaudatum* |
| Carnation ringspot | *X. diversicaudatum* |
| Cacao necrosis | *Longidorus* sp. (suspected) |
| Cherry leaf roll | *X. coxi, X. diversicaudatum, X. vuittenezi* |
| Cherry rasp leaf | *X. americanum* |
| Cowpea mosaic | *X. basiri* |
| Grapevine fanleaf | *X. index, X. italiae* |
| Grapevine chrome mosaic | *X. index, X. vuittenezi* (suspected) |
| Gladiolus notch leaf | |
|   California strain | *Paratrichodorus allius, P. porosus, P. minor* (syn. *P. christiei*) |
|   Dutch strain | *Trichodorus similis* |
|   English strain | *P. pachydermus, T. primitivus* |
|   German strain | *T. primitivus* |
|   Japanese strain | *P. minor* |
|   Oregon strain | *P. allius* |
|   Wisconsin strain | *P. minor* (syn. *P. christiei*) |
| Mulberry ringspot | *L. martini* |
| Myrobalan latent ringspot | *Longidorus* sp. (suspected) |
| Pea early browning | |
|   Dutch strain | *P. pachydermus, P. teres* |
|   English strain | *P. anemones, P. pachydermis, T. viruliferus* |
|   Italian strain | *T. viruliferus* |
| Peach rosette mosaic | *X. americanum, L. diadecturus*[206] |
| Prunus necrotic ringspot | *L. macrosoma* |
| Raspberry ringspot | |
|   Cherry strain | *L. macrosoma, X. diversicaudatum* |
|   English strain | *L. elongatus, L. macrosoma* |
|   Type strain | *L. elongatus* |
| Satellite tobacco ringspot | *X. americanum* |
| Strawberry latent ringspot | *X. coxi, X. diversicaudatum* |
| Tobacco rattle (Dutch strain) | *P. anemones, P. nanus, P. pachydermus, P. teres, P. allius, T. cylindricus, T. viruliferus, T. sparsus* |
| Tobacco ringspot | *X. americanum, X. coxi* |
| Tomato black ring | |
|   English strain | *L. attenuatus* |
|   Scottish strain | *L. elongatus* |
| Tomato ringspot | *X. brevicolle, X. revesi,*[207] *X. californicum,*[208] *X. americanum* |

Based on Taylor and Cadman,[202] Taylor,[203,204] Carter,[94] Harris,[93] and subsequent literature.

Longidoridae of suborder Dorylaimina, while *Trichodorus* and *Paratrichodorus* vectors of tobraviruses are placed in the family Trichodoridae of suborder Diphtherophorina. Both adult and larval stages of most nematode vectors transmit viruses equally efficiently. However, larvae of *Longidorus diadecturus* are more efficient vectors of peach rosette mosaic virus than adults in peach orchards in Ontario.[206] Nematodes do not retain viruses through molting or transmit them transovarially.

*3. Transmission*

Laboratory experiments indicate that nematode vectors retain viruses for long periods (Table 16). *Xiphinema* and *Trichodorus* spp. retain viruses for a considerable length of time

**Table 16**
**RETENTION PERIODS AND ACQUISITION FEEDING PERIODS OF SOME**
**NEMATODE-TRANSMITTED VIRUSES**

| Virus | Vector | Retention period | Acquisition feeding period |
|---|---|---|---|
| Arabis mosaic | *Xiphinema diversicaudatum* | 8 months | — |
| Gladiolus notch leaf | *Paratrichodorus allius* | 28 d | 1 h |
| Grapevine fanleaf | *X. index* | 8 months | 15 min (within 5 or 15 min) |
| Raspberry ringspot | *Longidorus elongatus* | 1—2 months | — |
| Strawberry latent ringspot | *X. diversicaudatum* | 84 d | — |
| Tobacco rattle, Dutch isolate | *Trichodorus pachydermis* | 36 d | — |
| Tobacco ringspot | *X. americanum* | 12 months | 8—24 h |
| Tomato blackring | *L. elongatus* | 1—2 months | — |
| Tomato ringspot | *X. americanum* | — | (Within 1 h) |

(usually several months), while *Longidorus* species retain them for a period of usually 1 to 2 months only. Arabis mosaic virus in *X. diversicaudatum*, strawberry latent virus in *X. diversicandatum*, grapevine fanleaf virus in *X. index*, and tobacco ringspot virus in *X. americanum* are specifically adsorbed as monolayers on the cuticle lining the lumina of the odontophore, anterior esophagus, and esophageal bulb of their respective nematodes.[209-212] Tobacco rattle virus particles are associated with pharyngeal walls and lining of esophageal lumen of *Trichodorus* and *Paratrichodorus* vectors.[213] In contrast, raspberry ringspot and tomato black ring viruses transmitted by *L. elongatus* are present within the odontostyle and cuticular stylet guiding sheath of the vector.[214] Particles of the English strain of raspberry ringspot virus and artichoke Italian latent virus are also associated with the inner surface of the odontostyle in *L. macrosoma* and *L. attenuatus*, respectively.[210,215] Thus, retention sites of viruses in *Xiphinema* and trichodorid nematodes (namely *Trichadorus* and *Paratrichodorus*) are similar in being associated with esophagus, but are markedly different from the retention sites of odontostyle and stylet guiding sheath in *Longidorus*.

Virus retention sites in the feeding apparatus of nematode vectors determines the difference in length of retention and transmission of these viruses by vectors.[204,209,213] Virus particles present in lumen of alimentary tract are held tightly and are later released slowly and gradually over long periods as the nematodes feed and the saliva passes into the plant cell through the esophagus lumen. Passage of saliva may change pH or ionic conditions within the esophagus lumen, resulting in altered surface charge of virus particles leading to their dissociation from retention sites.[204] Other factors may also be involved in this process. There is, however, no biological association between the viruses and vectors and they pass neither through molting nor eggs. The viruses are said to be retained, in analogy with aphid-transmitted viruses, in a nonpersistent manner, but are released slowly and intermittently from the extracellular retention sites of the feeding apparatus. In contrast, virus retention sites of odontostyle and stylet guiding sheath in *Longidorus* spp. accounts for the relatively short retention periods of these viruses and also supports the suggestion that the relationship between the two is of the mechanical type. Differences in virus retention sites parallel the differences in retention periods of viruses within the vector groups.

Viruses are acquired by nematode vectors, usually within short acquisition access periods which vary for different virus-vector combinations, from less than 5 min to within 1 h. It appears that nematode vectors acquire a virus soon after ingestion of sap from infected roots. Efficiency and frequency of virus transmission is low and variable if the acquisition access periods are small. However, both efficiency and frequency of transmission increase with increasing access periods to infected donor plants and to healthy bait plants. Thus, transmission efficiency of tomato ringspot virus by *X. americanum* was low, with up to a 48-h access period, but greatly increased with a 4-d access period. Similarly, transmission efficiency of tobacco ringspot virus by single *X. americanum* nematodes was low with a 24-h access period, but increased proportionately with time up to 10 d. The above situation is explained by the observation that nematodes feed intermittently so that, with increasing access time, an increasing number of nematodes conceivably feed on infected plants, leading to increased transmission efficiency of viruses by vectors.

There does not seem to be any incubation period of viruses within vector nematodes, and viruliferous vectors can possibly transmit viruses in a single brief feeding period.

*4. Virus-Vector Specificity*

Virus-vector specificity of nematode-transmitted viruses exhibits itself at different levels. The first level of specificity is that different virus groups are transmitted by different groups of nematode vectors: nepoviruses by *Xiphinema* and *Longidorus* spp. of family Longidoridae, and tobraviruses by *Trichodorus* and *Paratrichodorus* of family Trichodoridae. A second level of specificity exists within each nematode genus: a particular virus is generally transmitted by a particular nematode genus. A third level of specificity is easily discernable among nepoviruses: serologically distinct strains of a nepovirus are generally transmitted by closely related, but different species of the same nematode genus. Scottish strains of raspberry ringspot and tomato black ring viruses are transmitted by *L. elongatus*, while *L. macrosoma* and *L. attenuatus* transmit the English strains of the two respective viruses. However, strains showing slight serological differences may be transmitted by the same vector species: three strains of tomato ringspot virus are transmitted by *X. californicum*;[208] two tobacco ringspot virus strains are transmitted by *X. americanum*. However, contrary to the above generalizations, serologically unrelated viruses may have the same vector species: arabis mosaic and strawberry latent ringspot viruses are both transmitted by *X. diversicaudatum*.

The ability of viruses to be specifically adsorbed onto cuticle lining the lumina of odontophore and esophagus has been correlated with specificity of transmission by nematodes.[209] Passage of virus-containing sap from infected plant in nematode intestine during the ingestion phase of feeding may result in selective and specific adsorption of a transmissible virus at retention sites of a vector. A nontransmissible virus fails to be specifically adsorbed on the alimentary canal of a nematode. The nontransmissible arabis mosaic virus was found in intestine of *L. elongatus* (when it was fed on infected plants), but was not associated with any part of the feeding apparatus.[214] Capsid protein of the virus controls virus adsorption to and release from retention sites of the vector in the case of raspberry ringspot virus and tomato black ring virus.[216]

# REFERENCES

1. **Palti, J.,** *Cultural Practices and Infectious Crop Diseases,* Springer-Verlag, Berlin, 1981, 243.
2. **Mircetich, S. M. and Rowhani, A.,** The relationship of cherry leaf roll virus and blackline disease of English walnut trees, *Phytopathology,* 74, 423, 1984.
3. **Mircetich, S. M., Refsguard, J., and Matheron, M. E.,** Blackline of English walnut trees traced to graft transmitted virus, *Calif. Agric.,* 34, 8, 1980.
4. **Rao, A. L. N. and Varma, A.,** Transmission studies with cucumber green mottle mosaic virus, *Phytopathol. Z.,* 109, 325, 1984.
5. **Broadbent, L.,** Epidemiology and control of tomato mosaic virus, *Annu. Rev. Phytopathol.,* 14, 75, 1976.
6. **Allen, W. R.,** Dissemination of tobacco mosaic virus from soil to plant leaves under glasshouse conditions, *Can. J. Plant Pathol.,* 3, 163, 1981.
7. **Khalil, M. K. and Shalla, T. A.,** Detection and spread of potato virus S, *Plant Dis.,* 66, 368, 1982.
8. **Franc, G. D. and Banttari, E. E.,** The transmission of potato virus S by the cutting knife and retention time of infectious PVS on common surfaces, *Am. Potato J.,* 61, 253, 1984.

9. **Jones, R. A. C. and Fribourg, C. E.,** Beetle, contact, and potato true seed transmission of Andean potato latent virus, *Ann. Appl. Biol.,* 86, 123, 1977.
10. **Valverde, R. A.,** Brome mosaic virus isolates naturally infecting *Commelina diffusa* and *C. communis, Plant Dis.,* 67, 1194, 1983.
11. **Chiko, A. W.,** Reciprocal contact transmission of barley stripe mosaic virus between wild oats and barley, *Plant Dis.,* 67, 207, 1983.
12. **Chiko, A. W.,** Increased virulence of barley stripe mosaic virus for wild oats: evidence of strain selection by host passage, *Phytopathology,* 74, 595, 1984.
13. **Kleinhempel, H. and Kegler, G.,** Transmission of tomato bushy stunt virus without vectors, *Acta Phytopathol. Acad. Sci. Hung.,* 17, 17, 1982.
14. **Kegler, G. and Kegler, H.,** Beiträge zur Kenntnis der vektorlosen Übertragung pflanzenpathogener viren, *Arch. Phytopathol. Pflanzenschutz,* 17, 307, 1981.
15. **Hollings, M. and Stone, O. M.,** Red clover necrotic mosaic virus, *CMI/AAB Descriptions of Plant Viruses,* No. 181, 1977.
16. **Hiruki, C.,** The dianthoviruses: a distinct group of isometric plant viruses with bipartite genome, *Adv. Virus Res.,* 33, 257, 1987.
17. **Lanter, J. M., McGuire, J. M., and Goode, M. J.,** Persistence of tomato mosaic virus in tomato debris and soil under field conditions, *Plant Dis.,* 66, 552, 1982.
18. **Tomlinson, J. A., Faithfull, E. M., Webb, M. J. W., Fraser, R. S. S., and Seeley, N. D.,** *Chenopodium* necrosis: a distinctive strain of tobacco necrosis virus isolated from river water, *Ann. Appl. Biol.,* 102, 135, 1983.
19. **Avgelis, A.,** Occurrence of melon necrotic spot virus in Crete (Greece), *Phytopathol. Z.,* 114, 365, 1985.
20. **Tomlinson, J. A. and Faithfull, E. M.,** Studies on the occurrence of tomato bushy stunt virus in English rivers, *Ann. Appl. Biol.,* 104, 485, 1984.
21. **Koenig, R., Lesemann, D.-E., Huth, W., and Makkouk, M. K.,** Comparison of a new soilborne virus from cucumber with tombus-, diantho-, and other similar viruses, *Phytopathology,* 73, 515, 1983.
22. **Lynes, E. W., Teakle, D. S., and Smith, P. R.,** Red clover necrotic mosaic virus isolated from *Trifolium repens* and *Medicago sativa* in Victoria, *Aust. Plant Pathol.,* 10, 6, 1981.
23. **Shukla, D. D., Shanks, G. J., Teakle, D. S., and Behncken, G. M.,** Mechanical transmission of galinsoga mosaic virus in soil, *Austr. J. Biol. Sci.,* 32, 267, 1979.
24. **Lovisolo, O., Bode, O., and Völk, J.,** Preliminary studies on the soil transmission of petunia asteroid virus ( = petunia strain of tomato bushy stunt virus), *Phytopathol. Z.,* 53, 323, 1965.
25. **Gallitelli, D., Vovlas, C., and Avgelis, A.,** Some properties of cucumber fruit streak virus, *Phytopathol. Z.,* 106, 149, 1983.
26. **Smith, P. R., Campbell, R. N., and Fry, P. R.,** Root discharge and soil survival of viruses, *Phytopathology,* 59, 1678, 1969.
27. **Yarwood, C. E.,** Release and preservation of virus by roots, *Phytopathology,* 50, 111, 1960.
28. **Koenig, R.,** Plant viruses in rivers and lakes, *Adv. Virus Res.* 31, 321, 1986.
29. **Fletcher, J. T.,** Studies on the overwintering of tomato mosaic in root debris, *Plant Pathol.,* 18, 97, 1969.
30. **Broadbent, L., Read, W. H., and Last, F. T.,** The epidemiology of tomato mosaic. X. Persistence of TMV-infected debris in soil, and the effects of soil partial sterilization, *Ann. Appl. Biol.,* 55, 471, 1965.
31. **Tomlinson, J. A., Faithfull, E. M., Flewett, T. H., and Beards, G.,** Isolation of infective tomato bushy stunt virus after passage through the human alimentary tract, *Nature (London),* 300, 637, 1982.
32. **Roberts, F. M.,** The infection of plants by viruses through roots, *Ann. Appl. Biol.,* 37, 385, 1950.
33. **Paludin, N.,** Spread of viruses by recirculated nutrient solutions in soilless cultures, *Tidsskr. Planteavl,* 89, 467, 1985.
34. **Kegler, G., Kleinhempel, H., and Kegler, H.,** Untersuchung zur Bodenbürtigkeit des tomato bushy stunt virus, *Arch. Phytopathol. Pflanzenschutz,* 16, 73, 1980.
35. **Piazzolla, P., Castellano, M. A., and de Stradis, A.,** Presence of plant viruses in some rivers of southern Italy, *J. Phytopathol.,* 116, 244, 1986.
36. **Bennett, C. W.,** Seed transmission of plant viruses, *Adv. Virus Res.,* 14, 221, 1969.
37. **Mandahar, C. L.,** Vertical and horizontal spread of plant viruses through seed and pollen — an epidemiological view, in *Perspectives in Plant Virology,* Gupta, B. M., Singh, B. P., Verma, H. N., and Srivastava, K. M., Eds., Print House, Lucknow, 1985, 23.
38. **Mandahar, C. L.,** Virus transmission through seed and pollen, in *Plant Diseases and Vectors: Ecology and Epidemiology,* Maramorosch, K. and Harris, K. F., Eds., Academic Press, New York, 1981, 241.
39. **Shepherd, R. J.,** Transmission of viruses through seed and pollen, in *Principles and Techniques in Plant Virology,* Kado, C. I. and Agrawal, H. O., Eds., Van Nostrand Reinhold, New York, 1972, 267.
40. **Phatak, H. C.,** Seed-borne plant viruses — identification and diagnosis in seed health testing, *Seed Sci. Technol.,* 2, 3, 1974.
41. **Phatak, H. C.,** The role of seed and pollen in the spread of plant pathogens, particularly viruses, *Trop. Pest Manage.,* 26, 278, 1980.

42. **Neergaard, P.,** *Seed Pathology,* Vol. I, Macmillan, London, 1977, chap. 3.

43. **Bos, L.,** Seed-borne viruses, in *Plant Health and Quarantine in International Transfer of Genetic Resources,* Hewitt, W. B. and Chiarappa, L., Eds., CRC Press, Boca Raton, FL, 1977, 39.

43a. **Stace-Smith, R. and Hamilton, R. I.,** Inoculum thresholds of seedborne pathogens: viruses, *Phytopathology,* 78, 875, 1988.

44. **Hamilton, R. I., Leung, E., and Nichols, C.,** Surface contamination of pollen by plant viruses, *Phytopathology,* 67, 395, 1977.

45. **Yang, A. F. and Hamilton, R. I.,** The mechanism of seed transmission of tobacco ringspot virus in soybean, *Virology,* 62, 26, 1974.

46. **Carroll, T. W. and Mayhew, D. E.,** Anther and pollen infection in relation to the pollen and seed transmissibility of two strains of barley stripe mosaic virus in barley, *Can. J. Bot.,* 54, 1604, 1976.

47. **Massalski, P. R. and Cooper, J. I.,** The location of virus like particles in the male gametophyte of birch, walnut and cherry naturally infected with cherry leaf roll virus and its relevance to vertical transmission of the virus, *Plant Pathol.,* 33, 255, 1984.

48. **Childress, A. M. and Ramsdell, D. C.,** Detection of blueberry leaf mottle virus in highbush blueberry pollen and seed, *Phytopathology,* 76, 1333, 1986.

49. **Kelley, R. D. and Cameron, H. R.,** Location of prune dwarf and prunus necrotic ringspot viruses associated with sweet cherry pollen and seed, *Phytopathology,* 76, 317, 1986.

50. **Gaspar, J. O., Vega, J., Camargo, I. J. B., and Costa, A. S.,** An ultrastructural study of particle distribution during microsporogenesis in tomato plants infected with the Brazilian tobacco rattle virus, *Can. J. Bot.,* 62, 372, 1984.

51. **Brlansky, R. H., Carroll, T. W., and Zaske, S. K.,** Some ultrastructural aspects of the pollen transmission of barley stripe mosaic virus in barley, *Can. J. Bot.,* 64, 853, 1986.

52. **Mandahar, C. L. and Gill, P. S.,** The epidemiological role of pollen transmission of viruses, *Z. Pflanzenkr. Pflanzenschutz,* 91, 246, 1983.

53. **Alvarez, M. and Campbell, R. N.,** Transmission and distribution of squash mosaic virus in seeds of cantaloup, *Phytopathology,* 68, 257, 1978.

54. **Hamilton, R. I., Nichols, C., and Valentine, B.,** Survey for prunus necrotic ringspot and other viruses contaminating the exine of pollen collected by bees, *Can. J. Plant Pathol.,* 6, 196, 1984.

55. **Francki, R. I. B. and Miles, R.,** Mechanical transmission of sowbane mosaic virus carried on pollen from infected plants, *Plant Pathol.,* 34, 11, 1985.

56. **George, J. A. and Davidson, T. R.,** Pollen transmission of necrotic ringspot and sour cherry yellows viruses from tree to tree, *Can. J. Plant Sci.,* 43, 276, 1963.

57. **George, J. A. and Davidson, T. R.,** Further evidence of pollen transmission of necrotic ringspot virus and some cherry yellow viruses in sour cherry, *Can. J. Plant Sci.,* 44, 383, 1964.

58. **Gilmer, R. M. and Way, R. D.,** Pollen transmission of necrotic ringspot and prune dwarf viruses in sour cherry, *Phytopathology,* 50, 624, 1960.

59. **Gilmer, R. M. and Way, R. D.,** Evidence for tree to tree transmission of sour cherry yellows virus by pollen, *Plant Dis. Rep.,* 47, 1051, 1963.

60. **Cole, A., Mink, G. I., and Regev, S.,** Location of prunus necrotic ringspot virus on pollen grains from infected almond and cherry trees, *Phytopathology,* 72, 1542, 1982.

61. **Mink, G. I.,** The possible role of honeybees in long distance spread of prunus necrotic ringspot virus from California into Washington sweet cherry orchards, in *Plant Virus Disease Epidemiology,* Plumb, R. T. and Thresh, J. M., Eds., Blackwell Scientific, Oxford, 1983, 85.

62. **Childress, A. M. and Ramsdell, D. C.,** Bee-mediated transmission of blueberry leaf mottle virus via infected pollen in highbush blueberry, *Phytopathology,* 77, 167, 1987.

63. **Cole, A. and Mink, G. I.,** An agent associated with bee-stored pollen that degrades intact viruses, *Phytopathology,* 74, 1320, 1984.

64. **Nyeki, J. and Vertery, J.,** Effect of different ringspot viruses on the physiological and morphological properties of Montmorency sour cherry pollen, *Acta Phytopathol.,* 9, 23, 1974.

65. **Marenaud, C. and Llager, G.,** Étude de la diffusion de virus de type ILAR (taches annulaires nécrotiques) dans un verger de cerisier *(Prunus avium), Ann. Amelior. Plant.,* 26, 357, 1976.

66. **Haight, E. and Gibbs, A.,** Effect of viruses on pollen morphology, *Plant Pathol.,* 32, 369, 1983.

67. **Lister, R. M. and Murant, A. F.,** Seed transmission of nematode-borne viruses, *Ann. Appl. Biol.,* 59, 49, 1967.

68. **Mikel, M. A., D'Arcy, C. J., Rhodes, A. M., and Ford, R. E.,** Effect of maize dwarf mosaic virus infection on sweet corn pollen and silk, *Phytopathology,* 72, 428, 1982.

69. **Grasmick, M. E. and Slack, S. A.,** Effect of potato spindle tuber viroid on sexual reproduction and viroid transmission in true potato seed, *Can. J. Bot.,* 64, 236, 1986.

70. **Murdock, D. J., Nelson, P. E., and Smith, S. H.,** Histopathological examination of pelargonium infected with tomato ringspot virus, *Phytopathology,* 66, 844, 1976.

71. **Rader, W. E., Fitzpatrick, H. F., and Hildebrand, E. M.,** A seed-borne virus of musk-melon, *Phytopathology,* 37, 809, 1947.

72. **Vorra-urai, S. and Cockbain, A. J.,** Further studies on seed transmission of broad bean stain virus and Echtes Ackerbohnenmosaik-Virus in field beans *(Vicia faba),* Ann. Appl. Biol., 87, 365, 1977.

73. **Daubeny, H. A., Stace-Smith, R., and Freeman, J. A.,** The occurrence and some effects of raspberry bushy dwarf virus in red raspberry, *J. Am. Soc. Hortic. Sci.,* 103, 519, 1978.

74. **Murant, A. F., Chambers, J., and Jones, A. T.,** Spread of raspberry bushy dwarf virus by pollination, its association with crumbly fruit and problems of control, *Ann. Appl. Biol.,* 77, 283, 1974.

75. **Jones, A. T. and Woods, G. A.,** The virus status of raspberries *(Rubus idaeus* L.) in New Zealand, *N. Z. J. Agric. Res.,* 22, 173, 1979.

76. **Cameron, H. R., Milbrath, J. A., and Tate, L. A.,** Pollen transmission of prunus ringspot virus in prune and sour cherry orchards, *Plant Dis. Rep.,* 57, 241, 1973.

77. **Davidson, T. R.,** Field spread of prunus necrotic ringspot in sour cherries in Ontario, *Plant Dis. Rep.,* 60, 1080, 1976.

78. **Teakle, D. S.,** Transmission of plant viruses by fungi, in *Principles and Techniques in Plant Virology,* Kado, C. I. and Agrawal, H. O., Eds., Van Nostrand Reinhold, New York, 1972, 248.

79. **Teakle, D. S.,** Fungi, in *Vectors of Plant Pathogens,* Harris, K. F. and Maramrosch, K., Eds., Academic Press, New York, 1980, 417.

80. **Campbell, R. N.,** Fungal vectors of plant viruses, in *Fungal Viruses,* Molitoris, H. P., Hollings, M., and Woods, H. A., Eds., Springer-Verlag, Berlin, 1979, 8.

81. **Huth, W.,** Die Gelbmosaikvirose der Gerste in der Bundesrepublik Deutschland Beobachtungen seit 1978, *Nachrichtenbl. Dtsch. Pflanzenschutzdienst,* 36, 49, 1984.

82. **Guinchedi, L. and Langenberg, W. G.,** Beet necrotic yellow vein virus transmission by *Polymyxa betae* Keskin zoospores, *Phytopathol. Mediterr.,* 21, 5, 1982.

83. **Erasmus, D. S. and von Wechmar, M. B.,** The association of brome mosaic virus and wheat rusts. I. Transmission of BMV by uredospores of wheat stem and leaf rust, *Phytopathol. Z.,* 108, 26, 1983.

84. **Erasmus, D. S., Rybicki, E. B., and von Wechmar, M. B.,** The association of brome mosaic virus and wheat rusts. II. Detection of BMV in/on uredospores of wheat stem rust, *Phytopathol. Z.,* 108, 34, 1983.

85. **Mayo, M. A. and Reddy, D. V. R.,** Translation products of RNA from Indian peanut clump virus, *J. Gen. Virol.,* 66, 1347, 1985.

86. **Thouvenel, J. C. and Fauquet, C.,** Further properties of peanut clump virus and studies on its natural transmission, *Ann. Appl. Biol.,* 97, 99, 1981.

87. **Stobbs, L. W., Cross, G. W., and Manocha, M. S.,** Specificity and methods of transmission of cucumber necrosis virus by *Olpidium radicale* zoospores, *Can. J. Plant Pathol.,* 4, 134, 1982.

88. **Foxe, M. J.,** An investigation of the distribution of potato mop top virus in county Donegal, *J. Life Sci. R. Dublin Soc.,* 1, 149, 1980.

89. **Campbell, R. N.,** Longevity of *Olpidium brassicae* in air-dry soil and the persistence of the lettuce bigvein agent, *Can. J. Bot.,* 63, 2288, 1985.

90. **Westerlund, F. V., Campbell, R. N., Grogan, R. G., and Duniway, J. M.,** Soil factors affecting the reproduction and survival of *Olpidium brassicae* and its transmission of big vein agent to lettuce, *Phytopathology,* 68, 927, 1978.

91. **Yarwood, C. E. and Hecht-Poiner, E.,** Viruses from rusts and mildews, *Phytopathology,* 63, 1111, 1973.

92. **Temmink, J. H. M., Campbell, R. N., and Smith, P. R.,** Specificity and site of in vitro acquisition of tobacco necrosis virus by zoospores of *Olpidium brassicae, J. Gen. Virol.,* 9, 201, 1970.

93. **Harris, K. F.,** Arthropod and nematode vectors of plant viruses, *Annu. Rev. Phytopathol.,* 19, 391, 1981.

93a. **Ishihara, T.,** Families and genera of leafhopper vectors, in *Viruses, Vectors and Vegetation,* Maramorosch, K., Ed., Interscience, New York, 1969, 235.

93b. **Harris, K. F.,** Aphids, leafhoppers and planthoppers, in *Vectors of Plant Pathogens,* Harris, K. F. and Maramorosch, K., Eds., Academic Press, New York, 1980, 1.

94. **Carter, W.,** *Insects in Relation to Plant Disease,* 2nd ed., John Wiley & Sons, New York 1973, 759.

95. **Nielson, M. W.,** Taxonomic relationships of leafhopper vectors of plant pathogens, in *Leafhopper Vectors and Plant Disease Agents,* Maramorosch, K. and Harris, K. F., Eds., Academic Press, New York, 1979.

96. **Lockhart, B. E. L., Khaless, N., Lennon, A. M., and El Maataoui, M.,** Properties of bermuda grass etched line virus, a new leafhopper transmitted virus related to maize rayado fino and oat blue dwarf viruses, *Phytopathology,* 75, 1258, 1985.

97. **Greber, R. A.,** Ecological aspects of cereal chlorotic mottle virus, *Aust. Plant Pathol.,* 10, 29, 1981.

98. **Izadpanah, K., Ahmadi, A. A., Parvin, S., and Jafari, S. A.,** Transmission, particle size and additional hosts of the rhabdovirus causing maize mosaic in Shiraz, Iran, *Phytopathol. Z.,* 107, 283, 1983.

99. **Nagaraju, and Viswanath, S.,** Studies on the relationship of ragi streak virus and its vector *Cicadulina chinai, Indian Phytopathol.,* 34, 458, 1981.

100. **Xie, L.-H., Lin, J.-Y., and Guo, J. R.**, A new insect vector of rice dwarf virus, *Int. Rice Res. Newsl.*, 6, 14, 1981.

101. **Rivera, C. and Gamez, R.**, Multiplication of maize rayado fino virus in the leafhopper vector *Dalbulus maidis*, *Intervirology*, 25, 76, 1986.

102. **Nault, L. R., Gingery, R. E., and Gordon, D. T.**, Leafhopper transmission and host range of maize rayado fino virus, *Phytopathology*, 70, 709, 1980.

103. **Choudhury, M. M. and Rosenkranz, E.**, Vector relationship of *Graminella nigrifrons* to maize chlorotic dwarf virus, *Phytopathology*, 73, 685, 1983.

104. **Morinaka, T., Putta, M., Chettanachit, D., Parejarearn, A., Disthaporn, S., Omura, T., and Inoue, H.**, Transmission of rice gall dwarf virus by Cicadellid leafhoppers *Recilia dorsalis* and *Nephotettix nigropictus* in Thailand, *Plant Dis.*, 66, 703, 1982.

105. **Inoue, H. and Hirao, J.**, Effects of temperature on the transmission of rice waika virus by *Nephotettix cincticeps* Uhler (Homoptera: Cicadellidae), *Appl. Entomol. Zool.*, 15, 433, 1980.

106. **Conti, M.**, Vector relationships and other characteristics of barley yellow striate mosaic virus (BYSMV), *Ann. Appl. Biol.*, 95, 83, 1980.

107. **Chettanachit, D., Putta, M., Balaveang, W., Hongkajron, J., and Disthaporn, S.**, New rice grassy stunt virus (GSV) strain in Thailand, *Int. Rice Res. Newsl.*, 10, 10, 1985.

108. **Senboku, T., Shikata, E., Tiongco, E. R., and Ling, K. C.**, Transmission of rice ragged stunt disease by *Nilaparvata lugens* in Japan, *Int. Rice Res. Newsl.*, 3, 8, 1978.

109. **Morinaka, T., Putta, M., Chettanachit, D., Parejarearn, A., and Disthaporn, S.**, Transmission of rice ragged stunt disease in Thailand, *JARQ (Japan Agric. Res. Q.)*, 17, 138, 1983.

110. **Slykhuis, J. T.**, Mites, in *Vectors of Plant Pathogens*, Harris, K. F. and Maramorosch, K., Eds., Academic Press, New York, 1980, 325.

111. **Proeseler, G.**, Piesmids, in *Vectors of Plant Pathogens*, Harris, K. F. and Maramorosch, K., Eds., Academic Press, New York, 1980, 97.

112. **Harris, K. F.**, An ingestion-egestion hypothesis of noncirculative virus transmission, in *Aphids as Virus Vectors*, Harris, K. F. and Maramorosch, K., Eds., Academic Press, New York, 1977, 165.

113. **Hibino, H. and Cabautan, P. Q.**, Infectivity neturalization of rice tungro-associated viruses acquired by vector leafhoppers, *Phytopathology*, 77, 473, 1987.

114. **Adams, G. and Hsu, H. T.**, Comparison of structural proteins from two potato yellow dwarf viruses, *J. Gen. Virol.*, 65, 991, 1984.

115. **Whitcomb, R. F. and Davis, R. E.**, Mycoplasma and phytoreoviruses as plant pathogens persistently transmitted by insects, *Annu. Rev. Entomol.*, 15, 405, 1970.

116. **Madden, L. V. and Nault, L. R.**, Differential pathogenicity of corn stunting mollicutes to leafhopper vectors in *Dalbulus* and *Baldulus* species, *Phytopathology*, 73, 1608, 1983.

117. **Jackson, G. V. H. and Zettler, F. W.**, Sweet potato witches' broom and legume little-leaf diseases in the Solomon Islands, *Plant Dis.*, 67, 1141, 1983.

118. **Fletcher, J., Schultz, G. A., Davis, R. E., Eastman, C. E., and Goodman, R. M.**, Brittle root disease of horse radish: evidence for an etiological role of *Spiroplasma citri*, *Phytopathology*, 71, 1073, 1981.

119. **Dabek, A. J.**, Leafhopper transmission of *Rhynchosia* little leaf, a disease associated with mycoplasma-like organisms in Jamaica, *Ann. Appl. Biol.*, 103, 431, 1983.

120. **Dhingra, K. L. and Chenulu, V. V.**, Symptomatology and transmission of witches' broom disease of soybean in India, *Curr. Sci., India*, 52, 603, 1983.

121. **Maramorosch, K.**, Multiplication of aster yellows virus in its vector, *Nature*, 169, 4292, 1952.

122. **Liu, H.-Y., Gumpf, D. J., Oldfield, G. N., and Calavan, E. C.**, Transmission of *Spiroplasma citri* by *Circulifer tenellus*, *Phytopathology*, 73, 582, 1983.

123. **Liu, H.-Y., Gumpf, D. J., Oldfield, G. N., and Calavan, E. C.**, The relationship of *Spiroplasma citri* and *Circulifer tenellus*, *Phytopathology*, 73, 585, 1983.

124. **Purcell, A. H., Richardson, J., and Finlay, A.**, Multiplication of the agent of X-disease in a non-vector leafhopper *Macrosteles fascifrons*, *Ann. Appl. Biol.*, 99, 283, 1981.

125. **Lindsten, K.**, Planthopper vectors and plant disease agents in Fennoscandia, in *Leafhopper Vectors and Plant Disease Agents*, Maramorosch, K. and Harris, K. F., Eds., Academic Press, New York, 1979, 155.

126. **Morinaka, T., Chettanachit, D., Putta, M., Parejarearn, A., and Disthaporn, S.**, *Nilaparvata bakeri* transmission of rice ragged stunt virus, *Int. Rice. Res. Newsl.*, 6, 12, 1981.

127. **Ammar, E. D., Gingery, R. E., and Nault, L. R.**, Interactions between maize mosaic and maize stripe viruses in their insect vector, *Peregrinus maidis*, and in maize, *Phytopathology*, 77, 1051, 1987.

128. **Lindsten, K.**, Studies on virus diseases of cereals in Sweden. II. On virus diseases transmitted by the leafhopper, *Calligypona pellucida* (F.), Kungl, *Lantbrukshoegsk. Ann.*, 27, 199, 1961.

129. **Francki, R. I. B., Ryan, C. C., Hatta, T., Rohozinski, J., and Grivell, C. J.**, Serological detection of Fiji disease virus antigens in the planthopper *Perkinsiella saccharicida* and its inefficient ability to transmit the virus, *Plant Pathol.*, 35, 324, 1986.

130. **Nielson, M. W.**, *The Leafhopper Vectors of Phytopathogenic Viruses (Homoptera: Cicadellidae): Taxonomy, Biology, and Virus Transmission,* A.R.S. Tech. Bull. 1382, U.S. Department of Agriculture, Washington, D.C., 1968, 386.

131. **Sinha, R. C.**, Vertical transmission of plant pathogens, in *Vectors of Disease Agents, Interactions with Plants, Animals and Man,* McKelvey, J. J., Eldridge, B. E., and Maramorosch, K., Eds., Praeger, New York, 1981, 109.

132. **Inoue, H. and Omura, T.**, Transmission of rice gall dwarf by the green rice leafhopper, *Plant Dis.,* 66, 57, 1981.

133. **Falk, B. W. and Tsai, J. H.**, Serological detection and evidence for multiplication of maize mosaic virus in the planthopper, *Peregrinus maidis, Phytopathology,* 75, 852, 1985.

134. **Gingery, R. E., Nault, R., and Bradfute, O. E.**, Maize stripe virus: characteristics of a member of new virus class, *Virology,* 112, 99, 1981:

135. **Tsai, J. H. and Zitter, T. A.**, Transmission characteristics of maize stripe virus by the corn delphacid, *J. Econ. Entomol.,* 75, 397, 1982.

136. **Falk, B. W., Tsai, J. H., and Lommel, S. A.**, Differences in levels of detection for the maize stripe virus capsid and major non-capsid proteins in plant and insect hosts, *J. Gen. Virol.,* 68, 1801, 1987.

137. **Fukushi, T.**, Transmission of the virus through the eggs of an insect vector, *Proc. Imp. Acad. (Tokyo),* 9, 457, 1933.

138. **Fukushi, T.**, Studies on the dwarf disease of rice plant, *J. Fac. Agric. Hokkaido Univ.,* 37, 41, 1934.

139. **Black, L. M.**, A plant virus that multiplies in its insect vector, *Nature,* 166, 852, 1950.

140. **Black, L. M.**, Viruses that reproduce in plants and insects, *Ann. N.Y. Acad. Sci.,* 56, 398, 1953.

141. **Black, L. M.**, Occasional transmission of some plant viruses through the eggs of their insect vectors, *Phytopathology,* 43, 9, 1953.

142. **Yamada, W. and Yamamoto, H.**, Studies on the stripe disease of rice plant. I. On the virus transmission by an insect, *Delphacodes striatella* Fallen., *Spec. Bull. Okayama Prefect. Agric. Exp. Stn.,* 52, 93, 1955.

143. **Shinkai, A.**, Studies on insect transmission of rice virus diseases in Japan (in Japanese), *Nat. Inst. Agric. Sci. Bull. C.,* 14, 1, 1962.

144. **Fukushi, T. and Kimura, I.**, On some properties of the rice dwarf virus, *Proc. Jpn. Acad.,* 35, 482, 1959.

145. **Fukushi, T. and Kimura, I.**, Localization of rice dwarf virus in its insect vector, *Virology,* 21, 503, 1963.

146. **Sinha, R. C.**, Sequential infection and distribution of wound tumor virus in the internal organs of a vector after ingestion of virus, *Virology,* 26, 673, 1965.

147. **Sinha, R. C.**, Recovery of potato yellow dwarf virus from hemolymph and internal organs of an insect vector, *Virology,* 27, 118, 1965.

148. **Fukushi, T.**, Relationships between propagative rice viruses and their vectors, in *Viruses, Vectors and Vegetation,* Maramorosch, K., Ed., Wiley-Interscience, New York, 1969, 279.

149. **Nasu, S.**, Electron microscopy of the transovarial passage of rice dwarf virus, in *Viruses, Vectors and Vegetation,* Maramorosch, K., Ed., Wiley-Interscience, New York, 1969, 433.

150. **Nagraj, A. N. and Black, L. M.**, Hereditary variation in ability of a leafhopper to transmit unrelated viruses, *Virology,* 16, 152, 1962.

151. **Costa, A. S.**, Whiteflies as virus vectors, in *Viruses, Vectors and Vegetation,* Maramorosch, K., Ed., Wiley-Interscience, New York, 1969, 95.

152. **Costa, A. S.**, Whitefly-transmitted plant diseases, *Annu. Rev. Phytopathol.,* 16, 429, 1976.

153. **Bird, J. and Maramorosch, K.**, Viruses and virus diseases associated with whiteflies, *Adv. Virus Res.,* 22, 55, 1978.

154. **Muniyappa, V.**, Whiteflies, in *Vectors of Plant Pathogens,* Harris, K. F. and Maramorosch, K., Ed., Academic Press, New York, 1980, 39.

155. **Srivastava, K. M., Aslam, M., and Rao, B. L. S.**, A whitefly transmitted yellow vein mosaic disease of *Cosmos sulphureus* Cav., *Curr. Sci., India,* 54, 1126, 1985.

156. **Brown, J. K. and Nelson, M. R.**, Geminate particles associated with cotton leaf crinkle disease in Arizona, *Phytopathology,* 74, 987, 1984.

157. **Duffus, J. E., Larsen, R. C., and Liu, H. Y.**, Lettuce infectious yellow virus — a new type of whitefly-transmitted virus, *Phytopathology,* 76, 97, 1986.

158. **Iwaki, M., Thongmeearkom, P., Honda, Y., and Deema, N.**, Soybean crinkle leaf: a new whitefly-borne disease of soybean, *Plant Dis.,* 67, 546, 1983.

159. **Cohen, S. and Antignus, Y.**, A noncirculative whitefly-borne virus affecting tomatoes in Israel, *Phytoparasitica,* 10, 101, 1982.

160. **Amno-Nyako, F. C., Vetten, H. J., Allen, D. J., and Thottapilly, G.**, The relation between cowpea golden mosaic and its vector, *Bemisia tabaci, Ann. Appl. Biol.,* 102, 319, 1983.

161. **Reddy, K. S. and Yaraguntaiah, R. C.**, Virus-vector relationship in leaf curl disease of tomato, *Indian Phytopathol.,* 34, 310, 1981.

162. **Cohen, S., Duffus, J. E., Larsen, R. C., Liu, H. Y., and Flock, R. A.**, Purification, serology, and vector relationships of leaf curl virus, a whitefly-transmitted geminivirus, *Phytopathology,* 73, 1669, 1983.

163. **Walters, H. J.,** Beetle transmission of viruses, *Adv. Virus Res.,* 15, 339, 1969.
164. **Fulton, J. P., Scott, H. A., and Gamez, R.,** Beetles, in *Vectors of Plant Pathogens,* Harris, K. F. and Maramorosch, K., Eds., Academic Press, New York, 1980, 115.
165. **Fulton, J. P., Gergerich, R. C., and Scott, H. A.,** Beetle transmission of plant viruses, *Annu. Rev. Phytopathol.,* 25, 111, 1987.
166. **Reddy, D. R. R. and Varma, A.,** *Madurasia obscurella* Jacoby — a new vector of southern bean mosaic virus, *Curr. Sci., India,* 55, 109, 1986.
167. **Crestani, O. A., Kitajima, E. W., Lin, M. T., and Marinho, V. L. A.,** Passion fruit yellow mosaic virus, a new tymovirus found in Brazil, *Phytopathology,* 76, 951, 1986.
168. **Guy, P. L. and Gibbs, A. J.,** Further studies on turnip yellow mosaic tymovirus from an endemic Australian *Cardamine, Plant Pathol.,* 34, 532, 1985.
169. **Honda, Y., Iwaki, M., Thongmeearkom, P., Deema, N., and Srithongchai, W.,** Blackgram mottle virus occurring on mungbean and soybean in Thailand, *JARQ,* 16, 72, 1982.
170. **Greber, R. S.,** Some characteristics of *Solanum nodiflorum* mottle virus — a beetle-transmitted isometric virus from Australia, *Aust. J. Biol. Sci.,* 34, 369, 1981.
171. **Beniwal, S. P. S. and Bharathan, N.,** Beetle transmission of urdbean leaf crinkle virus, *Indian Phytopathol.,* 33, 600, 1980.
172. **Scott, H. A. and Fulton, J. P.,** Comparison of the relationships of southern bean mosaic virus and the cowpea strain of tobacco mosaic virus with the bean leaf beetle, *Virology,* 84, 197, 1978.
173. **Selman, B. J.,** Beetles — phytophagous Coleoptera, in *Viruses and Invertebrates,* Gibbs, A. J., Ed., North-Holland, Amsterdam, 1973, 157.
174. **Ghabrial, S. A. and Schultz, F. J.,** Serological detection of bean pod mottle virus in bean leaf beetles, *Phytopathology,* 73, 480, 1983.
175. **Kopek, J. A. and Scott, H. A.,** Southern bean mosaic virus in Mexican bean beetle and bean leaf beetle regurgitants, *J. Gen. Virol.,* 64, 1601, 1983.
176. **Freitag, J. H.,** Beetle transmission, host range, and properties of squash mosaic virus, *Phytopathology,* 46, 73, 1956.
177. **Hobbs, H. A. and Fulton, J. P.,** Beetle transmission of cowpea chlorotic mottle virus, *Phytopathology,* 69, 255, 1979.
178. **Lin, M. T., Hill, J. H., Kitajima, E. W., and Costa, C. L.,** Two new serotypes of cowpea severe mosaic virus, *Phytopathology,* 74, 581, 1984.
179. **Smith, C. E.,** Transmission of cowpea mosaic by bean leaf beetle, *Science,* 60, 268, 1924.
180. **Markham, R. and Smith, K. M.,** Studies on the virus of turnip yellow mosaic, *Parasitology,* 39, 330, 1949.
181. **Fulton, J. P. and Scott, H. A.,** Bean rugose mosaic virus and related viruses and their transmission by beetles, *Fitopathol. Bras.,* 2, 9, 1977.
182. **Gergerich, R. C., Scott, H. A., and Fulton, J. P.,** Regurgitant as a determinant of specificity in the transmission of plant viruses by beetles, *Phytopathology,* 73, 936, 1983.
183. **Monis, J., Scott, H. A., and Gergerich, R. C.,** Effect of beetle regurgitant on plant virus transmission using the gross wounding technique, *Phytopathology,* 76, 808, 1986.
184. **Gergerich, R. C., Scott, H. A., and Fulton, J. P.,** Evidence that ribonuclease in beetle regurgitant determines the transmission of plant viruses, *J. Gen. Virol.,* 67, 367, 1986.
185. **Gergerich, R. C., Scott, H. A., and Fulton, J. P.,** Some properties of beetle-transmitted viruses that may explain their transmissibility, *Phytopathology,* 76, 112, 1986.
186. **Slykhuis, J. T.,** Mites as vectors of plant viruses, in *Viruses, Vectors and Vegetation,* Maramorosch, K., Ed., Wiley-Interscience, New York, 1969, 121.
187. **Slykhuis, J. T.,** Transmission of plant viruses by eriophyid mites, in *Principles and Techniques in Plant Virology,* Kado, C. I. and Agrawal, H. O., Eds., Van Nostrand Reinhold, New York, 1972, 204.
188. **Oldfield, G. N.,** Mite transmission of plant viruses, *Annu. Rev. Entomol.,* 15, 343, 1970.
189. **Paliwal, Y. C. and Slykhuis, J. T.,** Localization of wheat streak mosaic virus in the alimentary canal of its vector, *Aceria tulipae* K., *Virology,* 32, 344, 1967.
190. **Paliwal, J. C.,** Fate of viruses in mite vectors and nonvectors, in *Vectors of Plant Pathogens,* Harris, K. F. and Maramorosch, K., Eds., Academic Press, New York, 1980, 357.
191. **Paliwal, J. C.,** Brome mosaic virus infection in the wheat curl mite *Aceria tulipae,* a nonvector of the virus, *J. Invertebr. Pathol.,* 20, 288, 1972.
192. **Roivainen, O.,** Mealybugs, in *Vectors of Plant Pathogens,* Harris, K. F. and Maramorosch, K., Eds., Academic Press, New York, 1980, 15.
193. **Rosciglione, B. and Castellano, M. A.,** Further evidence that mealybugs can transmit grapevine virus A (GVA) to herbaceous hosts, *Phytopathol. Mediterr.,* 24, 186, 1985.
194. **Anjaneyulu, A., Singh, S. K., Shukla, V. D., and Shenoi, N. M.,** Chlorotic streak, a new virus disease of rice, *Int. Rice Res. Newsl.,* 5, 12, 1980.
195. **Kaloostian, G. H.,** Psyllids, in *Vectors of Plant Pathogens,* Harris, K. F. and Maramorosch, K., Eds., Academic Press, New York, 1980, 87.

196. **Ananthakrishnan, T. N.,** Thrips, in *Vectors of Plant Pathogens,* Harris, K. F. and Maramorosch, K., Eds., Academic Press, New York, 1980, 149.

197. **Amin, P. W., Reddy, D. V. R., Ghanekar, A. M., and Reddy, M. S.,** Transmission of tomato spotted wilt virus, the causal agent of bud necrosis of peanut, by *Scirtothrips dorsalis and Frankliniella schultzei, Plant Dis.,* 65, 663, 1981.

198. **Kaiser, W. J., Wyatt, S. D., and Pesho, G. R.,** Natural hosts and vectors of tobacco streak virus in eastern Washington, *Phytopathology,* 72, 1508, 1982.

199. **Randles, J. W., Davies, C., Hatta, T., Gould, A. R., and Francki, R. I. B.,** Studies on encapsidated viroid-like RNA. I. Characterization of velvet tobacco mottle virus, *Virology,* 108, 111, 1981.

200. **Percival, J.,** The eelworm disease of hops, 'Nettle-headed' or 'shinkly' plants, *J. Southeast. Agric. Coll., Wye, Engl.,* 1, 5, 1895.

201. **Hewitt, W. B., Raski, D. J., and Goheen, A. C.,** Nematode vector of soil-borne fanleaf virus of grapevines, *Phytopathology,* 48, 586, 1958.

202. **Taylor, C. E. and Cadman, C. H.,** Nematode vectors, in *Viruses, Vectors and Vegetation,* Maramorosch, K., Ed., Wiley-Interscience, New York, 1969, 55.

203. **Taylor, C. E.,** Transmission of viruses by nematodes, in *Principles and Techniques in Plant Virology,* Kado, C. I. and Agrawal, H. O., Eds., Van Nostrand-Reinhold, New York, 1972, 668, 226.

204. **Taylor, C. E.,** Nematodes, in *Vectors of Plant Pathogens,* Harris, K. F. and Maramorosch, K., Eds., Academic Press, New York, 1980, 375.

205. **Wyss, U.,** Virus-transmitting nematodes: feeding behaviour and effect on root cells, *Plant Dis.,* 66, 639, 1982.

206. **Allen, W. R., Van Schagen, J. G., and Eveleigh, E. S.,** Transmission of peach rosette mosaic virus to peach, grape, and cucumber by *Longidorus diadecturus* obtained from diseased orchards in Ontario, *Can. J. Plant Pathol.,* 4, 16, 1982.

207. **Mountain, W. L., Powell, C. A., Forer, L. B., and Stouffer, R. F.,** Transmission of tomato ringspot virus from dandelion via seed and dagger nematode, *Plant Dis.,* 67, 867, 1983.

208. **Hoy, J. W., Mircetich, S. M., and Lownsbery, B. F.,** Differential transmission of prunus tomato ringspot virus strains by *Xiphinema californicum, Phytopathology,* 74, 332, 1984.

209. **Taylor, C. E. and Robertson, W. M.,** Sites of virus retention in the alimentary tract of the nematode vectors, *Xiphenema diversicaudatum* (Micol.) and *X. index* (Thorne and Allen), *Ann. Appl. Biol.,* 66, 375, 1970.

210. **Taylor, C. E. and Robertson, W. M.,** Acquisition, retention and transmission of viruses by nematodes, in *Nematode Vectors of Plant Viruses,* Lamberti, F., Taylor, C. E., and Seinhorst, J. W., Eds., Plenum Press, New York, 1975, 253.

211. **McGuire, J. M., Kim, K. S., and Douthit, L. M.,** Tobacco ringspot virus in the nematode *Xiphinema americanum, Virology,* 42, 212, 1970.

212. **Raski, D. J., Maggenti, A. R., and Jones, N. O.,** Location of grapevine fanleaf and yellow mosaic virus particles in *Xiphinema index, J. Nematol.,* 5, 208, 1973.

213. **Taylor, C. E. and Robertson, W. M.,** Location of tobacco rattle virus in the nematode vector, *Trichodorus pachydermis* Seinhorst, *J. Gen. Virol.,* 6, 179, 1970.

214. **Taylor, C. E. and Robertson, W. M.,** The location of raspberry ringspot and tomato black ring viruses in nematode vector, *Longidorus elongatus* (de Man), *Ann. Appl. Biol.,* 64, 233, 1969.

215. **Taylor, C. E., Robertson, W. M., and Roca, F.,** Specific association of artichoke Italian latent virus with the odontostyle of its vector *Longidorus attenuatus, Nematol. Mediterr.,* 4, 23, 1976.

216. **Harrison, B. D. and Murant, A. F.,** Nematode transmissibility of pseudo-recombinant isolates of tomato black ring virus, *Ann. Appl. Biol.,* 86, 209, 1977.

Chapter 9

# SPATIOTEMPORAL DETERMINANTS IN PLANT VIRUS EPIDEMICS: DEVELOPMENT OF A CONCEPTUAL MODEL

**Stewart M. Gray**

## TABLE OF CONTENTS

# I. INTRODUCTION

*A model is a lie which makes us know the truth.* R. MacArthur

Plant virus epidemics are directed by an ecologically complex system involving, but not limited to, viruses, vectors, host plants, the environment, and the interactions among these primary system components. Conceptualizing virus-vector-plant systems requires development of holistic ecological models that identify biological parameters likely to influence the system. Once the holistic system is acknowledged, simulation models can be developed to aid in the identification of important parameters and to prioritize research objectives. Biological and mathematical models do not replace the need for research, but rather stimulate and suggest research. Modeling is thus an initial step in plant virus epidemiology and should be used as a tool to develop strategies for research and disease management.

Current knowledge of the epidemiology of plant diseases caused by fungal pathogens and the successful development of control strategies for these diseases illustrate the usefulness of conceptual and simulation models.[1] The discipline of plant virus epidemiology has been slow to realize such benefits and remains largely a descriptive science. The reason is not a lack of appropriate methods, as evidenced by recent reports and reviews on modeling and quantitative analysis of plant virus epidemics.[2-8] I believe the basic problem is misconception and lack of educational pursuit of mathematical topics. As Thresh[9] stated, "with few notable exceptions plant virologists are not mathematically oriented". Many plant virologists view modeling as mathematical "hocus-pocus" used to generate data of questionable biological meaning. In addition, the majority of plant disease epidemiologists involved in teaching are reluctant to devote time and resources to the more complex virus-vector-host systems and instead emphasize the well-known systems and models of diseases caused by fungal pathogens.

The overall objective of this chapter is to present a conceptual model of virus-vector-host systems and discuss potential effects of several model components on the spatiotemporal dynamics of plant virus epidemics. This chapter discusses the complexities associated with virus-vector-plant systems and reinforces the concept that conceptual and simulation models are appropriate tools for the development of a research strategy. Subjects emphasized are biased toward plant viruses that require arthropod and nematode vectors for transmission and spread among host plants. Specific references are listed at the end of each section to direct the reader to alternative sources of information on subjects not covered or briefly covered in this chapter. For additional information on comparative plant virus epidemiology and plant virus ecology the reader is referred to a series of reviews by Thresh.[10-15]

## II. A CONCEPTUAL MODEL FOR PLANT VIRUS-VECTOR-HOST SYSTEMS

A sound understanding of the epidemiology of plant virus diseases is the key to rational control measures.[16] The expensive, long-term, labor-intensive field and laboratory studies required to gain the "sound understanding" emphasize the vital importance of developing conceptual ecological models as a first-step approach. The conceptual model defines the basic elements involved in a system and summarizes available information, experience, and hypotheses. The model organizes information in a biological way and may describe possible courses of investigation, suggest experiment design, identify voids, and prioritize research objectives. These models are improved stepwise by data from subsequent experiments. It is important to acknowledge that no complex system can be fully known in its interactive details. However, it is not possible to understand the epidemiology of a pathogen or attempt to control a pathogen without first being aware of the major factors influencing and driving the system. This awareness precedes the use of simulation models whose development and attributes have been reviewed recently.[17,18]

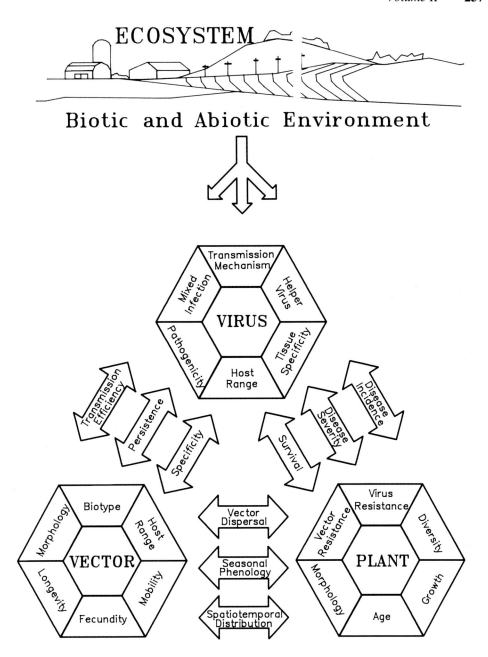

FIGURE 1. A conceptual, holistic ecological model of the intrinsic and interactive properties affecting the epidemiology of a vector-borne plant virus.

The conceptual model of plant virus epidemics is composed of three subject populations; the virus, its vector(s), and its host(s). The latter two in most cases are comprised of multiple species populations, each with individual intrinsic and interactive attributes. The environment is the total of biotic and abiotic agencies influencing each member of the population. The phenotype of the individuals in any of the populations is a product of the interaction between their genotype and the environment. Each subject population is comprised of a heterogeneous mix of individuals, each exerting an influence on the spatiotemporal dynamics of the virus epidemic. Figure 1 is a schematic model depicting some of the intrinsic and interactive properties of the three subject populations. The following sections of this chapter examine selected intrinsic and interactive characteristics of each of the three subject populations and discuss their potential effects on the spatiotemporal dynamics of plant virus epidemics.

## III. VIRUS CONTRIBUTIONS TO PLANT VIRUS EPIDEMICS

The virus is the simplest of the three subject populations. Plant viruses have small genomes relative to their plant hosts and vectors. The limited numbers of gene products, some of which are associated with multiple essential functions, restricts the phenotypic diversity of a virus population. Intrinsic properties of the virus will determine its mechanism of vector transmission, the host range, and the ability of the virus to infect, replicate, and move systemically in a compatible host.

### A. Mechanism of Vector Transmission

Plant virus transmission by vectors can be divided into three categories: circulative-propagative, circulative-nonpropagative, and noncirculative, based on the virus interaction within the vector.[19] However, during an epidemic the most important aspects of the transmission mechanisms are not the way the virus physically interacts with its vector, but the length of acquisition and inoculation times, the duration the virus is retained by an infectious vector, and whether the plant must be a host for both the virus and the vector or the virus alone. Host of the vector refers to the ability of the vector to develop a long-term feeding relationship and survive on the plant.

Circulative-propagative transmission refers to the ability of the virus to replicate and persist in a vector, a majority of which are leafhoppers or planthoppers. This type of transmission is restricted to five groups of plant viruses: phytoreoviruses, plant rhabdoviruses, fijiviruses, the rice stripe virus group, and the maize rayado fino virus group. These viruses generally have a narrow host range and are vectored by insect species whose plant host range is complementary to that of the virus.

Propagative transmission is advantageous as a mechanism of primary spread of the virus from mature or senescing crops to recently sown crops. The viruses are retained for extended periods of time by their vectors, e.g., up to 20 d for maize rayado fino virus,[20] or for the life of the vector in the case of Fiji disease virus.[21] The mobile insects are capable of long-distance movement to find suitable hosts or the insects can survive in the absence of a host by aestivation[22] or diapause.[23] The virus will persist in the dormant vector.[23,24] Since the virus and the vector have essentially the same host range, the virus is assured of transport to a new host plant if the vector survives.

Generally, the incidence of circulative-propagative viruses increases rapidly in recently sown crops due to the rapid influx of large numbers of adult vectors originating on mature or senescent crops.[21-23,25] Primary spread of the virus into the crop results in small clusters of diseased plants,[20,22] presumably the result of multiple infections of nearby plants by a single viruliferous vector.

Secondary spread of circulative-propagative viruses is important only if viruliferous insect populations can develop within the crop before the plants become physiologically inadequate hosts of the virus or the vector. Lengthy latent periods in the plant (i.e., the time from inoculation until the plant becomes a virus source) and in the vector will minimize the frequency of secondary spread.[22] Secondary spread of the disease within a crop becomes more important if transovarial transmission of the virus occurs, e.g., rice stripe virus. Three generations of the planthopper vector of rice stripe virus develop within the rice crop. The second-generation, viruliferous nymphs contribute to disease spread.[23] The third-generation nymphs go into a dormant period to overwinter and contribute to primary spread of the virus into plants growing in the spring.

Limitations of propagative transmission in sustaining epidemic development include low transmission efficiency, reduced vector fecundity, and increased vector mortality.[19,26,27] Also, the interaction between two propagative viruses doubly infecting a plant or insect host can delay or reduce transmission efficiency.[28] These factors alone or in combination can reduce the number of potential infection cycles in a crop as well as reduce the initial disease incidence.

Circulative transmission differs from propagative in that the virus does not replicate in its vector. Ingested virus is transported across the vector gut membrane into the hemolymph. Virus is selectively transported into the salivary glands from where it can be egested into plants.[29] Most circulative viruses are transmitted in the laboratory with varying degrees of efficiency by multiple vector species, but some appear to be vector specific.[30,31] In natural vector studies one species is typically found to be the predominant vector for each virus. Unlike the propagative viruses, mixed infections may increase the probability of transmission of some circulative viruses. Transencapsidation of the RNA from the MAV isolate of barley yellow dwarf virus (BYDV) by the capsid protein of the RPV isolate will break down the vector specificity common to these virus isolates, thereby increasing the number of vector species potentially transmitting the disease.[32]

Epidemiologically important features of circulative transmission include the long virus retention time of the vector, usually for the life of the vector. As with the propagative viruses, this allows for virus spread to new hosts from temporally or spatially distant sources. Also, a single vector is capable of infecting multiple plants during its life. Latent periods associated with the vector and host (measured in days) are sufficiently short for the primary infected plants to serve as a source for both virus and locally dispersing vectors. Primary spread of BYDV into a recently sown cereal crop was characterized by scattered and sparse infections, even when large numbers of vectors were involved. Secondary infection by nonviruliferous vectors acquiring virus from within the crop was attributed to progeny of the colonizing vector and was evident as clusters of infected plants slowly expanding outward from a primary source.[33] Similar trends have been observed for other aphid[34,35] and beetle-borne[36] circulative-nonpropagative viruses.

The requirement that the plant be a host for both the virus and the vector is the most salient epidemiological feature of both circulative and propagative transmission. The lengthy acquisition and/or inoculation times (i.e., minutes-hours) required for successful transmission of these viruses require the vector to establish a feeding relationship with the plant. Therefore, it is important that the host range of the virus and its vectors be similar. The long retention time associated with these viruses ensures that the virus will survive until its vector has located and established a feeding relationship with a suitable host. One possible exception to this feeding relationship involves the beetle-transmitted viruses, e.g., the como-, bromo-, tymo-, and sobemovirus groups. These viruses do circulate in their beetle vectors, although the pathway is different than described above.[37] Beetles can inoculate a plant with the first bite, but the transmission is very low. Presumably, inoculation of plants randomly tested, but rejected as a food source, by a viruliferous beetle would not play a significant role in the epidemiology of these viruses.

Noncirculative viruses are retained on the mouthparts, foregut, or midgut of their vectors by reversible binding involving structural or nonstructural virus proteins.[38] These viruses vary widely in the specificity of their vector associations. The best-known examples of noncirculative viruses are those transmitted by aphids, e.g., potyviruses, cucomoviruses, alfalfa mosaic virus, caulimoviruses, and carlaviruses. Vector specificity is uncommon, especially with viruses having a restricted host range, e.g., caulimoviruses and potyviruses. Acquisition and inoculation access times measured in seconds and the lack of a latent period within the vector allow noncolonizing vectors to contribute the the spread of the virus.[39-41] The probing behavior of aphids prior to feeding is enough for aphids to acquire virus or inoculate host and nonhost plants alike. These are all counterbalancing characteristics of the main disadvantage, i.e., the virus is generally retained by the vectors for only 4 to 6 h.

Dispersal of aphid-transmitted noncirculative viruses is limited more by time than by space because of the short virus retention period and the longevity of the fragile aphids. While long-distance transport of noncirculative aphid-borne viruses does occur,[42] dispersal of these viruses is more often local. Dispersal is favored by diverse agroecosystems or the continuous overlapping of cropping schedules. Secondary spread is usually attributed to nonviruliferous colonizing or noncolonizing alate aphids acquiring the virus and subsequently inoculating nearby healthy plants in their quest for suitable hosts.[43-46] Some secondary spread may result from local dispersal of apterous aphids developing on infected plants.[41]

Nepoviruses and tobraviruses are noncirculative viruses transmitted by specific species of nematodes.[47] These viruses are retained by the vector for weeks to months and therefore have the advantage of long-term survival either in the vector or in a host. The feeding behavior of nematodes is such that they will attempt to feed on any plant and therefore can transmit the virus to vector nonhosts as well as vector hosts.[48] Therefore, these viruses have the advantage of the circulative viruses in terms of persistence without the restriction that the host range of the virus and vector be similar.

Dispersal of nematode-borne viruses is limited more by space than by time because of the limited mobility and sedentary behavior of the vector. Nematode-transmitted viruses are normally found in clusters of infected plants.[49] The rapid appearance of clusters of disease soon after planting results from planting in soil infested with virus-carrying nematodes. Plant-parasitic nematode populations commonly have a clustered distribution.[50] Infection of multiple plants by a single nematode is of minimal importance due to the sedentary nature of the nematode once it has found a host. Secondary spread of the disease between adjacent plants will occur slowly either by dispersal of the nematode or, more likely, by expanding root systems of healthy plants coming in contact with infected roots and their associated nematodes.[51] Secondary spread is of minimal importance in annual crops due to the limited time the crop is available and susceptible to infection.

Irrespective of the vector, successful and continued transmission of noncirculative viruses relies on the acquisition of the virus by large numbers of vectors that continually disperse in search of new host plants. Host finding is a random event that may involve multiple attempts to feed on nonhost plants. The larger the number of viruliferous vectors, the greater the probability that a small number of vectors will actually survive the dispersal and inoculate a susceptible plant host.

Other mechanisms of transmission not directly associated with arthropod or nematode vectors, but contributing to the spatiotemporal characteristics of plant virus epidemics, include seed- and pollen-transmission. The role of seed and pollen in the spread of plant viruses has been reviewed[52,53] and is only briefly considered. Seed transmission provides mechanisms of long-range dispersal, long-term survival, and initial sources of inoculum subsequently spread by vectors or mechanical transmission. Hordeiviruses rely mainly on seed transmission for dispersal.

Transmission and spread by pollen is common among ilarviruses.[53] This provides a mechanism of wide dissemination of the virus. Pollen-collecting insects increase the efficiency of transmission by delivering the infected pollen and inoculating nearby and distant hosts of the same species.[54]

## B. Host Range

The ability of a virus to infect, replicate in, and survive in a host is at least partially determined by the viral genome or genomic products. The importance of the virus host range to the spatiotemporal characteristics of a virus epidemic was alluded to in the previous section. Two basic strategies can be described relating host range of the virus to its mechanism of transmission: restricted plant host range coupled with vector-specific transmission, or wide plant host ranges coupled with vector nonspecificity (Table 1).

Viruses with limited host ranges have evolved to ensure their continued survival through a persistent relation with specific vectors and by having a host range similar to that of the

**Table 1**
**RELATIONSHIP OF VIRUS GROUP, HOST RANGE, AND**
**MECHANISM OF TRANSMISSION**

| Virus group | Mechanism of host range | Transmission[a] | Vector |
| --- | --- | --- | --- |
| Comovirus | Narrow | C | Beetle |
| Tymovirus | Narrow | C | Beetle |
| Bromovirus | Narrow | C | Beetle |
| Maize rayado fino virus | Narrow | P | Leafhopper |
| Reoviridae | Narrow | P | Leafhopper |
| Rhabdoviridae | Narrow | P | Leafhopper/aphid |
| Geminivirus | Narrow | P | Leafhopper/whitefly |
| Luteovirus | Narrow/wide | C | Aphid |
| Pea enation mosaic virus | Narrow | C | Aphid |
| Caulimovirus | Narrow | NC | Aphid |
| Potyvirus | Narrow/wide | NC | Aphid |
| Carlavirus | Narrow | NC | Aphid |
| Cucumovirus | Wide | NC | Aphid |
| Alfalfa mosaic virus | Wide | NC | Aphid |
| Tobravirus | Wide | NC | Nematode |
| Nepovirus | Wide | NC | Nematode |

[a]  C = Circulative, P = propagative, NC = noncirculative.

vector. The virus is assured of transport to a new host, assuming the vector survives and the virus is retained. The vectors are mobile insects capable of long-distance migration and remain active throughout their life.

Viruses with wide host-plant ranges usually have nonspecific vector relationships. Aphid-borne viruses rely on large numbers of vectors dispersing over wide area to a multitude of plant species. The large number of viruliferous vectors increases the probability that a small proportion of the vectors will feed on a host of the virus. In the case of nematode vectors there is vector specificity, but the size of the vector population is large. Limited dispersal occurs, but the viruliferous nematodes can survive for extended periods of time to inoculate successive plant hosts in the same spatial area.

Exceptions to these generalizatioons are the aphid-borne caulimo-, poty-, and carlaviruses. Many of these viruses have narrow host ranges, but are transmitted efficiently by multiple species of aphids. They rely on a small percentage of a large vector population inoculating virus hosts.

## C. Summary

Other virus-related factors that will effect the spatiotemporal dynamics of virus epidemiology have been reviewed elsewhere andinclude the role of mixed infections[55] and helper-virus-dependent transmission.[56]

The vector-borne viruses cannot disperse independently of their vector; therefore, the spatiotemporal distribution of infected plants is dependent upon the dispersal behavior of the vector. The virus, or viruses in the case of mixed infections, contributes to the distribution of infected plants by determining the length of time and the specificity with which the virus is associated with a vector. Virus-related factors will also contribute to the efficiency with which virus is acquired by vectors. The virus-related factors are essential components of any descriptive or simulation model of plant virus epidemiology, yet they are given limited attention. A majority of the models being developed emphasize vector populations and diseased plant populations.[57] Models should include transmission parameters, i.e., acquisition, retention and inoculation times, as well as acknowledging interactions that influence the transmission parameters. The length of time and the efficiency of which a vector contributes to an epidemic is as important as the number of vectors potentially contributing to an epidemic.

## IV. VECTOR CONTRIBUTIONS TO PLANT VIRUS EPIDEMICS

A majority of plant viruses are dependent upon a vector for spread between available host plants. Therefore, understanding the spatial and temporal spread of a plant virus becomes an investigation of the population dynamics and dispersal of its vector. Population dynamics is a vast subject that will be discussed here in terms of factors involved directly in the epidemiology of plant virus diseases. Coverage of the subject will be biased toward insect population dynamics and more specifically to aphids, but many of the factors apply to other vector populations, e.g., nematodes and fungi. Emphasis will be on dispersal as a factor in population dynamics. Because of the difficulties in quantifying insect movement entomologists tend to underestimate the importance of colonization through dispersal.[58] As plant virus epidemiologists, we must realize that infection of host plants by most viruses is usually achieved only through dispersal of its vector. The more mobile vectors (e.g., aphids, leafhoppers, beetles) have the advantage of being able to disperse to a much larger spatial arena than the more sedentary vectors (e.g., nematodes, fungi, mites). Nevertheless, survival of the virus is dependent upon large numbers of the vectors acquiring the virus with a certain probability that a few vectors will survive and transmit the virus to a new host. The spatial aspects of vector dispersal are more important in the ecology of viruses with highly mobile vectors, whereas the temporal aspects of vector survival, i.e., the ability of the vector to survive in the absence of a host, are more important in the ecology of viruses with less mobile vectors.

For a detailed discussion of factors influencing vector population size, the reader is referred to Price.[58] For simplicity and brevity, we can assume that, given a suitable host and environment, vector populations will continue to increase in size until the carrying capacity of the system is reached. Host suitability and vector population density are the most significant determinants of vector population size and dispersal. As the host plant becomes unsuitable, vector reproduction will decrease and population mortality as well as vector dispersal will increase. As population density increases, feeding and reproductive sites become limited, resulting in decreased reproduction and increased dispersal.

### A. Spatial Characteristics of Vector Dispersal

The physiological and environmental factors that initiate vector dispersal will determine the extent of dispersal. Vector populations increasing beyond optimal densities on physiologically favorable hosts in a favorable environment are likely to disperse over short distances. In an agroecosystem this will result in dispersal to nearby plants of the same crop and exploitation of a homogeneous ecosystem. In aphid species that show alary polymorphism, the winged morphs colonize temporary, high-quality habitats which are then exploited by their apterous offspring. Overcrowding caused by rapid population buildup on favorable hosts may lead to the production of winged forms.[59] There is intraspecific variation in the duration of flights of alates produced in response to crowding,[60] and the variability has been linked to reproductive potential of the aphid.[61] Alate aphids developing on physiologically favorable hosts possess a large number of ovarioles and are reluctant to fly. When flight does occur, takeoff is at a shallow angle, resulting in short plant-to-plant flights and production of offspring within the same habitat.[61]

In terms of virus epidemiology, this type of dispersal contributes to the polycyclic disease cycle and secondary disease spread. This short-range dispersal is advantageous to the virus. It facilitates the exploitation of the host population by both its virus and the vector, assuming the virus can complete its infection cycle in the plant and be transmitted prior to dispersal of the vectors. The increase in vector population and the number of available virus sources increase the probability of virus-vector interactions and subsequent dispersal to new, uninfected hosts.

The secondary spread of a virus by short-distance dispersal of its vector need not result in an aggregation of adjacent virus-infected plants.[2] The result is likely to be loosely defined clusters of infected plants, because vector movement is not restricted to adjacent plants. Therefore, it is important not to limit spatial analysis to methods capable only of analyzing adjacent plants along rows, e.g., doublet analysis or runs analysis. Several alternative methods are available that will allow the investigator to quantify spatial patterns of diseased plants in any direction and over time.[2,3,62]

Vector populations that disperse in response to unfavorable environment and declining host suitability may behave very differently. In an agroecosystem, senescence of the host crop may trigger long-distance dispersal. "Long distance" implies movement out of the crop. In reality, the range of dispersal could be as close as an adjacent field of an emerging host crop or hundreds of kilometers to the nearest available host. A vector developing in an unfavorable environment has no way of knowing the proximity of its next available host; therefore, the actual distance to the next host should not influence the dispersal behavior. "Long-distance" dispersal often involves genetically controlled changes such as development of morphologically distinct forms more suited for long-distance migrations, as well as changes in development times and dispersal rates.[42,63] Aphids that develop on physiologically unfavorable hosts or in unfavorable environments have a lower reproductive capacity, but are better able to survive periods of starvation.[64] These aphids readily take off at steep angles that allow them to clear the vegetation and into the faster moving air currents.[61] Alate aphids with a reduced reproductive investment have a greater migratory urge, fly longer and more frequently, and are better able to survive starvation.[61] This should result in a differential dispersal pattern, i.e., "long distance", than for aphids with a greater reproductive investment (discussed above). The reduction in reproductive potential of migrating forms is also exhibited in leafhoppers and planthoppers. The ovaries of the long-winged, migrating form are reduced compared to the short-winged or the flightless reproductive forms.[65]

Epidemiologically, long-distance dispersal is a mechanism by which the vector and its associated virus can colonize and exploit new ecosystems. Long-distance spread is known to occur for viruses transmitted by the three mechanisms discussed previously, and by a wide range of vectors including aphids, whiteflies, and mites.[42] Long-distance dispersal of virus vectors results in the primary spread of viruses and may account for the complete epidemic in the case of monocyclic diseases.[66] Long-distance dispersal is also important in the establishment of disease foci. The spatial characteristics of these foci are dependent upon vector behavior and the crop environment. Random alighting in a field by the incoming vector will establish random foci. These may occur as clusters of plants if the vector is capable of moving to and infecting multiple plants. This is especially common with leafhopper, planthopper, and beetle vectors, which are strong fliers, and active insects that will frequently move between plants rather than settle on one plant. Aphids will attempt to visit several plants in proximity after locating a suitable host; however, their wing muscles will deteriorate rapidly at a rate proportional to the distance of flight.[61]

## B. Temporal Characteristics of Vector Dispersal

Swenson[67] commented that, all other things being equal, virus spread would be related to the number of vectors. The temporal pattern of virus spread has and continues to be explained on the basis of vector populations.[12,15] While the number of vectors is a primary factor, Swenson's "other things", which are the numerous intrinsic and interactive properties of the virus-vector-host system (Figure 1), must be defined and investigated and are never equal.

The temporal spread of a virus is a function of the number and species of vector, inoculation and acquisition times, the latent period in both the host and the vector, the average length of time an insect feeds on a plant, and the proportion of infected and uninfected plants.

The mechanism of transmission will determine, in part, the relative importance of each of these factors. Power[68] has developed a model that relates viruliferous vector movement into and within a field to the increase in disease of a monocyclic pathogen that is retained for the life of the vector. The model relates the time a viruliferous vector spends inoculating a plant to disease incidence. The highest infection rate and consequently the highest disease incidence results when a viruliferous vector remains on each plant visited for a period equal to the optimum inoculation feeding period. If the inoculation feeding period is less than optimum, transmission efficiency is decreased. Conversely, a longer than optimal inoculation feeding period reduces the number of plants the vector can visit and potentially inoculate. This model is useful for systems involving actively mobile vectors and propagative and perhaps circulative viruses, but assumes transmission efficiency does not decline during sequential inoculations and the infectability of the host does not change as a function of its physiological age. In addition, the infection rate will decrease as disease incidence increases due to multiple inoculations of already infected plants. Nevertheless, this model does prioritize the important factors associated with in-field, temporal spread of a monocyclic disease that persists in an active vector. The final disease incidence and rate of disease development will change based on the number of viruliferous vectors alighting on the crop over time and the feeding and dispersal behavior of the viruliferous vector once in the crop.

The situation becomes more complex with polycyclic virus epidemics and viruses not retained by their vectors for the life of the host or the vector. The seasonal phenology and behavior of the vector remain major factors, but the virus and host characteristics also play a major role in determining the temporal pattern of disease spread. An accurate model should take into account the seasonal phenology and numbers of alighting vectors, the probability of inoculation as a function of inoculation access period, the degree and success of colonization of the host by the vector, the latent period of the virus in an inoculated host, the acquisition efficiency and subsequent within-field dispersal of the viruliferous vectors, the probability of the vector retaining the virus as a function of the time the vector takes to move between host plants, and the time period that host will serve as an efficient source of the virus. The maximum rate of disease spread would require simultaneous optimization and maximization of several parameters. Any of the above-mentioned factors alone or in combination have the potential to reduce the rate of disease spread or the final incidence of disease.

### C. Summary

Vector-borne viruses are dependent upon their vectors for transmission; therefore, understanding the spatial and temporal dynamics of vector dispersal will provide insight into the spatial and temporal distribution of the viruses. However, as discussed in the previous sections, numerous factors contribute to the spatiotemporal characteristics of a plant virus epidemic. The complexity of the conceptual model discussed in this chapter emphasizes the importance of expanding virus epidemiological studies beyond the common virus incidence-vector presence type of investigation. While vector phenology and abundance are important parameters, there is a definite need to expand future studies to include all the parameters that affect successful transmission of a virus by its vector. Only then can the ultimate goal of using epidemiological data to develop virus control strategies be achieved.

Other vector-related factors influencing plant virus epidemiology have been reviewed elsewhere and include vector variation with respect to transmission efficiency,[69,70] host-finding behavior,[71] and methodology of vector population sampling.[72]

# V. HOST PLANT CONTRIBUTIONS TO PLANT VIRUS EPIDEMICS

The plant must not only serve as a host for the virus and perhaps the vector, but it must serve as an efficient link between the virus and its vector. A virus must be able to infect, replicate, and maintain itself in its hosts, but it also must be accessible to its vector. To ensure its continued survival a virus must be acquired, transported, and inoculated into new hosts periodically by its vector. Hypothetically, the optimum host would be easily inoculated and infected, be long-lived and unaffected by a virus and its vector, and be a constant, quality source of the virus. Any deviation from this optimum indicates the host may affect the epidemiology of the virus.

## A. Host Plant Age and Physiology

The host will have a pronounced effect on the number of infection cycles that can occur during an epidemic. The number of cycles will be determined by the length of the latent period in the host plant, i.e., the time from inoculation until the time the plant becomes a source of the virus, the length of time the infected plant serves as a source of the virus, and the time period the host is susceptible to the virus. Young plants of most crops are more susceptible to infection and systemic spread of viruses than older plants. Generally, younger or actively growing plants are better sources of the virus than are mature plants; however, exceptions do exist.[73] The quality of an infected plant as a source of a virus is not only dependent upon plant age, but also upon the time since inoculation. Once infected, the host is rarely able to recover from infection, but the virus titer or distribution does not remain constant.[73] A low virus titer or restricted virus distribution during disease development may reduce the probability that the host will contribute to further disease spread.[44,74] This is especially true of the noncirculative viruses.[44,86,88] When developing a model to describe virus spread, a parameter, analogous to the infectious or sporulation period of fungal disease models, may be necessary to identify the finite time when an infected host is a source of the virus.

## B. Host Plant Resistance

The various types and mechanisms of host plant resistance to viruses and their vectors have been reviewed,[75-77] but seldom are the epidemiological consequences of host plant resistance discussed. Total resistance or immunity of a plant to a virus or a vector would be the most absolute and desirable form of control. Unfortunately, this type of resistance is unavailable for most virus-vector systems. Partial resistance to viruses and their vectors has been identified in many crops and its benefit in reducing or delaying virus epidemics is now recognized.

Plant genotypes resistant to vectors of plant viruses can alter vector population size, activity, and feeding behavior, thereby altering the probability and efficiency of virus transmission by the resisted vector. The effect of plant resistance to vectors on transmission and spread of plant viruses is dependent upon the type and level of resistance (antixenosis, antibiosis, physical, antifeeding), the transmission characteristics (acquisition, latent, and inoculation periods), and the effect of virus infection on the expression of vector resistance.[75]

The effect of antixenosis on virus spread will depend upon the rapidity with which the vector recognizes the resistant plant as an unsuitable host relative to the inoculation and acquisition times of the virus. Interference with the initial setting and probing of aphid vectors is usually associated with physical attributes (pubescence, glandular hairs) of the leaf surface. If the resistance is strong enough to inhibit probing or feeding, then no acquisition or inoculation of virus will occur. Partial resistance may allow for some transmission, but may evoke a nonpreference response from the vector and result in reduced colonization or avoidance of the crop. Increased leaf pubescence on soybeans reduces the rate of spread and final disease incidence of soybean mosaic virus, although vector landing rates were similar to those in glabrous soybeans.[78]

Antibiosis acts to reduce the population of vectors developing on the host plant. Resistant muskmelon genotypes reduced the fecundity and longevity of colonizing aphid species and reduced the percentage of alates (winged forms) produced.[75] Any factor consistently reducing a vector population or dispersal of a vector population can be expected to have a similar or greater affect on virus spread.[67]

Another mechanism by which antibiosis may operate is by interfering with the ability of the vector to find its preferred feeding site.[79] This effect reduces the inoculation and acquisitions of circulative and propagative viruses.

Vector resistance has also been manifested as an alteration in the inoculation efficiency of specific vectors. The muskmelon genotypes Songwhan Charmi (SC) and aphid-resistant Top Mark (ARTM) possess a resistance to inoculation of some noncirculative viruses by the colonizing aphid, *Aphis gossypii*. Lecoq and Pitrat[80] found that this resistance (in SC) delayed epidemics of cucumber mosaic virus, resulting in higher yield compared to the susceptible variety 'Vedrantais'. Gray et al.[44] concluded that resistance to inoculation of watermelon mosaic virus 2 in the muskmelon genotype ARTM significantly reduced the rate of disease progress and the final disease incidence. In addition, there was a reduction in the amount of secondary spread of the virus, as indicated by reduced size of clusters of infected plants expanding from initial foci of infected plants.

The epidemiological effects of incomplete host plant resistance to viruses have recieved limited attention. A majority of the work on host plant resistance to viruses is conducted by plant breeders who select plants that yield and grow well under high inoculum pressure. The distinction between tolerance and resistance is seldom made, but is of great epidemiological importance if disease control is the ultimate objective. Partial virus resistance identified in crop plants and shown to reduce virus spread has been manifested as a reduction in virus titer, inhibition of systemic movement, and reduction in infection efficiency.

Resistant plants that are difficult to infect will affect polycyclic virus epidemics by delaying the onset of disease and reducing the initial amount of inoculum contributing to further cycles of disease spread.[81] This type of resistance is perhaps more effective against monocyclic epidemics; for example, rates of disease progress and final disease incidence of beet curly top virus were reduced in tomatoes.[82] The onset of disease was similar in resistant and susceptible plants, but the requirement of multiple inoculations significantly reduced the number of plants becoming infected and effectively slowed the epidemic.

Plants whose resistance is manifested as a delayed systemic invasion also delay the onset of disease.[83,84] Additionally, secondary infections are slowed because of the extended latent period prior to the infected plant becoming a source of the virus. Restrictions of systemic invasion reduce the number of potential virus-acquisition sites and reduce the transmission efficiency of vectors feeding on these plants. This is particularly effective in reducing secondary spread of the virus within the resistant host crop.

Suppressive virus resistance that results in a lower virus titer in the resistant host has been identified in several hosts.[85-88] This type of resistance reduces the acquisition efficiency of vectors and subsequently reduces the probability of secondary spread. Under conditions where secondary spread is the major contribution to disease incidence, suppressive resistance is an effective control measure.[44,88] Tables 2 and 3 list some reported instances of partial resistance of crop plants to viruses and their vectors.

Results of the relatively few epidemiological studies discussed in this section illustrate the importance of partial resistance. These types of resistance are often neglected or overlooked because of the incomplete protection. In many virus-vector-host systems, complete resistance or other forms of disease control have not been identified. Perhaps it is time that partial resistance to either the vector or the virus be recognized as an important potential

**Table 2**
**ASSOCIATION OF VECTOR RESISTANCE OF PLANTS**
**WITH THEIR DECREASED VIRUS INCIDENCE**

| Vector | Virus | Crop | Ref. |
|---|---|---|---|
| Nematode | Grape fanleaf | Grape | 89 |
| Mite | Wheat streak mosaic | Wheat | 90 |
| Thrip | Tomato spotted wilt | Groundnut | 91 |
| Planthopper | Rice ragged stunt | Rice | 92 |
| Leafhopper | Tungro | Rice | 93 |
| Aphid | Soybean mosaic | Soybean | 78 |
| Aphid | Watermelon mosaic 2 | Muskmelon | 44 |
| Aphid | Cucumber mosaic | Muskmelon | 80 |
| Aphid | Cucumber mosaic | Pepper | 94 |
| Aphid | Potato virus Y | Potato | 95 |
| Aphid | Barley yellow dwarf | Wheatgrasses | 79 |
| Aphid | Plum pox | Peach | 96 |
| Aphid | Raspberry virus complex | Raspberry | 97 |
| Beetle | Cowpea mottle | Cowpea | 98 |
| Whitefly | Tomato yellow curl | Tomato | 99 |

**Table 3**
**ASSOCIATION OF INCOMPLETE VIRUS RESISTANCE IN CROP PLANTS**
**WITH THEIR DECREASED VIRUS INCIDENCE**

| Virus | Crop | Type of resistance | Ref. |
|---|---|---|---|
| Barley yellow dwarf | Oats, barley | Reduced titer | 87 |
| Pepper mottle | Pepper | Reduced titer | 88 |
| Cowpea chlorotic mottle | Soybean | Reduced titer | 86 |
| Watermelon mosaic 2 | Muskmelon | Reduced titer and restricted distribution | 44, 85 |
| Southern bean mosaic | Soybean | Reduced titer and restricted distribution | 83 |
| Potato leaf roll | Potato | Restricted distribution | 74 |
| Maize dwarf mosaic | Maize | Resistance to inoculation | 81 |

control measure. Moreover, the benefits of partial resistance are increased when used with other cultural control practices.[80]

## C. Host Plant Diversity

The influence of host plant diversity on the transmission and spread of vector-transmitted plant pathogens was recently reviewed.[68] The use of multiple crops or multiline monocultures as a method of virus control has received minimal attention despite the success of such systems with other plant pathogens. The effects of these cropping strategies on virus disease spread or vector dispersal are not well understood, but a few generalities have been recognized. Plant community diversity decreases the amount of time a vector remains on a plant and increases the activity of interplant movement. Decreased feeding time reduces the probability of transmission of propagative and circulative viruses, but one may expect an increase in the transmission of noncirculative viruses. In fact, disease incidence tends to be lower in diverse plant communities for all types of viruses.[68] One possible explanation for this is that vectors of noncirculative viruses indiscriminately probe on hosts and nonhosts alike. The retention of noncirculative viruses is related not only to time, but to the number of feeding probes the viruliferous vector makes. Repeated probing on nonhost plants will cause the vector to lose the virus to plants which are not hosts of the pathogen. Clearly, the effectiveness of the diverse plant community in reducing virus incidence is dependent upon the ratio of host to nonhost plants of the virus rather than the actual number of plant species.

## D. Summary

Additional plant-related factors influencing virus epidemics have been reviewed, and include the role of weeds[100] and natural plant communities[101] and cultural practices used to manipulate the agroecosystem.[102]

The plant component of the conceptual model presented in this chapter has the greatest influence on the spatial and temporal characteristics of a virus epidemic. The characteristics of the plant community in time and space will influence the vector populations, virus populations, and the ability of the viruses and vectors to interact. The plant is the essential link between the virus and its vector, but the same plant need not be the source of both virus and vector. Descriptive and simulation models of plant virus epidemics must consider the intrinsic ability of the plant to resist the virus or its vector as well as the growth stages of the plant which are optimal for the virus and vector. In addition, the spatial and temporal characteristics of the plant ecosystem must also be considered with respect to the seasonal phenology of the vector population. The plant community cannot be considerd as a stable environment that contributes equally to an epidemic over time.

## VI. CONCLUSIONS

Understanding the epidemiology of a vectored plant virus disease involves an investigation of a complex ecological system. The components and characteristics involved are unique to each system, and this chapter has discussed but a few of the factors that are likely to have a major impact on the epidemic. As mentioned, many other factors equally as important have been reviewed elsewhere. Each of the system components and their attributes need to be recognized, but it is an impossible and unnecessary task to carefully study each one. Nevertheless, the essential investigations will require enormous amounts of resources, e.g., time, labor, money. Therefore, it is to the advantage of researchers to develop holistic conceptual models and utilize system analysis to direct and focus research efforts on the most important factors influencing their particular virus-vector-host system.

I have attempted to summarize the important intrinsic and interactive biological properties of a virus-vector-plant system and to generalize about how selected parameters may influence the spatiotemporal characteristics of an epidemic. There is an enormous amount of descriptive data on plant virus epidemiology available in the literature and numerous methods to describe these data. However, few systems have been studied in enough detail to develop unifying concepts or operational models. If virologists hope to provide adequate solutions and control strategies to current worldwide virus disease problems, we must move beyond descriptive epidemiology. Understanding and solving the problems will require a concerted effort of many disciplines and the use of all the analytical tools recently developed. The early success of the few vector-borne virus models being developed[7] should encourage a wider use of conceptual and simulation models.

## ACKNOWLEDGMENTS

I would like to thank my colleagues, Drs. C. L. Campbell, L. V. Madden, M. Milgroom, J. W. Moyer, A. G. Power, and A. J. Sawyer, for providing unpublished data and manuscripts and for their comments and suggestions on this manuscript.

# REFERENCES

1. **Leonard, K. J., and Fry, W. E.**, *Plant Disease Epidemiology*, Macmillan, New York, 1986, 370.
2. **Gray, S. M., Moyer, J. W., and Bloomfield, P.**, Two-dimensional distance class model for quantitative description of virus-infected plant distribution lattices, *Phytopathology*, 76, 243, 1986.
3. **Madden, L. V.**, Dynamic nature of within-field disease and pathogen distributions, in *The Spatial Component of Plant Disease Epidemics*, Jeger, M. J., Ed., Prentice-Hall, Englewood Cliffs, NJ, in press.
4. **Madden, L. V., Louie, R., and Knoke, J. K.**, Temporal and spatial analysis of maize dwarf mosaic epidemics, *Phytopathology*, 77, 148, 1987.
5. **Madden, L. V., Pirone, T. P., and Raccah, B.**, Temporal analysis of two viruses increasing in the same tobacco fields, *Phytopathology*, 77, 974, 1987.
6. **Madden, L. V., Pirone, T. P., and Raccah, B.**, Analysis of spatial patterns of virus-diseased tobacco plants, *Phytopathology*, 77, 1409, 1987.
7. **McLean, G. D., Garrett, R. G., and Ruesink, W. G.**, *Plant Virus Epidemics: Monitoring, Modelling and Predicting Outbreaks*, Academic Press, Sydney, 1986, 550.
8. **Nakasuji, F., Miyai, S., Kawamoto, H., and Kiritani, K.**, Mathematical epidemiology of rice dwarf virus transmitted by green rice leafhoppers, *Nephotettix cincticeps: a differential equation model*, *J. Appl. Ecol.*, 22, 839, 1985.
9. **Thresh, J. M.**, Plant virus epidemiology and control: current trends and future prospects, in *Plant Virus Epidemiology*, Plumb, R. T. and Thresh, J. M., Eds., Blackwell Scientific, London, 1983, 349.
10. **Thresh, J. M.**, The population dynamics of plant virus diseases, in *Populations of Plant Pathogens: Their Dynamics and Genetics*, Wolf, M. S. and Caten, C. E., Eds., Blackwell Scientific, Oxford, 1987, 135.
11. **Thresh, J. M,.** Plant virus dispersal, in *The Movement and Dispersal of Agriculturally Important Biotic Agents*, MacKenzie, D. R., Bradfield, C. S., Kennedy, G. G., Berger, R. D., and Taranto, D. J., Eds., Claitors Publishing, Baton Rouge, 1986, 51.
12. **Thresh, J. M.**, Progress curves of plant virus disease, *Adv. Appl. Biol.*, 8, 1, 1983.
13. **Thresh, J. M.**, Cropping practices and virus spread, *Annu. Rev. Phytopathol.*, 20, 193, 1982.
14. **Thresh, J. M.**, The epidemiology of plant virus diseases, in *Plant Disease Epidemiology*, Scott, P. R. and Bainbridge, A., Eds., Blackwell Scientific, Oxford, 1978, 79.
15. **Thresh, J. M.**, Temporal patterns of virus spread, *Annu. Rev. Phytopathol.*, 12, 111, 1974.
16. **Harrison, B. D.**, Epidemiology of plant virus diseases: a prologue, in *Plant Virus Epidemiology*, Plumb, R. T. and Thresh, J. M., Eds., Blackwell Scientific, Oxford, 1983, 1.
17. **Carter, N.**, Simulation modelling, in *Plant Virus Epidemics: Monitoring, Modelling and Predicting Outbreaks*, McLean, G. D., Garrett, R. G. and Ruesink, W. G., Eds., Academic Press, Sydney, 1986, 550.
18. **Kranz, J. and Hau, B.**, Systems analysis in epidemiology, *Annu. Rev. Phytopathol.*, 18, 67, 1980.
19. **Harrison, B. D.**, Plant virus transmission by vectors: mechanisms and consequences, in *Molecular Basis of Virus Disease*, Russel, W. C. and Almond, J. W., Eds., Cambridge University Press, 1987, 319.
20. **Gamez, R.**, The ecology of maize rayado fino virus in the American tropics, in *Plant Virus Epidemiology*, Plumb, R. T. and Thresh, J. M., Eds., Blackwell Scientific, Oxford, 1983, 267.
21. **Egan, B. T. and Hall, P.**, Monitoring the Fiji disease epidemic in sugarcane at Bundaburg, Australia, in *Plant Virus Epidemiology*, Plumb, R. T. and Thresh, J. M., Eds., Blackwell Scientific, Oxford, 1983, 287.
22. **Gamez, R. and Saavedra, F.**, Maize rayado fino: a model of a leafhopper-borne virus disease in the neotropics, in *Plant Virus Epidemics: Monitoring, Modelling and Predicting Outbreaks*, McLean, G. D., Garrett, R. G., and Ruesink, W. G., Eds., Academic Press, Sydney, 1986, 315.
23. **Kisimoto, R. and Yamada, Y.**, A planthopper-rice virus epidemiology model: rice stripe and small brown planthopper, *Laodelphax striatellus Fallen, in Plant Virus Epidemics: Monitoring, Modelling and Predicting Outbreaks*, McLean, G. D., Garrett, R. G., and Ruesink, W. G., Eds., Academic Press, Sydney, 1986, 327.
24. **Conti, M.**, Epidemiology of maize rough dwarf virus, *Phytoparasitica*, 12, 210, 1984.
25. **Autrey, L. J. C. and Ricaud, C.**, The comparative epidemiology of two diseases of maize caused by leafhopper-borne viruses in Mauritius, in *Plant Virus Epidemiology*, Plumb, R. T. and Thresh, J. M., Eds., Blackwell Scientific, Oxford, 1983, 277.
26. **Hsu, T. P. and Banttari, E. E.**, Dual transmission of the aster yellows mycoplasma-like organism and the oat blue dwarf virus and its effect on longevity and fecundity of the aster leafhopper vector *Macrosteles fascifrons*, *Phytopathology*, 69, 843, 1979.
27. **Sylvester, E. S.**, Circulative and propagative virus transmission by aphids, *Annu. Rev. Entomol.*, 25, 257, 1980.
28. **Ammar, E. D., Gingery, R. E., and Nault, L. R.**, Interactions between maize mosaic and maize stripe viruses in their vector *Peregrinus maidis* and in maize, *Phytopathology*, 77, 1, 1987.
29. **Gildow, F. E.**, Virus-membrane interactions involved in circulative transmission of luteoviruses by aphids, in *Current Topics in Vector Research*, Harris, K. F., Ed., Springer-Verlag, New York, 1987, 93.

30. **Rochow, W. F.,** Biological properties of four isolates of barley yellow dwarf virus, *Phytopathology,* 59, 1580, 1969.
31. **Johnson, R. A. and Rochow, W. F.,** An isolate of barley yellow dwarf virus transmitted specifically by *Schizaphis graminum, Phytopathology,* 62, 921, 1972.
32. **Rochow, W. F.,** Dependent transmission by aphids of barley yellow dwarf luteoviruses from mixed infections, *Phytopathology,* 72, 302, 1982.
33. **Plumb, R. T.,** Barley yellow dwarf virus — a global problem, in *Plant Virus Epidemiology,* Plumb, R. T. and Thresh, J. M., Eds., Blackwell Scientific, Oxford, 1983, 185.
34. **Heathcote, G. D.,** Virus yellows of sugar beet, in *Plant Virus Epidemics: Monitoring, Modelling and Predicting Outbreaks,* McLean, G. D., Garret, R. G., and Ruesink, W. F., Eds., Academic Press, Sydney, 1987, 399.
35. **Howell, W. E. and Mink, G. I.,** Role of aphids in the epidemiology of carrot virus diseases in central Washington, *Plant Dis. Rep.,* 61, 841, 1977.
36. **Gamez, R. and Moreno, R. A.,** Epidemiology of beetle-borne viruses of grain legumes in Central America, in *Plant Virus Epidemiology,* Plumb, R. T. and Thresh, J. M., Eds., Blackwell Scientific, Oxford, 1983, 103.
37. **Fulton, J. P., Gergerich, R. C., and Scott, H. A.,** Beetle transmission of plant viruses, *Annu. Rev. Phytopathol.,* 25, 111, 1987.
38. **Harrison, B. D. and Murant, A. F.,** Involvement of virus-coded proteins in transmission of plant viruses by vectors, in *Vectors in Virus Biology,* Mayo, M. A. and Harrap, K. A., Eds., Academic Press, London, 1984, 1.
39. **Raccah, B. and Singer, S.,** Role of colonizing and non-colonizing aphid species in the spread of non-persistent viruses in peppers, *Phytoparasitica,* 13, 161, 1985.
40. **Raccah, B. and Cal, O. A.,** The role of flying aphid vectors in the transmission of cucumber mosaic virus and potato virus Y to peppers in Israel, *Ann. Appl. Biol.,* 106, 451, 1985.
41. **Atiri, G. I., Enobakhare, D. A., and Thottappilly, G.,** The importance of colonizing and non-colonizing aphid vectors in the spread of cowpea aphid-borne mosaic virus in cowpea *Vigna unguiculata, Crop Prot.,* 5, 406, 1986.
42. **Thresh, J. M.,** The long-range dispersal of plant viruses, *Philos. Trans. R. Soc. London, Ser. B,* 302, 497, 1983.
43. **Garrett, R. G. and McLean, G. D.,** The epidemiology of some aphid-borne viruses in Australia, in *Plant Virus Epidemiology,* Plumb, R. T. and Thresh, J. M., Eds., Blackwell Scientific, 1983, 199.
44. **Gray, S. M., Moyer, J. W., Kennedy, G. G., and Campbell, C. L.,** Virus-suppression and aphid resistance effects on spatial and temporal spread of watermelon mosaic virus 2, *Phytopathology,* 76, 1254, 1986.
45. **Jayasena, K. W. and Randles, J. W.,** Patterns of spread of the nonpersistently transmitted bean yellow mosaic virus and the persistently transmitted subterranean clover red leaf virus in *Vicia faba, Ann. Appl. Biol.,* 104, 249, 1984.
46. **Scott, G. E.,** Nonrandom spatial distribution of aphid-vectored maize dwarf mosaic, *Plant Dis.,* 69, 893, 1985.
47. **Lamberti, F. and Roca, F.,** Present status of nematodes as vectors of plant viruses, in *Vistas on Nematology,* Veech, J. A. and Dickson, D. W., Eds., Society of Nematologists, Hyattsville, MD, 1987, 321.
48. **Madamba, C. P., Sasser, J. N., and Nelson, L. A.,** Some characteristics of the effects of *Meloidogyne* spp. on unsuitable host crops, *N.C. Agric. Exp. Stn. Tech. Bull.,* 169, 1, 1965.
49. **Harrison, B. D.,** Epidemiology of diseases caused by nematode-borne viruses: chairman's comments, in *Plant Disease Epidemiology,* Scott, P. R. and Bainbridge, A., Eds., Blackwell Scientific, London, 1978, 251.
50. **Noe, J. P. and Campbell, C. L.,** Spatial pattern analysis of plant-parasitic nematodes, *J. Nematol.,* 17, 86, 1985.
51. **Thresh, J. M. and Pitcher, R. S.,** The spread of nettlehead and related virus diseases of hop, in *Plant Disease Epidemiology,* Scott, P. R. and Bainbridge, A., Eds., Blackwell Scientific, London, 1978, 291.
52. **Phatak, H. C.,** The role of seed and pollen in the spread of plant pathogens particularly viruses, *Trop. Pest Manage.,* 26, 278, 1980.
53. **Mandahar, C. L.,** Virus transmission through seed and pollen, in *Plant Diseases and Vectors: Ecology and Epidemiology,* Maramorosch, K. and Harris, K. F., Eds., Academic Press, New York, 1981, 241.
54. **Mink, G. I.,** The possible role of honeybees in long-distance spread of prunus necrotic ringspot virus from California into Washington sweet cherry orchards, in *Plant Virus Epidemiology,* Plumb, R. T. and Thresh, J. M., Eds., Blackwell Scientific, London, 1983, 85.
55. **Rochow, W. F.,** The role of mixed infections in the transmission of plant viruses by aphids, *Annu. Rev. Phytopathol.,* 10, 101, 1972.

56. **Falk, B. W. and Duffus, J. E.,** Epidemiology of helper-dependent persistent aphid transmitted virus complexes, in *Plant Diseases and Vectors: Ecology and Epidemiology,* Maramorosch, K. and Harris, K. F., Eds., Academic Press, New York, 1981, 162.

57. **Garret, R. G., Ruesink, W. G., and McLean, G. D.,** Epilogue: a perspective, in *Plant Virus Epidemics: Monitoring, Modelling and Predicting Outbreaks,* McLean, G. D., Garrett, R. G., and Ruesink, W. G., Eds., Academic Press, Sydney, 1986, 525.

58. **Price, P. W.,** *Insect Ecology,* John Wiley & Sons, New York, 1975, 169.

59. **Shaw, M. J. P.,** Effects of population density on alienicolae of *Aphis fabae, Ann. Appl. Biol.,* 1970, 65, 197.

60. **Johnson, C. G.,** Lability of the flight system: a context for functional adaptation, in *Insect Flight,* Rainey, R. C., Ed., Academic Press, 1976, 217.

61. **Walters, K. F. A. and Dixon, A. F. G.,** Migratory urge and reproductive investment in aphids: variation within clones, *Oecologia (Berlin),* 1893, 58, 70.

62. **Gibbs, A.,** A simple convolution method for describing or comparing the distribution of virus-affected plants in a plant community, in *Plant Virus Epidemiology,* Plumb, R. T. and Thresh, J. M., Eds., Blackwell Scientific, London, 1983, 39.

63. **Dingle, H.,** Migration strategies of insects, *Science,* 175, 1327, 1972.

64. **Wellings, P. W., Leather, S. R., and Dixon, A. F. G.,** Seasonal variation in reproductive potential: a programmed feature of aphid life cycles, *J. Anim. Ecol.,* 49, 975, 1983.

65. **Taylor, R. A. J.,** Migratory behavior in the Auchenorrhyncha, in *The Leafhoppers and Planthoppers,* Nault, L. R. and Rodriguez, J. E., Eds., John Wiley & Sons, New York, 1985, 259.

66. **Martin, D. K.,** Studies on the main weed host and aphid vector of lettuce nectrotic yellows virus in South Australia, in *Plant Virus Epidemiology,* Plumb, R. T. and Thresh, J. M., Eds., Blackwell Scientific, London, 1983, 211.

67. **Swenson, K. G.,** Role of aphids in the ecology of plant viruses, *Annu. Rev. Phytopathol.,* 6, 351, 1968.

68. **Power, A. G.,** Cropping systems, insect movement, and the spread of insect transmitted diseases in crops, in *Research Approaches in Agricultural Ecology,* Glieuman, S. R., Ed., Springer-Verlag, New York, in press.

69. **Orlob, G. B.,** Further studies on the transmission of plant viruses by different forms of aphids, *Virology,* 16, 301, 1962.

70. **Rochow, W. F.,** Variation within and among aphid vectors of plant viruses, *Ann. N.Y. Acad. Sci.,* 105, 713, 1964.

71. **Kennedy, J. S., Booth, C. O., and Kershaw, W. J. S.,** Host finding by aphids in the field. II. *Aphis fabae* Scop. and *Brevicoryne brassicae* L.; with a re-appraisal of the role of host finding behavior in virus spread, *Ann. Appl. Biol.,* 1959, 47, 424.

72. **Irwin, M. E. and Goodman, R. M.,** Ecology and control of soybean mosaic virus, in *Plant Diseases and Their Vectors: Ecology and Epidemiology,* Maramorosch, K. and Harris, K. F., Eds., Academic Press, New York, 1981, 181.

73. **Gibbs, A. and Harrison, B.,** *Plant Virology: The Principles,* Edward Arnold, London, 1976, 137.

74. **Barker, H. and Harrison, B. D.,** Restricted distribution of potato leaf-roll virus antigen in resistant potato genotypes and its effect on transmission of the virus by aphids, *Ann. Appl. Biol.,* 109, 595, 1986.

75. **Kennedy, G. G.,** Host plant resistance and the spread of plant viruses, *Environmental Entomology,* 5, 827, 1976.

76. **Fraser, R. S. S.,** *The Biochemistry of Virus-Infected Plants,* John Wiley & Sons, New York, 1987, 103.

77. **Jones, A. T.,** Control of virus infection in crop plants through vector resistance: a review of achievements, prospects, and problems, *Ann. Appl. Biol.,* 111, 745, 1987.

78. **Gunasinghe, U. B., Irwin, M. E., and Kampmeier, G. E.,** Soybean leaf pubescence affects aphid vector transmission and field spread of soybean mosaic virus, *Ann. Appl. Biol.,* 112, 259, 1988.

79. **Shukle, R. H., Lampe, D. J., Lister, R. M., and Foster, J. E.,** Aphid feeding behavior: relationship to barley yellow dwarf virus resistance in *Agropyron* species, *Phytopathology,* 77, 725, 1987.

80. **Lecoq, H. and Pitrat, M.,** Field experiments on the integrated control of aphid-borne viruses in muskmelon, in *Plant Virus Epidemiology,* Plumb, R. T. and Thresh, J. M., Eds., Blackwell Scientific, London, 1983, 169.

81. **Louie, R.,** Effects of genotype and inoculation protocols on resistance evaluation of maize to maize dwarf mosaic virus strains, *Phytopathology,* 76, 769, 1986.

82. **Thomas, P. E. and Martin, M. W.,** Vector preference, a factor of resistance to curly top virus in certain tomato cultivars, *Phytopathology,* 61, 1257, 1971.

83. **Kuhn, C. W., Benner, C. P., and Hobbs, H. A.,** Resistance responses in cowpea to southern bean mosaic virus based on virus accumulation and symptomatology, *Phytopathology,* 76, 795, 1986.

84. **Simons, J. N.,** Factors affecting field spread of potato virus Y in south Florida, *Phytopathology,* 50, 424, 1960.

85. **Moyer, J. W., Kennedy, G. G., Romanow, L. R.,** Resistance to watermelon mosaic virus II multiplication in *Cucumis melo, Phytopathology,* 75, 210, 1985.
86. **Paguio, O. R., Boerma, H. R., and Kuhn, C. W.,** Disease resistance, virus concentration, and agronomic performance of soybean infected with cowpea chlorotic mottle virus, *Phytopathology,* 77, 703, 1987.
87. **Skaria, M., Lister, R. M., Foster, J. E., and Shaner, G.,** Virus content as an index of symptomatic resistance to barley yellow dwarf virus in cereals, *Phytopathology,* 75, 212, 1985.
88. **Zitter, T. A.,** Transmission of pepper mottle virus from susceptible and resistant pepper cultivars, *Phytopathology,* 65, 110, 1975.
89. **Bouquet, A.,** Resistance to grape fanleaf virus in Muscadine grape inoculated with *Xiphinema index, Plant Dis.,* 65, 791, 1981.
90. **Martin, T. J., Harvey, T. L., Bender, C. G., and Seifers, D. L.,** Control of wheat streak mosaic virus with vector resistance in wheat *Triticum aestivum, Phytopathology,* 74, 963, 1984.
91. **Reddy, D. V. R., Amin, P. W., McDonald, D., and Ghanekar, A. M.,** Epidemiology and control of groundnut bud necrosis and other diseases of legume crops in India caused by tomato spotted wilt virus, in *Plant Virus Epidemiology,* Plumb, R. T. and Thresh, J. M., Eds., Blackwell Scientific, London, 1983, 93.
92. **Parejarearn, A., Lapis, D. B., and Habino, H.,** Reaction of rice varieties to ragged stunt virus infection by three brown planthopper biotypes, *Int. Rice Res. Newsl.,* 9, 7, 1984.
93. **Hibino, H., Tiongco, E. R., Cabunagan, R. C., and Flores, Z. M.,** Resistance to rice tungro-associated viruses in rice under experimental and natural conditions, *Phytopathology,* 77, 871, 1987.
94. **Cohen, S.,** Resistance in peppers to cucumber mosaic virus transmission by aphids, *Phytoparasitica,* 10, 127, 1982.
95. **Lapointe, S. L., Tingey, W. M., and Zitter, T. A.,** Potato virus Y transmission reduced in an aphid-resistant potato species, *Phytopathology,* 77, 819, 1987.
96. **Massonie, G. and Maison, P.,** Peach resistance to *Myzus persicae* and to the transmission of the plum pox virus by this aphid, *Phytoparasitica,* 10, 127, 1982.
97. **Jones, A. T.,** Further studies on the effect of resistance to *Amphorophora idaei* in raspberry *Rubus idaeus* on the spread of aphid-borne viruses, *Ann. Appl. Biol.,* 92, 199, 1979.
98. **Allen, D. J., Thottappilly, G., and Rossel, H. W.,** Cowpea mottle virus field resistance and seed transmission in virus tolerant cowpea *Vigna unguiculata, Ann. Appl. Biol.,* 100, 331, 1982.
99. **Berlinger, M. J. and Dahan, R.,** Breeding for resistance to virus transmission by whiteflies in tomatoes, *Insect Sci. Appl.,* 8, 783, 1987.
100. **Duffus, J. E.,** Role of weeds in the incidence of virus diseases, *Annu. Rev. Phytopathol.,* 9, 319, 1971.
101. **Bos, L.,** Wild plants in the ecology of virus diseases, in *Plant Diseases and Vectors: Ecology and Epidemiology,* Maramorosch, K. and Harris, K. F., Eds., Academic Press, New York, 1981, 1.
102. **Zitter, T. A. and Simons, J. N.,** Management of viruses by alteration of vector efficiency and by cultural practices, *Annu. Rev. Phytopathol.,* 18, 289, 1980.

Chapter 10

DISEASE MANAGEMENT

**C. L. Mandahar, S. M. Paul Khurana, and I. D. Garg**

TABLE OF CONTENTS

# I. INTRODUCTION

The term "disease control" has been widely used in the literature. However, this term is restrictive in its connotation and suggests an accomplished event in which complete elimination of the pathogen is taken to have been achieved. In addition, reappearance of a "controlled" disease at a damaging level tends to indicate the failure of a control system. In contrast, the term "disease management" suggests four things: it conveys the working of a continuous system; suggests the necessity of carrying on continued adjustments and modifications in the system to face the changing situations; signifies the continued presence of a disease, under conditions in which the disease is managed in such a way that the disease losses or extent of damage to the crop are minimized and cut down to an economically acceptable level; and signifies the involvement of the economics of crop husbandry and consequently becomes an integral part of a crop production system. This brings in the concept of and need to define the economic threshold of a disease. It is that level of disease intensity that produces an incremental reduction in crop value greater than the cost of implementing a disease management strategy.[1]

Disease management acknowledges the dynamic nature of plant pathogens within agroecosystems, and developing a disease management system in turn becomes a dynamic process. Moreover, disease management is an integrated approach based on the judicious use of chemicals for protection and eradication, the use of natural genetic buffering mechanisms like resistant genes and tolerance, the use of crop-free periods, the use of virus-free propagative material and seed, the control of mechanical and vector transmission, and the application of cultural practices. Weed management must be regarded as an integral component of disease management. Such integrated disease management systems are ultimately the best and most promising approach for plant disease control and are much more effective than any single control method. Use of resistant varieties, crop scheduling, and insecticides together give good control of beet yellows virus complex in California. Such an integrated disease management program for this disease has also been successful in many European cooperating countries. An integrated six-point program has been employed for controlling beet curly top disease.[2] Early planting, use of resistant sugar beet cultivars, spraying of systemic insecticides at planting, and rangeland spraying have partially controlled this disease in California. Similarly, the only effective control measures for mycoplasmal and viral potato diseases are the use of disease-free seed, spraying the crop with systemic insecticides, crop inspection, and roguing of infected plants. Disease management is a holistic approach to disease control.

The role of the various individual components of disease management systems will be discussed in detail in order to understand the cumulative effect that the various control measures together have in managing plant virus diseases. Basically, a disease management strategy or any of its single components is aimed against the amount and efficacy of initial inoculum and its buildup by avoiding or reducing primary infection foci within or near a crop, against vector populations flying into and/or developing on a crop and their subsequent buildup, and against the rate of disease spread. In other words, control of various diseases means interference with virus ecology so that devising proper and effective management tactics against a virus disease is best based on epidemiology of that disease.

The efficacy of most control measures is related to levels of disease resistance and size of inoculum. Reduction in the amount of initial inoculum is particularly important because inoculum density is often correlated, although not always, in a linear fashion with disease intensity and also because some diseases engulf the crop dramatically once a certain minimum level of infection is exceeded. Moreover, there are particular thresholds of inoculum beyond which the chemical protectants fail to protect the crop adequately. It is therefore imperative to use certified seed and/or resistant cultivars accompanied by all the available cultural practices to limit inoculum to quantities the chemicals can handle. Moreover, the chemicals should be used early enough to limit buildup of inoculum to levels that will not overcome varietal resistance. For example, mineral oil was effective as long as the tobacco etch virus-diseased pepper plants fell below 3.5% in a field, but became completely ineffective when the diseased pepper plants exceeded 20% in a field.[3] The yield response of sugar beet to aphicide spraying depends on the amount of primary infection and virus incidence on un-sprayed areas.[4] The duration of the effect of 4-m high windbreaks on delaying the spread of cucumber mosaic virus (CMV) from infected melons to nearby vegetable plots of tomato, pepper, and squash in France in spring and early summer depended upon the level of virus infection in melon plots.[5]

## II. CULTURAL PRACTICES

Agricultural practices have a marked effect on the incidence and severity of virus diseases. Intensive crop management practices like irrigation, fertilization, use of high-yielding cultivars, monoculture, and higher crop plant densities have created new types of environments suitable for many weeds, vectors, and virus spread. This has made the use of cultural practices for controlling virus diseases all the more imperative. The role of certain cultural practices in avoiding or decreasing disease incidence is by now well recognized. Some of these cultural practices now commonly used as a matter of routine are removal of weeds, volunteers, groundkeepers, and other virus reservoirs, coupled with other field sanitation practices, roguing and eradication, use of cover crops and physical barriers, crop rotation and fallowing periods, etc. The effects of these and various other cultural practices on disease incidence and the possible ways of avoiding the outbreaks of virus diseases have been discussed in detail by Palti.[6] Only a few examples are given here.

A minor crop should be eliminated to save damage to a major crop by a virus affecting both the crop types. Peaches are of least importance in the northwestern U.S. and should be eradicated to check the spread of western beet yellows in sugar beets. Cucumbers act as reservoirs of CMV and are incompatible with the main crop of banana in Israel, which is also attacked by CMV.

Operations like manual or mechanical tilling, pruning, plucking, and harvesting result in spread of sap-borne and highly contagious viruses and viroids from plot to plot and plant to plant. Maximum spread of such viruses and viroids occurs when tubers of potato, fruits of coconut, cuttings of vines and ornamentals, and budwood of citrus, pome, and stone fruits, etc., are cut and/or grafted. The movement of man and equipment or leaf rubbing in rows of luxuriant yet tender foliage of potato, tomato, etc., by winds also help in virus spread. Such virus and viroid spread can be easily checked by disinfection of cutting tools and implements and by proper spacing between plants, which have proved helpful in reducing the spread of potato virus S and potato spindle tuber viroid (PSTV).[7]

One of the most widely practiced cultural control measure for managing virus disease is by altering planting dates. Success is ensured by adjusting sowing dates in such a way that vectors are least in number/less active during the vulnerable stage of plant growth. Seed potatoes in the northwestern Indian plains are raised during autumn (October to December) when aphid vector *Myzus persicae* populations are negligible.[8] Peanut bud necrosis disease in India, caused by tomato spotted wilt virus, can be reduced by altering the sowing dates in such a manner that these do not coincide with the peak flight activity of thrips vectors.[9] The incidence of maize rough dwarf virus in Israel was brought down from 45 to 3% by postponing the planting date from mid-April to the end of May. This was based on the finding that the virus fails to multiply in the planthopper vector *Laodelphax striatella* when the mean ambient temperature is above 24°C.[10] Several other such examples have been reported (Table 1).

**Table 1**
## DECREASING VIRUS DISEASE INCIDENCE BY CHANGING SOWING AND PLANTING DATES

| Crop and disease | Vector | Country | Restriction of peak vector populations to well-defined seasons and changed planting time |
|---|---|---|---|
| Bean golden mosaic | *Bemisia tabaci* | Brazil | Cooler temperatures reduce vector population; grow in late-dry season |
| Broad bean infected with a cowpea mosaic virus strain | *Sitona* sp. | France | Sow early in winter for plant to develop resistance against incoming vectors |
| Broad bean infected with sub-terranean clover red leaf | *Aulacorthum solani* | Tasmania | Vector populations peak in September; sow in May to July or November |
| Broad bean infected with Sudanese broad bean mosaic | *Aphis craccivora, Acyrthosiphon sesbaniae* | Sudan | Vector populations peak in January; sow in October to get adult plants in January |
| Carrot motley dwarf | *Cavariella aegopodii* | Australia | Sow after spring dispersal of vector is over |
| Groundnut rosette | *Aphis craccivora* | Nigeria | Sow in June rather than in July, when vector population is smaller |
| Maize rough dwarf | *Laodelphax striatella* | Israel | Vector population declines after late May; sow in late May |
| Sugar beet yellows | *Myzus persicae* | Worldwide | Sow early for plants to achieve less susceptible stage as vectors increase |
| Tomato spotted wilt | *Thrips tabaci, Frankliniella* spp. | Worldwide | Vector populations peak in warm humid seasons; avoid sowing or transplanting then |
| Tomato yellow leaf curl | *Bemisia tabacci* | Israel | Vector populations peak in summer; sow in spring |
| Tulips infected with tobacco rattle virus | Nematode | Netherlands | Plant in December instead of September because nematodes are less active in December |
| Winter wheat infected with wheat streak mosaic | Aphid | Alberta | Spring-sown wheat crop acts as virus and vector source for autumn-sown wheat crop; sow in September |
| Winter wheat infected with barley yellow dwarf | Aphid | Montana | Sow after September 10 by which time vector populations are on decline |

Modified from Palti, J., *Cultural Practices and Infectious Crop Diseases*, Springer-Verlag, Berlin, 1981, 243; with additions. With permission.

It is necessary and a common and successful practice in several countries to grow seed beds or have sites for producing virus-free propagative material like clones, tubers, and other vegetative propagative parts beyond a certain minimum distance from all the potential sources of infection. This is widely practiced in potato, sugar beet, cabbage, carrot, hop, and lettuce crops and fruit and ornamental plants. Several countries have enforced laws to compel the growers to practice such regional isolation. Beet seed crops cannot be grown in East Germany within 3 km of beet crops grown for processing.

## III. REPELLING AND LURING OF INSECT VECTORS

Repelling insect vectors, primarily aphids but also whiteflies, from fields by covering the soil with aluminum sheets or black or transparent polyethylene sheets, or by mulching the soil with some reflective material leads to delayed virus infection, decreased disease incidence, and greater yields.[11-14] Various types of mulches employed in this connection are aluminum foil, clear polyethylene, gray plastic, white plastic, yellow polyethylene, and organic mulches.

There was an 87 to 90% reduction in disease incidence caused by CMV and potato virus Y (PVY) in pepper plants in plots mulched with gray plastic in Israel. Watermelon mosaic virus-infected squash plants 11 weeks after planting in plots containing aluminum mulch, black plastic mulch, and unmulched plots in the U.S. were 4.1, 51.0, and 69.0%, respectively. The incidence of this virus in *Cucurbita pepo* in the Imperial Valley of California decreased by 94 and 77% in plots using aluminum foil and white plastic mulches, respectively. Spread of tomato yellow leaf curl virus in tomato crop in Israel was delayed for at least 20 d in plots mulched with yellow polyethylene sheets.

Mulching with silver polyethylene film greatly reduced bean yellow mosaic and pea seed-borne mosaic viruses in broad bean in Japan. Many growers in Florida now commonly use beds covered by black plastic mulch for establishing vegetable crops. Black plastic mulch is used in winter, when aphids are numerous and disease spreads rapidly, to increase soil temperature, while in summer the black plastic mulch is commonly sprayed with aluminum paint to reduce warm soil temperature. Aluminum foil is now being increasingly used in certain areas of the U.S. Watermelon mosaic virus occurred much less frequently in melon and watermelon plants in Australia mulched with reflective polyethylene, resulting in greatly increased yield in field trials.[15] Aluminum-surfaced plastic mulch reduced population of thrips vectors and of tomato spotted wilt virus in all crops tested and was superior to black plastic mulch.[16] use of organic mulches (straw, sawdust, rice husk) reduces aphid populations and virus incidence. Straw mulches have been successfully used in Israel to repel whiteflies and to protect cucumbers from cucumber yellow vein virus and tomato from tomato yellow leaf curl virus. Rice husk mulch offered some protection to citrus seedlings in seed beds against citrus tristeza virus (CTV) in a heavily nematode vector-infested area in Brazil.

The use of mulches has certain limitations, such as loss of reflectance with increase in size of plants and due to sandstorms, difficulty in laying aluminum foil, the problem of its disposal after the growing season, and the high cost involved. Moreover, field mulching is not effective in all cases. There was no reduction in the number of *M. persicae* in aluminum-mulched potato plots in northeastern Maine, while the number of *Aphis gossypii* was five times higher in aluminum-mulched squash plots than in nonmulched plots in other areas.[13]

The attraction of aphids towards the yellow color has been used to lure them away from the crop, resulting in far less disease incidence. The method, originally developed in Israel and subsequently recommended by the Israel Agricultural Extension Service, consists of 2 to 3 m long and 120 cm wide folded-over yellow polyethylene sheets placed around or at least along the windward side of a pepper field in such a way that there is a 2 to 3 m space between any two adjacent sheets, and their lower edge is 70 cm above ground.[14] The outer 60-cm-wide surface is covered with a thin layer of transparent glue that remains sticky for about 3 weeks. The sheets are fixed at a distance of 6 to 7 m from the border of the field a few days before germination of seeds. These yellow sticky polyethylene sheets lured incoming viruliferous aphids away from the field and trapped a very large number of them, resulting in a 51% reduction in cumulative infection by PVY and CMV. An identical experiment involving protection of seed potatoes against potato leaf roll virus (PLRV) reduced the number of infected harvested tubers from 17.2 to 2% in the first year and from 29 to 6% in the next year.[17,18] Sticky yellow polyethylene sheets reduced winged aphids by about 70% and PVY infection by about 38% in potato plots.[17]

The method also has its limitations. Sheets tend to tear in strong winds and they lose their efficiency as dust and sand cover their surface. Moreover, the sheets are effective as long as the vector populations remain moderate in size. The method is not always successful.

## IV. INDEXING AND CERTIFICATION PROGRAMS

The overwhelming importance of using healthy virus-free stocks, seeds, and other propagating material is self-evident and is the basic prerequisite for managing plant disease.[6,19-22] The use of certified seeds and stocks is the best way to ensure the growth of healthy and reasonably disease-free crops and has proved useful in almost all cases. Many countries of the world now have indexing and certification programs that routinely certify all planting materials and seeds of various plants to be healthy and virus free. Apple, cherry, citrus, grapes, hop, lettuce, plum, peach, potato, pear, quince, raspberry, soybean, strawberry, and sugar beet are some of the vegetables and fruits for which certification programs are in existence in different countries.

Using certified virus-free lettuce seed or seed containing less than 0.1% infected seed has been spectacularly successful for the last 17 years in controlling lettuce mosaic in the U.S., the U.K., and Israel.[23] It has been satisfactorily controlled in France by a tolerance limit of 1 infected seed per 1000 seeds. A certification scheme against beet yellows was initiated in the U.K. in 1951. It was based on the inspection of seed crop seedlings (stecklings) at the end of October after the autumn aphid migration. Steckling beds with more than 10% virus infection were ploughed up and those with less than 1% infection were certified for use as seed crops. A tolerance limit of less than 5% is allowed by the fruit tree certification programs in Washington and Montana in seed lots of *Prunus mahaleb* for prune dwarf and prunus necrotic ringspot viruses. The most prudent control measure for soybean mosaic is the use of virus-free seed. Virtual eradication of strawberry viruses from commercial plantings in Queensland, Australia, is due to the use of certified material along with a decrease in vector populations. Strawberry mottle virus has been eliminated from commercial strawberry orchards in Arkansas by employing virus-free plants.[24] The use of meristem cultures obtained from carnation necrotic fleck virus-free carnation stocks has almost completely eliminated this disease from Israel. The use of certified or disease-free potato and bean seed is the best control measure against potato viruses and bean common mosaic virus the world over. The destructive onion yellow dwarf virus has been eliminated from the main areas of garlic cultivation in France by using disease-free propagating material. Thousands of citrus trees are indexed annually in Israel and California to detect and eliminate tristeza-infected trees which otherwise would act as infection foci and aid in further natural spread of the disease.

Certification schemes have been developed in various countries for the production of disease-free certified propagating material from a few virus-free plants.[19] These virus-free plants are obtained by either of two ways: by selecting them from partially infected crop and then testing each individual selected plant for absence of virus, or by developing virus-free plants in any one of the various methods (heat therapy, apical meristem culture, etc.). These virus-free mother plants are maintained in insect-proof glass or screen houses to avoid reinfection.

Nuclear, basic, or foundation stock is first derived from these virus-free mother plants. The next step is to produce commercial or certified propagation material from the nuclear stock. This is usually done in normal cultural conditions, keeping in view the various cultural practices (like isolation, roguing, barriers, spraying, etc.) with particular reference to the epidemiology of the virus disease concerned. Inspection and certification is then conducted to assess the health of the final seeding material. The presence or absence of a virus from the seed lots is confirmed by serological techniques and infectivity assays of individual or bulk samples and by visual inspection for the absence of disease symptoms. Commonly certified propagative material is rarely absolutely free from virus infection; instead, a little or negligible amount of infected material may occur, but this infection is within the levels of tolerance so that for practical purposes such certified material is virus free.

In the U.S., a nonprofit Foundation Plant Material Service is located at the University of Californiat at Davis, while the U.S. Department of Agriculture and Washington State Department of Agriculture IR-2 repository of virus-free fruit trees is based at Prosser, WA. In India, the National Seeds Corporation is doing a good job in this connection.

Central Potato Research Institute, Shimla, India, has developed a ''seed plot technique'' for the northeastern plains of India for the production of healthy seeds.[25] It consists of starting with healthy seeds, growing the crop from October to December during the period of low aphid incidence, killing of haulms before the aphid population attains a critical buildup (20 per 100 leaves), and lifting of the crop as soon as tuber skin hardens. Use of systemic insecticides against leaf-sucking insects and aphids before removal of haulms, along with improved seed agronomic practices for seed production, has improved the technique still further. It became necessary to devise this technique since the area suitable for quality seed production in high hills above 2200 m was very limited and could only meet a fraction of the seed requirement of the country and at a very high cost.

## V. CROSS-PROTECTION (IMMUNIZATION)

Cross-protection denotes the suppression of symptoms upon inoculation with a virulent strain (the challenging or challenge virus) in a plant perviously infected/inoculated with a mild strain (protecting virus) of the same virus. Some other terms used in literature for this phenomenon are antagonism, immunization, induced immunity, interference, mutual antagonism, protection, etc.

This phenomenon was independently discovered during experiments involving tobacco mosaic virus (TMV) and potato virus X (PVX). Tobacco plants infected with a TMV strain exciting green mosaic symptoms did not show any change in symptoms on subsequent inoculation with TMV strain inducing yellow mosaic.[26] Out of the two TMV strains, one inducing white mosaic and the other green mosaic, the strain inoculated later could not be isolated from the inoculated tissue.[27] Mild strain of PVX gave immunity to *Datura stramonium* against subsequent infection with the severe strain.[28] Cross protection has since been found to be a common phenomenon and is operative in plants as well as protoplasts between strains of many viruses and viroids. Some of these are alfalfa mosaic, barley yellow mosaic, CTV, CMV, peanut mottle, PVX, PVY, raspberry ringspot, rice tungro, squash mosaic, TMV, tobacco ringspot, tomato aspermy, tomato spotted wilt, and other viruses. Strains of PSTV and citrus exocortis viroid also cross-protect each other.[29] A severe citrus exocortic viroid was persistently suppressed by a mild isolate of the same viroid in *Chrysanthemum morifolium*.[30] Cross-protection so far has been reviewed several times.[31-37]

Cross-protection may be complete (or of high degree) or imcomplete (or of low degree). A mild CMV strain protected tobacco, tomato, and squash plants against severe CMV strains which also failed to replicate and accumulate in protected plants.[38] This is complete cross-protection, since neither symptoms nor CMV replication could be detected. Mild PSTV strains could protect tomato and chrysanthemum plants against citrus exocortis viroid and severe PSTV strains only partially. Both viroids replicated in doubly infected plants, and severe symptoms were only delayed in protected chrysanthemum plants, but did not appear at all in protected tomato plants.[29] This is partial or incomplete cross-protection, since the challenge viroid did infect, multiply, and accumulate in protected plants and there was delayed or no symptom appearance. Partial cross-protection in maize plants occurs between strains A and B of maize dwarf mosaic virus.

Cross-protection may be reciprocal or unidirectional (unilateral). Cross-protection between P (severe) and S (mild) CMV strains was reciprocal in tomato plants. Each of the two strains protected against the other when inoculated prior to challenge inoculation, while coinoculation of the two strains resulted in mixed infections and reduction in synthesis of both strains.[39] In unidirectional or one-sided cross-protection, only one virus interferes with the other virus and not vice versa. Thus alfalfa mosaic virus cross-protects *Chenopodium amaranticolor* against CMV, while CMV fails to do so against alfalfa mosaic virus. Several other cases of reciprocal and unidirectional cross-protection are reported. Generally, however, cross-protection is reciprocal between related viruses, but is only in one direction between unrelated viruses.

The range of interference among strains can be very wide. It can vary from complete to intermediate to none, or a low degree of protection among strains of a virus.[40,41] This complete range of interference is shown among various PVX strains. Some PVX strains completely protect each other and others only partially, and cause merely reduced severity of symptoms, while still others show no protection at all. The degree of cross-protection among different strains of this virus was found to be correlated with their serological relationship. Serologically related PVX strains completely protect each other, while serologically unrelated strains cross-protect each other incompletely. This is also true of TMV. Many, but not all, of the closely related TMV strains give complete protection to the host against each other, while distantly related strains, like cucumber viruses 3 and 4, give no protection to cucumber plants against subsequent infection by TMV. However, superinfection of *Nicotiana sylvestris* by related TMV strains does occur. Similarly, the necrotizing TMV strains replicate in light green tissue of *N. sylvestris* even though this tissue contains large amounts of the related protecting TMV strain. Reports about such a situation prevailing among serologically related strains of other viruses are also available. Hence, serologically related strains may cross-protect against each other, but they may not do so as well, or some related strains of a virus may do so while other do not. Thus cross-protection, where it occurs, is an indication of the close relationship between the two viruses or even their being closely related strains of a single virus, while the lack of cross-protection does not necessarily mean that the two are not related to each other or are not the strains of a virus.

Nevertheless, exceptions do occur since a few cases are known where distinctly different viruses or completely unrelated viruses may cross-protect. Thus, tomato ringspot virus cross-protects tobacco plants against cherry leaf roll virus. Severe etch virus interferes with and decreases or suppresses multiplication of PVY in tobacco 'White Burley' plants in mixed inoculation or when PVY-infected plants were challenged after a month with severe etch virus. Maize mosaic virus or maize stripe virus, when inoculated in two maize cultivars, partially protect them against infection by the other virus.[42] Moreover, maize mosaic virus interfered with replication of maize stripe virus so that the concentration of the latter virus in doubly infected plants was less than in singly infected plants. Maize streak virus unilaterally protects maize plants against maize mosaic virus in Mauritius.[43] Even unrelated viroids cross-protect against each other.[29]

## A. Mechanism of Cross-Protection

Several theories have been propounded over the years to explain the mechanism of cross-protection.

### 1. Role of Capsid Protein

de Zoeten and Fulton[44] postulated that the genome of the incoming superinfecting virus is captured and encapsidated by the coat protein of the protecting virus (genomic masking), thereby preventing its replication. Some recent work supports de Zoeten and Fulton insofar as the coat protein seems to be the cause of cross-protection, but in a manner(s) which is different from the mechanism suggested by them. It can happen in different manners.

The addition of brome mosaic virus (BMV) coat protein to the reaction mixture inhibited BMV RNA synthesis by purified BMV replicase, and increasing coat protein concentrations increased the inhibitory effect.[45] Coat protein combined with RNA, and such an RNA was no longer available as a template for RNA synthesis. It was suggested on this basis that coat protein may be inhibiting RNA synthesis by merely blocking the replicase binding site on RNA rather than completely encapsidating the challenge RNA.[45] The capsid protein in this way may have a regulatory role in RNA replication of the incoming virus, leading indirectly to the manifestation of cross-protection phenomenon since the challenge virus fails to replicate.

Another hypothesis suggests that free coat protein from the protecting virus prevents uncoating of the challenging virus. The cross-protection observed in light green tissue of the *N. sylvestris*-TMV system was due to the inability of the incoming strain to uncoat and initiate infection, rather than to inhibition of virus replication in protected plants.[46] Some recent experimental evidence supports this suggestion. Extraneous homologous coat protein diminished the infectivity of virus inocula by inhibiting viral uncoating.[47] Wilson and Watkins[48] also suggested that cotranslational disassembly of TMV particles was preferentially inhibited in vitro by TMV coat protein. This happens because capsid protein prevents the initial release of a few coat protein subunits from the 5′ end of the challenge virus RNA or recoats the unmasked 5′ end of the challenge virus, leading in both cases to prevention of its translation and replication. This also explains the observation that a protected plant could be infected by RNA of a challenge virus if the former was encapsidated in coat protein of a virus which is not related to the protecting virus.[46,47]

Recent studies conducted on transgenic plants also suggest the participation of free coat protein in cross-protection phenomenon. Transgenic tobacco plants (*N. tabacum* 'Xanthi') expressing the TMV U1 coat protein gene were protected against disease development after inoculation with a serologically related severe TMV strain.[49] There was a dramatic decrease in chlorotic and necrotic lesions as well as in challenge virus accumulation in leaves of transgenic plants. The tobacco plants transformed with the coat protein gene of TMV showed delayed or no expression of viral symptoms when challenged with TMV.[50] The conclusions based on studies of transgenic plants can also explain the earlier reported results on providing cross-protection by coat protein-defective TMV mutants. Coat protein-defective TMV mutant PM1 was able to protect against subseqeunt infection by the common TMV strains,[51] possibly because this defective strain does produce coat protein although it is defective and insoluble. Similary, a *ts* TMV mutant Ni 118, which produces a nonfunctional denatured coat protein at the nonpermissive temperature, does cross-protect against the TMV common strain.[52]

The studies of Sherwood and Fulton,[46] Dodds et al.,[38] and Nelson et al.[49] found that virus-infected plants, on challenge inoculated with either virus or viral RNA of a closely related strain, exhibited greater protection against virus than against viral RNA as challenge inoculum. For example, tomato leaves protected by a mild CMV strain could be superinfected by RNA, but not by the virions of another CMV strain.[40] Thus, viral RNA as inoculum largely overcomes cross-protection in nontransformed and transgenic plants in all these cases. This tends to confirm the suggestion that free coat protein inhibits uncoating of particles of challenge virus.

It appears that coat protein in the above-mentioned cases is actively involved in the expression of cross-protection, but it is also equally likely that it has nothing to do with cross-protection in other cases because viroids, which generate no coat protein, also cross-protect.

*2. Blocking of Virus-Specific Sites*
The cells are postulated to contain finite sets of different receptors or infectible sites or replication sites, some of which are specific for one virus, while others are specific for another virus, and so on. The limited number of virus-specific infectible/replication sites of a cell(s) are all occupied by even a low concentration of the first virus. An incoming related virus will not be able to multiply, while an unrelated virus may be able to do so, since it has different infection sites. The studies of Salomon and Bar-Joseph[53] on competition between RNAs of two TMV strains for translation in rabbit reticulocyte lysate and wheat germ cell-free systems do suggest the presence of virus-specific binding/multiplication sites. The RNAs of the two strains were translated simultaneously if added to the translation system together at the beginning of the reaction. However, each of the two RNAs could inhibit translation of the other if the reaction (active translation) had already been initiated by the RNA of the first strain even when RNA of the other strain was added. On the other hand, translation of alfalfa mosaic virus RNA was not appreciably inhibited when its RNA was added after the

start of translation of TMV RNA. Addition of TMV RNA to an alfalfa mosaic virus RNA-initiated reaction also resulted in independent synthesis of protein products of the two viruses.

Some of the postulated theories concerning the mechanism of cross-protection call to attention the possible blocking of all or many of these virus-specific sites by the protecting virus.

Sherwood and Fulton[46] studied the interference between common (C) and petunia (P) strains of TMV on healthy and infected *N. sylvestris* leaves. Addition of TMV-C inoculum to TMV-P inoculum decreased the lesion numbers on healthy leaves proportionately to the concentration of TMV-C inoculum, but the decrease of lesions in infected leaves was only slight with increased TMV-C concentration. Moreover, addition of inactivated TMV-C, TMV coat protein, or bovine serum albumin to TMV-P inoculum decreased lesion numbers in both healthy and infected leaves in the same way as the addition of TMV-C did. It was concluded that nonspecific competition for infection sites was responsible for cross-protection.

Ross[54] hypothesized that the RNA of interfering virus may bind to ribosomes so that the subsequent binding of RNA of the challenge virus and its replication are prevented. Parallel to this is the exclusion hypothesis of Siegel,[55] which envisages that initiation of infection with a particle of one strain excludes the participation of particle of a second strain in the same infection.

Using up a material essential for virus replication, like RNA replicase, by the first strain of raspberry ringspot virus may be responsible for cross-protection of protoplasts against its challenge strain.[56] Similarly, cross-protection between serologically related CMV strains in *N. tabacum* 'Xanthi-nc' was primarily of a competitive nature presumably for virus-specific sites or virus-specific material essential for replication. This conclusion was based on the observation that chances of the challenging CMV virus to become established in a plant were essentially dependent upon the concentration of the CMV strain already existing in cells.

The results of Sterk and de Jager[57] do not support the blocking hypothesis, i.e., competition between virus particles or the RNAs of the two viruses for infectible sites. The top component of cowpea mosaic virus (CPMV) and inactivated CPMV RNA did not inhibit infection of cowpea by cowpea severe mosaic virus, while CPMV particles and infectious CPMV RNA did so. This rules out the possibility that interference is due to competition for virus attachment sites or for RNA receptor sites.

### 3. Role of Some Protecting/Inhibiting Substance(s)

The lower leaves of *Dianthus barbatus*, after inoculation with carnation mosaic virus, developed local lesions while upper leaves remained virus free but became resistant to infection by the same virus.[58] This was an unusual type of protection and was called acquired resistance or acquired immunity,[59] which was divided into two types, namely, the systemic and localized acquired resistances. Systemic acquired resistance is distinct from cross-protection because protection is induced by a substance(s) that passes through the graft union and protects the uninfected member of the graft against subsequent virus infection. The protecting substance(s) are called plant antibodies[59] or antiviral factor(s) (AVF).[60] The term "AVF" became popular because it does not convey any preconceived ideas or implications about its origin, chemical composition, and mechanism of protection, unlike the term "antibody". Later, the same phenomenon was investigated by several workers in greater detail using the TMV-infected Samsun NN and Xanthi-nc tobacco plants. Much work has now been conducted on acquired resistance and has been reviewed.[32,61-65]

It is recognized now that some general nonspecific mechanism, common to several host-virus combinations and which is also invoked by fungi, bacteria, and a number of natural and synthetic compounds, is involved in this nonspecific defense response of plants to pathogens inducing localized infections. Initially, formation of polyphenols and/or lignin and/or callose was suggested to be the cause of acquired resistance, but it is not believed to be so now. It is believed instead that formation of some inhibitory substance(s) is involved in this phenomenon. Much work has since been conducted on the identification of this inhibitory substance(s). It has been variously referred to as AVF, interferon (interferon-like substances), and b-proteins (pathogenesis-related proteins). It will not be out of place therefore to briefly examine the role of AVF, interferon-like substances, and b-proteins as the cause of acquired resistance. An identical process may be happening in cross-protection as well. Of these, pathogenesis-related proteins are treated in a separate chapter of this book.

Properties of the interfering agent produced in TMV-infected *Datura stramonium* were compared with those of the interferon.[61] Both lack virus specificity, are heat sensitive, are stable at pH 2.5 and thus nondializable, are inactivated or show reduced activity after treatment with trypsin, have identical molecular weight, are precipitated by zinc acetate and acetone, and the mode of action of both is indirect via the host. Recent evidence strongly supports this hypothesis. Human leukocyte interferon inhibited TMV multiplication in tobacco tissue and in tobacco protoplasts. The primary target of interferon action was TMV replication, although interferon also inhibited viral RNA translation and TMV accumulation. AVF is stable at pH 2, is a phosphorylated glycoprotein, has a molecular weight of about 22 kDa, and possesses some other features in common with interferon. So it may also be an interferon-like substance. It will not be far fetched, therefore, if an interferon-like substance is also involved in cross-protection.

### 4. Other Mechanisms

Some other mechanisms have also been propounded at various times. One mechanism envisages that RNA:RNA annealing between protecting and challenge virus RNA molecules may cause inhibition of replication or translation of superinfecting virus. No disease symptoms can thus be excited in a cross-protected plant. Successful infection of cross-protected plants by RNA of a challenge virus, as already mentioned above, shows that RNA:RNA annealing cannot be the cause of cross protection. In contrast to these cases, viral RNA from a necrotizing strain of sunn hemp mosaic virus failed to overcome protection by a nonnecrotizing strain.[47] Palukaitis and Zaitlin[35] postulated a "negative-strand capture" mechanism for cross-protection. The protecting viral RNA is visualized to bind to complementary minus-strand RNA of the challenge virus at the start of its replication.

Another hypothesis suggests that the virus, already present in cells, occurs in the form of aggregates that have specific "adsorptive" properties as a result of which the incoming virions of the same or a closely related virus become adsorbed on one of these aggregates.

### B. Cross-Protection as a Control Measure

In 1934, Kunkel[66] suggested the use of avirulent TMV strains to protect plants against more virulent strains. However, the first practical advantage of this phenomenon appears to have been taken in 1964 in the Isle of Wight in the U.K. where growers, in the absence of a wild TMV strain, used normally fairly severe TMV strain to inoculate tomato plants at the seedling stage. The results were encouraging and the percentage of unmarketable, poor-quality tomato fruits came down to under 3% from about 30% in the earlier year. Then in 1968 an avirulent TMV mutant, MII-16, a nitrous acid symptomless mutant obtained from TMV strain 1, was isolated in the Netherlands. However, the real breakthrough in managing tomato mosaic by immunization came only in 1972 when Rast[67] demonstrated the commercial feasibility and advantage of immunizing tomato crop with the avirulent MII-16 strain. This avirulent strain was manufactured commercially and became available by 1973 to 1974 in several countries as a preventive vaccine of tomato plants.

The control of tomato mosaic by immunization is now widely practiced commercially in several countries, including the U.K., the U.S., Japan, the Netherlands, Ireland, etc.[68-70] The avirulent strain is spray inoculated at the seedling stage by the gun method, which is better than manual inoculation. This method, along with the use of resistant cultivars, has almost completely wiped out this disease from commercial tomato crops in the Irish Republic.[71] Two slightly pathogenic strains (S-7 and V-69) are widely used in the U.S.S.R., by spraying the vaccines at cotyledon stage for vaccination of tomato crops against TMV.[72,73] The inoculated plants become resistant to highly virulent pathogenic strains after 7 to 8 d of inoculation. The resistance continues to manifest itself for several months and practically during the whole of the growing period so that the vaccinated plants give 30% increased yield. Vaccination is of great importance in the U.S.S.R., particularly, since no resistant tomato cultivars are available.

CTV is another disease in which cross-protection has shown great promise. CTV spread to a great majority of the Hassaku citrus trees in Japan during the 1950s. During a survey to locate a healthy source of budwood, an old, seemingly healthy Hassaku tree, which was actually later found to harbor a mild CTV isolate, was found. This tree served as the nucleus from which most of the Hassaku trees grown in Hiroshima were propagated. Unfortunately, however, the immunity conferred by the mild strain lasted for just 10 to 12 years so that fresh efforts have to be made to find a strain capable of imparting immunity for a longer period.[74] In contrast to this, preimmunization in Brazil of 'Pera' sweet orange, 'Galego' lime, and 'Ruby Red' grapefruit, with each of the 2 out of the 45 originally selected mild strains, permitted the immunized scions to grow satisfactorily on three tolerant rootstocks for 11 years. Moreover, three or more successive propagations from the initially immunized trees continued to show immunity so that the immunity conferred by each of the strains is considered lasting. More than 1,500,000 preimmunized 'Pera' sweet orange trees and 200,000 preimmunized 'Galego' lime trees have been planted in São Paulo, and no breakdown of protection has been observed.[75-77] Contrast this with the situation prevailing in the late 1950s when severe infection by CTV greatly discouraged cultivation of 'Pera' sweet orange trees. Similarly, in India mild strains of CTV gave fairly effective protection in the field to citron and acid lime against the severe strain.[78] Highly satisfactory use of protected clones has also been reported in grapefruit in Australia and in citrus trees in California.[37] However, such immunity in Israel has either failed or is incomplete and may be the possible reason for the recent spread of CTV after a long lag period.[79]

Papaya production is limited by papaya ringspot virus (PRV) in several countries of the world, since the host does not bear any resistance genes against the virus. Two avirulent nitrous acid mutants of the virus have been obtained, out of which one strain (PRV HA5-1) protected papaya plants in greenhouse tests against mechanical inoculation with a severe strain.[80] This strain could possibly be used to control the severe strain (PRV HA) in the field. It was necessary to challenge inoculate the severe strain after 18 to 26 d of inoculation with the mild (symptomless) mutant to obtain complete protection of papaya plants. Incomplete or no immunization was obtained if challenge inoculation was done earlier than 18 d. Conceivably, at least 18 d were needed for the symptomless mutant to buildup to a sufficient titer in plants to cross-protect them against the severe strain. This mutant also cross-protected *Cucumis metuliferus* against watermelon mosaic virus 1,[80] which is serologically identical to PRV.

In at least one case, natural occurrence of a protecting strain has been suggested to inhibit the spread of a disease in the field. The pattern of spread of rugose mosaic disease of 'Bing' sweet cherry, caused by prunus necrotic ringspot virus, suggested that a symptomless strain of the virus restricts the disease spread in nature.[81] Some other virus and viroid diseases where, primarily because of the availability of mild strains, cross-protection has been suggested to be applicable as a control measure are[37] apple blotch, apple leaf pucker, apple mosaic, avocado sun blotch, cocoa swollen shoot, cauliflower mosaic virus in Brussels sprouts, passionfruit woodiness, peach mosaic, and tomato aspermy.

As already mentioned, coat protein of a protecting virus may be the primary cause for phenotypic expression of cross-protection. Incorporation of the coat protein gene of such a virus in a host plant may provide a permanent protection against that virus. Reports of successful trnasformation of tobacco and tomato host plants through incorporation of the TMV coat protein gene into the host plant genome have appeared.[49,50,82] In addition, Beachy et al.[82] also incorporated the coat protein gene of alfalfa mosaic, PVX, and CMV in these plants. All these tobacco and tomato transgenic plants express the viral coat protein gene and are resistant to infection by these viruses.[82] Resistance is expressed as reduced number of infection sites and as decreased spread of virus from inoculated to other plant parts, and transgenic plants exhibit a high degree of resistance to TMV and tomato mosaic virus in greenhouse and field tests.[82] Probably, disassembly of the challenge virus is blocked in the cells transformed by TMV coat protein gene.[82] Tobacco engineered by incorporation of benign satellite RNA sequences resist the effects of CMV infection.[83]

However, the picture of cross-protection is not entirely rosy. Fulton[37] mentions some reasons against the use of cross-protection as a general means of virus disease control. The protection provided may be incomplete, and significantly, the amount of challenge inoculum needed for superinfection in several such cases is 100 to 1000 times more. A limiting factor in the use of cross-protection is that mild strains of some viruses may be mild only in some plants, but severe in other crop plants. So one has to be particularly cautious in the case of viruses with a wide host range coupled with vector transmission. Then there is always the possiblity that a mild strain may mutate to a severe strain, resulting in a catastrophe. Sometimes a protected crop may become more susceptible to other pathogens like fungi and bacteria.[83a] Moreover, it is often difficult to obtain desired mild strains and in inoculation of an entire crop. This is particularly so in annual crops. Fulton[37] mentions that the experience with tomato mosaic gives a measure of the difficulties encountered during implementation of cross-protection as a control measure for annual crops. These difficulties may discourage its use in annual crops, but it may be more useful for perennial crops as the success with CTV suggests.

Cross-protection as a control measure has been reviewed.[6,37]

## VI. DISEASE FORECASTING

Disease-predictive systems have largely been formulated on the basis of epidemiological studies and are of two types: systems that predict disease and systems that predict infection. The former (disease-predictive systems) predict the development and appearance of disease symptoms in relation to the occurrence of favorable environmental conditions. Consequently, such predictions are only made after favorable environmental and biological conditions have already occurred. Infection-prediction systems, on the other hand, predict many days in advance the occurrence of favorable environmental periods which can lead to infection of the crop. This is certainly a difficult proposition; hence most of the predictive systems in plant pathology are the disease-predictive systems.

Disease-predictive systems are of two further types on the basis of the rationale of their formulation. Empirical or deductive predictive systems are one type. They are formulated by studying and comparing the past records of disease occurrence and meteorological conditions in or near a particular locality. Such specific meteorological conditions are treated as "rules" of these systems which must be fulfilled for disease development to occur. No mathematics is involved in such systems except that concerning the meteorological data. Derived, fundamental, inductive, or logical predictive systems are the second type. They are developed from experimental data obtained in the laboratory or the field concerning the relationship of meteorological and biological conditions with disease development. Both the predictive systems have been developed and employed in plant pathology.

Availability of computers, calculators, modern statistical techniques, and sophisticated and sensitive environmental sensing equipment have made data recording, data processing, and accurate measurement of weather parameters easier, rapid, and accurate. All this has led to the development of better and more accurate prediction systems.

Predictive systems have proved very useful in fungal plant diseases, but progress in forecasting plant virus diseases has not been substantial. Epidemiology of a virus disease results from a complex of factors, but weather conditions early in the season that favor vector population development and their local or long-distance migration are the most important factors on which disease-predictive systems have been based. These predictive systems have generally been correlated with meteorological data collected over several years and with development and dispersal/migration of vectors from local or distant overwintering hosts to young crops. A more complicated system has been developed for rice dwarf virus disease in Japan. A theoretical mode, $At = 1 - \exp(-a \cdot N_2 \cdot L \cdot P)$, has been successfully verified in field and laboratoy tests where $At$ is the percentage of rice dwarf virus-infected rice hills, a is the number of hills infected successfully by the leafhopper vectors per day, $N_2$ is the number of adult leafhoppers per hill that have migrated, L is the mean longevity of adult hoppers in the field, and P is the percentage of viruliferous vectors.

Much of the spread of potato leaf roll virus (PLRV) in potatoes in Ireland and in the south of England occurs during early summer (i.e., from the end of May to early July) when alatae of *M. persicae* begin to disperse. Consequently, more potato plants were infected with PLRV and PVY during early summer (even though the number of alatae of *M. persicae* was quite low) than in late summer (mid-July to the end of August), even when numerous apterae were colonizing the crop.[84] PLRV incidence in seed potatoes in Scotland was much higher when winters preceding the seed crop were milder and allowed survival of the aphids and hence early migration to the crop.[85,86] A significant positive correlation has been found between winter temperatures and the dates of first catches of *M. persicae* in Scotland. This information has been successfully employed to forecast the spring migration of *M. persicae* and to advise growers whether or not the early control of aphids is necessary.[86]

*M. persicae* in the northwestern plains of India begins to appear on the potato crop from the end of November to the first week of December; vector populations build up from mid-December onward, and achieve their peak in April. A combination of high humidity (80 to 90%) and moderate temperature (25 to 30°C) increases transmission of PVY and PLRV by *M. persicae* by 30 to 35%, but the transmission of these viruses is almost half when relative humidity is low (50%).

Temperature also plays an important role in the activity and buildup of the vector aphid populations. Gabriel[87] proposed a model to forecast the incidence of PVY in tubers. It was based on the determination of two variables: the number of aphids over a period of 80 d after emergence of the crop, and the efficiency of tuber infection in relation to the time between the emergence of the crop and the appearance of aphids. The percentage of infected tubers could then be forecast with a multiple regression equation.

Smith et al.[88] used prevailing temperatures during a specified period of crop growth to predict PLRV incidence in Maine. The differences in PLRV incidence from 1954 to 1976 were linked with the linear combination of the transformed percentage of primary source of inoculum and thermal unit accumulation above 21°C from August 1 to 10.

Barley yellow dwarf virus, a serious disease problem in the north central states of the U.S., is spread by the aphid *Schizaphis graminum*. The source of both the virus and the vector is in the southern plains states of the U.S. Wallin and Loonan[89] compared the rainfall and temperature regimes of 75 years to determine the favorable ones for outbreaks of the vector in the southern plains states of the U.S. Rainfall less than normal during any 4 months from October through April was found to favor the development of moderate to high vector populations. This information can be used to predict outbreaks of the vector, which then migrates to the north-central states and causes high disease incidence. Thus a predictive system for this disease can be developed on this basis.

Weather data and the number of aphids have been used to predict the incidence of yellowing viruses of sugar beet crops in England for advising the farmers regarding application of insecticides.[90] Partial regression analysis was employed to relate the mean percentage of plants infected with yellowing viruses at the end of August to the number of (frost) days during January to March with minimum temperature below $-3°C$ and the mean temperature in April. Mean April temperature and the number of frost days accounted for the greatest percentage of variance in yellows incidence from year to year. The number of frost days were plotted against mean April temperatures on the graphs showing lines of equal percentage infection obtained by relating the weather variables to yellow incidence in August. The information about the expected incidence of yellows becomes available by the end of April, i.e., a few weeks before the vector appears on the crop. If the expected incidence is around 10%, farmers are advised to protect the crop by insecticidal sprays.

A severe epidemic of maize dwarf mosaic virus occurred on sweet corn during 1977 in Minnesota. Such an epidemic of this disease was unknown before, since the virus is endemic in the southern Great Plains of the U.S., where it perpetuates in the weed host *Sorghum halepense* and is spread to neighboring areas by the aphid vector *Schizaphis graminum*. Zeyen et al.[91] concluded that the 1977 epidemic could be associated with a weather pattern that could be linked to potential spread of the vector from the southern Great Plains of the U.S. to the northern states bordering Canada.

Parameters employed for forecasting some plant virus diseases are given in Table 2.

Knowledge about the onset of a favorable period for disease increase and/or vector development and migration can be used to our advantage. It enables us to apply chemicals at only certain times so as to achieve maximum disease and vector control as well as reduced financial input because of the reduced overall time period for which chemicals need to be applied. Such an application is likely to be more effective. Such spray warning schemes for the best time for pesticide application have been used in several countries for the sugar beet and potato viruses.[105,106] Predictive systems, for all these reasons, have become an integral part of plant disease control strategy.

## VII. QUARANTINE REGULATIONS

Quarantine, as a menas of exclusion of dangerous pests and pathogens of crop plants from a country, is an important step in disease management. There has always been movement of plant propagation material from one region/country to another region/country, and this movement has been increasing with time. Such a transfer of propagation materials has resulted in a number of disease and pest problems when such movement has taken place without proper checking/inspection at the receiving end. The result has been worldwide spread of bean common mosaic, lettuce mosaic, soybean mosaic, sugarcane mosaic, and several other viruses. This spread has largely occurred through the dissemination and movement of contaminated, infected, and/or infested plant material as well as through germplasm. Several viruses have been detected and intercepted in germplasm banks.[107]

Such dangers were widely recognized, and 125 countries of the world formulated quarantine regulations by which they prohibit and/or regulate the importation of particular plants and plant parts into particular countries. This controls and reduces the chances of inadvertant entry and spread of dangerous pathogens and pests/vectors. Almost all countries have established an official plant protection service for implementing domestic and foreign quarantine laws, for running plant health certification schemes, for conducting surveillance of phytosanitary situations within the country, and for controlling the entry of pathogens and pests/vectors. National quarantines have legal patronage and regulate the movement of materials between countries and within a country. The articles covered by national quarantine regulations obviously vary from country to country, depending upon its agricultural and horticultural crops and practices. All the national quarantine regulations share the following common features: Specification of prohibited material, regulation of movement of materials for scientific and educational purposes, specification of ports of entry and may insist on

## Table 2
## PARAMETERS EMPLOYED FOR FORECASTING OF PLANT VIRUS DISEASES

| Virus disease and place | Basis employed for disease prediction/dispersal of vectors | Ref. |
|---|---|---|
| Barley yellow dwarf | | |
| a. In New Zealand in winter | Sticky trap catches of vectors that have overwintered | 92 |
| b. In Minnesota and adjacent states | Incidence correlated with suitable weather conditions in April to May favoring vector influx following long-distance dispersal of aphid vectors from southern states | 89, 93 |
| Beet curly top in California | Main dispersal period of leafhoppers predicted on the basis of number of days with temperature above 45°C | 94 |
| Beet yellows in England | Mean percentage of infected plants at the end of August correlated with the number of days during January to March when temperature fell below −3°C ("frost days") and the mean temperature in April from 1951 to 1971 | 90 |
| Carrot motley dwarf in England | Disease intensity correlated with the number of aphid vectors caught on sticky trap catches | 95 |
| Cotton leaf curl in Sudan | Incidence correlated with severity of preceding dry season | 96 |
| Lettuce necrotic yellows in South Australia | Large number of *Hyperomyzus lactucae* trapped when mean weekly temperature fell in the 60 to 72°F range so that meteorological and aphid population data are used for predicting virus spread | 97 |
| Maize streak in Rhodesia (Zimbabwe) | Virus spread (vector dispersal) correlated with the amount of rain at the end of the preceding wet summer season | 98 |
| Pepper virus disease in Canada | Incidence of nonpersistent aphid-borne virus diseases in August correlated with temperature and sunshine data of April for 1970 to 1977 | 99 |
| Potato leaf roll | | |
| In England | Virus spread correlated with the number of alatae of *Myzus persicae* caught on sticky traps throughout the potato-growing season and the average health of potato stocks in the following year could be predicted from average trap data for several years | 100 |
| In Scotland | Virus spread correlated with average mean temperatures in April | 85, 101 |
| Rice dwarf in Japan | a. Prediction based on the amount of infection in the previous year and number of overwintering leafhopper vectors as amended by later vector populations | 102 |
| | b. A more complicated system based on the model $At = 1\text{-}exp\,(-\,a{\cdot}N_2{\cdot}L{\cdot}P)$ (see text) | 103 |
| Wheat streak mosaic in North America | Incidence in autumn crop correlated with infection in winter wheat crop | 104 |

phytosanitary certificates, fumigation of materials, and may prescribe safeguards such as isolation and refusal of entry.[108]

Viruses are difficult agents to detect and diagnose in propagative materials and need rapid, sensitive, and reliable techniques for their identification. A continued critical effort has to be made for updating the quarantine policy to effectively check the accidental transfer of viruses and viroids through true seed, tubers, bulbs, budwoods, rhizomes, etc.[108,109] Reddy[109] has selected 43 seed-borne viruses of quarantine significance in India. Countries promulgate plant protection quarantine regulations which list genera, species or common names of hosts of specific pathogens, viruses, viroids, vectors, etc., the entry of which is prohibited in a particular country if originating from one or more cited regions/countries.[108] In India, there are specific regulations against the import of barseem, castanea, cocoa, coffee, cotton, citrus, elm, flax, groundnut, Mexican jumping bean, pine, potato, rubber, sugarcane, sunflower, and tobacco. However, plant genera important from the quarantine risk point of view — such as banana, cassava, *Cydonia, Fragaria*, grapes, *Malus*, orchids, *Pyrus*, sweet potato, yam, and a large number of minor food, medicinal, and ornamental plants — are yet to be covered under the quarantine regulations.[109,110] The geographic distribution of some selected

## Table 3
### GEOGRAPHIC DISTRIBUTION IN WORLD OF THE IMPORTANT VIRAL AND VIROID DISEASES OF QUARANTINE IMPORTANCE IN DIFFERENT CROP PLANTS

| Disease | Host | Country |
|---|---|---|
| Banana Camheroon marbling | Banana | Camerons, Ivory Coast |
| Banana bunchy top | Banana | India, Sri Lanka |
| Cocoa Ceylon/Trinidad | Cocoa | Sri Lanka, Trinidad |
| Cocoa mottle leaf | Cocoa | Malaysia, Northern Sumatra, Togo |
| Cocoa swollen shoot | Cocoa | Ghana, Ivory Coast, Western Nigeria |
| Cassava mosaic/brown streak | Cassava | Africa, East Central Africa, India (Kerala), Sri Lanka, Thailand, Trinidad |
| Citrus exocortis | Citrus | Africa, India, U.S. |
| Citrus likubin | Citrus | Taiwan |
| Citrus psorosis | Citrus | U.S. |
| Citrus satsuma dwarf | Citrus | Japan |
| Citrus tristeza | Citrus | Africa, Asia, U.S. |
| Coffee ringspot | Coffee | Brazil, Costa Rica, Philippines |
| Groundnut bud necrosis | Groundnut | India |
| Groundnut stunt | Groundnut | Africa, Asia |
| Hop crinkle | Hops | Eastern Europe |
| Hop stunt | Hops | Japan |
| Pea seed-borne mosaic | Peas | Europe, India, U.S. |
| Raspberry ringspot | Raspberry | U.K. |
| Strawberry arabis mosaic | Strawberry | Europe, U.K. |
| Strawberry latent ringspot | Strawberry | Europe, U.K. |
| Sugar beet (latent) rosette | Sugar beet | Germany |
| Sugarcane Fiji | Sugarcane | Australia, Fiji |
| Sweet potato internal cork, russet crack, feathery mottle, and mosaic | Sweet potato | Africa, South America |
| Taro alomae and bobonae | Colocasia (Taro) | South Pacific |
| Tobacco rattle | Potato, tobacco | India, U.K., U.S. |
| Tomato black ring | Potato, tomato | Europe, India, Western Africa |

hosts of viruses that are readily transmitted through plant parts are summarized in Table 3.[110]

Several reviews on national and international quarantine laws have been published in journals, books, and official publications of various countries. Hewitt and Chiarappa,[111] Harris and Maramorosch,[112] and Kahn[108] give details.

## REFERENCES

1. **Apple, J. L.,** The theory of disease management, in *Plant Disease: An Advanced Treatise,* Vol. 1, Horsfall, J. G. and Cowling, E. B., Eds., Academic Press, New York, 1977, 79.
2. **Magyarosy, A. C.,** A new look at curly top disease, *Calif. Agric.,* 32, 13, 1978.
3. **Simons, J. N. and Zitter, T. A.,** Use of oil to control aphid-borne viruses, *Plant Dis.,* 64, 542, 1980.
4. **Hull, R. and Heathcote, G. D.,** Experiments on the time of application of insecticide to decrease the spread of yellowing viruses of sugar beet, 1954—1966, *Ann. Appl. Biol.,* 60, 469, 1967.
5. **Marrou, J., Quiot, J. B., Duteil, M., Labonne, G., Leclant, E., and Renoust, M.,** Ecologie et épidémiologie du Virus de la Mosaique du Concombre. VIII. Influence des brise-vents et des cultures environmentes sur la dissemination du Virus Mosaique du Concombre, *Ann. Phytopathol.,* 11, 375, 1975.
6. **Palti, J.,** *Cultural Practices and Infectious Crop Diseases,* Springer-Verlag, Berlin, 1981, 243.

7. **Singh, R. P.**, Control of viroid and contact transmitted virus diseases, in *Potato Pest Management in Canada,* Boiteau, G., Singh, R. P., and Parry, R. H., Eds., Canada Agriculture, New Brunswick, 1988, 309.

8. **Pushkarnath,** *Potato in the Subtropics,* Orient Longman, New Delhi, 1976, 289.

9. **Reddy, D. V. R. and Wightman, J. A.,** Tomato spotted wilt virus: thrips transmission and control, in *Current Topics in Vector Research,* Vol. 5, Harris, K. F., Ed., Springer-Verlag, New York, 1988, 203.

10. **Harpaz, I.,** *Maize Rough Dwarf,* Israel University Press, Jerusalem, 1972, 251.

11. **Zitter, T. A. and Simons, J. N.,** Management of viruses by alteration of vector efficiency and by cultural practices, *Annu. Rev. Phytopathol.,* 18, 289, 1980.

12. **Loebenstin, G. and Raccah, B.,** Control of non-persistently transmitted aphid-borne viruses, *Phytoparasitica,* 8, 221, 1980.

13. **Smith, F. F. and Webb, R. E.,** Repelling aphids by reflective surfaces, a new approach to the control of insect-transmitted viruses, in *Viruses, Vectors and Vegetation,* Maramorosch, K., Ed., Wiley-Interscience, New York, 1969, 631.

14. **Cohen, S.,** Control of whitefly vectors of viruses by colour mulches, in *Pathogens, Vectors, and Plant Diseases: Approaches to Control,* Harris, K. F. and Maramorosch, K., Eds., Academic Press, New York, 1982, chap. 3.

15. **McLean, G. D., Burt, J. R., Thomas, D. W., and Sproul, A. N.,** The use of reflective mulch to reduce the incidence of watermelon mosaic virus in Western Australia, *Crop Protect.,* 1, 491, 1982.

16. **Black, L. L., Bond, W. P., Story, R. N., and Batti, J. M.,** Tomato spotted wilt virus in Louisiana: epidemiological aspects, *Proc. Workshop Epidemiology of Plant Virus Diseases,* Orlando, FL, August 6 to 8, 1986.

17. **Zimmerman-Gries, S.,** Reducing the spread of potato leaf roll virus, alfalfa mosaic virus and potato virus Y in seed potatoes by trapping aphids on sticky yellow polyethylene sheets, *Potato Res.,* 22, 123, 1979.

18. **Marco, S.,** Reducing potato leaf roll virus (PLRV) in potato by means of baiting aphids by yellow surface and protecting crops by coarse nets, *Potato Res.,* 24, 21, 1981.

19. **Hollings, M.,** Disease control through virus-free stock, *Annu. Rev. Phytopathol.,* 3, 367, 1965.

20. **Hollings, M. and Stone, O. M.,** Techniques and problems in the production of virus-tested planting material, *Sci. Hortic.,* 20, 57, 1968.

21. **Cropley, R.,** The production and practical benefits of virus-free propagating material of fruit crops, in *Plant Health: The Scientific Basis for Administrative Control of Plant Parasites,* Ebbels, D. L. and King, J. E., Eds., Blackwell Scientific, Oxford, 1979, 121.

22. **Hollings, M. and Stone, O. M.,** Production and use of virus-free stocks of ornamental and bulb crops; some phytosanitary and epidemiological aspects, in *Plant Health: The Scientific Basis for Administrative Control of Plant Parasites,* Ebbels, D. L. and King, J. E., Eds., Blackwell Scientific, Oxford, 1979, 129.

23. **Grogan, R. G.,** Control of lettuce mosaic with virus-free seed, *Plant Dis.,* 64, 446, 1980.

24. **Fulton, J. P. and Moore, B. J.,** Strawberry virus dissemination in Arkansas, *Plant Dis.,* 66, 847, 1982.

25. **Nagaich, B. B., Puskarnath, Bhardwaj, V. P., Giri, B. K., Anand, S. R., and Upreti, G. C.,** Production of disease free seed potatoes in Indo-Gangetic Plains, *Indian J. Agric. Sci.,* 39, 238, 1969.

26. **McKinney, H. H.,** Mosaic diseases in the Canary Islands, West Africa and Gibraltar, *J. Agric. Res.,* 39, 557, 1929.

27. **Thung, T. H.,** Smetstof en plantencel bij enkele virus ziekten van de Tabaksplant, *Handel 6 Neid. Ind. Natuurwetensch. Congr.,* 450, 1931, (*Rev. Appl. Mycol.,* 11, 750, 1932).

28. **Salaman, R. N.,** Protective inoculation against a plant virus, *Nature,* 131, 468, 1933.

29. **Niblett, C. L., Dickson, E., Fernow, K. H., Horst, R. K., and Zaitlin, M.,** Cross-protection among four viroids, *Virology,* 91, 198, 1978.

30. **Schwinghamer, M. W. and Broadbent, P.,** Detection of viroids in dwarfed orange trees by transmission to chrysanthemum, *Phytopathology,* 77, 210, 1987.

31. **Price, W. C.,** Acquired immunity from plant virus diseases, *Q. Rev. Biol.,* 15, 338, 1940.

32. **Kassanis, B.,** Interactions of viruses in plants, *Adv. Virus. Res.,* 10, 219, 1963.

33. **Hamilton, R.,** Defences triggered by previous invaders, in *Plant Disease,* Vol. 1, Horsfall, J. G. and Cowling, E. B., Eds., Academic Press, New York, 1980, 279.

34. **Fulton, R. W.,** The protective effects of systemic virus infection, in *Active Defence Mechanisms in Plants,* Advanced Study Inst. Series, Wood, R. K. S., Ed., Plenum Press, New York, 1982, 381.

35. **Palukaitis, P. and Zaitlin, M.,** A model to explain the "cross protection" phenomenon shown by plant viruses and viroids, in *Plant Microbe Interactions: Molecular and Genetic Aspects,* Vol. 1, Kosuge, T. and Nester, E. W., Eds., Macmillan, New York, 1984, 420.

36. **Hamilton, R. I.,** *HortScience,* 20, 848, 1985.

37. **Fulton, R. W.,** Practices and precautions in the use of cross-protection for plant virus disease control, *Annu. Rev. Phytopathol.,* 24, 67, 1986.

38. **Dodds, J. A., Lee, S. Q., and Tiffany, M.,** Cross-protection between strains of cucumber mosaic virus; effect of host and type of inoculum on accumulation of virions and double-stranded RNA of the challenge strain, *Virology,* 144, 301, 1985.
39. **Dodds, J. A.,** Cross-protection and interference between electrophoretically distinct strains of cucumber mosaic virus in tomato, *Virology,* 118, 235, 1982.
40. **Bennett, C. W.,** Interference phenomenon between plant viruses, *Annu. Rev. Microbiol.,* 5, 295, 1951.
41. **Bennett, C. W.,** Interactions between viruses and virus strains, *Adv. Virus Res.,* 1, 39, 1953.
42. **Ammar, E. D., Gingery, R. E., and Nault, L. R.,** Interactions between maize mosaic and maize stripe viruses in their insect vector, *Peregrinus maidis,* and in maize, *Phytopathology,* 77, 1051, 1987.
43. **Autrey, L. J. C.,** Maize mosaic virus and other maize virus diseases in the islands of the Western Indian Ocean, In Proc. Int. Maize Virus Dis. Colloq. Worshop, Gordon, D. T., Knoke, J. K., Nault, L. R., and Ritter, R. M., Eds., August 2 to 6, 1982, Ohio State University, Wooster, 1983, 167.
44. **de Zoeten, G. A. and Fulton, R. W.,** Understanding generates possibilities, *Phytopathology,* 65, 221, 1975.
45. **Horikoshi, M., Nakayama, M., Yamaoka, N., Furusawa, I., and Shishiyama, J.,** Brome mosaic virus coat protein inhibits viral RNA synthesis in vitro, *Virology,* 158, 15, 1987.
46. **Sherwood, J. L. and Fulton, R. W.,** The specific involvement of coat protein in tobacco mosaic virus cross-protection, *Virology,* 119, 150, 1982.
47. **Zinnen, T. M. and Fulton, R. W.,** Cross-protection between sunn-hemp mosaic and tobacco mosaic viruses, *J. Gen. Virol.,* 67, 1679, 1986.
48. **Wilson, T. M. A. and Watkins, P. A. C.,** Influence of exogenous viral coat protein on the cotranslational disassembly of tobacco mosaic virus (TMV) particles in vitro, *Virology,* 149, 132, 1986.
49. **Nelson, R. S., Abel, P. P., and Beachy, R. N.,** Lesions and virus accumulation in inoculated transgenic tobacco plants expressing the coat protein gene of tobacco mosaic virus, *Virology,* 158, 126, 1987.
50. **Abel, P. P., Nelson, R. S., Hoffmann, B. De, N., Rogers, S. G., Fraley, R. T., and Beachy, R. N.,** Delay of disease development in transgenic plants that express the tobacco mosaic virus coat protein gene, *Science,* 232, 738, 1986.
51. **Zaitlin, M.,** Viral corss-protection: more understanding is needed, *Phytopathology,* 66, 382, 1976.
52. **Jockusch, H.,** Two mutants of tobacco mosaic virus temperature-sensitive in two different functions, *Virology,* 35, 94, 1968.
53. **Salomon, R. and Bar-Joseph, M.,** Translational competition between related virus RNA species in cell-free systems, *J. Gen. Virol.,* 62, 343, 1982.
54. **Ross, A. F.,** Interactions of viruses in the host, in *Virus Diseases of Ornamental Plants,* Tech. Commun. 36, Lawson, R. H. and Corbett, M. K., Eds., Int. Soc. Hort. Sci., The Hague, 1974, 247.
55. **Siegel, A.,** Mutual exclusion of strains of tobacco mosaic virus, *Virology,* 8, 470, 1959.
56. **Barker, H. and Harrison, B. D.,** Double infection, interference and super infection in protoplasts exposed to two strains of raspberry ringspot virus, *J. Gen. Virol.,* 40, 647, 1978.
57. **Sterk, P. and de Jager, C. P.,** Interference between cowpea mosaic virus and cowpea severe mosaic virus in cowpea host immune to cowpea mosaic virus, *J. Gen. Virol.,* 68, 2751, 1987.
58. **Gilpatrick, J. D. and Weintraub, M.,** An unusual type of protection with the carnation mosaic virus, *Science,* 115, 701, 1952.
59. **Yarwood, C. E.,** Acquired resistance to tobacco mosaic virus in bean, *Phytopathology,* 43, 490, 1953.
60. **Sela, I. and Applebaum, S. W.,** Occurrence of an antiviral factor in virus infected plants, *Virology,* 17, 543, 1962.
61. **Loebenstein, G.,** Localization and induced resistance in virus-infected plants, *Annu. Rev. Phytopathol.,* 10, 177, 1972.
62. **Loebenstein, G.,** Inhibition, interference and acquired resistance during infection, in *Principles and Techniques in Plant Virology,* Kodo, C. I. and Agrawal, H. O., Eds., Van Nostrand Reinhold, New York, 1972, 32.
63. **Kassanis, B.,** Some speculations on the nature of the natural defence mechanism of plants against virus infection, *Phytopathol. Z.,* 102, 277, 1981.
64. **Sela, I.,** Plant virus interactions related to resistance and localization of viral infections, *Adv. Virus Res.,* 26, 201, 1981.
65. **Sela, I.,** Interferon-like factor from virus-infected plants, *Perspect. Virol.,* 11, 129, 1981.
66. **Kunkel, L. O.,** Studies on acquired immunity with tobacco and aucuba mosaics, *Phytopathology,* 24, 437, 1934.
67. **Rast, A. T. B.,** MII-16, an artificial symptomless mutant of tobacco mosaic virus for seedling inoculation of tomato crops, *Neth. J. Plant Pathol.,* 78, 110, 1972.
68. **Rast, A. T. B.,** Variability of tobacco mosaic virus in relation to control of tomato mosaic in glasshouse tomato crops by resistance and cross-protection, *Agric. Res. Rept., Wageningen,* 834, 76, 1975.
69. **Fletcher, J. T. and Rowe, J. M.,** Observations and experiments on the use of an avirulent mutant strain of tobacco mosaic virus as a means of controlling tomato mosaic, *Ann. Appl. Biol.,* 81, 171, 1975.

70. **Fletcher, J. T.,** The use of avirulent strains to protect plants against the effects of virulent strains, *Ann. Appl. Biol.,* 89, 110, 1978.
71. **Staunton, W. P.,** Biological control of virus in tomato crops using cross-protection, in *Proc. Sem. Biol. Control,* Duggan, J. J., Ed., Royal Irish Academy, Dublin, 1978, 93.
72. **Kuznetsova, I. F.,** Vaktsinatsiya na domati, *Rastit. Zasht.,* 29, 21, 1981.
73. **Kuznetsova, I. F.,** in *Rev. Plant Pathol.,* 60 (Abstr.), 3338, 1981.
74. **Sasaki, A.,** Control of Hassaku dwarf by preimmunization with mild strain, *Rev. Plant Protect. Res.,* 12, 80, 1979.
75. **Muller, G. W. and Costa, A. S.,** Tristeza control in Brazil by preimmunization with mild strains, *Proc. Int. Soc. Citricult.,* 3, 868, 1977.
76. **Muller, G. W.,** Use of mild strains of citrus tristeza virus (CTV) to reestablish commercial production of 'Pera' sweet orange in Sao Paulo, Brazil, *Proc. Fla. State Hortic. Soc.,* 93, 62, 1980/1981.
77. **Costa, A. S. and Muller, G. W.,** Tristeza control by cross-protection: a U.S.-Brazil cooperative success, *Plant Dis.,* 64, 538, 1980.
78. **Balaraman, K. and Ramakrishnan, K.,** Cross-protection of acid lime with mild strains of tristeza, *Indian J. Agric. Res.,* 48, 741, 1978.
79. **Bar-Joseph, M.,** Cross-protection incompleteness: a possible cause for natural spread of citrus tristeza virus after a prolonged lag period in Israel, *Phytopathology,* 68, 1110, 1978.
80. **Yeh, S. D. and Gonsalves, D.,** Evaluation of induced mutants of papaya ringspot virus for control by cross-protection, *Phytopathology,* 74, 1086, 1984.
81. **Howell, W. E. and Mink, G. I.,** Control of natural spread of cherry rugose mosaic disease by a symptomless strain of prunus necrotic ringspot virus, *Phytopathology,* 74, 1139, 1984.
82. **Beachy, R. N, Register, J. C., Powell, P. A., Nelson, R. S., and Wisnieuski, L.,** Cross-protection against virus infection in transgenic plants, *5th Int. Congr. Plant Pathology,* Kyoto, Japan, 1988, 11.
83. **Baulcombe, D. C., Bevan, M., Harrison, B. D., and Mayo, M. A.,** Genetically engineered plants with resistance to cucumber mosaic virus, in *Biotechnology in Agriculture,* AFRC, London, 1987, 8.
83a. **Khurana, S. M. Paul, and Raychaudhuri, S. P.,** Interaction between plant pathogens with special reference to the viruses, *Int. J. Trop. Plant Dis.,* 6, 43, 1988.
84. **Heathcote, G. D. and Broadbent, L.,** Local spread of potato leaf-roll and Y viruses, *Eur. Potato J.,* 4, 138, 1961.
85. **Howell, P. J.,** Field studies on potato leaf-roll virus in south-eastern Scotland, 1962—1969, in relation to aphid populations and other factors, *Ann. Appl. Biol.,* 76, 187, 1974.
86. **Turl, L. A. D.,** An approach to forecasting the incidence of potato and cereal aphids in Scotland, *Eur. Plant Prot. Org. Bull.,* 10, 135, 1980.
87. **Gabriel, W.,** Essais d'amelioration de la prevision de l'infection des tubercules des pommes de terre par la virus Y, *Potato Res.,* 24, 301, 1981.
88. **Smith, O. P., Storch, R. H., Hepler, P. R., and Manzer, F. E.,** Prediction of potato leaf-roll virus disease in Maine from thermal unit accumulation and an estimate of primary inoculum, *Plant Dis.,* 68, 863, 1984.
89. **Wallin, J. R. and Loonan, D. V.,** Climatic factors affecting the development of high populations of the greenbug in Texas and Oklahoma, *Plant Dis. Rep.,* 61, 4, 1977.
90. **Watson, M. A., Heathcote, G. D., Lauckner, F. B., and Sowray, P. A.,** The use of weather data and counts of aphids in the field to predict the incidence of yellowing viruses of sugar beet crops in England in relation to the use of insecticides, *Ann. Appl. Biol.,* 81, 181, 1975.
91. **Zeyen, R. J., Stromberg, E. L., and Kuehnast, E. L.,** Long-range aphid transport hypothesis for maize dwarf mosaic virus: history and distribution in Minnesota, USA, *Ann. Appl. Biol.,* 111, 325, 1987.
92. **Close, R., Smith, H. C., and Lowe, A. D.,** Cereal virus warning system, *Commonw. Phytopathol. News,* 10, 7, 1970.
93. **Evans, D. A. and Medler, J. T.,** Flight activity of the corn leaf aphid in Wisconsin as determined by yellow pan trap collections, *J. Econ. Entomol.,* 60, 1088, 1967.
94. **Cook, W. C.,** The relation of spring movements of the beet lefhopper (*Eutettix tenellus*) in central California to temperature accumulations, *Ann. Entomol. Soc. Am.,* 38, 149, 1945.
95. **Watson, M. A. and Serjeant, E. P.,** The effect of motley dwarf virus on yield of carrots and its transmission in the field by *Cavariella aegopodii* Scop., *Ann. Appl. Biol.,* 53, 73, 1964.
96. **Boughey, A. S.,** The cause of variations in the incidence of cotton leaf curl in the Sudan Gezira, *Mycol. Paper Imp. Mycol. Inst.,* 22, 9, 1947.
97. **Randles, J. W. and Crowley, N. C.,** Epidemiology of lettuce necrotic yellows virus in South Australia. I. Relationship between disease incidence and activity of *Hyperomyzus lactucae, Aust. J. Agric. Res.,* 21, 447, 1970.
98. **Rose, D. J. W.,** Epidemiology of maize streak disease, *Annu. Rev. Entomol.,* 23, 259, 1978.
99. **Kemp, W. G. and Troup, P. A.,** A weather index to forecast potential of aphid-transmitted virus diseases of peppers in the Niagara peninsula, *Can. J. Plant Sci.,* 58, 1025, 1978.

100. **Broadbent, L.,** The correlation of aphid numbers with the spread of leaf roll and rugose mosaic in potato crops, *Ann. Appl. Biol.,* 37, 58, 1950.
101. **Howell, P. J.,** The relationship between winter temperatures and the extent of potato leaf roll virus in seed potatoes in Scotland, *Potato Res.,* 16, 30, 1973.
102. **Murumatsu, Y., Furuki, I., Kawaguchi, K., Sawaki, T., and Takeshima, S.,** Forecasting rice dwarf disease. II. Forecasting methods, *Rev. Plant Prot. Res. Jpn.,* 1, 83, 1968.
103. **Kiritani, K. and Sasaba, T.,** An experimental validation of the systems model for the prediction of rice dwarf virus infection, *Appl. Entomol. Zool.,* 13, 209, 1978.
104. **Fellows, H. and Sill, W. H.,** Predicting wheat streak mosaic epiphytotics in winter wheat, *Plant Dis. Rep.,* 39, 291, 1955.
105. **Hull, R.,** The spray warning scheme for control of sugar beet yellows in England. Summary of results between 1959—66, *Plant Pathol.,* 7, 1, 1968.
106. **Woodford, J. A. T., Shaw, M. W., McKinlay, R. G., and Foster, G. N.,** The potato aphid spray warning scheme in Scotland, 1975—1977, *Proc. 1977 Br. Crop Prot. Conf.,* 3, 247, 1977.
107. **Hampton, R., Waterworth, H., Goodman, R. M., and Lee, R.,** Importance of seedborne viruses in crop germplasm, *Plant Dis.,* 66, 977, 1982.
108. **Kahn, R. P.,** The host as a vector: exclusion as a control, in *Pathogens, Vectors, and Plant Diseases,* Harris, K. F. and Maramorosch, K., Eds., Academic Press, New York, 1982, 123.
109. **Reddy, O. R.,** Quarantine policy towards viruses and mycoplasma-like organisms of propagative plant material including seed, in *Current Trends in Plant Virology,* Singh, B. P. and Raychaudhuri, S. P., Eds., Today & Tomorrow Printers, New Delhi, 1982, 185.
110. **Raychaudhuri, S. P. and Khurana, S. M. Paul,** Plant quarantine problems in the introduction of viral, mycoplasmal and spiroplasmal diseases through seed and propagating materials in Asia, *Plant Prot. Bull.,* 36, 23, 1984.
111. **Hewitt, W. B. and Chiarappa, L., Eds.,** *Plant Health and Quarantine in International Transfer of Genetic Resources,* CRC Press, Cleveland, OH, 1977.
112. **Harris, K. F. and Maramorosch, K., Eds.,** *Pathogens, Vectors, and Plant Diseases: Approaches to Control,* Academic Press, New York, 1982.

Chapter 11

## TISSUE CULTURE

**R. Stace-Smith**

## TABLE OF CONTENTS

# I. INTRODUCTION

Virus diseases are present in virtually all food crop species and, depending on the virus and the severity of the host reaction, varying degrees of yield reduction may be attributed to infection with viruses. Some viruses remain localized near their point of infection, whether in the roots or in leaf tissue, and yield reductions associated with such infections are minimal. The majority of viruses do not remain confined to the infection site, but spread throughout the infected plant to become completely or nearly completely systemic. Seed that is produced by a systemically infected plant may or may not be infected and, depending upon the host-virus combination, varying degrees of seed transmission occur. Despite the fact that a low level of seed transmission may be of considerable epidemiological significance, virus-free progeny is attainable by indexing batches of seedlings from infected plants and selecting seedlings that show no evidence of virus infection.[1] This procedure is of little value in obtaining virus-free progeny from infected mother plants that are propagated vegetatively. When suckers, runners, corms, tubers, or cuttings are excised from an infected mother plant, the daughter plants are invariably infected. This means that virus diseases are of serious concern in perennial crops or crops that are propagated vegetatively.

Yield reductions attributed to virus infection in perennial crops are usually greatest in areas of the world that favor prolonged growing seasons, high populations of virus vectors, and an ample supply of infection sources. They are especially serious in developing countries in the tropics and subtropics, where the conditions are ideal for virus spread. Many of the staple crops that are grown in these regions are propagated vegetatively, with the result that virus-infected clones tend to become universally used. A similar situation existed with perennially propagated crops in the temperate zones until steps were taken to recover virus-free clones and develop strategies to propagate the virus-free clones under conditions where they remained virus free. Thus, one of the control measures that has been adopted for preventing or reducing virus-induced crop losses is aimed at the removal of virus sources and replacing infected sources with virus-free planting stocks. This process has been greatly assisted by developments in plant tissue culture, because tissue culture techniques have enabled plant propagators to use exceedingly small segments of tissue from an infected mother plant to establish new daughter plants. By utilizing tissue culture techniques, the problem of virus eradication from vegetatively propagated plants is comparable to virus eradication from seed-propagated plants in that in each case a proportion of the progeny is expected to be virus free.

This chapter will examine the role of plant tissue culture in virus disease control. From the above it is evident that a principal use of the tissue culture technique was to enable plant virologists to provide virus-free clones of a wide range of vegetatively propagated plants that were universally infected with one or more viruses. While virus eradication has been the dominant role of tissue culture, it is by no means the only role. It is a well-recognized fact that once a particular clone is rendered virus free, it remains virus free only until it is again exposed to virus infection. This means that virus-free clones must be propagated in an environment that precludes reinfection. An in vitro environment provides the ultimate degree of protection, hence tissue culture techniques may be used in at least the initial phases of clonal propagation. Further, they may be the method of choice for the storage and distribution of either virus-free clones or specific viruses. Finally, new advances in biotechnology offer the possibility of using transgenic plants to control specific virus diseases, and critical steps in the production of transgenic plants are performed in vitro. All of these aspects will be examined — some in more detail than others.

## II. VIRUS ERADICATION

As noted above, viruses are consistently transmitted from a mother plant to offspring whenever plants are propagated vegetatively, whether by bulbs, tubers, cuttings, grafting, or budding, and almost all new plants raised by these methods will be infected. Under good growing conditions, the infected progeny may grow indefinitely, yielding somewhat less than they would if they were not infected. Under suboptimal growing conditions or with a host-virus combination that is particularly severe, virus infection could be a major factor in economic loss. Further, vegetatively propagated plants tend to accumulate viruses, with the result that a new accession released by a plant breeder may be virus free upon release and yet become infected with two or more viruses if grown in an area where virus sources and vectors are prevalent. Accumulation of viruses has been a common problem with many cultivated crops, with the result that useful clonal material often became so crippled by viruses that its cultivation had to be abandoned. The need for "virus-free" stock therefore became essential. To supply healthy planting material, certification schemes are now operating for several kinds of plants, and the raising of stocks for propagation is often separated from the growing of a crop for its main purpose. Special stocks that form the basis of certification schemes are built up by propagating from single virus-free plants. Until recently, the production of virus-free stocks of plants that are grown as clonal varieties depended upon finding an uninfected plant of the variety to start the stock. Fortunately, this is no longer so, for methods have been developed whereby many plants can be freed from some or all of the viruses that commonly infect them. It can now be said that no useful clone need be abandoned because the whole clone is virus infected. The method that has found the widest application is meristem tip culture, usually from clonal material that has been subjected to a prolonged period of growth at a temperature that is considerably higher than would normally be used for plant production.

### A. Distribution of Viruses in Plants

Most viruses, when introduced into a susceptible host plant, spread throughout the plant, with the result that infectivity can be located in all affected tissue. This means that those viruses that are confined to phloem tissue can be detected in all the phloem cells, including phloem primordia at the growing tips of the shoots. The concentration of particles may vary, and several studies have demonstrated that virus concentration in shoot apices is dramatically lower than in comparable tissue beneath the shoot apices. In some virus-host combinations, no virus could be detected in the cells of the meristematic dome. It was this observation that led Morel and Martin[2] to postulate that virus-free plants, genetically identical to the mother plants, might be obtained by isolating the apical meristem from a systemically infected plant. Since this pioneering work, the technique has been used successfully by many plant pathologists to cure infected valuable cultivars of a wide range of plants.

The meristematic dome consists of a few layers of actively dividing cells located at the extreme tip of a shoot or primordial shoot. In a potato shoot, for instance, in addition to the extreme tip of the stem, meristematic domes are located at shoot primordia just beneath the growing point and also on shoot primordia in the axil of each leaf on the potato stem. It is in fact more convenient to excise meristematic tips from the shoot primordia located in the leaf axils than it is to dissect them from the shoot tip. The meristematic dome, containing about three layers of cells, is about 0.1 mm in diameter and about 0.25 mm long. These domes, if excised and placed in a suitable growing medium, almost invariably fail to grow. A somewhat larger unit, including the meristematic dome together with three or four leaf primordia and measuring from 0.3 to 0.8 mm in length, must be excised in order to achieve a reasonable proportion of units that ultimately produce established plantlets. The fact that all excised units include the meristematic dome has led to a plethora of terms to describe the tissue unit, including "meristem culture", "meristem tip culture", "tip culture'', ''shoot tip culture'', and ''axillary bud culture''. There is no implied difference in size — all units include several leaf primordia in addition to the meristematic dome.

Different viruses differ in their ability to invade the actively dividing region at the shoot apex. A few viruses are in such low concentration near the apex that excised cuttings several millimeters long may yield virus-free plantlets. Others approach the growing point so closely that little other than a few actively dividing cells at the apical dome are uninfected. The initial theory that virus-free plants result primarily from the total absence of infectious particles in the excised units is now being questioned. Infectivity tests to determine the extent of potato virus X (PVX) penetration into the apical region demonstrated that infective particles are present in meristematic tips.[3] PVX particles have also been observed by thin-section electron microscopy in the cytoplasm of apical dome cells of potato. Other viruses cannot be detected near the shoot tip. Potato leaf roll virus (PLRV), for example, was not detected in the dome and the first four leaf primordia of PLRV-infected plants, using autoradiograms of tips.[4]

## B. Meristem Culture

Meristematic buds from apices of stems or from leaf axils are usually selected for tissue culture. Their advantage is that the incipient shoot has already differentiated; to establish an independent plant, only elongation and root differentiation are required. It is of little practical value to culture other tissue, but it should be recognized that almost any part of a plant has the potential of producing propagules indistinguishable from the mother plant. Nonmeristematic tissue must undergo a development process involving the formation of callus tissue and subsequent differentiation into embryos and plantlets. For this reason, meristematic buds have become the tissue of choice used by virtually all workers as the starting material in the production of virus-free clones.

Meristem culture involves the dissection of a portion of the meristematic region and its culture on a nutrient medium for plantlet regeneration. The explant sometimes used is the apical dome, although in most cases one to several leaf primordia of the subapical region are also included (Figure 1). The number of leaf primordia included depends upon the plant species, the stability of the virus (or viruses) being eradicated, and the experience of the operator with respect to rooting success.

Some classical textbooks divide the apical meristem of a stem into two main regions — the ''promeristem'', which comprises the apical initials and neighboring cells, and the ''peripheral meristem'' below it in which the three basic meristems (the protoderm, the procambium, and the ground meristem) of the tissue systems can be distinguished. In other textbooks the apical meristem of the stem consists of both the apical dome portion of the shoot apex and the region with young leaf primordia below it. In this chapter meristem culture refers to the apical dome and the region of young leaf primordia below it, the unit measuring from 0.3 to 0.8 mm in length.

All explants originate from shoot tips, but they differ in size and the number of leaf primordia. Apical meristems, without any leaf primordia, have the highest probability of producing plantlets that are free from viruses. However, apical meristems without any leaf primordia have the lowest probability of surviving in the culture medium and eventually produce sufficient shoot and root tissue to ensure survival when the explant is transferred to soil. This is the practical reason why most workers settle for a compromise and excise a size that will eventually produce a reasonable proportion of established plantlets, but at the same time is not so large that most of the explants are still infected. There is no simple answer to the question of what is the best compromise, since it depends on many factors, most important of which is the preconditioning of the donor plant prior to excision and culture of the meristematic tip.

Another aspect that warrants consideration is the effect of explant size on the rate of contamination of cultured units. If the explant is not from a sterile seedling, a shoot containing

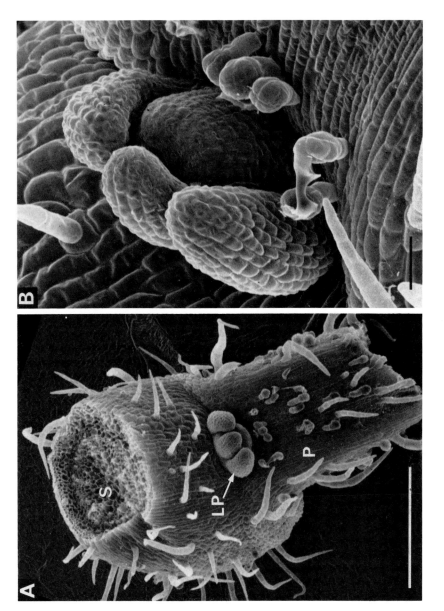

FIGURE 1. (A) Scanning electron micrograph of a bud in the leaf axil on a potato stem. LP = leaf primordia (three promordia visible in this photograph); P = petioles; S = stem. Bar represents 0.5 mm. (B) Close-up photograph of same bud as shown in (A), taken from a different angle and showing the meristematic dome in addition to the three leaf primordia. Bar represents 0.05 mm.

several axillary buds is usually surface sterilized in a dilute solution of sodium hypochlorite, often with a surfactant added, after a quick immersion in 95% ethanol. The shoot is rinsed a few times in sterile water before the axillary buds are excised under magnification in a laminar flow hood. Some explants do not survive surface sterilization, so the sterilization procedure should be as gentle as possible. In our experience with potato virus eradication,[5] contamination of explants was sometimes a problem when large tissue units (buds 1 to 3 mm long, containing a few small leaflets) were excised and cultured. When the small leaflets were peeled away, exposing the meristematic dome and leaf primordia for excision, contamination of explants was insignificant even if the sodium hypochlorite treatment was omitted from the procedure.

Pretreatment of the donor plant to enhance virus eradication will be discussed in the next section. However, it should be noted here that even when the donor plant is subjected to pretreatment, some viruses are recalcitrant, with the result that most or all of the regenerated explants are still infected. There are a number of ways of dealing with this problem, one of which is to reduce the size of the explant, even if its survival rate is lowered accordingly. This was successfully applied to the eradication of a complex of grapevine viruses, where shoot apices about 1 mm long and containing two or three leaf primordia were further fragmented to units of individual leaf primordia before being cultured.[6,7]

### 1. Culture Media and Environment

No aspect of meristem culture is as contentious as the composition of the culture medium that should be used to culture a particular species. The literature is rife with contradictions. This is understandable considering the fact that most research on the subject has been done by plant physiologists studying growth response. Their findings have been adapted by applied workers wanting tissue differentiation and root development. As a result, there is little agreement as to what constitutes the ideal meristem culture medium for a particular plant species. White's medium[8] was the most widely used during the early days of meristem culture, but since then many improvements in media have been made. A useful basic medium is the one devised by Murashige and Skoog,[9] and most workers today favor this medium or variations of this medium. It contains major and minor elements, sucrose as a carbon source, and a variety of vitamins and growth substances. A list of ingredients of some commonly used media in meristem culture is provided in the review paper prepared by Wang and Hu.[10] Workers are advised to consult original papers for details of additives that have been found particularly valuable in culturing meristem tips from individual species. Some of these are given in detail in the *Handbook of Plant Cell Culture*.[11]

Many of the numerous contributions in the field of plant tissue culture can be directly applied to meristem culture and micropropagation. The contribution of Murashige[12] is particularly notable. He developed the concept that the propagation of a plant through tissue culture must proceed through a sequence of stages and that it may be necessary to systematically explore the specific requirements of each stage. He identified three major stages, each with a different objective and possibly different requirements. The objective of stage I is simply to attain an aseptic tissue culture of the plant in question, and it is only necessary that the culture be free from obvious infection, that a suitable proportion of explants survive culture, and that there be rapid growth among the explants. The objective of stage II is a rapid increase of organs and other structures which can ultimately give rise to plants. The increase can be achieved in most instances by either inducing adventitious organ or embryo formation or by enhancing axillary shoot initiation. Stage III is intended to prepare the propagula for their successful transfer to soil, since successful tissue culture must result in reestablishment in soil of a high frequency of the tissue culture-derived plants. This stage involves rooting of shoot cuttings, hardening of plantlets to impart some tolerance to moisture stress, and possibly conferring a degree of resistance to certain pathogens. It is often possible to proceed through stages I and II by utilizing nutrient medium of the same composition, maximum success stage III media and culture conditions should be distinct from the other two.

The greatest success with the meristem culture technique has been with herbaceous horticultural species. This success is partially due to the weak apical dominance and strong root regeneration capacity of many herbaceous plants. Further, the economic importance of viruses in these crops has provided strong financial support for research in virus eradication and micropropagation. As a generalization, meristem tips excised from herbaceous plants develop and root more readily than comparable tissue from woody plants. However, this may reflect the state of the art with respect to tissue culturing and, as techniques improve, there may be little or no intrinsic difference between the two groups. With recent reports of success in culturing species that were once considered difficult or impossible to culture, it appears that virtually any plant species of economic significance can be cultured, provided expertise and effort are applied. Media composition is a critical factor in success, although considerable latitude is permitted with the concentration of the components of the basic medium. The most critical organic components are auxins and cytokinins, which will vary from species to species with respect to their kind and concentration. Root and shoot initiation is basically regulated by interaction betwen these two hormonal substances. A relatively high concentration of auxin favors root initiation while suppressing shoot formation. Conversely, a relatively high concentration of cytokinin favors shoot initiation and suppresses rooting. The control of root and shoot initiation by auxin-cytokinin balance appears to be a general phenomenon among plants.

Compared to herbaceous plants, the success rate in virus eradication from woody plant species has lagged far behind. The most difficulty is experienced at stage III, the root induction, especially when explants are taken from mature trees.

The choice between a liquid or a gel medium is usually made arbitrarily, the decision depending on the available facilities and the personal preference of the investigator. This haphazard approach, however, is not recommended since with some plants success or failure may depend on whether a liquid or an agar nutrient is employed. Further, the same species may require a different physical form of medium during each of the three stages in vitro. Some explants cultivated in a liquid medium perform better if supported by a filter paper bridge.[5] Problems like vitrification of the regenerated tissues or anaerobic growth of roots may result from oxygen deficiency associated with the use of agar-solidified medium. This problem may be solved by using an oxygen-enriched microenvironment for enhanced root and shoot growth, as was done by Gebhardt[13] using a foam substrate and nutrient solution to enhance the growth of raspberry shoot tip cultures in vitro. Of the three commercially available foam substrates that were tested, only a wide-pore phenol foam proved to be useful. Stationary liquid cultures can be used, but more commonly cultures in liquid nutrient are subjected to a degree of agitation. A rotating apparatus is used when gentle agitation is desired, with the rate of rotation usually being about 1 rpm.

The pH of the nutrient medium is a critical factor that is often neglected in tissue culture studies. The usual practice is to set the pH of the medium at some value within the range of 5.0 to 6.0 during medum preparation. However, most media are poorly buffered and the acidity is difficult to control. The widely used M-S medium developed by Murashige and Skoog,[9] when adjusted to pH 5.7 immediately after preparation, drops to pH 5.5 or 5.4 within a week, and even further when used for culture.[14] By the time rooted potato plantlets are ready to transfer from the culture tube to soil, the pH is usually between 4.5 and 5.0. Potato explants are able to tolerate this pH drift, but little is known about the influence of the actual pH values in the development of many other plant species.

The major factors of the tissue culture environment are light and temperature. The illumination of plant cultures must be considered in terms of intensity, length of the daily exposure period, and quality. Tissue culture failures are sometimes caused by the use of plant growth chambers when the light provisions have been intended for autotrophic plant development, which usually provides a higher light intensity than is desirable for tissue culture. Wang and Hu,[10] working with potato meristem cultures, found that the optimum light intensity for initiating cultures was 100 lx, increasing to 2000 lx after 4 weeks of culture (Stage II), and as the shoot reached a length of 1 cm (Stage III) the light intensity should be increased to 4000 lx. The high light intensity at Stage III was advantageous to the survival rate of the cultures.

There is little information on optimal temperature for meristem cultures, but it is known to vary from species to species. Meristems of bulbs of temperate plant origin require a rather low temperature, i.e., around 13°C,[10] while tropical plants grow well at temperatures as high as 30°C. In general practice, most meristem cultures are maintained in the temperature range 23 to 25°C. A day-night temperature fluctuation may be desirable for some plants, particularly those adapted to temperate or desert climates.

### 2. Pretreatment of Donor Plant

While the subject being discussed is virus eradication by means of meristem culture, it must be recognized that the meristem culture technique is relatively ineffective unless the donor plant is subjected to a pretreatment that enhances eradication. For this reason, most workers involved with virus eradication on a routine basis subject the donor plant to treatments aimed at reducing the virus concentration in the plant to the point where explants that are excised and cultured have a much better chance of being virus free than comparable explants excised from an untreated donor plant. Any manipulation that is done to the donor plant prior to excising explants that is advantageous in accomplishing the ultimate objective is time and money well spent. The most widely used pretreatment involves holding the donor plant in a high-temperature environment for a period of time before removing any plant tissue for meristem culture. This pretreatment is termed either "heat treatment" or "thermotherapy". Another procedure, less commonly used, involves exposing the plant to chemicals that inhibit virus multiplication, a treatment that is referred to as "chemotherapy". A third procedure, rarely used with virus-infected donor plants, but helpful with viroid-infected donor plants, involves holding the plant in a low-temperature environment, and is commonly called "low-temperature therapy". These three procedures will be discussed in greater detail.

### 3. Thermotherapy

Heat is an important therapeutic agent for treating disease in plants. As early as 1936, Kunkel[15] reported the eradication of peach yellows from infected peach trees by either dry heat or hot-water treatment. Although at the time it was thought that peach yellows was caused by a virus, it was later demonstrated that this and similar disease agents were mycoplasma-like organisms rather than viruses. The first report of a virus being eradicated from a plant by heat treatment was by Kassanis[16] in 1950. Since that time there have been several reports of eradication of viruses from whole plants by thermotherapy. Far more common, however, is the use of thermotherapy coupled with grafting, budding, or meristem culture of a portion of the treated donor plant as a means of obtaining an established explant that is genetically identical to the donor plant, but lacking the virus component carried by the donor plant.

Most viruses either do not replicate or replicate poorly in plants held above 30°C. Holding virus-infected plants continuously near the maximum temperature they can tolerate will often free them from infection. A few weeks at about 36°C has often been enough to cure plants of specific virus diseases. More frequently, however, the virus is not completely eradicated and, although the virus concentration is drastically reduced, virus replication resumes when the plant is returned to normal growing conditions. Nevertheless, heat treatment is still a valuable technique in establishing a virus-free clone. After long periods at high temperature, some parts of the plant may be virus free, even though others are still infected. New shoots produced during treatment are especially likely to be uninfected, so when cuttings are taken from plants and rooted, one or more may produce a virus-free plant.

Although applicable to some viruses, heat therapy followed by excision and rooting of shoots does not work with most viruses, hence alternative techniques must be used. The most widely used technique involves coupling thermotherapy with meristem tip culture. The smaller the tissue unit used to establish an explant, the less likely it will be infected. Carried to the extreme, the minute shoot tip constitutes the ideal tissue unit. The small size required to be virus free is incapable of surviving in an exposed environment, but it is able to survive in the protected environment of a culture tube. For this reason, the most common technique for obtaining virus-free clones is to subject the mother plant to a prolonged period of heat therapy and, immediately after treatment, to excise and culture meristematic tips from the plants. Thus, heat treatment is frequently combined with meristem culture to eradicate those viruses that are not readily eliminated by either treatment alone.

One problem with thermotherapy is the survival of the donor plant while undergoing the prolonged treatment that is required to ensure that the majority of explants are likely to be virus free. This can usually be solved for a particular species provided the investigator is prepared to devote as much time and effort to a study of plant survival during thermotherapy as is devoted to cultural requirements of the excised unit. With the potato plant, for instance, Mellor and Stace-Smith[5] found that the limit of survival of potato tubers was about 3 weeks at 37°C, while at a slightly lower temperature survival was greatly increased. When the air temperature alternated daily from 33 to 37°C and the soil temperature from 30 to 32°C, tubers survived up to 10 weeks, but rooted cuttings survived more than 6 months. Further, the shoots that developed during heat treatment were slender and axillary buds were small with few rudimentary leaves, as compared to the tuber pieces which had robust shoots with stout axillary buds, each with many rudimentary leaves to be removed before the bud could be excised. Not only were the buds from cuttings easier to excise, but the total volume of buds of similar length taken from cuttings was less than that of buds excised from plants derived from tuber pieces. This size difference was reflected in virus eradication.

Survival of potato plants during thermotherapy is a relatively simple problem as compared with survival of woody plants such as cherry or grapevine. Janecková and Blattný[17] determined that cherry plants had to be treated for at least 28 d at 37°C before green ring mottle virus was eliminated. However, after such treatment, most of the vegetative tips that were excised for grafting failed to survive. Lenz et al.[18] had a similar problem in attempts to eliminate a complex of prunus necrotic ringspot and prune dwarf viruses which normally resist thermotherapy at 36 to 38°C for 3 weeks. They succeeded in prolonging the survival of plants during heat treatment by growing them in a gas-tight chamber with a soil temperature of 20°C and an air temperature of 40°C, with the carbon dioxide content increased to 900 ppm and oxygen lowered to 5%, and with a high light intensity. Other workers have successfully eradicated prunus necrotic ringspot from cherry cultivars by acclimatizing the plants at 25°C before subjecting them to 37°C for 4 to 5 weeks.[19] Grapevine is another woody host with many viruses, some of which are difficult to eradicate. Monette[20] successfully eradicated two nepoviruses by an in vitro treatment of 6 h at 39°C and 18 h at 22°C for a duration of 40 d. The technique of applying heat therapy to rooted grapevine shoots during in vitro culture has been successfully used to eliminate fanleaf, leaf roll, and fleck.[21]

Some investigators solve the heat-stress problem by closely observing the progress of the donor plant and removing shoots for axillary bud culture just before the total collapse of the donor plant. Bjarnason et al.[22] attempted to eliminate viruses from rose cultivars grown in hydroponics by combining heat treatment with meristem culture and found that the plant survival time was 8 to 20 d, depending on the cultivar. They succeeded in eradicating viruses from the donor plants by excising and culturing axillary buds just before collapse of the donor plant, but many of the axillary buds did not survive in culture.

In summary, the advantages of coupling a heat-treatment facility with a tissue-culture facility are recognized throughout the world. At the Vancouver Research Station, for example, we routinely apply thermotherapy to all potato clones that are included in our potato virus eradication program. This program now includes some 250 cultivars and advanced selections, with 15 to 20 added each year. These are made available upon request to agencies in Canada and elsewhere as nuclear material for certification programs.[23,24]

*4. Chemotherapy*

There are many reports in the literature of substances that suppress symptoms or reduce the virus concentration in plants and it is logical to assume that some of these substances may be usefully applied to a virus-infected plant to obtain virus-free explants. However, until recently the results were disappointing. Some of the analogs of purines and pyrimidines, used at concentrations that do not destroy plant tissue, inhibit the multiplication of viruses. Some authors have actually claimed virus eradiction using these analogs, but the reports have been based on small numbers of plants or the results have been open to other interpretations.[5,25] Despite the apparent lack of success in virus eradication, antiviral compounds could augment the effectiveness of virus eradication by meristem tip culture, particularly with those viruses that are difficult to eradicate by the standard heat treatment-meristem culture combination.

In a review of chemotherapy on plant viruses, Tomlinson[26] noted that knowledge of the subject was fragmentary and many uncertainties remained in interpreting the results that had been published. Despite the uncertainty of interpreting results, several workers have continued searching for the effective chemotherapeutant. A recent review on antiviral chemicals for plant disease control[27] indicates that the literature on this subject has expanded considerably since the previous review.[26] The most promising substance is ribavirin (Virazole), a chemical that has been extensively tested against human viruses and one which appears to be effective against both RNA and DNA viruses. Cassells and Long[28] demonstrated that incorporation of ribavirin into potato explant and meristem culture media resulted in an increased percentage of virus-free progeny. Viruses eliminated singly or as complexes included potato viruses X, Y, S, and M. Similarly, Hansen and Lane[29] successfully eliminated apple chlorotic leafspot virus from apple shoot cultures by growing them in a culture medium containing ribavirin at rates of 10, 20, 40, and 80 $\mu M$ for a 4-week period. The shoots remained virus free after transfer to ribavirin-free medium, to greenhouse conditions, and to the field. Phytotoxicity was observed in treatments with 40 and 80 $\mu M$ ribavirin. Use of ribavirin may therefore be a practical alternative to thermotherapy, particularly in situations where adequate heat-treatment facilities are unavailable.

Another compound, vidarabine, which has been used as an antiviral agent against certain DNA animal viruses, may also be a useful additive in culture media used for the elimination of plant viruses. Stone[30] reported improved eradication of four viruses in *Ullucus tuberosus* by adding vidarabine (1 mg/l) to the culture medium.

*5. Low-Temperature Therapy*

In 1962, Selsky and Black[31] speculated that the application of low-temperature treatment to many virus-diseased plants which have been refractory to high-temperature treatments presents interesting possibilities for future investigations. They based their comments on their experimental work on the eradication of wound tumor virus from sweet clover plants, where growth of cuttings at a low temperature (14°C) markedly increased the frequency of virus-free plants. In the intervening years, the challenge to investigate low-temperature treatment has been explored further by Barnett et al.,[32] who found that stolon tip cuttings from white clover plants at 10°C were free from clover yellow vein virus (CYVV), while cuttings from plants held at 40°C were free from alfalfa mosaic, white clover mosaic, and CYVV. These results demonstrated that high-temperature therapy is in general far superior to low-temperature therapy for virus eradication, and low-temperature therapy has not been explored further.

While low-temperature therapy is of no particular value as a virus eradication technique, it does appear to be helpful as part of a viroid eradication program. Early attempts to eradicate viroids using the standard heat treatment-meristem culture technique were notably unsuccessful, resulting in eradication from a low percentage of surviving plantlets.[33,34] In a few cases, viroids were eliminated from shoot apical meristems ranging in size from 0.2 to 0.5 mm long, but not from larger meristems where the plantlet survival rate was higher. It was later shown[35] that potato spindle tuber viroid (PSTV) is discernible in gels as UV-absorbing peaks at temperature above 24°C and that it replicates best above 30°C. This observation suggested the advisability of using low-temperature therapy for viroid eradication. Lizárraga et al.[36] successfully eliminated PSTV from a potato clone by a combination of low-temperature treatment (5 to 8°C) for 6 months and subsequent meristem culture. A similar treatment used in attempts to eradicate hop stunt viroid, where infected hop plants were held at 10 ± 2°C for periods of up to 4 months, did not enhance the efficiency of eradication. More recently, Paduch-Cichal and Kryczyński[37] reported on the value of prolonged low-temperature therapy (longer than 3 months) in eradication of PSTV and three viroids in chrysanthemum (chrysanthemum stunt, chrysanthemum chlorotic mottle, and cucumber pale fruit). The PSTV-free plants were obtained from meristem tips cut from sprouts grown from infected tubers after 6 months of therapy at 6 to 7°C in the dark. A 3-month therapy period proved to be too short for efficient eradication of viroids from either potato or chrysanthemum.

## C. Mechanism of Virus Eradication

The exact mechanism by which thermotherapy assists in eradicating viruses from plant tissue is poorly understood. Most of the early successes in the use of thermotherapy were achieved in plants infected with mycoplasmalike organisms, suggesting that this group of organisms had low thermal inactivation points. Later, when a number of true viruses were eradicated from plants by means of thermotherapy, it was recognized that some viruses known to have a high temperature coefficient of heat inactivation could be eradicated. Kassanis[38] was the first to recognize that success in heat treatment depends not solely on the intrinsic properties of the virus, but on some interaction between the virus and the metabolism of the host cells. The hypothesis was developed that heat therapy results from a shift in balance between virus synthesis and degradation in which synthesis is reduced at temperatures sublethal for both the virus and the host. At higher temperatures, synthesis does not occur, and inactivation of the virus results from the heat. Nyland and Goheen[39] feel that the most plausible hypothesis for the mechanism of heat therapy is that high temperature causes the destruction of essential chemical activities in both virus and host, but that the host is better able to recover from the damage. Stated another way, the temperature coefficient of thermal inactivation for the host exceeds that of the virus at certain temperatures. The inactivation process may be purely physical, or it may be aided by biological changes induced in the host metabolism by high temperatures. Wang and Hu[10] suggest the following ways in which heating in vivo may inactivate viruses, limit virus production, and produce virus-free plants or plant parts: (1) inactivation of intact virus particles present in a cell by breaking of their RNA; (2) disruption of the virus particles with subsequent enzymatic degradation of their components; (3) inactivation of accessory viral enzymes; (4) prevention of virus particle assemblies, beause the coat protein may not be able to assume the correct packing configuration at high temperatures; and (5) slowing down or preventing the movements of a virus through a plant while allowing apical growth to continue, thereby having the newly grown parts free of virus when meristem tips are excised and cultured.

With regard to the mechanism of eradication, there is conclusive evidence with a number of viruses that infectious particles are present in the meristematic tissue and that excised tips contain virus at the time they are placed on the culture medium. Electron microscopic evidence is limited to relatively few viruses, but the fact that particles have been detected whenever a thorough search has been made suggests that the phenomenon is the rule rather than the exception. Further, a portion of these tips develop into plantlets that are free from virus, providing conclusive evidence that eradication takes place during culture.

The mechanism of virus eradication during culture is not known, but a number of explanations have been proposed. Hollings and Stone[40] suggested that an inactivation system exists within the apex whose action is helped by removing the mature portion of the plant. Ingram[41] suggested that eradication may be due either to some inactivating factor produced by the explant or to the effect of some constituent in the culture medium. Quak[42] speculated that the disappearance of virus particles may be attributed to contact of the meristem with the culture medium. Mellor and Stace-Smith[5] believed that viral replication requires enzymes that are normally available to the cells near the meristematic dome and that excision of the tip temporarily disorganizes the growth process, with the result that enzymes required for one or more steps of viral replication are unavailable. Without viral replication, the normal process of viral degradation would continue, but no new particles would be formed.

None of the above explanations is fully satisfactory. They all have a weakness in making the assumption that viruses exist in the plant extremities as fully formed particles. The viroids, which do not have a coat protein, but exist as an RNA strand, are not only heat stable, but also survive in excised meristematic shoots. With viroids in mind, one might speculate that those viruses that are difficult to eradicate by combined thermotherapy-meristem culture exist in the plant extremities not as particles, but as free RNA. Further, recent advances in molecular biology of viruses indicate the possibility of the viral RNA being incorporated in the plant genome, or a replicative RNA or a specialized form of genomic RNA surviving in the meristematic tissue.

## D. Technique Options

The above outline of the techniques that are commonly employed by those workers in the process of obtaining virus-free clones clearly shows that a wide range of options is available. The particular option selected for a specific host-virus combination has depended upon a number of factors, among them being the state of knowledge of virus eradication at the time the problem was being investigated, published information available on eradication of the virus in question, the technical training and experience of the investigator, survival problems encountered with the host when held at high temperatures, whether or not a suitable culture medium had already been devised for the host species, and the propagation technique that was routinely applied for the commercial propagation of the plant being investigated.

During the 1950s, few plant pathologists involved with field problems relating to virus disease control had any knowledge or experience with plant tissue culture. It was therefore natural that practically all of the virus eradication attempts during this period were directed toward the application of heat therapy. By 1957, Kassanis[38] undertook to prepare a review on the effects of changing temperature on plant virus diseases and, in this review, he listed 33 viruses which had been inactivated in plants by heat. At this time, the science of plant virus identification was nowhere near what it is today, so some of the viruses included in the list were unidentified. Further, close to half of the ''viruses'' in the list are no longer classified with the viruses, but belong to the group of mycoplasmalike organisms, a group of organisms that is extremely heat labile. Only after thermotherapy was attempted with a wide range of host-virus combinations did it become obvious that this technique alone was of limited value, because the unit that was excised to establish an independent plant was too large to be completely free of virus infectivity. In 1969, Nyland and Goheen[39] compiled a list of 26 viruses that had not been inactivated by heat treatment in vivo. This list, which included many economically important viruses, emphasized the fact that thermotherapy was not adequate to meet the challenge and that more sophisticated techniques were essential.

In reviewing the literature today, some 2 decades after Nyland and Goheen compiled their review, one is struck by the realization that there have been dramatic advances in the field of plant virus eradication. Many new and ingenious techniques, or combinations of techniques, have been devised. A summary of the viruses that have been successfully eradicated is compiled in Table 1, and the techniques that have been used are listed below, with the abbreviation for each technique used in Table 1.

**Natural escape from virus infection (NE)** — Some viruses occur erratically in their host, and opportune removal and propagation of small pieces of tissue may yield clones that are free from the virus. See Hollings[25] for several examples of this phenomenon.

**Whole-plant thermotherapy (WPT)** — This is the least sophisticated of all techniques. It consists of placing a potted plant in a heat chamber held at approximately 37°C and removing the plant after a few weeks.

**Natural thermotherapy (NT)** — This involves the susceptibility of viruses to inactivation during growth in a natural high-temperature environment. The technique was investigated in the Imperial Valley in California for strawberry virus eradication.[43]

**Thermotherapy, followed by excision of cuttings (TC)** — Small cuttings (5 to 20 mm long) taken immediately after heat treatment from the shoot tips of treated plants are often free from viruses. Hollings[25] listed many examples of success using this technique.

**Thermotherapy, followed by grafting (TG)** — This technique is commonly used with stone fruit, apple, and pear cultivars. The trees to be treated are grown in pots and budded before treatment. After varying treatment periods, samples are removed from the treated trees and inserted into suitable rootstocks.[44-46]

**In vitro micrografting (IVM)** — This procedure combines standard grafting techniques with meristem culture methods, in that the meristematic tip (the scion) is excised and grafted on a sterile rootstock seedling. For virus elimination, the meristem must be small enough to be virus free, and the rootstock must also be virus free. The procedure was first used to eradicate citrus viruses[47,48] and a virus from apple.[49]

**Meristem culture (MC)** — This widely used technique has already been discussed.

**Fragmented shoot tip culture (FSTC)** — This modification of the standard meristem shoot-tip culture involves the in vitro propagation of fragmented shoot apices. With grapvine shoot tips, the fragments first grow into leaflike structures and then produce adventitious shoots at the basal end of the leaves. The shoots can be rooted to form whole plants. The technique has been used successfully to regenerate plants that are free from several viruses that affect grapevine.[6,7]

**Thermotherapy, followed by meristem culture (TMC)** — As noted earlier, this is the method of choice for the majority of virus eradication being done today.

**Diurnal alternating temperatures (DAT)** — Thermotherapy usually involves exposing the infected plant to a continuous temperature of about 37°C for several weeks, a treatment that causes some deterioration of the tissues of the treated plants. In an attempt to minimize tissue damage, several workers have investigated holding the treated plant at diurnal temperatures alternating between a period at high temperatures that is restrictive to virus multiplication and a period that is optimal for plant growth and virus multiplication. PLRV was eradicated from infected potato plants by Hamid and Locke[50] using diurnal treatments at 40°C (4 h) and 16 to 20°C (20 h). A similar regime was found to be effective for inactivation or eradication of viruses from apple,[51] tobacco,[52,53] and cowpea.[53] In practice, this modification of thermotherapy is usually coupled with meristem culture.

**In vitro therapy (IVT)** — Heat treatment can be applied to a plantlet in a culture tube instead of being applied to a potted plant. Walkey and Cooper[54] reported that cucumber mosaic virus and alfalfa mosaic virus were eradicated or their concentration greatly diminished when infected meristem tips were grown at 32 to 34°C, whereas similar cultures grown at 22°C remained infected.

**Low-temperature therapy (LTT)** — As noted above, this technique is helpful in obtaining viroid-free clones.

## Table 1
## SUMMARY OF PLANT VIRUSES THAT HAVE BEEN ERADICATED FROM INFECTED PLANTS, CATEGORIZED INTO VIRUS GROUPS, AND THE PROCEDURE USED TO ACHIEVE ERADICATION

| Virus group[a] | Virus | Eradication procedure[b] | Ref.[c] |
|---|---|---|---|
| Alfalfa mosaic | Alfalfa mosaic | TMC, MC, TC | 25, 39 |
| Bromovirus | Cowpea chlorotic mottle | DAT | 53 |
| Carlavirus | Alstroemeria carla | MC | 59 |
| | Carnation latent | TMC | 25, 39 |
| | Chrysanthemum B | TMC, DAT | 25, 39 |
| | Hop latent | TMC, MC | 42, 60 |
| | Hop mosaic | MC | 60 |
| | Lily symptomless | TMC, MC | 42, 61 |
| | Narcissus latent | MC | 61 |
| | Nerene latent | TMC | 61 |
| | Passion fruit latent | TMC, MC | 62 |
| | Poplar mosaic | TC | 39 |
| | Potato M | TMC | 42 |
| | Potato S | TMC | 42 |
| | Shallot latent | TMC | 63 |
| Caulimovirus | Carnation etched ring | MC, TMC | 25, 42 |
| | Cauliflower mosaic | MC, TMC | 42 |
| | Dahlia mosaic | TC | 25, 39, 42 |
| | Strawberry vein banding | TC, TMC | 58 |
| Closterovirus | Apple chlorotic leaf spot | IVM, TG, CMC | 29, 39, 64 |
| | Apple stem grooving | IVM | 65 |
| | Apple stem pitting | DAT | 39 |
| | Cherry green ring mottle | TG | 17, 39 |
| | Citrus tristeza | MC | 25, 39 |
| | Grapevine leaf roll | FSTC, TMC | 6, 66 |
| | Grapevine stem pitting | FSTC | 7 |
| Comovirus | Broad bean mottle | MC, TMC | 25 |
| | Ullucus virus C | CMC | 30 |
| Cucumovirus | Cucumber mosaic | NE, TC, DAT | 30, 39 |
| | Peanut stunt | LTT, MC | 32 |
| | Tomato aspermy | NE, TC | 25, 42 |
| Dianthovirus | Carnation ringspot | TMC, TC | 25, 42 |
| Geminivirus | Abutilon mosaic | TC | 25, 39 |
| | African cassava mosaic | MC, TMC | 67, 68 |
| Ilarvirus | Apple mosaic | NE, TG | 25, 39 |
| | Asparagus 2 | MC, TMC | 69 |
| | Prune dwarf | TC, MC, IVM | 25, 64 |
| | Prunus necrotic ringspot | TC, MC, IVM | 19, 22, 39 |
| | Rose mosaic | TC, MC | 25, 39 |
| | Tobacco streak | NE, TC, MC | 58, 69 |
| Luteovirus | Potato leaf roll | MC, DAT, NT | 39, 42 |
| | Strawberry mild yellow edge | MC, TMC | 58 |
| Nepovirus | Arabis mosaic | TMS, MC | 39, 42 |
| | Cherry leaf roll | TMC | 42 |
| | Grapevine fanleaf | TMC | 39, 42 |
| | Raspberry ringspot | TMC | 70 |
| | Strawberry latent ringspot | TMC | 42 |
| | Tobacco ringspot | TMC | 39 |
| | Tomato black ring | TMC | 42 |
| | Tomato ringspot | TMC | 39, 42 |
| Plant reovirus | Wound tumor | LTT | 31 |
| Plant rhabdovirus | Raspberry vein chlorosis | TC, TMC | 58 |
| | Strawberry crinkle | NT, TMC, MC, WPT | 58 |

**Table 1 (continued)**
**SUMMARY OF PLANT VIRUSES THAT HAVE BEEN ERADICATED**
**FROM INFECTED PLANTS, CATEGORIZED INTO VIRUS GROUPS, AND**
**THE PROCEDURE USED TO ACHIEVE ERADICATION**

| Virus group[a] | Virus | Eradication procedure[b] | Ref.[c] |
|---|---|---|---|
| Potexvirus | Banana mosaic | TMC | 71 |
| | Cassava brown streak | TMC | 68 |
| | Cymbidium mosaic | TMC | 42 |
| | Hydrangea ringspot | TMC, TC | 25 |
| | Lily virus X | MC | 61 |
| | Narcissus mosaic | MC | 61 |
| | Papaya mosaic | TMC | 30 |
| | Potato aucuba | TMC | 5 |
| | Potato X | TMC, MC | 42 |
| | White clover mosaic | TC, LTT | 32 |
| Potyvirus | Alstroemeria mosaic | MC | 59 |
| | Artichoke latent | MC | 72 |
| | Asparagus 1 | TMC | 69 |
| | Bean yellow mosaic | LTT, TMC, MC | 32, 39 |
| | Carnation vein mottle | TMC | 25, 42 |
| | Clover yellow vein | LTT, TC | 25, 32 |
| | Dasheen mosaic | MC, TMC | 42, 73 |
| | Freesia mosaic | MC, TMC | 42 |
| | Hippeastrum mosaic | MC | 42 |
| | Hyacinth mosaic | MC | 42 |
| | Iris mild mosaic | MC, CMC | 61 |
| | Iris severe mosaic | MC | 61 |
| | Leek yellow stripe | MC, TMC | 63 |
| | Narcissus yellow stripe | MC | 61 |
| | Onion yellow dwarf | MC, TMC | 63 |
| | Papaya mosaic | CMC | 30 |
| | Pea seed-borne mosaic | MC | 42 |
| | Plum pox | MC, TG, IVM | 74, 75 |
| | Potato A | MC, TMC | 42 |
| | Potato V | MC, TMC | 42 |
| | Potato Y | MC, TMC | 42 |
| | Ryegrass mosaic | MC | 76 |
| | Sugarcane mosaic | MC | 42 |
| | Sweet potato feathery mottle | MC | 42 |
| | Tulip breaking | MC | 61 |
| | Turnip mosaic | MC, TMC | 42 |
| | Ullucus mosaic | CMC | 30 |
| | Yam mosaic | TMC | 77 |
| Tobamovirus | Odontoglossum ringspot | SIC | 55 |
| | Tobacco mosaic | TMC | 42 |
| | Ullucus tobamovirus | CMC | 30 |
| Tobravirus | Tobacco rattle | MC | 61 |
| Tombusvirus | Carnation Italian rinsgpot | TC, WPT | 78 |
| | Narcissus tip necrosis | MC | 61 |
| | Pelargonium leaf curl | NE | 25 |
| | Tomato bushy stunt | TC | 25 |
| | Turnip crinkle | TMC | 25 |
| Tomato spotted wilt | Tomato spotted wilt | NE | 25 |
| Tymovirus | Poinsettia mosaic | TC, TMC, CSC | 56, 79 |
| Ungrouped | Apple chlorotic rigspot | TMC | 39 |
| | Blackberry calico | TC | 58 |
| | Black currant reversion | TG | 58 |
| | Black raspberry necrosis | WPT, TC | 58 |
| | Carnation mottle | MC, TMC | 25 |

**Table 1 (continued)**
**SUMMARY OF PLANT VIRUSES THAT HAVE BEEN ERADICATED**
**FROM INFECTED PLANTS, CATEGORIZED INTO VIRUS GROUPS, AND**
**THE PROCEDURE USED TO ACHIEVE ERADICATION**

| Virus group[a] | Virus | Eradication procedure[b] | Ref.[c] |
|---|---|---|---|
| Ungrouped | Cherry necrotic rusty mottle | TC | 39 |
| | Citrus psorosis | IVM | 39, 80 |
| | Citrus xylopsorosis | IVM | 80 |
| | Fig mosaic | MC | 81 |
| | Ginger mosaic | MC | 10 |
| | Gooseberry vein banding | MC | 58 |
| | Grapevine corky bark | FSTC | 7 |
| | Grapevine asteroid mosaic | TMC | 39 |
| | Grapevine yellow mosaic | TC | 82 |
| | Grapevine vein mosaic | WPT | 82 |
| | Pear stony pit | WPT | 39 |
| | Poinsettia cryptic | TMC, CSC | 56, 79 |
| | Raspberry bushy dwarf | TC, TMC | 58 |
| | Raspberry leaf mottle | WPT, TC | 58 |
| | Raspberry leaf spot | WPT, TC | 58 |
| | Red currant vein banding | TG | 58 |
| | Rubus yellow net | TC | 58 |
| | Strawberry chlorotic fleck | TMC | 58 |
| | Strawberry feathery leaf | TMC | 58 |
| | Strawberry latent C | MC | 58 |
| | Strawberry mottle | WPT | 58 |
| | Strawberry pallidosis | MC | 58 |
| | Sweet potato mild mottle | MC, TC | 83 |
| Viroids | Chrysanthemum chlorotic mottle | LTT, MC | 37, 84 |
| | Chrysanthemum stunt | LTT, MC | 37, 84 |
| | Citrus exocortis | MC, IVM | 39, 80 |
| | Cucumber pale fruit | LTT, MC | 37 |
| | Grapevine viroid | MC | 85 |
| | Hop stunt | LTT | 37 |
| | Potato spindle tuber | LTT, MC | 33, 35, 37 |

a   Based on the Fourth Report of the International Committee on Taxonomy of Viruses.[57] Viroids are
   included for comparative purposes.
b   Refer to text for an explanation of abbreviations used in this column.
c   To keep citations to a minimum, only recent references are cited. For original references concerning
   small fruit crops, readers are referred to a recent handbook[58] and, for most references prior to 1977, to
   review articles by Hollings,[25] Nyland and Goheen,[39] and Quak.[42]

**Chemotherapy during meristem culture (CMC)** — The application of this technique
has been discussed in detail above.

**Serological inactivation before culturing (SIC)** — This ingenious technique was developed by Inouye[55] to assist in the eradication of odontoglossum ringspot (ORSP) from
virus-infected cymbidium. Very few virus-free plants could be obtained by meristem culture,
but following immersion of meristem tips for 1 h in ORSP antiserum more than half of the
plantlets derived from the cultures were virus free.

**Cell suspension culture (CSC)** — This technique was devised to assist in the elimination
of poinsettia mosaic virus and ponsettia cryptic virus from infected cultivars.[56] Shoot tips
from infected plants were stimulated to form calli which were transferred to a liquid medium
and shaken continuously for 20 or more days. Single cells and small cell aggregates which
were removed from the suspension by filtration differentiated to rooted plantlets when plated

on a solid medium. Most of the plantlets were free of both viruses. This is an effective method for virus elimination from infected plants provided that homohistonts, which give uniform regenerates, are used as donor plants.

Several workers have speculated on the correlation between the size and shape of plant viruses in relation to how readily they could be eradicated by heat therapy. On the basis of the limited experimental data available in 1957, Kassanis[38] speculated that viruses with rod-shaped particles are less likely to be inactivated in vivo by heat treatment. At that time, all the viruses known to have rod-shaped particles had survived in plants exposed to a temperature of 36°C, whereas most, but not all, of those with spherical particles had been inactivated. Hollings[25] supported Kassanis' generalization, noting that many of the mechanically transmitted viruses with isometric particles could be eliminated by heat. He found that rigid rods of the tobamovirus group withstood heat treatment, as had most of the viruses in the potexvirus group. He also found that the aphid-borne flexuous rods (potyviruses) proved generally difficult to eliminate.

Nyland and Goheen[39] attempted to group viruses by tabulating the known facts published on heat therapy and by relating these facts to other properties of the viruses. Unfortunately, at the time they undertook their tabulation, the grouping of viruses based upon their properties was still in a state of chaos so that their attempt to relate heat therapy with virus properties was premature. Their tabulation did clearly show that "viruses of the yellow group", now classified as mycoplasmalike organisms, were readily inactivated. Another group, primarily ilarviruses, were hot-water labile. A third group, which included many potyviruses, closteroviruses, ilarviruses, and nepoviruses, was designated as easily inactivated on the arbitrary basis that thermotherapy of less than 29 d was sufficient to permit propagation of explants free of infection. Another group, designated as difficult to inactivate, included the carlavirus and potexvirus groups.

Plant virus nomenclature and classification has advanced considerably in the past 2 decades, to the point where there is now a sound basis of classification that is acceptable to the majority of workers.[57] This, together with a steady flow of new literature on virus eradication, prompted the tabulation (Table 1) of those plant viruses where successful eradication had been reported in relation to recognized plant virus groups. With some 15 eradication options being used, some of which are more effective than others, it is difficult to analyze the available information and formulate unequivocal statements. However, the following generalizations can be made with some degree of assurance:

1. Viruses that are included in a particular group behave similarly with respect to ease of eradication.
2. Viruses with isometric particles are more readily inactivated than those with rod-shaped particles. This indicates that the generalization proposed by Kassanis[38] remains valid.
3. For those viruses with isometric particles, the ilarviruses are most readily inactivated and the comoviruses are the most difficult.
4. For those viruses with rod-shaped particles, the long flexuous rods (potyviruses and carlaviruses) are more readily eradicated than viruses of intermediate length (potexviruses and carlaviruses).
5. The most difficult viruses to eradicate belong to the tobamovirus group. These viruses appear to be even more refractory than the viroids.

## III. IN VITRO MICROPROPAGATION

Plant cell culture became generally achievable following the identification of indole acetic acid as auxin and the discovery of kinetin as the first cytokinin. Current methods of regenerating plants for rapid in vitro propagation are based on manipulating the auxin-cytokinin balance to achieve adventitious organogenesis, removing apical dominance with higher levels of cytokinin.[86] Orchids were the first plants to be regenerated by clonal propagation in tissue culture and, due to the impressive successes by commercial orchidologists, extension to other crops followed. These crops included ferns and other ornamentals and later vegetables, fruits, field crops, and forest trees. The number of species that are now propagatable by tissue culture is approaching 1000, although not all have been tested for commercial feasibility.[87]

The method is not without serious flaws, the main one being that it is too laborious and slow to compete economically with standard propagation procedures. Successful transfer to soil is not assured and pathogen exclusion by current practices is largely coincidental. Some new approaches, aimed at automated technology, may well improve the economics of plant tissue culture.[88]

This chapter is not concerned with tissue culture for mass propagation, but rather with the use of in vitro micropropagation as a means of excluding plant viruses, especially in the early phases of multiplication. This technique is based on the principle that any viruses present in a particular clone are capable of being eradicated, although once eradication is accomplished the clone is susceptible to recontamination by viruses. Hence, in vitro micropropagation provides a means by which a few or many plants may be produced in an environment that virtually precludes contamination. In some crops where adequate technology is developed and where there is economic justification, in vitro micropropagation may be extended to a few thousand plantlets. Two crops where extended in vitro micropropagation is used, and where exclusion of viruses is a dominant reason for using tissue culture procedures, are potatoes and strawberries. With these crops, as indeed with any crops, it is important to recognize that virus exclusion and testing steps must precede the plant multiplication process.

## A. Case Study: Potato

Heat treatment, meristem culture, and micropropagation techniques have enabled the development of virus-free collections of potato cultivars that can be used as initial starting material for seed certification programs.[24] When virus-free clones were first made available 2 decades ago, there was a reluctance in some areas to replace existing clones, many of which were universally infected with PVX and PVS. The primary concerns were that mild strains of PVX and PVS might be useful in protecting plants against severe strains, that all clones of a cultivar might not be equally productive, and that the virus eradication treatment might induce or cause deleterious mutations. These concerns appear to be unwarranted, and comparisons of separate clones of several cultivars which have undergone grower selection prior to virus eradication have provided reassuring evidence of cultivar stability.[89]

The starting material of many potato seed programs consists of in vitro-grown plantlets arising from a pathogen-tested nuclear plant. The initial stages of seed potato programs utilize varying levels of in vitro micropropagation depending on the size and location of the program. The objective of in vitro-derived microtubers or plantlets is to provide a quality product for use in seed potato programs. The propagation units directly derived from in vitro procedures are restricted to the early phases of production, with later phases consisting of glasshouse- or screenhouse-produced units. Propagation usually involves excising single nodes with leaves from small in vitro plantlets and placing each node onto the surface of agar-solidified medium. In 3 to 4 weeks the axillary bud grows into a plantlet with six or seven more nodes and becomes available for subculture. If larger numbers of plantlets are required, larger stem cuttings containing three or four nodes are placed in liquid medium in shake culture. After a few weeks of rapid growth, each flask contains 60 to 70 nodes. When a suitable number of small plantlets have been produced, they are transplanted to beds or pots for tuber production.

In recent years, interest on microtuber production under in vitro conditions has developed in many countries. A rapid, cost-effective method involving the addition of benzylamino-purine, chlorocholine chloride, and sucrose to the propagation medium has been developed at the International Potato Center.[90]

Several important breakthroughs in the 1960s and 1970s have led to the implementation of rapid multiplication and tissue culture methods in Third World countries.[91] The first of these was the ability to free seed potato stocks of systemic pathogens via thermotherapy and tissue culture, permitting the continuous use of pathogen-free seed stocks and thereby reducing virus contamination of fields allocated to the production of seed potatoes. Improved virus detection methods, principally the widespread application of the ELISA technique, further reduced viral contamination. Rapid multiplication methods in an environment where contamination was absent or slight were adapted to meet the needs created by varietal or climatic variations. Advances were made in the production of tuberlets, which permits the production of uniform field or bed plantings without the inherent disadvantages associated with staggered transplanting and harvesting dates of other methods. Finally, the use of tissue culture as a rapid multiplication method[92] permits the production of a large number of virus-free plantlets in a small area. These plantlets can be used directly for production or as mother plants which can be used as a source of cuttings.

As with Third World countries, micropropagation as a tool for production of nuclear potato seed stocks has made giant steps in developed countries in the last decade.[93,94] Most seed-producing areas of North America and Europe have either built or modified existing structures to accomodate this type of propagation either in vitro or in protected environments. The trend is for seed growers, government agencies, or private companies to start small in vitro programs and, depending on their initial success, perseverance, and control of costs, to expand their facilities and capabilities. With a full range of pathogen testing being incorporated into developing in vitro systems as well as a reasonable degree of good management, viruses and other pathogens will decrease. It is expected that there will be an expansion of in vitro programs throughout North America and Europe. Transplanting plantlets to the field will increase as demand outpaces greenhouse space and mechanization is expanded. In vitro production of microtubers will increase, particularly if methods can be devised so that microtubers are cost competitive with minitubers that are produced in glasshouse facilities. Further, reliable ways must be found to break dormancy of microtubers, since they often undergo long periods of dormacy. This problem has been partially solved by the simple procedure of cutting the microtubers in half.[95]

## B. Case Study: Strawberry

The basic principle of the method for micropropagation of strawberry is simple. The growth of dormant axillary buds is induced by the presence of cytokinin in the culture medium, and subsequent elimination of cytokinin permits the development of leaves and roots. The factors that optimize this procedure to the point where it has become suitable as a mass-production method were first published by Nishi and Oosawa.[96] The procedure has since been revised and improved.[97,98] It offers three valuable features that are not possible by conventional methods: (1) rapid increase of new or scarce germplasm at any time of the year; (2) availability of nursery stock plants that are absolutely free of diseases, insectes, and nematodes; and (3) a compact, disease-free repository.

Virus elimination from strawberry cultivars is an essential first step. Posnette[99] was the first to eliminate strawberry viruses by hot-air treatment, but since then it has been found that thermotherapy in conjunction with tissue culture or excised apical buds dramatically increases the success rate. To maximize recovery of the largest number of virus-free plants, meristems 0.5 to 0.9 mm long are most often used. Boxus[100] feels that previous heat treatment is not necessary if one works with very small explants (0.2 to 0.4 mm). He succeeded in eliminating unidentified viruses from 73 meristem clones of 25 infected cultivars. Scott and

Zanzi[101] started with heat-treated plants from indexed *Fragaria vesca* and *F. virginiana* indicator clones. These plants were grown in a greenhouse to produce runner tips for meristems. Each year the propagation cycle was started from new, 0.5 to 0.7 mm long meristems excised from vigorous runner tips during May to August. Cultures were held in a growth room at 26°C with a 16-h light period. After 4 weeks, meristems had sufficient growth to transfer them to a multiplication medium, where stem tufts developed. These tufts were transferred to fresh medium every 4 to 5 weeks, but only three successive transfers were made in the multiplication stages in order to avoid or reduce the possibility of mutations arising during multiplication.

The possibility of deleterious mutations arising during in vitro micropropagation is a concern to all commercial laboratories involved in the multiplication of strawberries. The seriousness of the problem is difficult to assess, since there are conflicting reports in the literature. Scott and Zanzi[101] observed two leaf variegations in potted plants derived from tissue-cultured plantlets. One was a yellow-green mosaic or a solid yellow color in sections of the leaf blade in the cultivar Belrubi. Affected plants were discarded prior to planting, and no new variegations were detected later in a planting of 1700 plants. The second variegation was a narrow white streak in the leaf blades of the cultivar Aliso. This symptom was apparent on 72 of 3550 plants after they were put in a fiberglass house. The plants were removed and no others were detected. Although both types of variegation can be rogued easily, their appearance did emphasize the requirement of nurserymen to check their stocks at an early stage of propagation to eliminate those that are not true to type.

Boxus et al.[98] have evaluated several strawberry fields established with micropropagated plants which included disease-free control plants of the same genotype. Some of the micropropagated plants grew better than the control plants, although precise statistical comparisons were not made. The authors commonly observed a dwarf habit associated with a micropropagated clone of cv. Gorella. Fruit size was decreased and the plants had more leaves with more petioles and reduced leaf surface. Cold storage of the plants at −2°C for several months caused the first-generation Gorella plants to grow to normal size. Boxus et al.[98] recognize that there may be difficulties associated with micropropagation, and they feel it is important to evaluate different cultivars with regard to duration of cold treatment required and the number of subcultures that can be made without an undue risk of mutations. In 1988 Boxus[102] reported that the field behavior of tissue-cultured plants was very good, with no statistical differences for fruit size and fruit weight, provided excessive subculturing is avoided. He recommended less than 12 subcultures and noted that the increase in flowering and fruit deformation observed in some tissue-cultured clones is related not only to the number of subcultures, but also to the day length.

Kinet and Parmentier[103] found that the qualitative responses to daylight of micropropagated strawberry plants, cv. Gorella, were identical to those of material propagated by conventional runnering. Flowering was induced by short days, while inflorescence response was promoted by long photoperiods. Inflorescence structure was markedly influenced by day length in that the proportion of basal branching was increased by prolonged short-day treatment. Highly floriferous plants occurred more frequently when the number of subcultures preceding root development was high, and particularly when reproductive structures developed in long days.

In the micropropagation system used in France, in vitro plants taken from test tubes are submitted to a conventional 2-year multiplication system with runnering before being used for fruit production. It was found that plants were reasonably homogeneous during the multiplication phase in nurseries, but heterogeneity between plants occurred in the following 2 years with the appearance of some physiological disorders which increased as fruit set and matured.[104] Many plants collapsed at that time and, since no specific disease was identified as the cause, in vitro techniques were suspected of generating off-type material. Highly subcultured clones were the most heterogeneous, sensitive to mildew, had poor rooting systems, and decreased fruit size and weight. By halving the concentration of cytokinetin and auxin to 0.5 mg/l instead of the 1 mg/l that had been used and by limiting the number of subcultures to ten, the problem appears to have been solved. The micropropagation method will be retained in the French procedure for certifying plants with the above modifications incorporated.

A private biotechnology company in Brazil has been using tissue culture techniques for the production of certified strawberry plants.[105] In 1986, a total of 200,000 stock plants was produced in vitro and hardened in styrofoam containers of soil-free medium for 1 month in a greenhouse. Survival rate during hardening was over 95% and the occurrence of off-type plants was insignificant. There were sufficient in vitro plants to establish a 70-ha nursery where 16 million commercial runner plants were generated in one 6-month culture cycle.

While strawberry micropropagation was pioneered in Europe and is being used commercially in France, Italy, Belgium, and Germany, it has not replaced conventional propagation in many countries. The strawberry-producing areas of the U.S. and Canada, for instance, rely almost completely on conventional propagation to supply their nursery fields. The reasons for this practice are twofold: first, cost; and secondly, a concern about the stability of some cultivars during long-term micropropagation. Estimates of the cost of propagating strawberry plants by tissue culture have been difficult to calculate because of rapidly changing techniques as new information develops and as the skills of technicians improve. Micropropagation appears to be considerably more expensive than conventional nursery propagation for direct fruit production so that, in those countries with micropropagation programs, it is used only to provide planting stock for strawberry nurseries rather than for commercial strawberry production. However, there are valuable advantages of strawberry micropropagation that suggest a possible expansion, namely: (1) rapid multiplication of new cultivars or virus-free clones of cultivars; (2) provision of basic nursery stock that is assuredly free of pathogens, insects, and nematodes; (3) rapid multiplication of everbearing cultivars; (4) propagation at any season of the year; (5) propagation of many small plants in a small space; (6) storage of small plants in limited storage space; (7) retention of a repository for variety collections; (8) retention of a repository for virus isolates; and (9) provision of an acceptable propagation unit for the international exchange of germplasm.

The concern about genetic stability is more difficult to allay. Certainly, some of the earlier reports on the field evaluation of tissue-cultured strawberry were discouraging. Customers expect to receive true-to-type plants and fruit production equivalent to control plants that have not been micropropagated. Following many years of experience with tissue-culture micropropagation of strawberry in Belgium,[102] Italy,[106] France,[104] and Brazil,[105] it appears that genetic instability is no longer considered to be a serious problem.

## IV. REGENERATION OF TRANSGENIC PLANTS

For a detailed treatment of the subject of virus disease control through transgenic plants, readers are referred to Chapter 13. This section is concerned with a small segment of the broad subject, namely, regeneration of transgenic plants.

Cross protection, the procedure whereby plants are deliberately infected with a mild strain of a virus to protect them against infection with more virulent strains, has been successfully applied to a few host-virus combinations. It is assumed that the coat protein of the mild strain prevents the expression of the coat protein gene of the more virulent strains, thereby reducing virus accumulation and symptom expression. Carrying this approach one step further, there is encouraging evidence that genetically engineering plants using expression vectors carrying the coat protein gene may result in transgenic plants that are protected against virus infection.

The general concept of genetic engineering is relatively simple. It is usually considered as a three-step process. First, a gene is physically isolated in large quantities as purified DNA. The second step involves its stable introduction into a living cell by transformation of the modified gene. The third step is the regeneration of a single transformed cell into a whole transformed plant. Initially, plants were generated from calli derived from protoplasts transformed by cocultivation with *Agrobacterium tumefaciens* cells. However, protoplast cultures has certain limitations, such as: (1) not all species of plants can be regenerated readily from protoplasts; (2) the entire process takes up to 6 months from protoplast to plant; (3) plants derived from protoplasts are subject to mutations or chromosomal abnormalities; and (4) protoplast culture technology can be difficult to reproduce from one experiment to the next. An alternative procedure involving transformation of stem or root explants in vitro proved to be laborious for large-scale experiments and not easy to use with modified Ti plasmids that lacked the tumor-inducing gene. To overcome these limitations, Horsch et al.[107] have devised a technique whereby discs punched from surface-sterilized leaves are submerged overnight in a culture of *A. tumefaciens* and then incubated upside down on nurse culture plates containing a medium that induces regeneration of shoots of the species being transformed. Shoot regeneration may occur within 2 to 4 weeks, and transformants can be confirmed by their ability to form roots in a medium containing kanamycin.

*Agrobacterium* is a powerful tool for gene transfer to dicotyledonous plants, but it was considered less valuable for monocotyledonous plants because they were less susceptible to infection by the bacterium.[108] The failure of *Agrobacterium* to transform monocots appears to be associated with insufficient amounts of specific wound substances produced in these plants. When *A. tumefaciens* was treated with wound exudates of the dicot *Solanum tuberosum* and used as a vector to discs of yam bulbil tissue, transformation was accomplished.[109] This may expand the scope of the technique to include a wide range of monocots and dicots. Further experimental work will be required in order to regenerate plants from the transformed cells.

# REFERENCES

1. **Stace-Smith, R. and Hamilton, R. I.,** Inoculum thresholds of seedborne pathogens: viruses, *Phytopathology,* 78, 875, 1988.
2. **Morel, G. and Martin, C.,** Guérison de dahlias atteints d'une maladie à virus, *C. R. Acad. Sci. Paris,* 235, 1324, 1952.
3. **Pennazio, S. and Redolfi, P.,** Potato virus X eradication in cultured potato meristem tips, *Potato Res.,* 17, 333, 1974.
4. **Faccioli, G., Rubies-Autonell, C., and Resca, R.,** Potato leafroll virus distribution in potato meristem tips and production of virus-free plants, *Potato Res.,* 31, 511, 1988.
5. **Mellor, F. C. and Stace-Smith, R.,** Virus-free potatoes by tissue culture, in *Applied and Fundamental Aspects of Plant Cell, Tissue and Organ Culture,* Reinert, J. and Bajaj, Y. P. S., Eds., Springer-Verlag, Berlin, 1977, 616.
6. **Barlass, M., Skene, K. G. M., Woodham, R. C., and Krake, L. R.,** Regeneration of virus-free grapevines using in vitro apical culture, *Ann. Appl. Biol.,* 101, 291, 1982.
7. **Barlass, M.,** Elimination of stem pitting and corky bark diseases from grapevine by fragmented shoot apex culture, *Ann. Appl. Biol.,* 110, 653, 1987.
8. **White, P. R.,** *A Handbook of Plant Tissue Culture,* Jacques Cattell Press, Lancaster, PA, 1943.
9. **Murashige, T. and Skoog, F.,** A revised medium for rapid growth and bioassays with tobacco tissue cultures, *Physiol. Plant.,* 15, 473, 1962.
10. **Wang, P. J. and Hu, C. Y.,** Regeneration of virus-free plants through in vitro culture, in *Advances in Biochemical Engineering,* Vol. 18, Fiechter, A., Ed., Springer-Verlag, Berlin 1980, 61.
11. **Evans, D. A., Sharp, W. R., and Ammirato, P. V., Eds.,** *Handbook of Plant Cell Culture,* Vol. 4, Macmillan, New York, 1986.
12. **Murashige, T.,** Plant propagation through tissue cultures, *Annu. Rev. Plant Physiol.,* 25, 135, 1974.
13. **Gebhardt, K.,** Development of a sterile cultivation system for rooting of shoot tip cultures (red raspberries) in duroplast foam, *Plant Sci.,* 39, 141, 1985.

14. **Mellor, F. C. and Stace-Smith, R.**, Development of excised potato buds in nutrient culture, *Can. J. Bot.*, 47, 1617, 1969.
15. **Kunkel, L. O.**, Heat treatments for the cure of yellows and other virus diseases of peach, *Phytopathology*, 26, 809, 1936.
16. **Kassanis, B.**, Heat inactivation of leaf-roll virus in potato tubers, *Ann. Appl. Biol.*, 37, 339, 1950.
17. **Janecková, M. and Blattný, C.**, [Thermotherapy of sour cherry cultivar Fanal infected by green ring mottle] *Ochr. Rostlin*, 16, 161, 1980.
18. **Lenz, F., Baumann, G., and Kornkamhaeng, P.**, High temperature treatment of *Prunus avium* L. "F12/1" for virus elimination, *Phytopathol. Z.*, 106, 373, 1983.
19. **Bogush, L. Y., Chernets, A. M., and Bondarenko, S. S.**, [In vitro propagation of sour cherry cultivars after heat treatment] *Sadovod. Vinograd. Vinodel. Mold.*, 6, 33, 1984.
20. **Monette, P. L.**, Elimination *in vitro* of two grapevine nepoviruses by an alternating temperature regime, *J. Phytopathol.*, 116, 88, 1986.
21. **Galzy, R.**, Technique de thermothérapie des viruses de la vigne, *Ann. Epiphyt.*, 15, 245, 1964.
22. **Bjarnason, E. N., Hanger, B. C., Moran, J. R., and Cooper, J. A.**, Production of prunus necrotic ringspot virus-free roses by heat treatment and tissue culture, *N.Z. J. Agric. Res.*, 28, 151, 1985.
23. **Mellor, F. C. and Stace-Smith, R.**, Virus-free potatoes through meristem culture, in *Biotechnology in Agriculture and Forestry*, Vol. 3, Bajaj, Y. P. S., Ed., Springer-Verlag, Berlin, 1987, 30.
24. **Wright, N. S.**, Assembly, quality control and use of a potato cultivar collection rendered virus-free by heat therapy and tissue culture, *Am. Potato J.*, 65, 181, 1988.
25. **Hollings, M.**, Disease control through virus-free stock, *Annu. Rev. Phytopathol.*, 3, 367, 1965.
26. **Tomlinson, J. A.**, Chemotherapy of plant viruses and virus diseases, in *Pathogens, Vectors, and Plant Diseases: Approaches to Control*, Harris, K. F. and Maramorosch, K., Eds., Academic Press, New York, 1982, 23.
27. **Hansen, A. J.**, Anti-viral chemicals for plant disease control, *Crit. Rev. Plant Sci.*, 8, 45, 1989.
28. **Cassells, A. C. and Long, R. D.**, The elimination of potato viruses X, Y, S and M in meristem and explant cultures of potato in the presence of Virazole, *Potato Res.*, 25, 165, 1982.
29. **Hansen, A. J. and Lane, W. D.**, Elimination of apple chlorotic leafspot virus from apple shoot cultures by ribavirin, *Plant Dis.*, 69, 134, 1985.
30. **Stone, O. M.**, The elimination of four viruses from *Ullucus tuberosus* by meristem-tip culture and chemotherapy, *Ann. Appl. Biol.*, 101, 79, 1982.
31. **Selsky, M. I. and Black, L. M.**, Effect of high and low temperatures on the survival of wound-tumor virus in sweet clover, *Virology*, 16, 190, 1982.
32. **Barnett, O. W., Gibson, P. B., and Seo, A.**, A comparison of heat teatment, cold treatment, and meristem tip-culture for obtaining virus-free plants of *Trifolium repens*, *Plant Dis. Rep.*, 59, 834, 1975.
33. **Stace-Smith, R. and Mellor, F. C.**, Eradication of potato spindle tuber virus by thermotherapy and axillary bud culture, *Phytopathology*, 60, 1857, 1970.
34. **Hollings, M. and Stone, O. M.**, Attempts to eliminate chrysanthemum stunt from chrysanthemum by meristem-tip culture after heat-treatment, *Ann. Appl. Biol.*, 65, 311, 1970.
35. **Sänger, H. L. and Ramm, K.**, Radioactive labelling of viroid RNA, in *Modifications of the Information Content of Plant Cells*, Markham, R., Davies, D. R., Hopwood, D. A., and Horne, R. W., Eds., North-Holland/American Elsevier, Amsterdam, 1975, 229.
36. **Lizárraga, R. E., Salazar, L. F., Roca, W. M., and Schilde-Rentschler, L.**, Elimination of potato spindle tuber viroid by low temperature and meristem culture, *Phytopathology*, 70, 754, 1980.
37. **Paduch-Cichal, E. and Kryczyński, S.**, A low temperature therapy and meristem-tip culture for eliminating four viroids from infected plants, *J. Phytopathol.*, 118, 341, 1987.
38. **Kassanis, B.**, Effects of changing temperature on plant virus diseases, *Adv. Virus Res.*, 4, 221, 1957.
39. **Nyland, G. and Goheen, A. C.**, Heat therapy of virus diseases of perennial plants, *Annu. Rev. Phytopathol.*, 7, 331, 1969.
40. **Hollings, M. and Stone, O. M.**, Investigation of carnation viruses. I. Carnation mottle, *Ann. Appl. Biol.*, 53, 103, 1964.
41. **Ingram, D. S.**, Growth of plant parasites in tissue culture, in *Plant Tissue and Cell Culture*, Street, H. E., Ed., Blackwell Scientific, Oxford, 1973, 392.
42. **Quak, F.**, Meristem culture and virus-free plants, in *Applied and Fundamental Aspects of Plant Cell, Tissue, and Organ Culture*, Reinert, J. and Bajaj, Y. P. S., Eds., Springer-Verlag, Berlin, 1977, 598.
43. **Frazier, N. W., Voth, V., and Bringhurst, R. S.**, Inactivation of two strawberry viruses in plants grown in a natural high-temperature environment, *Phytopathology*, 55, 1203, 1965.
44. **Nyland, G.**, Heat inactivation of stone fruit ringspot virus, *Phytopathology*, 50, 380, 1960.
45. **Fridlund, P. R.**, The national and international roles of the IR-2 virus-free fruit tree repository, *Acta Hortic.*, 67, 83, 1976.
46. **Németh, M.**, *Virus, Mycoplasma and Rickettsia Diseases of Fruit Trees*, Akadémiai Kiadó, Budapest, 1986, 135.

47. **Murashige, T., Bitters, W. P., Rangan, T. S., Nauer, E. M., Roistacher, C. N., and Holliday, P. B.,** A technique of shoot apex grafting and its utilization towards recovering virus-free *Citrus* clones, *HortScience,* 7, 118, 1972.
48. **Navarro, L., Roistacher, C. N., and Murashige, T.,** Improvement of shoot-tip grafting in vitro for virus-free citrus, *J. Am. Soc. Hortic. Sci.,* 100, 471, 1975.
49. **Huang, S.-C. and Millikan, D. F.,** *In vitro* micrografting of apple shoot tips, *HortScience,* 15, 741, 1980.
50. **Hamid, A. and Locke, S. B.,** Heat inactivation of leafroll virus in potato tuber tissues, *Am. Potato J.,* 38, 304, 1961.
51. **Larsen, E. C.,** Heat inactivation of viruses in apple CLMM 109 by use of daily temperature cycles, *Tidsskr. Planteavl.,* 78, 422, 1974.
52. **Walkey, D. G. A. and Freeman, G. H.,** Inactivation of cucumber mosaic virus in cultured tissues of *Nicotiana rustica* by diurnal alternating periods of high and low temperature, *Ann. Appl. Biol.,* 87, 375, 1977.
53. **Lozoya-Saldaña, H. and Dawson, W. O.,** Effect of alternating temperature regimes on reduction or elimination of viruses in plant tissues, *Phytopathology,* 72, 1059, 1982.
54. **Walkey, D. G. A. and Cooper, V. C.,** Effect of temperature on virus eradication and growth of infected tissue cultures, *Ann. Appl. Biol.,* 80, 185, 1975.
55. **Inouye, N.,** [Effect of antiserum treatment on the production of virus-free *Cymbidium* by meristem culture] *Nogaku Kenkyu,* 60, 123, 1983.
56. **Preil, W., Koenig, R., Engelhardt, M., and Meier-Dinkel, A.,** [Elimination of poinsettia mosaic virus (PoiMV) and poinsettia cryptic virus (PoiCV) from *Euphorbia pulcherrima* Willd. by cell suspension culture], *Phytopathol. Z.,* 105, 193, 1982.
57. **Matthews, R. E. F.,** Classification and nomenclature of viruses, *Intervirology,* 17(1—3), 1, 1982.
58. **Converse, R. H., Ed.,** *Virus Diseases of Small Fruits,* Agric. Handbook No. 631, Department of Agriculture, Washington, D.C., 1987.
59. **Hakkaart, F. A. and Versluijs, J. M. A.,** Virus elimination by meristem-tip culture from a range of Alstroemeria cultivars, *Neth. J. Plant Pathol.,* 94, 49, 1988.
60. **Adams, A. N.,** Elimination of viruses from the hop (*Humulus lupulus*) by heat therapy and meristem culture, *J. Hortic. Sci.,* 50, 151, 1975.
61. **Brunt, A. A., Stone, O. M., and Phillips, S.,** Recent progress in the production, propagation and distribution of virus-free flower bulbs, *Glasshouse Crops Research Institute Annual Report for 1983,* 1984, 109.
62. **Hakkaart, F. A. and Versluys, J. M. A.,** [Virus in *Passiflora caerulea* eliminated by meristem culture] *Vakbl. de Bloemisterij,* 36, 24, 1981.
63. **Walkey, D. G. A., Webb, M. J. W., Bolland, C. J., and Miller, A.,** Production of virus-free garlic (*Allium sativum* L.) and shallot (*A. ascalonicum* L.) by meristem-tip culture, *J. Hortic. Sci.,* 62, 211, 1987.
64. **Navarro, L., Llácer, G., Cambra, M., Arregui, J. M., and Juárez, J.,** Shoot-tip grafting in vitro for elimination of viruses in peach plants (*Prunus persica* Batsch), *Acta Hortic.,* 130, 185, 1982.
65. **G'bova, R.,** [Tissue culture in developing virus-free fruit species], *Rastenievud. Nauki,* 24, 3, 1987.
66. **Iri, M., Shimura, T., Togawa, H., and Ueno, K.,** [Elimination of grapevine leafroll virus by heat treatment and meristem tip culture], *Ann. Phytopathol. Soc. Jpn.,* 48, 685, 1982.
67. **Kartha, K. K. and Gamborg, O. L.,** Elimination of cassava mosaic disease by meristem culture, *Phytopathology,* 65, 826, 1975.
68. **Kaiser, W. J. and Teemba, L. R.,** Use of tissue culture and thermotherapy to free East African cassava cultivars of African cassava mosaic and cassava brown streak diseases, *Plant Dis. Rep.,* 63, 780, 1979.
69. **Yang, H.-J. and Clore, W. J.,** Obtaining virus-free plants of *Asparagus officinalis* L. by culturing shoot tips and apical meristems, *HortScience,* 11, 474, 1976.
70. **Sweet, J. B., Constantine, D. R., and Sparks, T. R.,** The elimination of three viruses from *Daphne* spp. by thermotherapy and meristem excision, *J. Hortic. Sci.,* 54, 323, 1979.
71. **Gupta, P. P.,** Eradication of mosaic disease and rapid clonal multiplication of bananas and plantains through meristem tip culture, *Plant Cell, Tissue Organ Cult.,* 6, 33, 1986.
72. **Pécaut, P.,** [Improvement of artichoke varieties: vegetatively propagated and seed propagated varieties, virus-free clones from multiplication *in vitro*], *C.R. Seances Acad. Agric. Fr.,* 69, 69, 1983.
73. **Zettler, F. W. and Hartman, R. D.,** Dasheen mosaic virus as a pathogen of cultivated aroids and control of the virus by tissue culture, *Plant Dis.,* 71, 958, 1987.
74. **Mosella, C. L., Signoret, P. A., and Jonard, R.,** [Development of apical micrografting methods to eliminate two types of virus particles in peach (*Prunus persica* Batsch)], *C. R. Heb. Seances Acad. Sci. Ser. D,* 290, 287, 1980.
75. **Jordović, M.,** [Practical aspects of investigations of plum pox virus], *Zast. Bilja,* 33, 445, 1982.
76. **Dale, P. J.,** The elimination of ryegrass mosaic virus from *Lolium multiflorum* by meristem tip culture, *Ann. Appl. Biol.,* 85, 93, 1977.

77. **Mantell, S. H., Haque, S. Q., and Whitehall, A. P.,** Apical meristem tip culture for eradication of flexuous rod viruses in yams *(Dioscorea alata), Trop. Pest Manage.,* 26, 170, 1980.

78. **Hollings, M., Stone, O. M., and Bouttell, G. C.,** Carnation Italian ringspot virus, *Ann. Appl. Biol.,* 65, 299, 1970.

79. **Paludan, N. and Begtrup, J.,** Inactivation of poinsettia mosaic virus and poinsettia cryptic virus in *Euphorbia pulcherrima* using heat-treated mini-cuttings and meristem-tip culture, *Tidsskr. Planteavl,* 90, 283, 1986.

80. **González, M., Peña, I., González Rego, J., Zamora, V., and Rodríguez, I.,** [Introduction in Cuba of in vitro grafting of shoot apices of the genus *Citrus* and related genera as means of obtaining virus-free plants], *Agrotecnia de Cuba,* 9, 61, 1977.

81. **Muriithi, L. M., Rangan, T. S., and Waite, B. H.,** In vitro propagation of fig through shoot tip culture, *HortScience,* 17, 86, 1982.

82. **Pop, I. V.,** [Results of research on heat treatment of some virus diseases of grapevine], *An. Inst. Cercet. Prot. Plant.,* 19, 27, 1986.

83. **Hollings, M., Stone, O. M., and Bock, K. R.,** Purification and propertries of sweet potato mild mottle, a white-fly borne virus from sweet potato *(Ipomoea batatas)* in East Africa, *Ann. Appl. Biol.,* 82, 511, 1976.

84. **Paludan, N.,** [Chrysanthemum stunt and chlorotic mottle. Establishment of healthy chrysanthemum plants and low temperature storage of chrysanthemum, carnation and pelargonium in test-tubes], *Planteavl,* 84, 349, 1980.

85. **Duran-Vila, N., Juárez, J., and Arregui, J. M.,** Production of viroid-free grapevines by shoot tip culture, *Am. J. Enol. Vitic.,* 38, 217, 1988.

86. **Wickson, M. E. and Thimarn, K. V.,** The antagonism of auxin and kinetin in apical dominance, *Physiol. Plant.,* 11, 62, 1958.

87. **Murashige, T. and Huang, L.-C.,** Cloning plants by tissue culture: early years, current status and future prospects, *Acta Hortic.,* 212, 35, 1987.

88. **Levin, R., Gaba, V., Tal, B., Hirsch, S., and De Nola, D.,** Automated plant tissue culture for mass propagation, *Biotechnology,* 6, 1035, 1988.

89. **Wright, N. S.,** Uniformity among virus-free clones of ten potato cultivars, *Am. Potato J,* 60, 381, 1983.

90. **Tovar, P., Estranda, R., Schilde-Rentschler, L., and Dodds, J. H.,** Induction and use of in vitro potato tubers, *CIP Circ.,* 13, 4, 1, 1985.

91. **Bryan, J. E.,** Implementation of rapid multiplication and tissue culture methods in Third World countries, *Am. Potato J.,* 65, 199, 1988.

92. **Roca, W. M., Espinoza, N. O., Roca, M. R., and Bryan, J. E.,** A tissue culture method for the rapid propagation of potatoes, *Am. Potato J.,* 55, 691, 1978.

93. **Jones, E. D.,** A current assessment of in vitro culture and other rapid multiplication methods in North America and Europe, *Am. Potato J.,* 65, 209, 1988.

94. **Addy, N. A.,** Opportunities and challenges for private industry, *Am. Potato J.,* 65, 221, 1988.

95. **Ewing, L. L., McMurry, S. E., and Ewing, E. E.,** Cutting as a method of breaking dormancy in microtubers produced in vitro, *Am. Potato J.,* 64, 329, 1987.

96. **Nishi, S. and Oosawa, K.,** Mass production method of virus-free strawberry plants through meristem callus, *Jpn. Agric. Res. Q.,* 7, 189, 1973.

97. **Boxus, P., Quoirin, M., and Laine, J. M.,** Large scale propagation of strawberry plants from tissue culture, in *Applied and Fundamental Aspects of Plant Cell, Tissue, and Organ Culture,* Reinert, J. and Bajaj, Y. P. S., Eds., Springer-Verlag, Berlin, 1977, 130.

98. **Boxus, P., Damiano, C., and Brasseur, E.,** Strawberry, in *Handbook of Plant Cell Culture,* Vol. 3, Ammirato, P. V., Evans, D. A., Sharp, W. R., and Yamada, Y., Eds., Macmillan, New York, 1984, 17.

99. **Posnette, A. F.,** Heat inactivation of strawberry viruses, *Nature (London),* 171, 312, 1953.

100. **Boxus, P.,** Rapid production of virus-free strawberry by "in vitro" culture, *Acta Hortic.,* 66, 35, 1976.

101. **Scott, D. H. and Zanzi, C.,** Rapid propagation of strawberries from meristems, in *The Strawberry,* Childers, N. F., Ed., Horticultural Publications, Gainesville, FL, 1981, 213.

102. **Boxus, P.,** Review of in vitro strawberry mass production, Program and Abstracts, International Strawberry Symposium, Cesena, Italy, 1988, 50.

103. **Kinet, J.-M. and Parmentier, A.,** Flowering behavior of micropropagated strawberry plants, Program and Abstracts, Interantional Strawberry Symposium, Cesena, Italy, 1988, 52.

104. **Rancillac, M. and Nourrisseau, J.-G.,** Micropropagation and strawberry plant quality, Program and Abstracts, International Strawberry Symposium, Cesena, Italy, 1988, 55.

105. **Paiva, M., Mecê, M. C. L. B., Sá, J. C. M., and Roes, B.,** Mass strawberry production via tissue culture, Program and Abstracts, International Strawberry Symposium, Cesena, Italy, 1988, 53.

106. **Faedi, W., D'Ercole, N., Turci, P., Bazzocchi, C., and Siroli, M.,** Effect of different propagation systems on strawberry plants performance, Program and Abstracts, International Strawberry Symposium, Cesena, Italy, 1988, 51.

107. **Horsch, R. B., Fry, J. E., Hoffmann, N. L., Eichholtz, D., Rogers, S. G., and Fraley, R. T.,** A simple and general method for transferring genes into plants, *Science,* 227, 1229, 1985.
108. **De Cleene, M.,** The susceptibility of monocotyledons to *Agrobacterium tumefaciens, Phytopathol. Z.,* 113, 81, 1985.
109. **Schafer, W., Gorz, A., and Kahl, G.,** T-DNA integration and expression in a monocot crop plant after induction of *Agrobacterium, Nature (London),* 327, 529, 1987.

Chapter 12

## DISEASE RESISTANCE MECHANISMS

### R. S. S. Fraser

## TABLE OF CONTENTS

# I. INTRODUCTION

## A. Control of Diseases Caused by Plant Viruses

Plant resistance is the most important means of controlling diseases and crop losses caused by viruses. This is mainly because other strategies are not available or have not been fully developed.

Chemical treatments, which are widely used against microbial pathogens of plants, are either not available for diseases caused by plant viruses or are unsuitable for application at the crop level because of expense or phytotoxicity. There have been numerous reports of compounds with antiviral activity in plants,[1-3] but so far, their use has mostly been restricted to elimination of viruses from infected breeding or propagation lines.[3] It is perhaps unlikely that chemicals with directly antiviral activity will be developed for field use, given current concern about pesticide usage and consequences for human and animal consumers. However, chemicals which stimulate natural resistance mechanisms in plants[4] may show promise.

Disease avoidance mechanisms include use of virus-free seed or planting material for vegetatively propagated crops and methods to prevent virus spread, such as good crop hygiene or control of virus vectors. Such methods offer poor prospects for disease control once a virus becomes established in the crop.

In the 1930s, it was demonstrated that plant resistance to viruses could be inherited simply in a Mendelian manner.[5] This, together with the ability to characterize and detect viruses which was emerging at about that time, provided the spur to include breeding for virus resistance in crop improvement programs. Since then, virus-resistant cultivars have been produced for innumerable crop species.[6] However, there are gaps in this defense. For many viruses, the host species have no known source of genetic resistance. In other cases, forms of virus which overcome the host resistance gene have evolved, sometimes with considerable rapidity. Some resistance mechanisms have been amply studied under laboratory conditions and have been shown to be more or less effective there, but it has not yet proved possible to translate them to use on a field scale.

The investigation of resistance mechanisms is currently receiving a great stimulus from the application of molecular biological methods. Initially, these have been applied primarily to the virus, but increasingly they are being used on the host. And linked to this, methods for genetic manipulation of plants are now being used to develop new types of mechanism of resistance to virus diseases. My objectives in this chapter are to review what is known of natural mechanisms of resistance to plant viruses, to question whether they might be more effectively developed for crop protection, and to explore briefly some of the new molecular approaches. The study of resistance mechanisms is of interest not only because of its importance in practical crop protection, but because of the insight it gives into how plants and their simplest pathogens interact and co-evolve.

## B. Types of Resistance Mechanism

Resistance mechanisms vary in terms of visual manifestation, effects on virus multiplication and symptom formation, the stage in the viral replicative cycle which serves as the target of the resistance, and in the depth of our understanding of the mode of action. Resistance undoubtedly involves many different types of mechanism, and these types are difficult to classify — especially when they are incompletely understood. However, a workable classification can be arrived at by considering the level of complexity of the host population at which the resistance operates.[7]

**Resistance at the species level: nonhost immunity** — All individuals of a species are completely unaffected by a particular virus, in that no detectable multiplication occurs and no symptoms are formed after attempted inoculation.

**Resistance at the cultivar level** — This is the type of resistance used by the plant breeder.

Resistant cultivars or breeding lines contain a gene or genes conferring resistance against a virus which is *pathogenic* to susceptible cultivars of the species

**Resistance at the level of the individual plant** — Resistance may be conferred on an individual plant of a susceptible species or cultivar by a prior infection or by environmental or chemical treatment. This type of resistance is often called "induced" or "acquired" resistance,[8,9] and there appear to be several types of mechanism. The resistance induced is not normally heritable and must be conferred afresh on each generation. However, in the special case of resistance conferred on susceptible individuals by transformation with DNA which is integrated into the host chromosome, this type of induced resistance is heritable.

### C. Interactions of Viruses with Plant Resistance Mechanisms

In true cases of nonhost immunity, there is no interaction, or the virus replicative cycle fails before anything can be detected. The virus is *nonpathogenic* to the nonhost species. However, the qualification about the ability to detect early pathogenesis or not introduces a gray area: closer examination may reveal that what was previously thought to be a completely nonpathogenic noninteraction may in fact have become blocked at an early stage of the replicative cycle, with at least some of the very earliest stages having been accomplished. This distinction has implications for possible mechanisms of apparent nonhost immunity, which will be developed in Section III.

In cultivar resistance, the interesting question is how the behavior of the virus in the resistant host differs from that in the susceptible host: how does the host resistance mechanism interfere with the virus replicative cycle? Several examples will be considered below, in terms of genetical and biochemical control.

A frequent occurrence when genetically resistant cultivars are introduced is the emergence of isolates of the virus which are able to overcome the particular resistance gene. These are designated as *virulent*, as opposed to the *avirulent* original isolates which are inhibited by the resistance gene. The genetics and possible biochemical mechanisms of virulence must therefore be discussed as a counterpart to resistance.

For induced resistance mechanisms, the means by which viral pathogenesis is inhibited — or apparently inhibited — are again of interest, and will be discussed. Do viruses evolve to overcome induced resistance mechanisms as they do with cultivar resistance? The evidence is not yet clear, partly because many types of induced resistance mechanism have so far been studied only in the laboratory. The selection pressure for evolution of virulence against induced resistance which would be offered by the continuous presence of that mechanism in nature or in agriculture has not been available, but the question of evolution of virulence against antiviral genetic elements in transgenic plants is highly relevant.

## II. POSSIBLE MODELS FOR RESISTANCE MECHANISMS

Before examining what is known of resistance mechanisms, it is useful to pose various questions about what *types* of mechanism might be involved and to consider some possible models. These can be used to make predictions which can then be tested against experiment and observations.

### A. Possible Targets of Resistance Mechanisms in the Viral Replicative Cycle

In the broadest sense, the replicative cycle of a virus includes not only the biochemical processes leading to the production of progeny particles within the infected host cell, but also processes leading to the dissemination of the virus to other plants, and affecting its persistence in time. Resistance mechanisms can operate at these broader levels as well as within the individual plant (Figure 1). The likelihood of matching virulence in the virus varies with the level at which the resistance operates. We can distinguish five main types of resistance target.

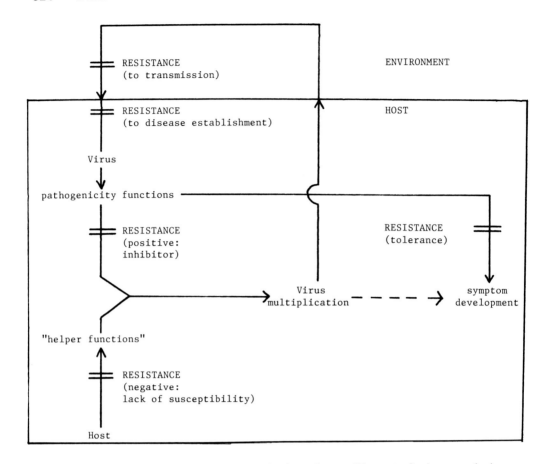

FIGURE 1.    The replicative cycle of a plant virus, showing various possible targets of resistance mechanisms.

## 1. Resistance to Transmission

For viruses which are transmitted mechanically, and thus in nature or in agriculture depend on chance meeting with a new host to perpctuate the infection cycle, there is little or no scope for this type of resistance: factors affecting virus spread and persistence will be only indirectly involved. In contrast, for viruses which are transmitted by some specific mechanism, such as by an insect vector or through seed of the plant, there are possibilities for resistance mechanisms which influence virus transmission more directly.

One comparatively well-understood mechanism concerns resistance to aphid transmission of beet yellows virus (BYV) and potato virus Y (PVY) in the wild potato *Solanum berthaultii*. This has special epidermal hairs not found on cultivated potatoes, which release the aphid alarm pheromone (E)-β-farnesene and prevent aphids from settling.[10-12] In wheat, plant resistance to the wheat curl mite reduces transmission of wheat streak mosaic virus (WSMV).[13] Useful virus resistance may also be observed when the host plants are not actively resistant to the vector, but are disfavored in comparison with other hosts of the insect, for reasons such as color or texture of the foliage.[14,15]

In contrast to these successful cases, aphid resistance in cowpeas did not provide resistance to infection with cowpea aphid-borne mosaic virus (CAMV).[16] The reason was that although aphids were less numerous on resistant than on susceptible cowpeas, they did make more numerous, although shorter, feeding probes on the resistant hosts, and thus were likely to have been highly effective in transmission of the virus. Acquisition of the virus from aphid-resistant plants was, however, reduced.[17]

A single recessive gene in barley controls seed transmission of barley stripe mosaic virus (BSMV),[18] and there are other examples of varietal differences in seed transmissibility.[19,20] It is not always clear whether these mechanisms involve specific interference with the ability of the virus to invade the seed, or whether they are further expressions of mechanisms generally reducing virus spread and multiplication in the plant as a whole.

### 2. Resistance to Disease Establishment

This includes true nonhost immunity, where attempts can be made to deliver the virus to the plant, but no evidence of infection is obtained. Possible mechanisms are considered in Section III. Then there are some examples of resistance in normally susceptible species, where the mechanism appears to operate against the earliest stages of infection. These mechanisms may be correlated with physical factors such as a thicker epidermis or cuticle, preventing mechanical inoculation, or with a reduced number of ectodesmata in the epidermis.[21,22] Such mechanisms tend to be nonspecific, operating against several viruses affecting the host species.

### 3. Resistance to Virus Spread within the Plant

Numerous cases of cultivar resistance operate by inhibiting the spread of the virus from the point of infection: the virus can be localized in a single infected cell,[23] in a small group of infected cells,[24] in a local lesion which forms after an early period of virus multiplication and cell-to-cell spread,[25-28] or in the inoculated leaf.[29] Localization in single cells is difficult to detect. Some species were traditionally regarded as nonhost for particular viruses until careful examination revealed single- or few-cell infection. This has been referred to as "subliminal" infection.[23,30] Recently, an example of cultivar resistance has also been shown to involve single-cell localization.[31] Localization in lesions is by far the commonest known mechanism. It is normally associated with necrosis, but chlorotic and cryptic lesions are also known.[27]

### 4. Nonlocalized Resistance to Virus Multiplication

In this target area, the cell-to-cell and systemic movement of the virus is not inhibited, but multiplication is inhibited. Infected plants of the resistant cultivar contain smaller amounts of virus than susceptible ones.[32]

### 5. Resistance to Symptom Formation

Sometimes resistant cultivars are classified as such because they have less severe symptoms, but nevertheless can be shown to contain virus.[33,34] This effect, also known as *tolerance*, might be a secondary consequence of a nonlocalized resistance to virus multiplication, or might represent an independent inhibition of the pathways leading to visible symptom formation. In either case, measurement of virus concentrations in resistant and susceptible plants is required to clarify the situation.[35-37]

## B. Is the Resistance Mechanism Positive or Negative?

Figure 1 shows some possible configurations of resistance mechanisms. Within the plant, the viral genome is assumed to specify certain pathogenicity functions which are required for replication and which are involved in formation of disease symptoms. The host supplies "helper" functions which are required by the virus. These may be at a general level — energy and precursor supplies, and a protein synthesizing system, for example — or at a more specific level, such as a host-specified component of the replicase.[38] Resistance might operate either as a *positive* mechanism, in which the host produces an inhibitor of some stage of pathogenesis, or as a *negative* mechanism, in which the host fails to produce a normally required helper function, or produces one which is unsuitable for virus multiplication.

When the virus throws up a virulent isolate overcoming resistance, it achieves different things in the two models. Virulence against the positive model is a failure to interact with the host-specifed inhibitor. Virulence in the negative model is an ability to multiply without the host-specified helper function. This may be a more difficult evolutionary step for the virus than avoidance of an inhibitor. Negative resistance mechanisms would therefore be predicted to have higher durability than positive mechanisms.

### C. Recognition Events: All-or-Nothing and Quantitative Interactions

In resistance mechanisms where the virus comes into direct contact with the plant, we can distinguish a spectrum of types of interaction between host- and virus-encoded molecules. One concept is that there is a *recognition event* between the two components which determines the outcome of the interaction. For positive models, the recognition leads to resistance. For negative models, where resistance is lack of a host helper function, recognition leads to susceptibility, and failure of recognition leads to resistance.

In genetic terms, the prediction is that resistance alleles should be dominant in the positive model, but recessive in the negative model. These conclusions imply that the outcome of the recognition event is an all-or-nothing response. However, the amount of resistance may also vary quantitatively, depending on the nature and concentrations of the host- and virus-specified reactants. In this case, resistance is predicted to be incompletely dominant, i.e., gene dosage dependent.

### D. Is the Resistance Mechanism Constitutive or Induced?

The components of the resistance mechanism might be present in the plant and operative from the time of infection, or they might be induced after infection. The former model could involve positive mechanisms constitutively expressed, but also include negative mechanisms which by definition are constitutive. In the latter model, the recognition event between host- and virus-specified molecules would be the first reaction in a chain leading to synthesis of the antiviral moiety. In this case, the prediction is that early virus multiplication in resistant plants will proceed as in susceptible plants, until such time as resistance becomes activated.

### E. Are there Horizontal and Vertical Resistances to Plant Viruses?

These concepts were introduced by Vanderplanck[39] and expounded by Robinson,[40] mainly in the context of fungal pathogens. Briefly, and somewhat oversimplified, vertical resistance is controlled by single or few major genes, and isolates of the pathogen develop which can overcome specific resistance genes. Horizontal resistance tends to slow disease development, is thought to be polygenic, and is effective against all isolates of the pathogen.

It should be clear from this chapter that vertical resistance to viruses is well proven in numerous species, in that differential reactions[41] between pathogen and host lines do occur. Robinson[40] was of the opinion that viruses had such a capacity to mutate to overcome major resistance genes that vertical resistance, if it existed, would be quickly broken down. This is clearly not always the case.

The evidence for horizontal resistance to viruses is much less clear: polygenic resistance present in wild species may well have been lost in breeding for other attributes, but the problem has not been directly investigated, and would not be a simple task. Some indirect evidence for horizontal resistance is considered elsewhere.[42]

### III. NONHOST IMMUNITY

There are three possible models. In the positive model, the nonhost plant contains an inhibitor or inhibitors which are 100% effective against all isolates of the virus in question. Holmes[43] suggested that this might be a result of additive effects of a large number of resistance genes — say 20 to 40 — against the virus, but there does not appear to be any direct evidence for this. In the negative model, the nonhost lacks some helper function which is essential to the virus and which is found in host species. This is the basis of the hypothesis advanced by Bald and Tinsley[44-46] to explain host range and nonhost immunity. Finally, nonhost immunity might be due to more trivial reasons such as physical barriers to infection, unsuitable delivery systems, or unfavorable cell pH or ionic conditions.

There is no great amount of direct experimental evidence for any of these theories. For the positive model, the antiviral inhibitor, or consortium of inhibitors, have not been characterized. Studies of the negative model have concentrated on areas of metabolism where host helper components might be involved. Some work has been done to investigate possible virus-specific receptors; for example, by mixed reconstitution of virus particles containing the RNA of a hosted type and the coat protein of a nonhosted type, and vice versa.[47-49] The interpretation of such experiments is complex, but generally they do not provide strong support for a role of virus-specific receptors, or rather their absence, in nonhost immunity. The difficulties of research on virus-specific receptors and early interactions are perhaps best expressed by citing a recent report that tobacco mosaic virus (TMV) is uncoated and the RNA expressed when injected into *Xenopus laevis* (toad) oocytes![50] Another possible area for operation of the negative model is replicase activity. There is now some evidence that replicase can involve host- and virus-coded subunits,[38] but much more information is needed on how this might control nonhost immunity, if at all.

Some of the best experimental evidence tends to favor the third model, in that some viruses can infect and multiply well in protoplasts prepared from nonhost species.[51-53] This suggests that one barrier to infection at the whole-plant level might be at the cell wall or against the delivery mechanism.

## IV. CULTIVAR RESISTANCE

### A. Genetic Control

*1. Resistance*

Numerous crop species have been bred for resistance to numerous viruses; consequently, the genetics of resistance is comparatively well understood. Generally, resistance is simple in genetic terms, being controlled by single or only a few genes.[6] However, quite complex systems of resistance to a particular virus can be built up by combining genes at different loci, if these are available.[54] Little is known of the genetics of resistance in wild species: it is possible that quite different systems of genetic control may have evolved under natural as opposed to artificial selection.

Table 1 summarizes some of the features of genetic controls from a survey of 63 combinations of crop species and their viruses.[6,7,55] The small number of cases of possibly oligogenic resistance refers to reports implying an *obligate* cooperation between genes at different loci, as opposed to simple additive effects of independently acting genes. The complex recessive system of resistance to bean common mosaic virus (BCMV) in *Phaseolus vulgaris* is one where obligate cooperation is well established.[29,54] Resistance depends on the alleles at at least two loci being homozygous recessive. The *bc-u* locus is required in all cases, together with at least one from the *bc-1, bc-2,* or *bc-3* loci. The latter three loci are virus strain specific; the first is not. Alone, *bc-u* has no effect on virus multiplication; the other loci alone may have some effect, but less than when in combination with *bc-u*.[29]

Some cases interpreted as showing oligogenic resistance involving epistatic or hypostatic interactions between resistance genes[56,57] were later shown to be under simple monogenic control when the observations were repeated under controlled environmental conditions.[58,59] The apparent genetic complexity of the earlier work was actually a product of genotype × environment interaction.

Table 1 also contains examples of monogenic resistance, where the expression of the resistance phenotype is indirectly modified by the host genetic background. For example, the *Yd₂* gene for barley yellow dwarf virus (BYDV) resistance is highly effective in Ethiopian

**Table 1**
## GENETICS OF RESISTANCE TO VIRUSES IN CROP SPECIES AND SOME FEATURES OF RESISTANCE GENE ACTION AND VIRULENCE

| Genetic basis | Number of host-virus combinations |
|---|---|
| Single dominant gene | 29 |
| Incompletely dominant (gene dosage dependent) | 10 |
| Apparently recessive | 11 |
| Subtotal: monogenic | 50 |
| Possibly oligogenic | 5 |
| Monogenic, with possible modifier genes or effects of host genetic background | 8 |
| Subtotal: oligogenic (?) | 13 |
| Total number of host-virus combinations in sample | 63 |

| Localization | Immune | Yes | Partial | No | Not known | Total |
|---|---|---|---|---|---|---|
| Dominant alleles | 0 | 19 | 0 | 2 | 8 | 29 |
| Incompletely dominant | 0 | 0 | 4 | 8 | 0 | 12 |
| Apparently recessive | 5 | 1 (?) | 1 | 2 | 4 | 13 |

Immune = no virus detectable; Yes = normally involving lesion formation; No = resistance permitting some systemic spread; Not known = not tested, or not reported in the literature.

| Temperature response | ts | tr | Not known | Total |
|---|---|---|---|---|
| Dominant alleles | 7 | 2 | 20 | 29 |
| Incompletely dominant | 1 | 1 (?) | 10 | 12 |
| Apparently recessive | 2 | 2 | 9 | 13 |

ts = Temperature sensitive; tr = temperature resistant.

| Virulent isolates reported | Yes | No | Not known | Total |
|---|---|---|---|---|
| Dominant alleles | 16 | 1 | 12 | 29 |
| Incompletely dominant | 8 | 3 | 1 | 12 |
| Apparently recessive | 4 | 1 | 8 | 13 |

From Fraser, R. S. S., *Biochemistry of Virus-Infected Plants*, Research Studies Press/John Wiley & Sons, Chichester, U.K., 1987. With permission.

barley, but lost its effects when transferred to European lines. This was shown to be associated with slow plant growth rate; effectiveness was recovered when the gene was transferred back to rapidly growing lines.[60-62]

Turning to the resistances controlled monogenically, it is clear from Table 1 that most genes in the sample are dominant, with smaller numbers of gene dosage-dependent and recessive cases. In fact, the number of gene dosage-dependent types is probably an underestimate. Much evaluation of allelic relationships in the literature has been done on the basis of visible symptoms, which tend to give illusory clear-cut indications of dominance or recessiveness. When the effects of resistance genes on virus multiplication are measured, however, gene dosage-dependence is commonly found. Two cases can be cited as examples. In tomato, the *Tm-1* gene for TMV resistance completely prevents formation of mosaic symptoms in the heterozygote and thus appears dominant. However, virus multiplication is more strongly inhibited in *Tm-1/Tm-1* plants than in *Tm-1/ +* plants.[32] Resistance to BCMV in *P. vulgaris* controlled by the complex *bc* system referred to above appears to be recessive at all loci.[54] However, measurement of virus multiplication in plants heterozygous and

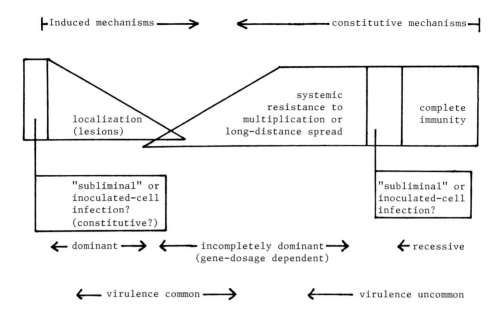

FIGURE 2. Some general conclusions about mechanisms of resistance to viruses and the nature of genetic controls.

homozygous for the *bc* genes clearly shows a gene dosage dependence of the inhibition of virus multiplication.[29] In terms of the models discussed above, it appears that many interactions between resistance genes and viruses may be influenced by quantitative interactions.

Table 1 shows that dominant genes are strongly associated with resistance mechanisms which operate by virus localization: this is consistent with a recognition event which leads to induction of an all-or-nothing response, such as lesion formation. The gene dosage-dependent mechanisms tend to allow spread of virus through the plant, but to inhibit multiplication or symptom severity. Recessive resistance is generally not associated with localization, but in a number of cases causes complete immunity. These would be consistent with the negative model of resistance; if the host helper function which is missing in resistant plants affects an early stage of the viral replicative cycle, then no sign of pathogenesis should be detected, and the resistance should be fully recessive. The single case of recessive resistance associated with localization in Table 1 is anomalous, as the alternative reaction was complete immunity rather than susceptibility.[63]

In a small percentage of reports, the temperature sensitivity of resistance mechanisms has been investigated. Such data as are available suggest that resistance controlled by dominant genes tends to break down at high temperature. It is possible that the localization mechanisms involved share some common feature in different plants which tends to confer temperature sensitivity. Recessive resistance operating by the "negative" model should not show temperature sensitivity, as there should be no gene product to undergo thermal denaturation. The two temperature-sensitive cases of recessive resistance shown in Table 1 are not in disagreement with this contention. One involved apparent selection of variants of the virus adapted to high temperature, while the other concerned the anomalous case discussed above.

Figure 2 summarizes some of the properties of resistance mechanisms in relation to the genetic controls and to some of the predictions made in Section II. Some general conclusions can be drawn, although much of the evidence is from a limited number of cases. Dominant resistance tends to involve localization, probably a positive inhibitor model, and an induced mechanism. Gene dosage-dependent mechanisms tend to allow spread through the plant and may be constitutive. Finally, there is evidence for a small number of cases of fully recessive mechanisms which may operate on the negative model and which by definition are constitutive. From the limited evidence available, it appears that dominant, localizing mechanisms do not operate when isolated protoplasts are inoculated,[26] whereas systemically effective mechanisms are effective in protoplasts.[64]

## 2. Virulence

Table 1 shows that overall, virulence is comparatively common; most genes in the survey have been overcome. The large number of cases where virulent isolates are classified as "not known" concern resistance genes which, from the literature, do not appear to have been tested against a wide range of virus isolates. Those classified as "no virulence reported" have been found to be durable during long exposure to a selection of isolates. Clearly, extreme durability is rare. This is well illustrated by the *N* gene for TMV resistance in *Nicotiana*. First reported almost 50 years ago,[5] it appeared effective against all isolates of this virus under normal conditions, until overcome by a pepper isolate in 1984.[65]

It has been difficult or impossible to study the genetics of virulence in viruses by classical genetical means, because crossing and segregation studies are impossible. Furthermore, the limited size of the viral genome — enough to code for a few to a dozen proteins — implies that virulence cannot be the sole property determined by a particular gene. Rather, virulence is likely to be a pleiotropic effect of a viral gene which has some other function in pathogenicity and which must retain that function along with virulence.

Some mapping of virulence to particular regions of the virus genome was possible for those viruses with multicomponent genomes. The genome components can be separated, and *pseudorecombinants* prepared using components from virulent and avirulent strains. A number of examples, reviewed in detail elsewhere,[6] suggest that virulence maps in different genomic segments in different viruses. However, until the functions specified by the different segments are known and until they can be compared between virus groups, the interpretation of these data is limited.

Recently, recombinant DNA methods have allowed much more detailed mapping of virulence and are allowing a quantum leap in our understanding of the phenomenon. cDNA clones are prepared from virulent and avirulent isolates, cut with a suitable restriction enzyme, and religated to form an artificial recombinant. Infectious RNA can be derived by use of an expression vector. By use of a range of restriction enzymes cutting at different sites, a large number of recombinants can be produced and screened for biological activity. Fine-scale genetic mapping of the virulence determinant can then be completed by sequencing the small regions of the genome identified as controlling virulence. It is advantageous to compare spontaneous or artificial mutants with the parent strain, rather than comparing naturally occurring isolates, to minimize the background variation in sequence which is not directly related to the determination of virulence. Figure 3 summarizes the location of some virulence determinants which have been identified in TMV. It is clear that for different host resistance genes, virulence maps in different viral functions, including all of the major TMV-specified proteins. As the different host resistance genes probably involve different mechanisms (*N'* causes local lesions,[70,71] *Tm-2* causes localization in the infected cell,[31] and *Tm-1* permits spread throughout the plant[32]) it is perhaps not surprising that different viral functions may be altered in the acquisition of virulence. It is particularly interesting that *Tm-2* resistance, which probably acts against movement of the virus from cell to cell, is overcome by a change in the 30K protein which controls cell-to-cell spread.[72] However, it should also be noted that any of these types of virulence could operate at the RNA level rather than involve altered protein functions.

## 3. Gene-for-Gene Interactions

The frequency of evolution of virulent isolates of viruses has caused formation of gene-for-gene relationships between host resistance and viral virulence. These interacting genetic

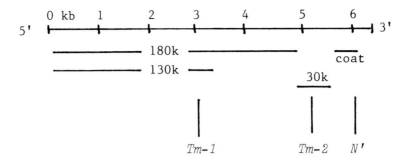

FIGURE 3.   Genetic map of TMV, showing the location of base changes determining virulence/avirulence against three host resistance genes: $N'$ in tobacco[66,67] and $Tm-1$ and $Tm-2$ in tomato.[68,69]

systems are by no means as complex as those evolved by plants and their fungal pathogens,[73,74] where some hosts can have a large number of genes or alleles for disease resistance, and the pathogen has developed a correspondingly large number of genes for virulence. Few plants have yet been found to have more than about two or three genes for resistance to a particular virus. On the other hand, the situation is more complex than represented by Robinson,[40] who argues that the ability of viruses to evolve virulence is so great that gene-for-gene interactions would break down completely. Virulence is common as Table 1 shows, but some resistance genes have shown good durability, and virulent isolates of a virus have not necessarily been present in all geographical areas where a resistance gene is deployed; these factors have helped the persistence of reasonably stable gene-for-gene interactions.

Three main examples may be cited to illustrate some of the features of gene-for-gene interactions. In tomato, resistance to TMV is controlled by the $Tm-1$, $Tm-2$, and $Tm-2^2$ genes;[75] the latter two are allelic. $Tm-1$ was overcome within about 1 year of introduction in commercial horticulture.[76] $Tm-2$ proved rather more robust, but has been overcome enough by virulent isolates to render it of little use in practical crop protection.[77] $Tm-2^2$ has been the mainstay of TMV-resistant varieties in the U.K. for over a decade. Rare isolates of TMV are known which overcome this gene, but they appear to be defective in infectivity and have not become widespread in tomato crops.[78] TMV isolates containing virulence genes against $Tm-1$ and either $Tm-2$ or $Tm-2^2$ have been reported;[76,77] recently we have found isolates capable of overcoming both $Tm-2$ and $Tm-2^2$.[78a] A gene-for-gene "checkerboard" for the tomato-TMV interaction is presented elsewhere.[42]

Resistance to BCMV in *P. vulgaris* is one of the most complex gene-for-gene systems to have been elucidated.[29,54] A partial "checkerboard" is presented by Fraser[6] for one of the possible pairs of recessive resistance genes. It is calculated that with the known genes in the recessive *bc* system, there should be 32 host resistance phenotypes (excluding the heterozygous forms with no or limited effectiveness) and 32 corresponding virulent forms of the virus. Not all of these have been found. Thus, no isolate of virus has yet overcome the *bc-3* resistance, and all isolates overcoming *bc-1²* also overcome the allelic *bc-1*. Therefore, isolates overcoming *bc-1²* alone must either be extremely rare or virulence against *bc-1²* may involve a biochemical property which also confers the ability to overcome *bc-1*. The most complex resistance-breaking strains of BCMV yet found contain virulence against three resistance genes or alleles.

Table 2 shows a form of a partial "checkerboard" for resistance to TMV in *Capsicum annuum*.[79-82] The data for virus multiplication confirm the gene-for-gene interaction inferred from the examination of visible symptoms. Pepper strains of TMV are thought to have developed from infections on other crops, and probably mostly from tomatoes. In developing virulence against the pepper resistance genes and adapting to the different host, they seem to have lost the ability to multiply successfully in the original host,[80] and changes in particle

**Table 2**
**A GENE-FOR-GENE "CHECKERBOARD" FOR STRAINS OF TMV-SUSCEPTIBLE AND RESISTANT PLANTS OF *CAPSICUM ANNUUM***

| Virus isolate | FO | Ob | SL |
|---|---|---|---|
| Virulence genotype | 0 | 1 | 1,3 |
| Host resistance genotype | | | |
| $L^+/L^+$ | Mosaic, a little necrosis (S) 3.12 | Mosaic, some necrosis (S) 0.89 | Mosaic (S) 1.94 |
| $L^1/L^1$ | — (R) 0.25 | Mosaic, some necrosis (S) and abscission 0.56 | Mosaic (S) 1.40 |
| $L^3/L^3$ | — (R) 0.09 | — (R) 0.11 | Mosaic (S) 1.13 |

*Note:* — Indicates no symptoms, (S) indicates a susceptible reaction, (R) resistance. Numbers indicate virus concentration (mg/g fresh weight). All data are for systemically infected leaves.

length have also been reported.[83] The "Samsun latent" strain in particular[84] multiplies very poorly in 'Samsun' tobacco,[82] although this variety contains no known genes for TMV resistance.

## B. Mechanisms of Resistance and Virulence: Case Histories

The discussion above of models and genetic systems has already indicated some of the mechanisms which are involved in resistance to viruses and has hinted at the considerable diversity encountered in different hosts and viruses. My objective in this section is to consider a small number of individual cases in more detail. These are chosen to represent the best-understood cases and to illustrate the diversity further.

### 1. Mechanisms Involving Local Lesion Formation

Infection is thought to become established in single cells after mechanical inoculation, and may be initiated by a single virus particle.[85] Thereafter, the virus appears to multiply and spread from cell to cell for a time, then further spread is prevented in some way. The early rate of multiplication can be at the same rate as in a susceptible, nonlocalizing host.[25,28]

Most examples of localization involve necrosis, although there is little evidence that the necrosis itself is the primary deterrent to virus spread. This is because the full expression of necrosis only occurs rather late after infection, and virus particles can be seen by electron microscopy or detected by immunological methods outside the necrotic area.[86,87] Furthermore, some localization mechanisms do not cause necrosis, but give chlorotic lesions, or cryptic lesions, which are only detectable by staining the leaf for starch.[88,89] These exceptions are rare, however. An interesting question which remains is why localization mechanisms are so often accompanied by necrosis. It may be that the mechanism which inhibits virus spread involves changes in cell membranes which inevitably trigger necrosis as a secondary event. Alternatively, necrosis may truly be a second-line defense mechanism which has evolved in response to a resistance-related selection pressure. The quinones and other compounds which form during necrogenesis can inhibit viral infectivity and may prevent secondary infections.[90,91]

Numerous workers have investigated changes in many aspects of metabolism during lesion formation and virus localization. Sadly, none of these reports really gives firm information about the mechanism which directly inhibits virus spread or multiplication. The nature of the presumed recognition event between virus- and host-coded molecules which may initiate the whole cascade of ensuing metabolic changes is unknown. Finally, the product of the resistance gene and its mode of action are not yet known in any case. However, the metabolic studies which have been carried out do give some clues about possible mechanisms of resistance and much more information about secondary consequences.

The earliest changes detectable after inoculation occur within a few to 24 hours,[91-95] and involve increased electrolyte leakage, implying membrane damage or altered permeability. These changes may be causally related to the increase in ethylene biosynthesis, which is detectable shortly before lesion appearance.[96,97] The increase in ethylene is itself thought to be the trigger for synthesis of the "pathogenesis-related" (PR) proteins,[98] which accumulate some time after lesions first appear.[99]

By about 24 to 48 h after inoculation, the first signs of inhibition of virus multiplication become apparent.[100] The early stages of induction of resistance and necrosis require cell-to-cell contact, but this requirement is lost at about the same time.[101] Also just before lesion appearance, there is evidence for increased activity of the phenylpropanoid pathway.[102-104] This is undoubtedly involved in supply of precursors for the synthesis of lignin and other cell wall changes which occur in the later period of lesion growth. Whether these thickened walls have any significance in virus resistance is open to question.[105-107] They may represent a secondary defense to possible fungal or bacterial infection. Compounds synthesized from intermediates in the phenylpropanoid pathway may also have direct antimicrobial activity. This, together with the suggested antimicrobial activities of the β-1-3 glucanases and chitinases among the induced PR proteins, emphasizes that later metabolic events accompanying lesion formation involve the activation of secondary defenses against chance invasion by other pathogens.[104,108]

Many other metabolic changes have been reported at about the time of lesion appearance and during early lesion development. They include increases in mitochondria,[109] ribosomes,[86] and the phytohormones abscisic acid[110] and indoleacetic acid.[111] It is doubtful whether any of these are directly involved in resistance.

In the later stages of lesion development, there is evidence of major wall changes, especially thickening and deposition of callose, and ultrastructural evidence of cell collapse and disruption or degradation of the organelles.[112] Lesion growth generally continues for 5 to 15 d, then stops.[7,27]

In contrast to experiments which record metabolic changes during localization and lesion formation, other workers have tried to demonstrate the synthesis of compounds with demonstrable antiviral activity. Sela and his co-workers have investigated an "antiviral factor" (AVF) from TMV-infected *N. glutinosa* which appears to show certain parallels with interferon. AVF is a phosphorylated glycoprotein.[113,114] Antiviral activity has generally been assayed by mixing with TMV and showing that AVF will inhibit lesion formation on hypersensitive tobacco, or by treating leaf disks. Sela[115] and Sela et al.[116] have proposed complex schemes for the metabolism and mode of action of AVF, including a role in the induction of some antiviral state in uninfected tissue. Most of the proposed schemes of action have still to be supported by experimental evidence and elucidation of the steps involved.

The parallels between AVF and interferon were further strengthened when it was reported that human interferon was a potent inhibitor of TMV multiplication in tobacco[117,118] and indeed was more effective than tobacco AVF. However, several other groups have now reported that mammalian interferons have no effect on TMV and other plant viruses.[119-121] There have also been other reports suggesting inhibition of the multiplication of plant viruses.[122,123]

Loebenstein and Gera[124,125] have isolated a different antiviral agent, which they call inhibitor of viral replication (IVR), from protoplasts of TMV-infected 'Samsun NN' tobacco plants. IVR consists of a pair of proteins of molecular weights 26 and 57 kDa. It inhibits virus multiplication in protoplasts, leaf disks, or intact plants, and appears to be present in the intercellular spaces of infected leaves.[126] IVR was also effective against other viruses and in other hosts, but its effects on host cell metabolism do not seem to have been reported. The lack of specificity for TMV raises questions about how it may be related to the operation of the resistance mechanism specified by the *N* gene in tobacco.

The *N* gene of tobacco is undoubtedly the most studied of the genes controlling hypersensitivity towards a plant virus (TMV), and tobacco has been among the earliest plants to be exposed to genetic manipulation by recombinant DNA techniques. It is not surprising, then, that attempts to clone and study a plant gene for virus resistance have focused on this system. Zaitlin and co-workers prepared genomic and cDNA libraries from *N* gene tobacco and have screened these with probes derived from the polyA mRNAs of TMV-infected *N* gene and susceptible tobacco varieties. The hope was that clones would be found which would react with *N* gene probes, but not with susceptible plant probes, and that these differentially expressed mRNAs would be related to hypersensitivity.[127,128] Unfortunately, although over 110,000 clones were screened, there was little indication of specificity for hypersensitivity.[128] Another problem in this type of approach is that genes involved in secondary events are also induced during necrotization, and these provide a background of noise which has to be screened and eliminated before the *N* gene product can be found. The induction of PR proteins follows this model.[129]

Although progress to isolating genes controlling localization is slow, there is now more information on the molecular biology of the virus side of the interaction with the supposed gene product. These studies of the virus may eventually provide strong clues about the nature of the resistance gene product and assist its isolation.

The location of the virulence determinant for local lesion formation on *N'* gene tobacco plants in the coat protein gene of TMV has already been mentioned.[66,67] Knorr and Dawson[66] noted that the response might be determined by the TMV RNA or by the capsid protein: certainly the interaction of different strains of TMV with the resistance mechanism specified by the *N'* gene appears to be related to capsid protein properties.[71,130] Knorr and Dawson[66] speculated that the capsid protein of lesion-forming strains might elicit the hypersensitive response; alternatively, the protein of systemically spreading strains might suppress a hypersensitive reaction which is normally elicited by virus multiplication. Although six independent local lesion-forming mutants all had the same single alteration in base sequence, causing a change from serine to phenylalanine at position 148 on the capsid protein, another mutant causing lesions did not have this change. Partial sequencing of this mutant suggested that it might be possible to induce necrosis by mutations occurring outside the coat protein gene. Furthermore, mutants with deletions in the coat protein gene were shown to be able to induce necrosis in *n* gene tobacco plants, which do not normally react hypersensitively to TMV.[131] Thus, there may be many pathways for the induction of necrosis, but it seems clear that the coat protein gene is crucial in determining the virulence/avirulence reaction with the *N'* gene.

In *N* gene plants, TMV mutants lacking the complete coat protein gene all caused lesions,[131] and so did a mutant which does not synthesize coat protein.[132] Thus, there is a clear contrast between *N* and *N'* in that the coat protein seems to be critical in determining the outcome of interaction with the latter, but not with the former — even though the genes are thought to be alleic.[133]

## 2. Resistance to Cowpea Mosaic Virus in Cowpea

In a survey of over 1000 lines of cowpea, Beier and co-workers[134] found 65 which were "immune" to cowpea mosaic virus (CPMV). However, protoplasts from almost all these lines could replicate the virus effectively. One immune line, 'Arlington', gave protoplasts which would only support very low levels of CPMV multiplication.[135,136] The two CPMV RNAs are translated to form polyproteins, which are then cleaved by a specific virus-coded protease to form a number of end products such as capsid protein and the putative replicase.[137,138] Extracts of 'Arlington' cowpeas were shown to inhibit the in vitro translation of CPMV RNA, but not that of the related cowpea severe mosaic virus (CPSMV). CPSMV also overcomes the resistance of 'Arlington' cowpea.[139] Furthermore, extracts of 'Arlington' leaves were shown to contain a protease inhibitor which specifically inhibits processing of the CPMV polyproteins, but not that of the CPSMV polyproteins. This provides a convincing explanation of the mechanism of resistance to CPMV and of the apparent virulence of CPSMV.

### 3. Resistance to TMV in Tomato

The genes at the two loci conferring resistance to TMV in tomato clearly operate in quite different ways. *Tm-2* and *Tm-2²* both restrict the virus to single cells in the epidermis; presumably the cells which have been mechanically inoculated.[31] Virus multiplication within the inoculated cells of resistant plants does not appear to be inhibited, and protoplasts from resistant plants will support virus multiplication.[64,140] This strongly suggests that the resistance operates by preventing cell to cell spread of the virus. This conclusion is supported by the finding that a strain of TMV which overcomes the *Tm-2* resistance has a mutation in the gene coding for the 30-kDa protein:[68] this protein is involved in cell-to-cell spread of TMV.[72,141]

*Tm-1* resistance, in contrast, is effective in protoplasts,[140] but does not prevent virus spread throughout the plant.[32] Watanabe and co-workers[142] showed that the synthesis of all known viral-coded proteins and RNAs was inhibited in protoplasts containing the *Tm-1* gene, but that the resistance could be partly overcome in heterozygous plants if TMV RNA at high concentration was used as inoculum. The former observation may suggest that an early step in the viral replication cycle is the target of the resistance mechanism. Meshi et al.[69] analyzed virulent and avirulent strains of TMV and found that those which overcame *Tm-1* had two base changes in the 130- and 180-kDa proteins, corresponding to alterations in the amino acids at positions 984 and 979. These two proteins, one of which is the read-through product of the other, are thought to contribute to the putative replicase activities. The significance of these changes was confirmed by construction of artificial recombinants between avirulent and virulent types using cDNA techniques. The recombinants showed that both positions could confer virulence, but that the one at amino acid position 979 was more effective than that at position 984. Interestingly, the effect of all mutations to virulence examined was to decrease the local net charge of the proteins, suggesting that there is an electrostatic interaction between the replicase proteins and the supposed resistance factor derived from the *Tm-1* gene. Meshi et al.[69] also raised the possibility that the host resistance factor might be an altered replicase component which interacts variably with the 130- and 180-kDa proteins, depending on their local net charge. They argued against an interaction of the *Tm-1* gene product with the TMV RNA as such, because variation in base sequence between strains is rather common in this region, even though biological behavior vis-à-vis *Tm-1* falls into only two classes.

In earlier work on the temperature response of *Tm-1* activity, we showed that the suppression of symptom formation by avirulent strains was effective at all growth temperatures used, while the inhibition of virus multiplication was strongly temperature dependent and declined over the range from 20 to 33°C.[143] It is conceivable that an electrostatic interaction between TMV-specified proteins and a supposed resistance gene product which inhibits replicase activity could show this type of temperature dependence. However, the symptom results are more difficult to explain, even though the link between virus multiplication and symptom severity can be fairly tenuous.[144] A recent finding is that *Tm-1* plants contain somewhat higher concentrations of abscisic acid than susceptible plants, and that exogenous abscisic acid does suppress symptom formation, although rather high concentrations are required.[145] This might suggest that the two effects of the *Tm-1* gene are brought about by at least partly separate mechanisms.

## V. INDUCED RESISTANCE

This class of resistances is very diverse: it includes mechanisms which are highly effective and those whose effectiveness is questionable. Some mechanisms are highly specific for an individual virus, or particular isolates of that virus, while other mechanisms are effective against several types of pathogen and even insect pests. Some mechanisms have been utilized in practical crop protection, while others have only been studied as laboratory phenomena.

### A. Acquired Systemic and Local Resistance

In plants responding to virus infection by formation of necrotic local lesions, local and sometimes systemic "resistance" to a second inoculation is induced, in that the lesions formed after the challenge inoculation are smaller and/or less numerous than those formed on previously untreated control plants.[146,147] The resistance is induced by numerous biotic and abiotic agents and is highly unspecific.[148,149] There is considerable doubt as to whether virus multiplication is actually reduced by this type of resistance, or whether it is merely an expression of a reduced number of infections, and smaller amount of necrosis per lesion.[150]

Most research on acquired resistance has centered on the association with the pathogenesis-related (PR) proteins, and it has often been suggested that they are somehow involved in virus resistance,[53,98,99] although others have argued against such a link.[151,152] No direct antiviral effect of the proteins has yet been demonstrated,[153] and recent discoveries, showing that certain PRs have enzyme or inhibitor activities which may be effective against insect pests or microbial pathogens, also argue against this.[104] Some PR proteins have as yet no known function and could conceivably still turn out to be antiviral. However, in the light of the other evidence, it would not be relevant to give an extended treatment of the properties of PR proteins in this review. Fuller discussions of the evidence for and against antiviral functions of PRs are given elsewhere.[99,154]

### B. Green Islands

Leaves infected systemically with a plant virus often show a distinctive light green/dark green mosaic, in which the virus is located in the light green areas. The dark green areas have a higher chlorophyll concentration than healthy leaves and contain little or no virus. These areas can be difficult to challenge inoculate with the same virus. This resistance tends to be specific to the inducing virus.[155]

Whenham et al.[156] showed that dark green areas contained higher concentrations of abscisic acid than healthy or light green, infected tissue, for TMV-infected tobacco leaf. Exogenous abscisic acid was shown to increase chlorophyll concentration of healthy tissue and to make it partly resistant to challenge inoculation.[151] Thus, the TMV-stimulated increase in ABA concentration might be involved in this resistance mechanism.

More recent work has also shown changes in cytokinins in dark green tissues. In particular, infection reduced the concentration of zeatin, and increased the concentrations of zeatin-*O*-glucoside and zeatin riboside-*O*-glucoside.[157] These changes were shown to occur as secondary consequences of the TMV-induced increase in abscisic acid, but may also be of importance in view of several reports of inhibition of virus multiplication by compounds with cytokinin activity.[158]

Mechanisms based on the dark green island effect have not yet found application in crop protection. If similar mechanisms govern the partial failure of viruses to infect meristematic regions and embryos, there could be a possible means to influence seed transmission of some viruses.

## C. Cross-Protection

This involves a specific interference between two virus strains which are usually closely related. In practical crop protection, the plants are deliberately inoculated with a mild or attenuated strain of the virus, which gives a systemic infection. The plants then show a degree of resistance against chance or experimental infection with normal or severe strains.[9,159] Over the last 2 decades there was considerable argument about the mechanisms involved, and in particular whether or not viral coat protein was an important factor.[154,160,161] One problem was that the plant virus systems used for comparisons were often quite different. "Cross-protection" is likely to involve a variety of mechanisms, and conclusions reached in one system may not be applicable in another.

Recently, the ground of the argument has shifted dramatically, with the production of transgenic plants containing and expressing the coat protein genes of various viruses.[162,163] These have indeed been shown to have resistance to viruses, and the mechanism appears similar to that in "whole virus" cross protection. The mechanisms of cross protection remain to be fully explained, as does the resistance of coat protein transgenic plants. However, it seems likely that in both, the virus coat protein is inhibiting the early stages of uncoating the viral RNA prior to establishment of the infection.[164]

## D. Virus Inhibitors and Inducers of Resistance

Many plants contain substances which will inhibit viruses of other plants, especially when applied at the stage of initial infection. Other substances appear to induce a form of resistance in the donor plant, which may act against establishment or against virus multiplication. These mechanisms have recently been comprehensively reviewed by Verma and Prasad.[4] They are not considered in detail here because they generally involve man-made combinations, and may not have direct relevance to types of resistance mechanism operating within a single plant, which is the topic of the present review. Furthermore, there does seem to be a need to demonstrate that the inhibitors, or the resistances induced, are specific for the viruses affected, rather than just being general inhibitors of a range of aspects of metabolism. Having said this, some of the mechanisms might still offer useful avenues to practical crop protection.

## VI. CONCLUSION: PROSPECTS FOR RESISTANCE MECHANISMS IN CROP IMPROVEMENT

The mechanisms discussed in this review are very diverse, and of varying effectiveness in long-term crop protection. How might they be improved, and what are the possibilities for expanding the approach?

Nonhost resistance is sometimes mentioned as a suitable target for genetic engineering. It is by definition so completely effective that it should be of value to transfer it to host plants. There are two problems. If the mode of action is the positive model, involving a number of host genes, the problems of transfer by conventional or nonconventional breeding methods are formidable. If the nonhost immunity operates by a negative mechanism, involving lack of a susceptibility function, there is nothing to transfer to the host. Rather, the gene for the susceptibility function in the host has to be deleted or the function disabled insofar as it interacts with the virus.

There are several problems with genetically controlled cultivar resistance. For many crops the genetic base of resistance is very narrow, and there is a need to seek further sources in old cultivars or wild relative species. Tomlinson[163a] has calculated that in U.K. horticulture alone, there are 22 crops in which no resistance is available against a total of 25 viruses. Another problem is virulence: the results considered in this review suggest that most resistance genes are eventually overcome and that single virus isolates can sometimes overcome several host resistance genes.

Nevertheless, breeding for resistance has given meaningful advances in practical crop protection against virus diseases. It will be useful to expand the effort to create oligogenic systems which are probably more robust. Individual genes may also be tested for robustness, under quarantine conditions, against numerous strains and mutants of the virus. A gene which proved to have low durability under these conditions should only then be released in commercial varieties in combination with other resistance genes. The *Tm-1* gene for TMV resistance in tomato, which has low durability, was arguably "wasted" by being introduced on its own.[76] It may have been better to reserve it until it could have been introduced in cultivars also containing genes at the *Tm-2* locus.

Genetic engineering of virus resistance genes is often spoken of as the way forward in crop protection. There are several problems, not least being the isolation of the genes in a form suitable for manipulation. Until the gene products can be identified, attempts at isolation have to rely on methods such as transposon mutagenesis or hybridization enrichment, which are by nature random and therefore require screening of large numbers of samples. These approaches have problems of background noise giving large numbers of "false positives". Nevertheless, from the amount of researcher effort being devoted to this goal, it is likely that resistance genes will be isolated in the not too distant future.

Transfer of these genes to other species might confer virus resistance, but there may also be problems of function in an alien genetic background or where resistance gene function depends on a host-specific elicitor. Furthermore, virulent isolates of the virus might also overcome a gene transferred by nonconventional methods as easily as one transferred by conventional breeding.

Another aspect of the isolation of resistance genes is that it should give a quantum leap in our understanding of how resistance mechanisms work. Stemming from this, site-directed mutagenesis may allow the development of artificial allelic series of resistance genes which can be inserted into plants in combination to form efficient and durable oligogenic systems.

In the shorter term, meaningful progress is coming from transgenic plants expressing parts of the genetic material of the virus itself. In theory, and increasingly in practice, it is possible to express a small part of the virus genetic information in such a way as to interfere with the functioning of the whole. Furthermore, modification of these small parts of information to enhance their interference potential is a distinct possibility. Thus, any viral function which can be expressed in a defective form may potentially inhibit.

Some approaches which have already shown promise include expression of the coat protein gene, which has given useful resistance against TMV, alfalfa mosaic virus (AlMV), and other viruses.[162,163,165] Attempts to inhibit translation or transcription of viral RNA by expressing antisense sequences[166,167] appear to have had less success so far in plant systems. For those viruses with satellite RNAs which modify symptom expression and virus multiplication, there have been encouraging developments.[168,169] Transgenic plants expressing satellite to cucumber mosaic virus (CMV), for example, show reduced disease severity when inoculated with CMV.

There are qualifications and dangers to some of these approaches. These types of genetically engineered novel resistances might still be overcome by "virulent" isolates of the virus. The possibility that the novel resistance mechanism itself might cause symptoms or a deleterious interaction with another attacking virus also needs to be considered. For example, with the satellite mechanism, there are also satellite/virus combinations which can cause more severe symptoms: this danger will need careful evaluation. It will also be necessary to check the extent to which the transferred genetic element may be capable of spreading in the host plant population — as with transgenic satellite sequences — and whether or not the new selection pressures put on the viral population might lead to evolution of forms capable of causing greater damage on susceptible or conventionally resistant plants.

On the other hand, the prospect is that these novel approaches to resistance will enhance our ability to control or even eliminate plant viruses. The diversity of mechanisms available should make construction of oligogenic systems a reality, and derivation of the genes from the virus rather than the host overcomes the problems of nonavailability or narrowness of host genetic base.

# REFERENCES

1. **Fraser, R. S. S. and Whenham, R. J.**, Inhibition of the multiplication of tobacco mosaic virus by methyl benzimidazol-2yl-carbamate, *J. Gen. Virol.*, 39, 191, 1978.
2. **Schuster, G.**, Improvement in the antiphytoviral chemotherapy by combining ribavirin (virazole) and 2,4-dioxo-hexahydro-1,3,5-triazine (DHT), *Phytopathol. Z.*, 103, 323, 1982.
3. **Cassells, A. C.**, Chemical control of virus diseases of plants, in *Progress in Medicinal Chemistry*, Vol. 20, Ellis, G. P. and West, G. B., Eds., Elsevier, Amsterdam, 1983, 119.
4. **Verma, H. N. and Prasad, V.**, Virus inhibitors and inducers of resistance: potential avenues for biological control of viral diseases, in *Biocontrol of Plant Diseases*, Vol. 3, Mukerji, K. G. and Tewari, J. P., Eds., CRC Press, Boca Raton, FL, in press.
5. **Holmes, F. O.**, Inheritance of resistance to tobacco mosaic disease in tobacco, *Phytopathology*, 28, 553, 1938.
6. **Fraser, R. S. S.**, Genes for resistance to plant viruses, *CRC Crit. Rev. Plant Sci.*, 3, 257, 1986.
7. **Fraser, R. S. S.**, *Biochemistry of Virus-Infected Plants*, Research Studies Press/John Wiley & Sons, Chichester, U.K., 1987.
8. **Fraser, R. S. S.**, Mechanisms of induced resistance to virus diseases, in *Mechanisms of Resistance to Plant Diseases*, Fraser, R. S. S., Ed., Martinus Nijhoff/Dr. W. Junk, Dordrecht, 1985, 373,
9. **Sequiera, L.**, Cross protection and induced resistance: their potential for plant disease control, *Trends Biotechnol.*, 2, 25, 1984.
10. **Gibson, R. W. and Pickett, J. A.**, Wild potato repels aphids by release of aphid alarm pheromone, *Nature*, 302, 608, 1983.
11. **Dawson, G. W., Gibson, R. W., Griffiths, D. C., Pickett, J. A., Rice, A. D., and Woodcock, C. M.**, Aphid alarm pheromone derivative affecting settling and transmission of plant viruses, *J. Chem. Ecol.*, 8, 1377, 1982.
12. **Gibson, R. W., Pickett, J. A., Dawson, G. W., Rice, A. D., and Stribley, M. F.**, Effects of aphid alarm pheromone derivatives and related compounds on non- and semi-persistent plant virus transmission by *Myzus persicae*, *Ann. Appl. Biol.*, 104, 203, 1984.
13. **Martin, T. J., Harvey, T. L., Bender, C. G., and Seifers, D. L.**, Control of wheat streak mosaic virus with vector resistance in wheat, *Phytopathology*, 74, 963, 1984.
14. **Pitrat, M. and Lecoq, H.**, Inheritance of resistance to cucumber mosaic virus transmission by *Aphis gossypii* in *Cucumis melo*, *Phytopathology*, 70, 958, 1980.
15. **Hibino, H., Tiongco, E. R., Cabunagan, R. C., and Flores, Z. M.**, Resistance to rice tungro-associated viruses in rice under experimental and natural conditions, *Phytopathology*, 77, 871, 1987.
16. **Atiri, G. I., Ekpo, E. J. A., and Thottapilly, G.**, The effect of aphid resistance in cowpea on infestation and development of *Aphis craccivora* and the transmission of cowpea aphid borne mosaic virus, *Ann. Appl. Biol.*, 104, 339, 1984.
17. **Atiri, G. I. and Thottappilly, G.**, *Aphis craccivora* settling behaviour and acquisition of cowpea aphid-borne mosaic virus in aphid-resistant cowpea lines, *Entomol. Exp. Appl.*, 39, 241, 1985.
18. **Carroll, T. W., Gossel, P. L., and Hockett, E. A.**, The inheritance of resistance to seed transmission of barley stripe mosaic virus in barley, *Phytopathology*, 69, 431, 1979.
19. **Davis, R. F. and Hampton, R. O.**, Cucumber mosaic virus isolates seedborne in *Phaseolus vulgaris:* serology, host-pathogen relationships and seed transmission, *Phytopathology*, 76, 999, 1986.
20. **Morales, F. J. and Castano, M.**, Seed transmission characteristics of selected bean common mosaic virus strains in differential bean cultivars, *Plant Dis.*, 71, 51, 1987.
21. **Holmes, F. O.**, Concomitant inheritance of resistance to several virus diseases in tobacco, *Virology*, 13, 409, 1961.
22. **Thomas, P. and Fulton, R. W.**, Correlation of the ectodesmata number with nonspecific resistance to initial virus infection, *Virology*, 34, 459, 1968.
23. **Sulzinski, M. A. and Zaitlin, M.**, Tobacco mosaic virus replication in resistant and susceptible plants: in some species virus is confined to a small number of initially infected cells, *Virology*, 121, 12, 1982.
24. **Hussain, M. H., Melcher, U., Whittle, T., Williams, A., Brannan, C. M., and Mitchell, E. D.**, Replication of cauliflower mosaic virus DNA in leaves and suspension culture protoplasts of cotton, *Plant Physiol.*, 83, 633, 1987.

25. **Taniguchi, T.,** Similarity in the accumulation of tobacco mosaic virus in systemic and local necrotic infection, *Virology,* 19, 237, 1963.
26. **Otsuki, Y., Shimomura, T., and Takebe, I.,** Tobacco mosaic virus multiplication and expression of the *N* gene in necrotic responding tobacco varieties, *Virology,* 50, 45, 1972.
27. **Fraser, R. S. S.,** Mechanisms involved in genetically controlled resistance and virulence. virus diseases, in *Mechanisms of Resistance to Plant Diseases,* Fraser, R. S. S., Ed., Martinus Nijhoff/Dr. W. Junk, Dordrecht, 1985, 143.
28. **Fraser, R. S. S.,** Virus recognition and pathogenicity: implications for resistance mechanisms and breeding, *Pestic. Sci.,* 23, 267, 1988.
29. **Day, K. L.,** Resistance to Bean Common Mosaic Virus in *Phaseolus vulgaris* L., Ph.D. thesis, University of Birmingham, U.K., 1984.
30. **Cheo, P. C. and Gerard, J. S.,** Differences in virus-replicating capacity among plant species inoculated with tobacco mosaic virus, *Phytopathology,* 61, 1010, 1971.
31. **Nishiguchi, M. and Motoyoshi, F.,** Resistance mechanisms of tobacco mosaic virus strains in tomato and tobacco, in *Plant Resistance to Viruses, Ciba Foundation Symposium 133,* Evered, D. and Harnett, S., Eds., John Wiley & Sons, Chichester, U.K., 1987, 38.
32. **Fraser, R. S. S. and Loughlin, S. A. R.,** Resistance to tobacco mosaic virus in tomato: effects of the *Tm-1* gene on virus multiplication, *J. Gen. Virol.,* 48, 87, 1980.
33. **McCreight, J. D., Kishaba, A. N., and Mayberry, K. S.,** Lettuce infectious yellows tolerance in lettuce, *J. Am. Soc. Hortic. Sci.,* 111, 788, 1986.
34. **Loi, N., Osler, R., Snidaro, M., Ardigo, A., and Lorenzoni, C.,** Tolerance to BYDV (barley yellow dwarf virus) in inbreds and hybrids of maize, *Maydica,* 31, 307, 1986.
35. **Allen, D. J.,** Identification of resistance to cowpea mottle virus, *Trop. Agric.,* 57, 325, 1980.
36. **Cisar, G., Brown, C. M., and Jedlinski, H.,** Effect of fall or spring infection and sources of tolerance of barley yellow dwarf of winter wheat, *Crop Sci.,* 22, 474, 1982.
37. **McKenzie, R. I. H., Burnett, P. A., Gill, C. C., Comeau, A., and Brown, P. D.,** Inheritance of tolerance to barley yellow dwarf virus in oats, *Euphytica,* 34, 681, 1985.
38. **Mouches, C., Candresse, T., and Bove, J. M.,** Turnip yellow mosaic virus RNA-replicase contains host and virus-encoded subunits, *Virology,* 134, 78, 1984.
39. **Vanderplanck, J. E.,** *Disease Resistance in Plants,* 2nd ed., Academic Press, Orlando, FL, 1984.
40. **Robinson, R. A.,** *Plant Pathosystems,* Springer-Verlag, Berlin, 1976.
41. **Person, C.,** Gene-for-gene relationships in host:parasite systems, *Can. J. Bot.,* 37, 1101, 1959.
42. **Fraser, R. S. S.,** Genetics of host resistance to viruses and of virulence, in *Mechanisms of Resistance to Plant Diseases,* Fraser, R. S. S., Ed., Martinus Nijhoff/Dr. W. Junk, Dordrecht, 1985, 62.
43. **Holmes, F. O.,** Additive resistances to specific viral diseases in plants, *Ann. Appl. Biol.,* 42, 129, 1955.
44. **Bald, J. G. and Tinsley, T. W.,** A quasi-genetic model for plant virus host ranges. I. Group reactions within taxonomic boundaries, *Virology,* 31, 616, 1967.
45. **Bald, J. G. and Tinsley, T. W.,** A quasi-genetic model for plant virus host ranges. II. Differentiation between host ranges, *Virology,* 32, 321, 1967.
46. **Bald, J.G. and Tinsley, T. W.,** A quasi-genetic model for plant virus host ranges. III. Congruence and relatedness, *Virology,* 32, 328, 1967.
47. **Hiebert, E., Bancroft, J. B., and Bracker, C. E.,** The assembly *in vitro* of some small spherical viruses, hybrid viruses and other nucleoproteins, *Virology,* 34, 492, 1968.
48. **Atabekov, J. G., Novokov, V. K., Vishnichenko, V. K., and Javakhia, V. G.,** A study of the mechanisms controlling the host range of plant viruses. II. The host range of hybrid viruses reconstituted *in vitro* and of free viral RNA, *Virology,* 41, 108, 1970.
49. **Atabekov, J. G.,** Host specificity in plant viruses, *Annu. Rev. Phytopathol.,* 13, 127, 1975.
50. **Turner, P. C., Watkins, P. A. C., Zaitlin, M., and Wilson, T. M. A.,** Tobacco mosaic virus particles uncoat and express their RNA in *Xenopus laevis* oocytes: implications for early interactions between plant cells and viruses, *Virology,* 160, 515, 1987.
51. **Furusawa, I. and Okuno, T.,** Infection with BMV of protoplasts derived from five plant species, *J. Gen. Virol.,* 40, 489, 1978.
52. **Huber, R., Hontilez, J., and van Kammen, A.,** Infection of cowpea protoplasts with both the common strain and the cowpea strain of TMV, *J. Gen. Virol.,* 55, 241, 1981.
53. **Van Loon, L. C.,** Disease induction by plant viruses, *Adv. Virus Res.,* 33, 205, 1987.
54. **Drijfhout, E.,** Genetic interaction between *Phaseolus vulgaris* and bean common mosaic virus with implications for strain identification and resistance breeding, *Agric. Res. Rep. (Wageningen),* 872, 1978.
55. **Fraser, R. S. S.,** Genetics of plant resistance to viruses, in *Plant Resistance to Viruses, Ciba Foundation Symposium 133,* Evered, D. and Harnett, S., Eds., John Wiley & Sons, Chichester, U.K., 1987, 6.
56. **Shifris, O., Myers, C. H., and Chupp, C.,** Resistance to mosaic virus in the cucumber, *Phytopathology,* 32, 773, 1942.

57. **Baggett, J. R. and Frazier, W. A.,** The inheritance of resistance to bean yellow mosaic virus in *Phaseolus vulgaris, J. Am. Soc. Hortic. Sci.,* 70, 325, 1957.

58. **Wasuwat, S. L. and Walker, J. C.,** Inheritance of resistance in cucumber to cucumber mosaic virus, *Phytopathology,* 51, 423, 1961.

59. **Provvidenti, R. and Schroeder, W. T.,** Resistance in *Phaseolus vulgaris* to the severe strain of bean yellow mosaic virus, *Phytopathology,* 63, 196, 1973.

60. **Jones, A. T. and Catherall, P. L.,** The relationship betwen growth rate and the expression of tolerance to barley yellow dwarf virus in barley, *Ann. Appl. Biol.,* 65, 137, 1970.

61. **Catherall, P. L. and Hayes, J. D.,** Assessment of varietal reaction and breeding for resistance to the yellow dwarf virus in barley, *Euphytica,* 15, 39, 1966.

62. **Catherall, P. L., Hayes, J. D., and Boulton, R. E.,** Breeding cereals resistant to virus diseases in Britain, *Ann. Phytopathol.,* 9, 241, 1977.

63. **Thompson, A. E., Lower, R. L., and Thornberry, H. H.,** Inheritance in beans of the necrotic reaction to tobacco mosaic virus, *J. Hered.,* 53, 89, 1962.

64. **Motoyoshi, F. and Oshima, Y.,** Infection with tobacco mosaic virus of leaf mesophyll protoplasts from susceptible and resistant lines of tomato, *J. Gen. Virol.,* 29, 81, 1975.

65. **Csillery, G., Tobias, I., and Rusko, J.,** A new pepper strain of tobacco mosaic virus, *Acta Phytopathol. Acad. Sci. Hung.,* 18, 195, 1984.

66. **Knorr, D. A. and Dawson, W. O.,** A point mutation in the tobacco mosaic virus capsid protein gene induces hypersensitivity in *Nicotiana sylvestris, Proc. Natl. Acad. Sci. U.S.A.,* 85, 170, 1988.

67. **Saito, T., Meshi, T., Takamatsu, N., and Okada, Y.,** Coat protein gene sequence of tobacco mosaic virus encodes a host response determinant, *Proc. Natl. Acad. Sci. U.S.A.,* 84, 6074, 1987.

68. **Meshi, T., Motoyoshi, F., Maeda, S., Yoshioka, H., Watanabe, H., and Okada, Y.,** Responsible mutations in the 30K protein gene of TMV for overcoming the *Tm-2* resistance, *Abstr. 5th Intl. Conf. Plant Pathol., Kyoto,* 1988, 38.

69. **Meshi, T., Motoyoshi, F., Adachi, A., Watanabe, Y., Takamatsu, N., and Okada, Y.,** Two concomitant base substitutions in the putative replicase genes of tobacco mosaic virus confer the ability to overcome the effects of a tomato resistance gene, *Tm-1, EMBO J.,* 7, 1575, 1988.

70. **Weber, P. V. V.,** Inheritance of a necrotic lesion reaction to a mild strain of tobacco mosaic virus, *Phytopathology,* 41, 593, 1951.

71. **Fraser, R. S. S.,** Varying effectiveness of the *N'* gene for resistance to tobacco mosaic virus in tobacco infected with virus strains differing in coat protein properties, *Physiol. Plant Pathol.,* 22, 109, 1983.

72. **Ohno, T., Takamatsu, N., Meshi, T., Okada, Y., Nishiguchi, M., and Kiho, Y.,** Single amino acid substitution in 30 K protein of TMV defective in transport function, *Virology,* 131, 255, 1983.

73. **Flor, H. H.,** The complementary genetic systems of flax and flax rust, *Adv. Genet.,* 8, 29, 1956.

74. **Crute, I. R.,** The genetic bases of relationships between microbial parasites and their hosts, in *Mechanisms of Resistance to Plant Diseases,* Fraser, R. S. S., Ed., Martinus Nijhoff/Dr. W. Junk, Dordrecht, 1985, 80.

75. **Pelham, J.,** Strain-genotype interaction of tobacco mosaic virus in tomato, *Ann. Appl. Biol.,* 71, 219, 1972.

76. **Pelham, J., Fletcher, J. T., and Hawkins, J. H.,** The establishment of a new strain of tobacco mosaic virus resulting from the use of resistant varieties of tomato, *Ann. Appl. Biol.,* 65, 293, 1970.

77. **Hall, T. J.,** Resistance at the *Tm-2* locus in the tomato to tomato mosaic virus, *Euphytica,* 29, 189, 1980.

78. **Fraser, R. S. S., Gerwitz, A., and Payne, J. A.,** Resistance to tobacco mosaic virus in tomato, Rep. National Vegetable Research Station for 1986/87, Wellbourne, 1987, 20.

78a. **Fraser, R. S. S.,** unpublished data.

79. **Betti, L., Tanzi, M., and Canova, A.,** Pepper mosaic virus strains and their adaptation to the host. 1. Biological and serological behaviour, *Phytopathol. Mediterr.,* 27, 7, 1988.

80. **Betti, L., Tanzi, M., and Canova, A.,** Evolutionary changes in TMV pepper strains as a result of repeated host passages, *Phytopathol. Mediterr.,* 25, 39, 1986.

81. **Tobias, I., Rast, A. T. B., and Maat, D. Z.,** Tobamoviruses of pepper, eggplant and tobacco: comparative host reactions and serological relationships, *Neth. J. Plant Pathol.,* 88, 257, 1982.

82. **Tobias, I., Fraser, R. S. S., and Gerwitz, A.,** The gene-for-gene relationship between *Capsicum annuum* L. and tobacco mosaic virus: effects on virus multiplication, ethylene synthesis, and accumulation of pathogenesis-related protein, *Physiol. Mol. Plant Pathol.,* 35, 271, 1989.

83. **Tanzi, M., Betti, L., Bertaccini, A., and Canova, A.,** Pepper mosaic virus strains and their adaptation to the host. II. Morphological behaviour, *Phytopathol. Mediterr.,* 27, 28, 1988.

84. **Greenleaf, W .H., Cook, A. A., and Heyn, A. N. J.,** Resistance to tobacco mosaic virus in *Capsicum* with special reference to the Samsun latent strain, *Phytopathology,* 54, 1367, 1964.

85. **Garcia-Arenal, F., Palukaitis, P., and Zaitlin, M.,** Strains and mutants of tobacco mosaic virus are both found in virus derived from single lesion-passaged inoculum, *Virology,* 132, 131, 1984.

86. **Da Graca, J. V. and Martin, M. M.,** An electron microscope study of hypersensitive tobacco infected with tobacco mosaic virus at 32°C, *Physiol. Plant Pathol.,* 8, 215, 1976.

87. **Konate, G., Kopp, M., and Fritig, B.,** Studies on TMV multiplication in systemically and hypersensitively reacting tobacco varieties by means of radiochemical and immunoenzymatic methods, *Agronomie,* 3, 95, 1983.

88. **Kim, K. S.,** Subcellular responses to localized infection of *Chenopodium quinoa* by pokeweed mosaic virus, *Virology,* 41, 179, 1970.

89. **Cohen, J. and Loebenstein, G.,** An electron microscope study of starch lesions in cucumber cotyledons infected with tobacco mosaic virus, *Phytopathology,* 65, 32, 1975.

90. **Tanguy, J. and Martin, C.,** Phenolic compounds and the hypersensitivity reaction in *Nicotiana tabacum* infected with tobacco mosaic virus, *Phytochemistry,* 11, 19, 1972.

91. **Weststeijn, E. A.,** Peroxidase activity in leaves of *Nicotiana tabacum* var. Xanthi nc. before and after infection with tobacco mosaic virus, *Physiol. Plant Pathol.,* 8, 63, 1976.

92. **Weststeijn, E. A.,** Permeability changes in the hypersensitive reaction of *Nicotiana tabacum* cv. Xanthi-nc after infection with tobacco mosaic virus, *Physiol. Plant Pathol.,* 13, 253, 1978.

93. **Pennazio, S. and Sapetti, C.,** Electrolyte leakage in relation to viral and abiotic stresses inducing necrosis in cowpea leaves, *Biol. Plant,* 24, 218, 1982.

94. **Pennazio, S., Appiano, A., and Redolfi, P.,** Changes occurring in *Gomphrena globosa* leaves in advance of the appearance of tomato bushy stunt virus necrotic local lesions, *Physiol. Plant Pathol.,* 15, 177, 1979.

95. **Kasamo, K. and Shimomura, T.,** Response of membrane-bound $Mg^{2+}$-activated ATPase of tobacco leaves to tobacco mosaic virus, *Plant Physiol.,* 62, 731, 1978.

96. **De Laat, A. M. M. and Van Loon, L. C.,** The relationship between stimulated ethylene production and symptom expression in virus-infected tobacco leaves, *Physiol. Plant Pathol.,* 22, 261, 1983.

97. **De Laat, A. M. M., Van Loon, L. C., and Vonk, C. R.,** Regulation of ethylene biosynthesis in virus-infected tobacco leaves. I. Determination of the role of methionine as the precursor of ethylene, *Plant Physiol.,* 68, 256, 1981.

98. **Van Loon, L. C.,** The induction of pathogenesis-related proteins by pathogens and specific chemicals, *Neth. J. Plant Pathol.,* 88, 265, 1983.

99. **Antoniw, J. F. and White, R. F.,** Changes with time in the distribution of virus and PR protein around single local lesions of TMV-infected tobacco, *Plant Mol. Biol.,* 6, 145, 1986.

100. **Takahashi, T.,** Studies on viral pathogenesis in plant hosts. IV. Comparison of early processes of tobacco mosaic virus infection in the leaves of 'Samsun NN' and 'Samsun' tobacco plants, *Phytopathol. Z.,* 77, 157, 1973.

101. **Kalpagam, C., Foglein, F. J., Nytrai, A., Premecz, G., and Farkas, G. L.,** Expression of the *N* gene in plasmolysed leaf tissues and isolated protoplasts of *Nicotiana tabacum* cv Xanthi-nc infected by TMV, in *Current Topics in Plant Pathology,* Kiraly, Z., Ed., Akademiai Kiado, Budapest, 1977, 395.

102. **Massala, R., Legrand, M., and Fritig, B.,** Effect of α-aminoacetate, a competitive inhibitor of phenylalanine ammonia lyase, on the hypersensitive resistance of tobacco to tobacco mosaic virus, *Physiol. Plant Pathol.,* 16, 213, 1980.

103. **Legrand, M., Fritig, B., and Hirth, L.,** O-Diphenol O-methyl transferases of healthy and tobacco mosaic virus-infected hypersensitive tobacco, *Planta,* 144, 101, 1978.

104. **Fritig, B., Kauffmann, S., Dumas, B., Geoffroy, P., Kopp, M., and Legrand, M.,** Mechanism of the hypersensitivity reaction of plants, in *Plant Resistance to Viruses, Ciba Foundation Symposium 133,* Evered, D. and Harnett, S., Eds., John Wiley & Sons, Chichester, U.K., 1987, 92.

105. **Appiano, A., Pennazio, S., D'Agostino, G., and Redolfi, P.,** Fine structure of necrotic local lesions induced by tomato bushy stunt virus in *Gomphrena globosa* leaves, *Physiol. Plant Pathol.,* 11, 327, 1977.

106. **Favali, M. A., Conti, G. G., and Bassi, M.,** Modifications of the vascular bundle ultrastructure in the 'resistance zone' around necrotic lesions induced by tobacco mosaic virus, *Physiol. Plant Pathol.,* 13, 247, 1978.

107. **Russo, M., Martelli, G. P., and di Franco, A.,** The fine structure of local lesions of beet necrotic yellow vein virus in *Chenopodium amaranticolor, Physiol. Plant Pathol.,* 19, 237, 1981.

108. **Hooft van Huijsduijnen, R. A. M., Kauffmann, S., Brederode, F. T., Cornelissen, B. J. C., Legrand, M., Fritig, B., and Bol, J. F.,** Homology between chitinases that are induced by TMV infection of tobacco, *Plant Mol. Biol.,* 9, 411, 1987.

109. **Weintraub, M., Ragetli, H. W. J., and Lo, E.,** Mitochondrial content and respiration in leaves with localized virus infections, *Virology,* 50, 841, 1972.

110. **Whenham, R. J. and Fraser, R. S. S.,** Effect of systemic and local lesion-forming strains of tobacco mosaic virus on abscisic acid concentration in tobacco leaves: consequences for the control of leaf growth, *Physiol. Plant Pathol.,* 18, 267, 1981.

111. **Van Loon, L. C. and Berbee, A. T.,** Endogenous levels of indoleacetic acid in leaves of tobacco reacting hypersensitively to TMV, *Z. Pflanzenphysiol.,* 89, 373, 1978.

112. **Israel, H. W. and Ross, A. F.**, The fine structure of local lesions induced by tobacco mosaic virus in tobacco, *Virology,* 33, 272, 1967.

113. **Mozes, R., Antignus, Y., Sela, I., and Harpaz,I.**, The chemical nature of an antiviral factor (AVF) from virus-infected plants, *J. Gen. Virol.,* 38, 241, 1978.

114. **Antignus, Y., Sela, I., and Harpaz, I.**, Further studies on the biology of an antiviral factor (AVF) from virus-infected plants and its association with the *N*-gene of *Nicotiana* species, *J. Gen. Virol.,* 35, 107, 1977.

115. **Sela, I.**, Plant virus interactions related to resistance and localization of viral infections, *Adv. Virus Res.,* 26, 201, 1981.

116. **Sela, I., Grafi, G., Sher, N., Edelbaum, O, Yagev, H., and Gerassi, E.**, Resistance systems related to the *N* gene and their comparison with interferon, in *Plant Resistance to Viruses, Ciba Foundation Symposium 133,* Evered, D. and Harnett, S., Eds., John Wiley & Sons, Chichester, U.K., 1987, 109.

117. **Orchansky, P., Rubinstein, M., and Sela, I.**, Human interferons protect plants from virus infection, *Proc. Natl. Acad. Sci. U.S.A.,* 79, 2279, 1982.

118. **Reichman, M., Devash, Y., Suhadolnik, R. J., and Sela, I.**, Human leukocyte interferon and the antiviral factor (AVF) from virus-infected plants stimulate plant tissues to produce nucleotides with antiviral activity, *Virology,* 128, 240, 1983.

119. **Antoniw, J. F., White, R. F., and Carr, J. P.**, An examination of the effect of human α-interferons on the infection and multiplication of tobacco mosaic virus in tobacco, *Phytopathol. Z.,* 109, 367, 1984.

120. **Huisman, M. J., Broxterman, H. J. G., Schellekens, H., and Van Vloten-Doting, L.**, Human interferon does not protect cowpea plant cell protoplasts against infection with alfalfa mosaic virus, *Virology,* 143, 622, 1985.

121. **Loesch-Fries, L. S., Halk, E. L., Nelson, S. E., and Krahn, K. J.**, Human leukocyte interferon does not inhibit alfalfa mosaic virus in protoplasts or tobacco tissue, *Virology,* 143, 626, 1985.

122. **Orgakov, V. I., Kaplan, I. B., Taliansky, M. E., and Atabekov, J. G.**, Suppression of the multiplication of potato viruses by human leukocyte interferon, *Dokl. Akad. Nauk. USSR,* 276, 743, 1984.

123. **Carter, W. A., Swartz, H., and Gillespie, D. H.**, Independent evolution of antiviral and growth-modulating activities of interferon, *J. Biol. Response Modif.,* 4, 447, 1985.

124. **Loebenstein, G. and Gera, A.**, Inhibitor of virus replication released from tobacco mosaic virus infected protoplasts of a local lesion responding tobacco cultivar, *Virology,* 114, 132, 1981.

125. **Gera, A. and Loebenstein, G.**, Further studies of an inhibitor of virus replication from tobacco mosaic virus-infected protoplasts of a local lesion-responding tobacco cultivar, *Phytopathology,* 73, 111, 1983.

126. **Loebenstein, G.**, Inhibitor of viral replication, in *Plant Resistance to Viruses, Ciba Foundation Symposium 133,* Evered, D. and Harnett, S., Eds., John Wiley & Sons, Chichester, U.K., 1987, 116.

127. **Smart, T. E., Dunigan, D. D., and Zaitlin, M.**, *In vitro* translation products of mRNAs derived from TMV-infected tobacco exhibiting a hypersensitive response, *Virology,* 158, 461, 1987.

128. **Dunigan, D. A., Golemboski, D. B., and Zaitlin, M.**, Analysis of the *N* gene of *Nicotiana,* in *Plant Resistance to Viruses, Ciba Foundation Symposium 133,* Evered, D. and Harnett, S., Eds., John Wiley & Sons, Chichester, U.K., 1987, 120.

129. **Bol, J. F., Hooft van Huijsduijnen, R. A. M., Cornelissen, B. J. C., and van Kan, J. A. L.**, Characterization of pathogenesis-related proteins and genes, in *Plant Resistance to Viruses, Ciba Foundation Symposium 133,* Evered, D. and Harnett, S., Eds., John Wiley & Sons, Chichester, U.K., 1987, 72.

130. **Van Regenmortel, M. H. V.**, Serological studies on naturally occurring strains and chemically induced mutants of tobacco mosaic virus, *Virology,* 31, 467, 1967.

131. **Dawson, W. O., Bubrick, P., and Grantham, G. L.**, Modifications of the tobacco mosaic virus coat protein gene affecting replicaiton, movement and symptomatology, *Phytopathology,* 78, 783, 1988.

132. **Sarkar, S. and Smitamana, P.**, A proteinless mutant of tobacco mosaic virus: evidence against the role of a viral coat protein for interference, *Mol. Gen. Genet.,* 184, 158, 1981.

133. **Valleau, W. D.**, The relative positions of the *N* and *N'* factors on *Nicotiana tabacum* chromosomes, *Phytopathology,* 33, 14, 1943.

134. **Beier, H., Siler, D. J., Russell, M. L., and Bruening, G.**, Survey of susceptibility to cowpea mosaic virus among protoplasts and intact plants from *Vigna sinensis* lines, *Phytopathology,* 67, 917, 1977.

135. **Beier, H., Bruening,G., Russell, M. L., and Tucker, C. L.**, Replication of cowpea mosaic virus in protoplasts isolated from immune lines of cowpeas, *Virology,* 95, 165, 1979.

136. **Kiefer, M. C., Bruening, G., and Russell, M. L.**, RNA and capsid accumulation in cowpea protoplasts that are resistant to cowpea strain SB, *Virology,* 137, 71, 1984.

137. **Goldbach, R. and Krijt, J.**, Cowpea mosaic virus-encoded proteinase does not recognize primary translation products of mRNAs from other comoviruses, *J. Virol.,* 43, 1151, 1982.

138. **Franssen, H., Moerman, M., Rezelman, G., and Goldbach, R.**, Evidence that the 32,000-Dalton protein encoded by bottom component of cowpea mosaic virus is a proteolytic processing enzyme, *J. Virol.,* 50, 183, 1984.

139. **Bruening, G., Ponz, F., Glascock, C., Russell, M. L., Rowhani, A., and Chay, C.,** Resistance of cowpeas to cowpea mosaic virus and to tobacco ringspot virus, in *Plant Resistance to Viruses, Ciba Foundation Symposium 133,* Evered, D. and Harnett, S., Eds., John Wiley & Sons, Chichester, U.K., 1987, 23.

140. **Motoyoshi, F. and Oshima, N.,** Expression of genetically controlled resistance to tobacco mosaic virus in isolated tomato leaf mesophyll protoplasts, *J. Gen. Virol.,* 34, 499, 1977.

141. **Watanabe, Y., Morita, N., Nishiguchi, M., and Okada, Y.,** Attenuated strains of tobacco mosaic virus: reduced synthesis of a viral protein with a cell-to-cell movement function, *J. Mol. Biol.,* 194, 699, 1987.

142. **Watanabe, Y., Kishibayashi, N., Motoyoshi, F., and Okada, Y.,** Characterization of *Tm-1* gene action on replication of common isolates and a resistance-breaking isolate of TMV, *Virology,* 161, 527, 1987.

143. **Fraser, R. S. S. and Loughlin, S. A. R.,** Effects of temperature on the *Tm-1* gene for resistance to tobacco mosaic virus in tomato, *Physiol. Plant Pathol.,* 20, 109, 1982.

144. **Fraser, R. S. S., Gerwitz, A., and Morris, G. E. L.,** Multiple regression analysis of the relationships between tobacco mosaic virus multiplication, the severity of mosaic symptoms, and the growth of tobacco and tomato, *Physiol. Mol. Plant Pathol.,* 29, 239, 1986.

145. **Fraser, R. S. S. and Whenham, R. J.,** Abscisic acid metabolism in tomato plants infected with tobacco mosaic virus: relationships with growth, symptoms and the *Tm-1* gene for TMV resistance, *Physiol. Mol. Plant Pathol.,* 34, 215, 1989.

146. **Ross, A. F.,** Localized acquired resistance to plant virus infection in hypersensitive hosts, *Virology,* 14, 329, 1961.

147. **Ross, A. F.,** Systemic resistance induced by localized virus infections in plants, *Virology,* 14, 340, 1961.

148. **Kuc, J.,** Expression of latent genetic information for disease resistance in plants, in *Cellular and Molecular Biology of Plant Stress,* Key, J. L. and Kosuge, T., Eds., Alan R. Liss, New York, 1985, 303.

149. **McIntyre, J. L., Dodds, J. A., and Hare, J. D.,** Effects of localized infections of *Nicotiana tabacum* by tobacco mosaic virus on systemic resistance against diverse pathogens and an insect, *Phytopathology,* 71, 297, 1981.

150. **Fraser, R. S. S. and Clay, C. M.,** Pathogenesis-related proteins and acquired systemic resistance: causal relationship or separate effects?, *Neth. J. Plant Pathol.,* 89, 283, 1983.

151. **Fraser, R. S. S.,** Are pathogenesis-related proteins involved in acquired systemic resistance of tobacco plants to tobacco mosaic virus?, *J. Gen. Virol.,* 57, 305, 1982.

152. **Dumas, E. and Gianinazzi, S.,** Pathogenesis-related (b) proteins do not play a central role in TMV localization in *Nicotiana rustica, Physiol. Plant Pathol.,* 28, 243, 1986.

153. **Kassanis, B. and White, R. F.,** Effect of polyacrylic acid and b proteins on TMV multiplication in tobacco protoplasts, *Phytopathol. Z.,* 91, 269, 1978.

154. **Fraser, R. S. S.,** Mechanisms of induced resistance to virus diseases, in *Mechanisms of Resistance to Plant Diseases,* Fraser, R. S. S., Ed., Martinus Nijhoff/Dr. W. Junk, Dordrecht, 1985, 373.

155. **Loebenstein, G., Cohen, J., Shabtai, S., Coutts, R. H. A., and Wood, K. R.,** Distribution of cucumber mosaic virus in systemically infected tobacco leaves, *Virology,* 81, 117, 1977.

156. **Whenham, R. J., Fraser, R. S. S., Brown, L. P., and Payne, J. A.,** Tobacco mosaic virus-induced increase in abscisic acid concentration in tobacco leaves: intracellular location in light and dark green areas, and relationship to symptom development, *Planta,* 168, 592, 1986.

157. **Whenham, R. J.,** Effect of systemic tobacco mosaic virus infection on endogenous cytokinin concentration in tobacco *(Nicotiana tabacum* L.) leaves: consequences for the control of resistance and symptom development, *Physiol. Mol. Plant Pathol.,* 35, 85, 1989.

158. **Fraser, R. S. S. and Whenham, R. J.,** Plant growth regulators and virus infection: a critical review, *Plant Growth Regul.,* 1, 37, 1982.

159. **Rast, A. T. B.,** Variability of tobacco mosaic virus in relation to control of tomato mosaic in glasshouse crops by resistance breeding and cross protection, *Agric. Res. Rep. (Wageningen),* 843, 1, 1975.

160. **De Zoeten, G. A. and Fulton, R. W.,** Understanding generates possibilities, *Phytopathology,* 65, 221, 1975.

161. **Zaitlin, M.,** Virus cross protection: more understanding is needed, *Phytopathology,* 66, 382, 1976.

162. **Tumer, N. E., O'Connell, K. M., Nelson, R. S., Sanders, P. R., Beachy, R. N., Fraley, R. T., and Shah, D. M.,** Expression of alfalfa mosaic virus coat protein gene confers cross-protection in transgenic tobacco and tomato plants, *EMBO J.,* 6, 1181, 1987.

163. **Nelson, R. S., Abel, P. P., and Beachy, R. N.,** Lesions and virus accumulation in inoculated transgenic tobacco plants expressing the coat protein gene of tobacco mosaic virus, *Virology,* 158, 126, 1987.

163a. **Tomlinson, J. A.,** personal communication.

164. **Wilson, T. M. A. and Watkins, P. A. C.,** Influence of exogenous virus coat protein on the cotranslational disassembly of tobacco mosaic virus (TMV) particles, *in vitro, Virology,* 149, 132, 1986.

165. **Loesch-Fries, L. S., Merlo, D., Zinnen, T., Burhop, T., Hill, K., Krahn, K., Jarvis, N., Nelson, S., and Halk, E.,** Expression of alfalfa mosaic virus RNA 4 in transgenic plants confers virus resistance, *EMBO J.,* 6, 1845, 1987.

166. **Izant, J. G. and Weintraub, H.,** Constitutive and conditional suppression of exogenous and endogenous genes by antisense RNA, *Science,* 229, 345, 1985.
167. **Morch, M. D., Joshi, R. L., Denial, T. M., and Haenni, A. L.,** A new 'sense' approach to block viral RNA replication *in vitro, Nucleic Acids Res.,* 15, 4123, 1987.
168. **Gerlach, W. L., Llewellyn, D., and Haseloff, J.,** Construction of a plant disease resistance gene from the satellite RNA of tobacco ringspot virus, *Nature,* 328, 802, 1987.
169. **Harrison, B. D., Mayo, M. A., and Baulcombe, D. C.,** Virus resistance in transgenic plants that express cucumber mosaic virus satellite RNA, *Nature,* 328, 799, 1987.

Chapter 13

GENETICALLY ENGINEERED RESISTANCE: TRANSGENIC PLANTS

**C. Hemenway, L. Haley, W. K. Kaniewski, E. C. Lawson, K. M. O'Connell, P. R. Sanders, P. E. Thomas, and N. E. Tumer**

TABLE OF CONTENTS

# I. INTRODUCTION

Host resistance is one of several classical approaches to virus disease control. The others are cure of the disease (i.e., by chemotherapy, thermotherapy, or meristem culture), maintenance and use of pathogen-free seed or propagules (i.e., by use of seed certification and tissue culture techniques), restriction of long-distance movement of the pathogen into new areas (quarantine), restriction of local dissemination into and among crop plants, and restriction of deleterious effects of disease. Despite these approaches, many viruses still are causing serious economical problems. Among all of these approaches, plant resistance is the most desirable and practical. Resistance in plants may be inherent (genetic) or induced. Genetic resistance has long been considered the ideal approach to disease control. Once achieved, it has the potential of replacing all other approaches, is easy and inexpensive to apply, has no undesirable environmental impact, and requires no induction. However, suitable resistance genes frequently are not available in genetically compatible germplasm. When they are, many years may be required for their incorporation into cultivars with other desired characteristics. Transfer of genes from related species is sometimes possible, employing ploidy manipulation, embryo rescue, protoplast fusion, and other difficult technologies, but these genes frequently are inseparably linked to undesirable characteristics.[1] Furthermore, most plant virus resistance genes are not fully effective or universally applicable against all isolates of the virus.[2]

Induced resistance is conferred on a susceptible plant by a prior event. It is not heritable, but ultimately has a genetic basis. Of several types known,[3-6] the most effective was termed "cross-protection" by McKinney.[7] His discovery that infection with one strain of a virus conferred resistance to subsequent infection by a closely related strain was soon confirmed by others.[8] This approach proved to be applicable to many different viruses[3] and was recognized as a potentially effective means of controlling the effects of severe strains of virus.[9]

Because of significant progress in genetic engineering and plant transformation techniques, a new type of induced resistance based on expression of viral genes or sequences in plants has been developed. This genetically engineered protection against virus infection was first demonstrated in plants developed to express the tobacco mosaic virus (TMV) coat protein (CP) gene.[10] The applicability of this approach to control other RNA viruses was demonstrated with plants expressing the CP genes from alfalfa mosaic virus (AlMV),[11-13] cucumber mosaic virus (CMV),[14] potato virus X (PVX),[15] and tobacco streak virus (TSV).[16] Protection from virus infection has also been demonstrated in transgenic plants developed to express antisense transcripts to CP genes of TMV,[17] CMV,[14] and PVX,[15] or viral satellite RNAs from CMV or tobacco ringspot virus.[18,19] It is likely that the genetically engineered approaches described above will benefit agriculture. In addition, analysis of plants that express various viral genes or sequences will promote our understanding of virus infection.

In this chapter, various aspects of both classical cross-protection and genetically engineered protection will be described in order to emphasize the similarities and differences between these approaches. Further, the significance and relative effectiveness of these approaches will be discussed.

# II. CLASSICAL CROSS-PROTECTION

A variety of terms have been applied to the phenomenon of cross-protection, but the original term coined by McKinney[7] is now widely accepted. A potential area of confusion arises because the term is now applied to describe both the presumed cause (suppression of viral functions) and effect (suppression of symptoms) of the phenomenon. Essentially the same problem exists in the terminology for genetic resistance. Originally, the term evolved among growers to imply any quality of a plant that precludes or supresses the deleterious effects of disease. Cooper and Jones[20] applied the term to causes (performance of viral functions) in their proposal for terminology for a more extensive definition of plant/virus interactions. By their proposal, a plant, in response to inoculation, is either infectible or immune (not infectible). If infectible, specific viral functions either proceed with relative efficiency (termed "susceptibility") or with relative difficulty (termed "resistance"), and the infection may cause little (termed "tolerance") or much injury (termed "sensitivity") to the plant. We apply the terminology of Cooper and Jones[20] with reference to specific mechanisms of both genetic and induced resistance.

In general, viruses must be closely related to cross-protect, but more is required than relatedness. For example, isolates of the same virus vary in the degree to which they cross-protect, and cross-protection between isolates of some viruses sometimes cannot be demonstrated.[21,22] Protection between pairs of isolates may not be reciprocal, and some pairs of isolates will not protect in either direction, although other isolates of the same virus do cross-protect.[23] Some of these variations may reflect the degree to which virus isolates vary in their capacity to infect all cells of a plant, for there is much evidence that a cell must be infected to be protected.[24] Availability of uninfected cells could provide a means of initial infection for a challenge virus.

The degree of resistance afforded by cross-protection is seldom, if ever, complete. Superinfection of protected plants can usually be achieved when inoculum concentrations are increased 100- to 1000-fold more than required to infect healthy plants.[3] Since inoculum concentrations to which plants are exposed in the field rarely reach these levels, superinfection may never be achieved in the field environment among plants with only moderate levels of protection. A case in point is that of protection of potato with mild strains of PVX. Superinfection with severe strains is achieved experimentally,[21,25,26] but is never observed in the field. In contrast, severe PVX symptoms are observed among potato plants free of a protecting mild strain of PVX. Since superinfection is highly concentration dependent, it would be interesting to know more about efficiency of cross-protection against insect transmission, which involves only minute quantities of virus compared with the amounts involved in mechanical transmission. Classical cross-protection against aphid transmission has been demonstrated.[27]

It is clear that cross-protection suppresses initial infection because it is inoculum concentration dependent and affects the numbers of local lesions after inoculation. To some extent, the reductions and delays in virus accumulation and systemic invasion observed in protected plants could reflect merely the suppression of initial infection. However, the reduction in systemic virus content in protected plants, observed in tissues produced long after initial infection, may also represent suppression in virus replication and transport.

Several mechanisms have been proposed to account for classical cross-protection (Chapter 10). Most have focused on viral RNA or CP as the mediating factor. Gibbs[28] proposed that excess replicase of the protecting virus might bind, but not transcribe RNA of the challenge virus, and thus create an unproductive complex of the challenge RNA that would stop the infection process. By this theory, superinfection would never occur, but in fact it does. Ross[29] proposed that replicase of the challenging virus might be insufficiently selective and copy the protecting viral RNA, by virtue of its greater abundance, more often than that of its own. This theory could account for limited replication of the challenge virus, depending upon the affinity between RNA and replicase. In support of this theory, experiments with pseudorecombinants show that related strains of a virus can utilize the same replicase.[30] Palukaitis and Zaitlin[31] proposed that protection against infection by RNA viruses could involve the direct hybridization of the challenge viral ($-$) RNA with the excess ($+$) RNA of the protecting virus. Also in support of an RNA hypothesis, Niblett et al.[32] found that cross-protection could be obtained with viroids, which are not encapsidated and do not code for any known proteins.

Prominent among theories involving CP is that of de Zoeten and Fulton,[33] which states that the RNA of the challenge virus is re-encapsidated by excess CP of the protecting virus, thus preventing replication of the challenge virus. The process would involve a form of genetic masking in which some of the RNA of one virus is encapsidated in the CP of a related strain.[34] However, experimental evidence has not consistently confirmed this hypothesis. Zaitlin[35] found that plants inoculated with a mutant of TMV incapable of encapsidating TMV RNA were somewhat protected. As an alternative, Sherwood and Fulton[36] proposed that cross-protection results because stripping or uncoating of the original capsid is blocked. They found that the RNA of a necrotic strain of TMV or virions of the same strain treated with bentonite infected light green areas of *Nicotiana sylvestris* plants previously infected with a systemic strain of TMV. The same challenge RNA re-encapsidated in TMV capsid protein would not infect the same tissue, but it did when encapsidated in brome mosaic virus (BMV) capsid protein. The fact that infection was not suppressed when the TMV RNA was encapsidated with the protein of a different virus suggested that capsid protein of the challenger determines specificity of the reaction. Dodds et al.[37] confirmed with CMV that protection was largely overcome when RNA was used as inoculum.

In vitro experiments by Wilson[38] indicated that TMV particles "loosened" by brief treatment at elevated pH served as good templates for translation reactions. Further, Wilson and Watkins[39] showed that addition of high concentrations of TMV capsid protein blocked the in vitro translation reaction. They proposed that translation of the challenge virus RNA is blocked by re-encapsidation with the protecting viral capsid protein during disassembly of the challenge virus. Further, work by Zinnen and Fulton[40] suggests that an additional mechanism may be involved in protection. They found that plants infected with a systemic strain of sunn hemp mosaic virus (SHMV) were completely protected, not only against virions, but also against the RNA of a necrotic strain of the same virus.

The hypotheses outlined above are all based upon suppression of initial infection. Work by Dodds[41] and Dodds et al.[37] could indicate yet another mechanism that would involve suppression of viral invasion. A challenging CMV strain replicated well in, but was confined to, the leaves into which it was inoculated.

Although many of the mechanisms that have been proposed are indirectly supported, none of them can explain in full the experimental data. It seems likely that there are different mechanisms for different virus-host systems, or multiple mechanisms with different preferences that are system dependent.

Cross-protection has been used in several instances to control virus disease (Chapter 10), but there are several potential disadvantages to using this approach.[4] A mild protecting strain may (1) be difficult to isolate or produce, (2) cause some loss of yield, (3) spread to other hosts in which it causes damage, (4) fail to protect against strains found in the field in all areas, (5) select in favor of more damaging strains, (6) mutate to a more damaging variant, (7) induce susceptibility to other viruses via synergistic interactions, and (8) increase susceptibility to nonviral pathogens. Furthermore, the plant remains susceptible for some time after inoculation with the protecting strain until systemic invasion is complete. The protecting strain may also be difficult and costly to apply, especially in the case of annual crops where uniform, annual inoculation would be required.

## III. RESISTANCE IN TRANSGENIC PLANTS THAT EXPRESS VIRAL COAT PROTEIN GENES

By developing plants to express specific viral genes and sequences, protection against virus infection has been achieved without many of the disadvantages typical of classical protection. The CP genes engineered into plants thus far were derived from several different RNA viruses — TMV, AlMV, TSV, CMV, and PVX. The levels of expression of these genes in transgenic plants vary somewhat for each CP system. This variation, in part, may be due to differences in control sequences used to express the genes. Gene copy number and chromosomal location of the T-DNA insert may also influence expression.[42,43] Variation is also due to differences in detection methods; levels of CP detected by ELISA generally are lower than those detected by Western analysis. In addition, sensitivity depends on the quality of the antibody preparation. Therefore, the levels reported below should not be considered as absolute values.

Although these viruses differ in their particle morphology, genome organization, and pathology, the CP for each is translated from a subgenomic RNA. Transgenic plants expressing any one of these CPs have been analyzed for protection from infection by the corresponding viruses. In general, symptoms and corresponding virus accumulation were partially or completely suppressed and/or delayed in these transgenic plants.

## A. Tobacco Mosaic Virus

In structural terms, TMV is one of the best-characterized plant viruses. The virus particle is a very stable, rigid rod that contains a single-stranded, positive-sense RNA genome.[44] The TMV CP gene was the first plant viral CP gene expressed in tobacco plants.[10,45]

Transgenic tobacco plants were developed by Bevan et al.[45] to express the CP gene of the OM strain of TMV. The CP coding sequence of the OM strain differs by two amino acids from that of the U1 strain utilized by Powell Abel et al.[10] Both groups used a plant expression vector containing the cauliflower mosaic virus (CaMV) 35S promoter and the nopaline synthase (NOS) polyadenylation signal, but the precise sequences surrounding the promoter differed somewhat.[46]

Bevan and Harrison[47] detected CP levels of 0.001% of soluble leaf protein and did not observe protection in their transgenic tobacco plants. In contrast, plants developed by Powell Abel et al.[10] contained CP at 0.1% of the total extractable protein and were protected against TMV U1 infection; symptom development was significantly delayed in these transgenic plants. This difference in protection observed by the two groups is probably due to the difference in levels of CP expression, which may be a function of the somewhat different sequences surrounding the genes. These sequence differences could affect the rate of transcription, translational efficiency, or the stability of the chimeric mRNAs. Nelson et al.[48] found that transgenic tobacco plants expressing TMV U1 CP were protected not only against the U1 strain, but also against the severe PV230 strain. Both the number of lesions and virus accumulation in inoculated leaves were reduced in transgenic tobacco leaves, and systemic disease was delayed.

Two field tests of transgenic tomato plants expressing the TMV U1 CP confirmed the efficacy of genetically engineered protection to TMV.[49,50] Nelson et al.[50] reported that a significant percentage of transgenic tomato plants expressing CP were protected against the TMV U1 strain. The incidence of symptom development was reduced among the field-grown, transgenic tomato plants after inoculation with this strain, and virus accumulation was reduced in the protected plants. In a recent field test, 100% of the transgenic tomato plants inoculated with U1 and PV230 strains of TMV were completely protected.[49] They contained no detectable or infectious virus at harvest, as determined by ELISA and by transmission to local lesion hosts.

The capacity of the TMV CP to protect against a closely related virus, tomato mosaic virus (ToMV), was recently investigated.[49,50] The CP of the L strain of ToMV has 88% homology at the amino aid level with that of TMV U1.[51] In general, transgenic tomato plants expressing the TMV U1 CP were protected against infection by several isolates of ToMV. A percentage of these plants exhibited complete protection or a delay in symptom development. In contrast, all of the transgenic plants challenged with TMV U1 were completely protected. These data agree with the classical cross-protection results of Broadbent[52] and indicate that the degree of protection afforded by expression of the CP gene of one virus is dependent on the extent of its relationship with the challenge virus.

## B. Alfalfa Mosaic Virus

AlMV is a multicomponent, bacilliform-shaped virus that contains three genomic RNAs (RNA1 to RNA3) and a subgenomic RNA (RNA4). AlMV infection can only occur if RNA1 through RNA3 and RNA4 or its CP are present.[53]

Tumer et al.[11] reported expression of AlMV CP in both tobacco (*N. tobacum* cv. Samsun) and tomato (*Lycopersicon esculentum* cv. VF36) plants that were transformed with vectors containing the CP gene under the control of the CaMV 35S promoter and the nopaline synthase polyadenylation signal. The levels of CP expression in tobacco and tomato leaves ranged from 0.1 to 0.4% and from 0.1 to 0.8%, respectively, of the total extractable leaf protein, as determined by Western blot analysis. van Dun et al.[13] also reported AlMV CP expression in leaves of 'Samsun NN' tobacco, using a similar vector with the CaMV 35S promoter and the NOS polyadenylation signal. Their plants expressed CP at 0.01 to 0.05% of total extractable protein by Western blot analysis. Previous attempts by this group to regenerate transgenic plants expressing the AlMV CP under the control of the E9 pea small subunit promoter and the rbcS-E9 3' end had not resulted in detectable levels of CP in the plants.[54]

Using an expression vector containing the AlMV CP driven by the CaMV 19S promoter and NOS polyadenylation signal, Loesch-Fries et al.[12] obtained transformed tobacco plants that expressed AlMV CP. Levels of CP expression in these plants ranged from 0.004 to 0.08% of the total extractable plant protein, as detected by ELISA. These lower levels of CP expression could be attributed to the use of the CaMV 19S rather than the CaMV 35S promoter. In addition, differences in detection levels are typically observed between Western and ELISA methods of analysis. Expression was found to vary within different leaves of a single plant.[12,55] This variation in expression of CP within plants may be due to leaf age or simply to a random distribution of CP in the plant tissues.

Transgenic plants produced by all three groups were protected against AlMV. In general, symptom development on inoculated and upper leaves of transgenic plants were reduced or absent.[11-13] No virus was detectable by ELISA in either inoculated or upper leaves of symptomless CP+ plants 2 weeks after inoculation with AlMV, whereas the corresponding leaves of control plants contained high levels of virus.[55] Loesch-Fries et al.[12] reported a correlation between the level of CP expression and symptom severity, where plants with higher CP levels exhibited fewer symptoms than those expressing lower levels. van Dun et al.[13] found that leaf extracts from previously inoculated transgenic CP+ plants were not infectious on a local lesion host. By this test, they determined that symptomless transgenic plants contained no transmissible virus. This protection was effective against related strains of AlMV, but not against TMV[12,16] or potato virus Y (PVY).[55]

When plants or protoplasts expressing CP were challenged with AlMV viral RNA1 to RNA3, successful infections occurred.[16,56] In contrast, control plants or protoplasts did not become infected when inoculated with viral RNA1 to RNA3. These results indicated that the endogenous CP in transgenic plants was biologically active and could function to activate AlMV infection.

van Dun et al.[16] also analyzed protection against AlMV in plants engineered to express a truncated CP gene. They inserted 4 nucleotides (nts) at 293 nts from the 5' end of the cDNA to RNA4. Transcripts from this construct code for 87 amino acids at the N terminus, plus an additional 25 amino acids due to the frame shift. The appropriate-sized transcripts were observed in transgenic tobacco plants, but the corresponding polypeptides were not detected using a polyclonal antibody to AlMV. Plants expressing the truncated transcripts were not protected from infection by AlMV or AlMV RNA1 to RNA4 and did not activate the AlMV genome when inoculated with RNA1 to RNA3.

## C. Tobacco Streak Virus

TSV, the type member of the Ilarvirus group, is similar to AlMV in that it has a tripartite genome and requires RNA4 or the CP to activate RNA1 through RNA3 for infection.[53] In addition, the CP genes from AlMV and TSV have been found to be interchangeable for activation of the genome.[57,58] However, these two viruses differ in several aspects, and their CP genes share no amino acid sequence homology.[58-60]

Tobacco plants (cv. Samsun NN and cv. Xanthi nc) that expressed TSV CP were developed using a plant transformation vector driven by the CaMV 35S promoter and the NOS polyadenylation signal.[16] When these plants were challenged with AlMV RNA1 through RNA3, infection occurred. Thus, the endogenous TSV CP was able to activate the AlMV genome. When TSV CP+ plants were challenged with TSV, they were protected; but when these plants were challenged with AlMV, no protection was observed. Similarly, although AlMV CP+ plants were protected against AlMV, no protection was observed against TSV.[16,55] Plants expressing TSV CP also were not protected against the unrelated virus, TMV,[16] which is consistent with results obtained with transgenic plants expressing other viral CP genes.[12,55]

## D. Cucumber Mosaic Virus

The demonstration of CP-mediated protection against the cucumovirus, CMV, is agronomically very significant, as the virus infects over 750 species of plants. CMV is a multicomponent virus with icosahedral particles comprised of three genomic RNAs (RNA1 to RNA3) and one subgenomic RNA (RNA4). As with the unrelated AlMV, RNA3 contains a copy of the CP gene; however, the subgenomic RNA4 that is derived from RNA3 serves as the template for CP synthesis. RNA1 and RNA2 are encapsidated in separate particles, and RNA3 and RNA4 are packaged together. Several strains of CMV also contain a satellite RNA (RNA5 or CARNA5).[61]

The CP gene from the D strain of CMV was expressed under the control of the CaMV 35S promoter and the pea rbcS E9 3' end in tobacco plants.[14] The levels of CP in these transgenic tobaccos constituted up to 0.002% of total leaf protein as detected by ELISA. The CMV C strain, whose CP differs from the D strain CP by only 3 of 218 amino acids, was mechanically inoculated onto the transgenic plants over a fiftyfold range in inoculum concentration. A significant percentage of CP+ plants did not develop symptoms after challenge. Those plants that did not develop symptoms initially still were symptomless after 1 month, at which time all control plants were chlorotic and stunted. In addition, the virus levels in the symptomless CP+ plants were greatly reduced compared to levels in control plants. This reduction in virus levels was similar to the protection observed in classical cross-protection experiments of Dodds et al.[37,41]

## E. Potato Virus X

PVX, the type member of the potexvirus group, is a flexuous rod-shaped particle with a single genomic RNA. Unlike all of the other viruses discussed in this section, the PVX RNA genome is polyadenylated and does not have a tRNA-like structure at its 3' end.[62]

Tobacco plants transformed with the PVX CP gene under the control of the CaMV 35S promoter and the pea rbcS E9 3' end have been analyzed for protection against PVX infection.[15] Progeny from several different transformed plant lines were analyzed for CP expression and then were inoculated with several concentrations of PVX. Fewer lesions developed on inoculated leaves of transgenic CP+ plants than on vector control plants, and systemic symptoms were delayed or absent. Virus accumulation was also significantly reduced in transgenic plants as detected by ELISA. Plants expressing CP above a certain threshold level (at least 0.02% of total extractable protein) were protected, in terms of virus accumulation and symptom attenuation, over a two log range in inoculum concentration. In contrast, plants expressing signficantly lower levels of CP were protected only at lower inoculum concentrations.

To determine if the virus present at significantly reduced levels in transgenic plants was modified in any way, Kaniewski et al.[63] passaged extracts from these plants onto both transgenic and control tobacco plants. Concurrent with each serial passage, symptoms in transgenic plants decreased. Infectivity of extracts from inoculated leaves of transgenic tobacco plants progressively regressed with each passage until, after four passages, they no longer infected transgenic plants. However, these extracts did infect control plants, on which symptoms produced were similar to those induced by virus that had not been passaged through transgenic plants. These data indicated that virus was not modified during passage through transgenic plants. Rather, the recoverable infectivity from inoculated leaves was reduced at each serial passage by CP-mediated resistance.

Extracts from the systemic leaves of PVX-inoculated transgenic plants were not infectious on either transgenic or wild-type tobacco plants. If these extracts contained any virus, it was at concentrations below the dilution endpoint of the virus.

Protection afforded by expression of a truncated PVX CP gene also has been analyzed.[15] The CP gene was truncated by removing sequences downstream of a convenient restriction site located at 135 nt from the 5' end, and then was inserted into an expression vector containing the CaMV 35S promoter and the pea rbcS E9 3' end. Transcripts from this truncated PVX CP gene were detected by Northern analysis in these plants, but the corresponding polypeptides were not detected using a polyclonal antibody made to the virus. The levels of transcripts were approximately sixfold lower than those of full-length sense transcripts in the plants that express high levels of CP. The plants expressing the truncated transcripts were not protected, even with inoculum concentrations at which protection was observed in transgenic plants expressing low levels of CP.

## IV. POTENTIAL MECHANISMS INVOLVED IN COAT PROTEIN-MEDIATED PROTECTION

The effectiveness of the CP-mediated approach to virus resistance has been well documented, but the precise mechanisms involved have not been defined. Although there is a correlation between levels of CP expression and the extent of protection, the inhibition of virus infection could be due to either the CP per se or to the CP transcript. The CP transcript could inhibit virus infection by binding to the challenge virus minus strand during replication. As all of the CP constructs described above contain regions at the 3' end that are involved in replicase binding, the CP transcripts could bind challenge virus replicase or components necessary for replication. This question was recently resolved for AlMV and TMV by van Dun et al.[16] and Powell et al.,[64] respectively. When transgenic tobacco plants expressing a translationally defective TMV transcript were challenged with virus, no protection was observed. In contrast, plants that express CP from transcripts lacking the 3' replicase binding site were protected.[64] Similarly, transgenic plants expressing translationally defective AlMV transcripts were not protected.[16] Thus, protection was dependent upon the presence of CP rather than the CP transcript.

A more difficult problem to resolve is how CP expression in transgenic plants can inhibit virus infection. The reduction in lesion numbers (initial sites of infection) observed on inoculated leaves of protected plants indicates that the endogenous CP blocks virus infection at an early stage. Protection in transgenic plants expressing either TMV[48] or AlMV CP[12,13] was largely overcome by inoculation with the corresponding viral RNAs, suggesting that endogenous CP inhibits virus infection by preventing uncoating or uptake of the challenge virus.

To further investigate this possibility, Register and Beachy[65] analyzed protection in protoplasts isolated from tobacco plants that expressed TMV CP. They observed that transgenic protoplasts were protected from infection with virus, as evidenced by reduction in virus and viral RNA accumulation. However, protection was overcome when these protoplasts were inoculated with TMV RNA. These data suggest that infection was blocked prior to or concomitant with replication. To address the possibility that endogenous CP might inhibit disassembly of the challenge virus, they inoculated protoplasts with TMV virions that were "loosened" by incubation at pH 8.0. This treatment has been demonstrated to remove some of the CP molecules from the 5′ ends of the TMV rods and to enhance translation in vitro.[38] Register and Beachy[65] found that protoplasts inoculated with these pH 8.0 treated rods were no longer protected against initial infection. Thus, endogenous CP in transgenic protoplasts could not block infection with the altered TMV. These data further support the hypothesis that endogenous CP inhibits uncoating, either by binding to the virion to block disassembly or by blocking some site for disassembly in the cell.

In contrast, protection against PVX in transgenic tobacco plants expressing PVX CP was not overcome with PVX RNA.[15] Both lesion numbers on inoculated leaves and virus levels in inoculated and systemic leaves were reduced, and the extent of reduction was the same in both virus and RNA inoculated plants.

This different response to RNA inoculation between transgenic plants expressing TMV or PVX may be due to differences in assembly and/or disassembly of these viruses. Assembly of TMV proceeds bidirectionally, but primarily in the 3′ to 5′ direction, from an origin of assembly sequence (OAS) located near the 3′ end of the RNA.[61] PVX assembly has not been well characterized, but assembly of the potexvirus, papaya mosaic virus (PMV), was demonstrated to proceed 5′ to 3′ from an OAS located near the 5′ end.[66] The disassembly of TMV proceeds from 5′ to 3′ and appears to be cotranslational *in vitro*[38] and *in vivo*.[67] *In vitro* cotranslational disassembly experiments involve pH 8.0 treatment of the TMV particles. This treatment removes some CP molecules from the 5′ end, which results in enhanced translation of the viral RNA. Such experiments have not been successful with PMV, possibly because the pH 8.0 treatment is not sufficient to loosen a potexvirus particle.[68] Alternatively, potexviruses may disassemble differently. Although disassembly of PVX has not been analyzed, the *in vitro* disassembly of PMV at alkaline pH was from 3′ to 5′.[69]

Although PVX CP might inhibit uncoating when the inoculum is virus, some other mechanism must be involved when the inoculum is PVX RNA. Perhaps the endogenous PVX CP binds to some region of the challenge viral RNA to block infection. One region of predicted strong binding is the OAS. Assuming that the OAS for PVX is located near the 5′ end as in PMV, it is possible that the PVX CP in transgenic plants could bind to this region and prevent translation of the replicase and/or inhibit the replication process per se.

## V. PROTECTION IN TRANSGENIC PLANTS EXPRESSING VIRAL ANTISENSE RNA

Gene expression has been successfully regulated by antisense RNA in *Xenopus* oocytes,[70,71] bacteria,[72,73] eukaryotic cells,[74,75] *Drosophila*,[76,77] and in plants.[78-81] Several recent reviews have been published on the potential and application of antisense RNA for gene regulation and protection against viral infection in bacteria, animals, and plants.[31,82-84] Resistance to virus infection by cells expressing high levels of antisense RNA to the target virus has been demonstrated in bacterial systems[85,86] and animal systems.[87] Recently, transgenic plants expressing antisense RNA to several viral CP genes or to other regions of viral RNA genomes have been analyzed for resistance to infection.[14-17,88,88a]

### A. Expression of Antisense RNA to Viral Coat Protein Genes

Protection against virus infection in plants expressing CP antisense RNA has been reported to be moderately successful for PVX,[15] CMV,[14] and TMV.[17] Each of these CP genes were inserted in the antisense orientation into the same expression vectors used for the sense constructs described above.

Tobacco plants expressing antisense RNA to PVX CP were significantly protected at 0.05 μg/ml PVX inoculum, as indicated by reduction in symptom development and in virus accumulation.[15] This experiment indicated that virus levels in inoculated leaves of plants expressing antisense RNA were only 20% of those observed in control plants, and were similar to levels detected in leaves of plants expressing low levels of PVX CP and CP transcript; symptoms were also suppressed. In contrast, plants expressing higher levels of PVX CP contained less than 1% of the virus levels observed in control plants. When the inoculum concentration was increased to 0.5 μg/ml, the antisense plants were no longer protected, whereas plants expressing low levels of CP were significantly protected.

Reduced virus levels were also observed when transgenic plants expressing CMV CP antisense RNA were inoculated with CMV.[14] In these plants, transcript levels were similar to the sense transcript levels in the best CP-expressing plants. As observed with PVX, the reduction in virus accumulation or disease symptoms was effective with antisense RNA only at low levels of challenge virus inoculum. Transgenic tobacco plants expressing TMV CP antisense transcripts were also protected against low inoculum concentrations of TMV, as evidenced by the absence of symptom development in a significant percentage of these plants.[17]

The antisense CP constructs described above contain sequences complementary to the 3' ends of the corresponding viral RNAs. As these regions are important for replicase binding, it is possible that the protection observed with these constructs results from inhibition of replication of the challenge virus. Thus, replication could be blocked by RNA/RNA interactions or by binding of replicase or other components required for replication to the endogenous antisense transcripts.

With these possibilities in mind, Powell et al.[17] analyzed protection in transgenic tobacco plants that expressed antisense RNA to a CP gene that lacked 117 nts from the 3' end. These plants were not protected, even at low inoculum concentrations. Therefore, these data suggest that the TMV CP antisense transcripts block virus infection by annealing to the viral RNA to prevent replicase from binding. Thus, the resistance in plants expressing antisense RNA is overcome by high virus inoculum, because the ratio of inhibiting antisense RNA to virus sense RNA is reduced and viral replication is able to proceed.

An alternative type of mechanism that has been reported for some bacterial systems and mammalian cells is inhibition of translation by RNA/RNA interactions.[72-74] Izant and Weintraub[74] found that RNA complementary to only the 5' nontranslated leader sequence was sufficient to block translation.[74] In contrast, Hemenway et al.[15] found that transgenic tobacco plants expressing RNA complementary to the 5' 135 nts of th PVX CP gene were not protected against virus infection at moderate inoculum concentrations. However, the levels of the antisense transcripts were threefold lower than transcripts made from the full-length constructs. It might be possible to achieve some protection with these truncated transcripts by lowering inoculum concentrations, by removing the CaMV 35S leader sequence, by including more sequences upstream of the authentic PVX CP gene, or by increasing transcript levels. However, the degree of translational inhibition probably would not be comparable to that observed in other systems because high levels of subgenomic CP RNAs are generated during infection.

## B. Expression of Antisense Sequences to Other Regions of Viral Genomes

Rezaian et al.[88] generated transgenic tobacco plants that express antisense RNA to the three genomic RNAs of the Q strain of CMV. In all cases, the antisense sequences were inserted into expression vectors containing the CaMV 35S promoter and the NOS 3' end. One construct contained sequences complementary to the 5' end of a cDNA to RNA1, which includes part of the nontranslated leader and coding sequence of the putative replication proteins. Included in the second construct was a sequence complementary to the 3' end of RNA2, including the end of the coding region, the 3' nontranslated sequence and part of the very 3' end that is conserved among all three genomic RNAs. The third construct contained sequence complementary to the 5' end of RNA3 that encompasses part of the nontranslated leader and some coding sequence for a protein possibly involved in cell to cell movement of CMV. The relative levels of transcripts expressed from these three constructs in transgenic plant were 1:15:500, respectively. As the same expression vector was used for all three, these differenes in expression could be due to structural effects on expression or stability, or to position effects during insertion into the plant genome.

Primary transformants expressing these antisense RNAs were propagated by subculturing, and then were challenged with different concentrations of CMV. Only plants expressing antisense RNA to the 5' end of RNA1 were protected, as evidenced by reduced virus accumulation and reduced infectivity of extracts from these plants compared to controls. However, protection was not observed in plants derived from other primary transofmants that expressed similar or higher levels of the same antisense RNA. Thus, the protection observed in plants from the one primary transformant cannot be directly correlated with expression of this antisense RNA. These experiments are different from those described above with plants expressing CMV, PVX, or TMV CP antisense RNAs, in which progeny from self-fertilized, primary transformants were analyzed for protection.

Transgenic tobacco plants expressing sequences complementary to the 3' half (nts 3395 to 6395) of the TMV genome have been analyzed for protection against TMV or TMV RNA.[17] Although transcripts detected in these plants contain the same sequences found in the full-length TMV CP antisense transcripts described previously, they did not confer protection, even at low inoculum concentrations. Powell et al.[17] proposed that the longer antisense RNA may not hybridize well to the plus strand RNA of challenge virus because of structural constraints. As plants containing this antisense RNA also appear to develop disease symptoms sooner than control plants, it is possible that subgenomic RNAs are being produced from the antisense transcripts. Expression of the 30K protein and CP from these subgenomic RNAs may account for the enhanced infection.

Recently, transgenic tobacco plants have been developed to express sense and antisense transcripts to the 5' nontranslated leader region (nts 7 to 55) or to the 5' coding region of the 120K protein (nts 64 to 115) of TMV.[88a] These sequences were inserted between the CaMV 35S promoter and NOS 3' end or into the nopaline synthase NOS gene of a plant expression vector. Only plants transformed with antisense constructs under the control of the 35S promoter were protected against infection, as evidenced by the absence of symptoms. In addition, a larger percentage of progeny plants expressing antisense transcripts to the leader region were protected than plants expressing antisense transcripts to the 5' coding region of the 120K protein. This protection was overcome at higher inoculum concentrations, as was observed with plants expressing CP antisense transcripts.[14,15,17]

In general, the existing antisense approaches to plant protection have shown some potential for reduction of virus infection and may be an important tool for engineering resistance.

## VI. EXPRESSION OF VIRAL SATELLITE RNA TO CONTROL VIRUS INFECTION

Satellites of plant viruses are entities that can replicate only with the aid of specific helper viruses. Two classes of satellites have been described, the satellite viruses and the satellite RNAs. Satellite viruses appear to have evolved from viruses and encode their capsid protein, but no other known proteins; they replicate only in the presence of the helper virus. The satellite RNAs also replicate only in the presence of their helper virus, but do not apparently code for any proteins. They are encapsidated by the helper virus capsid protein, but do not exhibit any sequence homology to the viral genome. They may have evolved from plant RNAs.[90] Recent reviews describe the known plant virus satellites and the biology and molecular characterization of these entities.[90-91a]

The satellites have various effects on their helper viruses and have been referred to as "molecular parasites" of the helper virus.[92] Of particular interest to the phytopathologist is the phenomenon of suppression or attenuation of symptoms when some satellites are in association with their helper virus. This phenomenon has been used to reduce the disease severity of CMV in pepper.[93] However, symptoms are not suppressed in all hosts for the virus and satellite. The expression of symptoms induced by the helper virus in combination with either satellite RNA or satellite virus is influenced by the host plant. In some plant hosts, satellites may increase the severity of symptoms and induce necrosis or have no effect on disease modulation. Site-directed mutagenesis of two satellites of CMV, one protective strain and one necrotic strain, showed that protein production was not necessary for the protective function of the S-CARNA5 RNA or the necrotic activity of the D-CARNA5 RNA.[94]

Harrison et al.[18] demonstrated that tobacco plants transformed with cDNA to CMV satellite RNA5 (CARNA5) expressed CARNA5 RNA on infection of these plants with CMV. Both replication of the helper CMV and symptom expression were reduced. These effects were not observed in the inoculated or the first few upper leaves, but were apparent in subsequent leaves. This is in contrast to transgenic plants expressing viral CP genes, in which protection was observed on inoculated and upper leaves. Harrison et al.[18] also found that a related virus, tomato aspermy virus (TAV), induced expression of the CARNA5 RNA, which resulted in milder symptoms on the transgenic plants. However, in TAV case, reduction in symptoms was not accompanied by any apparent reduction in virus replication.

Gerlach et al.[19] reported that transgenic tobacco plants expressing tobacco ringspot virus (TobRV) satellite RNA were protected from infection by the corresponding virus. In these experiments, satellite RNA was amplified after inoculation of the transgenic plants, and virus replication and symptom development were inhibited. Similar results were observed with transgenic tobacco plants expressing sequences complementary to the satellite RNA to TobRV. However, there was a delay before protection was observed in this case.

The mechanism by which satellite RNA expression reduces helper virus replication and/or symptom severity has not been determined. Current theories include competition for cellular machinery necessary for replication between the helper virus and satellite. Possible ribozyme activity of the satellite RNAs and interaction of this activity with the RNA of the helper may also be a mechanism for reduction of helper virus replication.[95-98] The "hammerhead" structure proposed in the structural analysis of satellite RNAs may be involved in self-cleavage of the RNA and may play an as yet undetermined role in interaction with the helper virus RNA. The use of the satellites for disease control may have limited use because of the specificity of the interactions of the satellites to particular helper viruses and the limited number of viruses that have satellites. Mutations may occur in engineered satellite sequences that may serve to increase disease severity.[99] In cases where the symptoms are reduced, but virus titers are not decreased, satellite protection may not reduce the spread of the virus to unprotected plants. Research on satellite RNAs may yield some answers to fundamental questions of RNA/RNA interactions and evolution of infectious viral and virus-like agents.

## VII. TRANSGENIC PLANTS THAT EXPRESS NONSTRUCTURAL VIRAL GENES

Expression of nonstructural viral genes in plants will be important for molecular analysis of virus infection and may help to define additional approaches for protection. Deom et al.[100] demonstrated that tobacco plants expressing the TMV 30-kDa protein were able to complement the temperature-sensitive mutant of TMV, Ls1. At the nonpermissive temperature, this mutant cannot move from cell to cell in control plants. In contrast, Ls1 was able to spread throughout inoculated and systemic leaves of the transgenic tobacco plants expressing the 30-kDa protein. These plants were not protected from infection by the U1 strain of TMV.

Other cDNAs corresponding to RNAs of multicomponent RNA viruses have also been expressed in plants.[101,102] Garcia et al.[101] were able to detect full-length transcripts corresponding to the entire mRNA of cowpea mosaic virus (CPMV) in transformed cowpea callus tissue. Tobacco plants transformed with cDNAs to either AlMV RNA1 or RNA2 were described by van Dun et al.[102] Although the appropriate sized transcripts were detected in plants, no translation products reacted with antibodies made to synthetic peptides. However, they were able to demonstrate that transgenic protoplasts containing cDNA1 complemented infection by RNA2 and RNA3, indicating that biologically active protein from RNA1 was accumulating in the plant. Plants expressing transcripts corresponding to either RNA1 or RNA2 were not protected against either of two strains of AlMV.

Recently, transgenic tobacco plants were engineered to express a protein involved in replication of the DNA virus, tomato golden mosaic virus (TGMV).[103] A single viral protein (AL1) encoded by the A component of this virus was detected in transgenic plants by Western analysis. Further, plants from several transgenic lines were able to complement TGMV A components containing mutations in the AL1 open reading frame in infectivity assays. Thus, the AL1 protein expressed in these plants appears to be biologically active. Such plants will be used to further analyze the role of AL1 in replication and to identify critical cis-acting regulatory sequences.

## VIII. CONCLUSIONS

The resistance conferred on plants by CP-mediated protection is similar to that conferred by classical cross-protection. It is highly specific for virus strains closely related to the inducing strain. It can be overcome in some cases by increasing the inoculum concentration. It retards initial infection, subsequent virus accumulation and systemic invasion, and severity of symptoms. Its effectiveness is greatest when high levels of viral product are available in the challenged cells, and it seems to vary mechanistically in the same ways as classical cross-protection; i.e., it may be effective only against virions or, in other instances, against both intact virions and RNA. However, the precise mechanisms involved in CP-mediated protection and classical cross-protection may not be the same.

Ultimately, CP-mediated protection may prove to be more effective than classical cross-protection and may be applicable to a wider range of viruses. CP-mediated protection would almost certainly be effective against those viruses that fail to cross-protect classically due to their inability to infect all cells of the host. Whereas only mild strains are suitable for classical protection, any strain may be utilized for CP-mediated protection. The specificity of classical cross-protection could be circumvented in CP-mediated protection by expressing more than one CP gene in plants. Furthermore, it may be possible to engineer constructs that are more effective and less specific than those involved in classical cross-protection.

At this point, we have only limited knowledge concerning the field effectiveness of CP-mediated protection. The effectiveness of any resistance depends on how it impacts critical epidemiological factors. Important among these factors is the degree to which the resistance retards dissemination of the pathogen within the crop. We know that CP-mediated protection has the potential to reduce incidence of initial infection, delay or completely inhibit systemic invasion, and suppress virus accumulation and disease symptoms. Even with a relatively weak resistance, the combined effect of these suppressions to viral functions could dramatically reduce field disease.

An interesting feature of CP-mediated protection is that it transforms induced resistance to inherent genetic resistance. This approach not only avoids the disadvantages of classical cross-protection, but it has the advantages of both induced and genetic resistance. In contrast to genetic resistance, CP-mediated resistance is relatively easily introduced into an existing cultivar without altering other characteristics. By introducing CP-mediated protection, it may be possible to rescue old susceptible cultivars and to concentrate on developing other desirable characteristics in breeding programs. While genetic resistance genes are often linked with undesirable characteristics, CP genes have no apparent negative effects on growth or productivity of the transgenic plant.

Thus far, the use of viral antisense sequences to control virus infection has not been as effective as the CP-mediated approach. This difference is undoubtedly due to the different mechanisms involved. In transgenic plants expressing CP, the block to challenge virus infection appears to be prior to or concomitant with replication. In some cases, there is evidence indicating that the endogenous CP blocks disassembly of the challenge virus, either by interfering with the disassembly process per se or by binding to some site in the cell for disassembly. To block infection with antisense CP transcripts, it seems that RNA/RNA annealing must occur to prevent replicase from binding. High levels of endogenous CP antisense RNA would be required to achieve inhibition in this type of system where replication is underway. Thus, with existing levels of expression of antisense CP RNAs, protection has been effective only at low inoculum concentrations.

Protection against virus infection can also be achieved by expression of viral satellite RNAs in plants. This approach is not generally applicable to as many different RNA viruses as the CP-mediated protection because it is limited to those viruses that contain satellites. In addition, plants expressing satellite RNA are tolerant to viral infection, whereas plants expressing CP are resistant or immune in many cases.

In summary, genetically engineered resistance has tremendous potential, either as a sole means of virus control or in combination with other methods.

## REFERENCES

1. **Martin, M. W.,** Developing tomatoes resistant to curly top viruses, *Euphytica,* 19, 243, 1970.
2. **Fraser, R. S. S.,** Resistance to plant viruses, *Oxford Sur. Plant Mol. Cell Biol.,* 4, 1, 1987.
3. **Fulton, R. W.,** The protective effect of systemic virus infection, in *Active Defence Mechanisms in Plants,* Wood, R. K. S., Ed., Plenum Press, London, 1980, 231.
4. **Fulton, R. W.,** Practices and precautions in the use of cross protection for plant virus disease control, *Annu. Rev. Phytopathol.,* 24, 67, 1986.
5. **Hamilton, R. I.,** Defence triggered by previous invaders: viruses, in *Plant Disease: An Advanced Treatise,* Vol. 5, Horsefall, J. G. and Cowling, E. B., Eds., Academic Press, New York, 1980, 279.
6. **Ponz, F. and Bruening, G.,** Mechanisms of resistance to plant viruses, *Annu. Rev. Phytopathol.,* 24, 355, 1986.
7. **McKinney, H. H.,** Mosaic diseases in the Canary Islands, West Africa, and Gibraltar, *J. Agric. Res.,* 39, 557, 1929.
8. **Thung, T. H.,** Smetstof en plantentencel bij enkele virusziekten van de tabaksplant, *Z. Ned. Indisch Natuurwetensch. Congr. Bandoeng Java,* 450, 1931.
9. **Johnson, J.,** An acquired partial immunity to the tobacco streak disease, *Trans. Wis. Acad. Sci. Arts Lett.,* 30, 27, 1937.
10. **Powell Abel, P., Nelson, R. S., De, B., Hoffman, H., Rogers, S. G., Fraley, R. T., and Beachy, R. N.,** Delay of disease development in transgenic plants that express the tobacco mosaic virus coat protein gene, *Science,* 232, 738, 1986.
11. **Tumer, N. E., O'Connell, K. M., Nelson, R. S., Sanders, P. R., Beachy, R. N., Fraley, R. T., and Shah, D. M.,** Expression of alfalfa mosaic virus coat protein gene confers cross-protection in transgenic tobacco and tomato plants, *EMBO J.,* 6, 1181, 1987.
12. **Loesch-Fries, L. S., Merlo, D., Zinnen, T., Burhop, L., Hill, K., Drahn, K., Jarvis, N., Nelson, S., and Halk, E.,** Expression of alfalfa mosaic virus RNA 4 in transgenic plants confers resistance, *EMBO J.,* 6, 1845, 1987.
13. **van Dun, C. M. P., Bol, J. F., and van Vloten-Doting, L.,** Expression of alfalfa mosaic virus and tobacco rattle virus coat protein genes in transgenic tobacco plants, *Virology,* 159, 299, 1987.
14. **Cuozzo, M., O'Connell, K. M., Kaniewski, W., Fang, R.-X., Chua, N.-H., and Tumer, N. E.,** Viral protection in transgenic plants expressing the cucumber mosaic virus coat protein or its antisense RNA, *Bio/Technol.,* 6, 549, 1988.

15. **Hemenway, C., Fang, R.-X., Kaniewski, W. K., Chua, N.-H., and Tumer, N. E.,** Analysis of the mechanism of protection in transgenic plants expressing the potato virus X coat protein or its antisense RNA, *EMBO J.,* 7, 1273, 1988.

16. **van Dun, C. M. P., Overduin, B., van Vloten-Doting, L., and Bol, J. F.,** Transgenic tobacco expressing tobacco streak virus or mutated alfalfa mosaic virus coat protein does not cross-protect against alfalfa mosaic virus infection, *Virology,* 164, 383, 1988.

17. **Powell, P. A., Stark, D. M., and Beachy, R. N.,** Protection against tobacco mosaic virus in transgenic plants that express TMV antisense RNA, *Proc. Natl. Acad. Sci. U.S.A.,* 86, 6949, 1989.

18. **Harrison, B. D., Mayo, M. A., and Baulcombe, D. C.,** Virus resistance in transgenic plants that express cucumber mosaic virus satellite RNA, *Nature,* 328, 799, 1987.

19. **Gerlach, W. L., Llewellyn, D., and Haseloff, J.,** Construction of a plant disease resistance gene from the satellite RNA of tobacco ringspot virus, *Nature,* 328, 802, 1987.

20. **Cooper, J. I. and Jones, A. T.,** Responses of plants to viruses: proposals for the use of terms, *Phytopathology,* 73, 127, 1983.

21. **Matthews, R. E. F.,** Criteria of relationship between plant virus strains, *Nature,* 163, 175, 1949.

22. **Duffus, J. E.,** Host relationships of beet western yellow virus strains, *Phytopathology,* 54, 736, 1964.

23. **Fulton, R. W.,** Superinfection by strains of tobacco streak virus, *Virology,* 85, 1, 1978.

24. **Kunkel, L. L.,** Studies on acquired immunity with tobacco and aucuba mosaics, *Phytopathology,* 24, 437, 1934.

25. **Bercks, R.,** Serologische beitrage zur frage der abwehr von zweitinfektionen bei X-viren, *Phytopathol. Z.,* 24, 54, 1948.

26. **Bald, J. G.,** Potato virus X: effectiveness of acquired immunity in older and younger leaves, *J. Counc. Sci. Ind. Res.,* 21, 247, 1948.

27. **Costa, A. S. and Muller, G. W.,** Tristeza control by cross-protection: a U.S. Brazilian cooperative success, *Plant Dis.,* 64, 538, 1980.

28. **Gibbs, A.,** Plant virus classification, *Adv. Virus Res.,* 14, 263, 1969.

29. **Ross, A. F.,** Interaction of viruses in the host, *Acta Hortic.,* 36, 247, 1974.

30. **Barker, H. and Harrison, B. D.,** Double infection, interference, and superinfection in protoplasts exposed to two strains of raspberry ringspot virus, *J. Gen. Virol.,* 40, 647, 1978.

31. **Palukaitis, P. and Zaitlin, M.,** A model to explain the "cross-protection" phenomenon shown by plant viruses and viroids, in *Plant-Microbe Interactions: Molecular and Genetic Perspectives,* Kosuge, T. and Nester, E. W., Eds., Macmillan, New York, 1984, 420.

32. **Niblett, C. L., Dickson, E., Fernow, K. H., Horsch, R. K., and Zaitlin, M.,** Cross-protection amongst four viroids, *Virology,* 91, 198, 1978.

33. **de Zoeten, G. A. and Fulton, R. W.,** Understanding generates possibilities, *Phytopathology,* 65, 221, 1975.

34. **Dodds, J. A. and Hamilton, R. I.,** Structural interactions between viruses as a consquence of mixed virus infections, *Adv. Virus Res.,* 20, 33, 1976.

35. **Zaitlin, M.,** Viral cross-protection: more understanding is needed, *Phytopathology,* 66, 382, 1976.

36. **Sherwood, J. L. and Fulton, R. W.,** The specific involvement of coat protein in tobacco mosaic virus cross-protection, *Virology,* 119, 150, 1982.

37. **Dodds, J. A., Lee, S. Q., and Tiffany, M.,** Cross-protection between strains of cucumber mosaic virus: effect of host and type of inoculum on accumulation of virions and double-stranded RNA of the challenge strain, *Virology,* 144, 301, 1985.

38. **Wilson, T. M. A.,** Cotranslational disassembly of tobacco mosaic virus in vitro, *Virology,* 137, 255, 1984.

39. **Wilson, T. M. A. and Watkins, P. A. C.,** Influence of exogenous viral coat protein on the cotranslational disassembly of tobacco mosaic virus (TMV) particles in vitro, *Virology,* 149, 132, 1986.

40. **Zinnen, T. M. and Fulton, R. W.,** Cross-protection between sun-hemp mosaic and tobacco mosaic viruses, *J. Gen. Virol.,* 67, 1679, 1986.

41. **Dodds, J. A.,** Cross-protection and interference between electrophoretically distinct strains of cucumber mosaci virus in tomato, *Virology,* 118, 235, 1976.

42. **Sanders, P. R., Winter, J. A., Barnason, A. R., Rogers, S. G., and Fraley, R. T.,** Comparison of cauliflower mosaic virus 35S and nopaline synthase promoters in transgenic plants, *Nucleic Acids Res.,* 15, 1543, 1987.

43. **Jones, J. D. G., Dunsmuir, P., and Bedbrook, J.,** High level expression of introduced chimaeric genes in regenerated transformed plants, *EMBO J.,* 4, 2411, 1985.

44. **Gibbs, A.,** Tobamovirus classification, in *The Plant Viruses,* Vol. 2, von Regenmortel, M. H. V. and Fraenkel-Conrat, H., Eds., Plenum Press, New York, 1986, 59.

45. **Bevan, M. W., Mason, S. E., and Goelet, P.,** Expression of tobacco mosaic virus coat protein by a cauliflower mosaic virus promoter in plants transformed by *Agrobacterium, EMBO J.,* 4, 1921, 1985.

46. **Beachy, R. N.,** Virus cross-protection in transgenic plants, in *Plant Gene Research. Temporal and Spatial Regulation of Plant Genes,* Verma D. P. and Goldberg, R. B., Eds., Springer-Verlag, New York, 1988, 313.

47. **Bevan, M W. and Harrison, B. D.,** Genetic engineering of plants for tobacco mosaic virus resistance using the mechanisms of cross-protection, in *Molecular Strategies for Crop Protection,* Arntzen, C. J. and Ryan, C., Eds., Alan R. Liss, New York, 1986, 215.
48. **Nelson, R. S., Powell Abel, P., and Beachy, R. N.,** Lesions and virus accumulation in inoculated transgenic tobacco plants expressing the coat protein gene of tobacco mosaic virus, *Virology,* 158, 126, 1987.
49. **Sanders, P. R., Kaniewski, W., Haley, L., Layton, J., LaVallee, B. J., Delanney, X., and Tumer, N.,** unpublished data.
50. **Nelson, R. S., McCormick, S. M., Delanney, X., Dube, P., Layton, J., Anderson, E. J., Kaniewska, M., Proksch, R. K., Horsch, R. B., Rogers, S. G., Fraley, R. T., and Beachy, R. N.,** Virus tolerance, plant growth, and field performance of transgenic tomato plants expressing coat protein from tobacco mosaic virus, *Bio/Technol.,* 6, 403, 1988.
51. **Nozu, Y., Ohno, T., and Okada, Y.,** Amino acid sequences of some common Japanese strains of tobacco mosaic virus, *J. Biochem.,* 68, 39, 1970.
52. **Broadbent, L.,** The epidemiology of tomato mosaic. VII. The effect of TMV on tomato fruit yield and quality under glass, *Ann. Appl. Biol.,* 54, 209, 1964.
53. **Francki, R. I. B., Milne, R. G., and Hatta, T.,** *Atlas of Plant Viruses,* CRC Press, Boca Raton, FL, 1985.
54. **van Dun, C. M. P.,** Expression of Viral cDNA in Transgenic Tobacco, thesis, Rijksuniveriteit te Leiden, 1988.
55. **Kaniewski, W., Haley, L., and Tumer, N.,** unpublished data.
56. **Loesch-Fries, L. S., Halk, E., Merlo, D., Jarvis, N., Nelson, S., Krahn, K., and Burhop, L.,** Expression of alfalfa mosaic virus coat protein gene and anti-sense cDNA in transformed tobacco tissue, in *Molecular Strategies for Crop Protection,* Arntzen, C. J. and Ryan, C., Eds., Alan R. Liss, New York, 1987, 221.
57. **Gonsalves, D. and Garnsey, S. J.,** Infectivity of heterologous RNA-protein mixtures from alfalfa mosaic, citrus leaf rugose, citrus variegation, and tobacco streak virus, *Virology,* 67, 319, 1975.
58. **van Vloten-Doting, L.,** Coat protein is required for infectivity of tobacco streak virus: biological equivalence of the coat proteins of tobacco streak and alfalfa mosaic virus, *Virology,* 65, 215, 1975.
59. **Zuidema, D. and Jaspars, E. M. J.,** Comparative investigations on the coat protein binding sites of the genomic RNAs of alfalfa mosaic and tobacco streak viruses, *Virology,* 135, 43, 1984.
60. **Cornelissen, B. J. C., Janssen, H., Zuidema, D., and Bol, J. F.,** Complete nucleotide sequence of tobacco streak virus RNA3, *Nucleic Acids Res.,* 12, 2407, 1984.
61. **Francki, R. I. B.,** *The Plant Viruses,* Vol. 1, Plenum Press, New York, 1985.
62. **Milne, R. G.,** *The Plant Viruses,* Vol. 4, Plenum Press, New York, 1988.
63. **Kaniewski, W. K., Hemenway, C., Haley, L., and Tumer, N. E.,** Analysis of PVX infection in transgenic plants after serial passage, submitted.
64. **Powell, P. A., Sanders, P. R., Tumer, N. E., Fraley, R. T., and Beachy, R. N.,** Protection against tobacco mosaic virus infection in transgenic plants requires accumulation of coat protein rather than coat protein RNA sequences, *Virology,* 175, 124, 1990.
65. **Register, J. C., III and Beachy, R. N.,** Tobacco mosaic virus infection of transgenic plants is blocked at an early event in infection, *Virology,* 166, 524, 1988.
66. **Abou Haidar, M. G. and Erickson, J. W.,** Structure and in vitro assembly of papaya mosaic virus, in *Molecular Plant Virology,* Vol. 1, Davies, J. W., Eds., CRC Press, Boca Raton, FL, 1985, 85.
67. **Shaw, J. G., Plaskitt, K. A., and Wilson, T. M. A.,** Evidence that tobacco mosaic virus particles disassemble cotranslationally in vivo, *Virology,* 148, 326, 1986.
68. **Wilson, T. M. A. and Shaw, J. G.,** Does TMV uncoat cotranslationally in vivo?, *Trends Biochem. Sci.,* 10, 57, 1985.
69. **Lok, S. and Abou Haidar, M. G.,** The polar alkaline disassembly of papaya mosaic virus, *Virology,* 113, 637, 1981.
70. **Melton, D. A.,** Injected anti-sense RNAs specifically block messenger RNA translation in vivo, *Proc. Natl. Acad. Sci. U.S.A.,* 82, 144, 1985.
71. **Kawasaki, E. S.,** Quatitative hybridization arrest of mRNA in *Xenopus* oocytes using single-stranded complementary DNA or oligonucleotide probes, *Nucleic Acids Res.,* 13, 4991, 1985.
72. **Mizuno, T., Chou, M., and Inouye, M.,** A unique mechanism regulating gene expression: translational inhibition by a complementary RNA transcript (mic RNA), *Proc. Natl. Acad. Sci. U.S.A.,* 81, 1966, 1984.
73. **Simon, R. W. and Kleckner, N.,** Translational control of IS10 transposition, *Cell,* 34, 682, 1983.
74. **Izant, J. G. and Weintraub, H.,** Constitutive and conditional suppression of exogenous and endogenous genes by anti-sense RNA, *Science,* 229, 345, 1985.
75. **Kim, S. K. and Wold, B. J.,** Stable reduction of thymidine kinase activity in cells expressing high levels of anti-sense RNA, *Cell,* 42, 129, 1985.
76. **Rosenberg, U. B., Preiss, A., Seifert, E., Jackle, H., and Knipple, D C.,** Production of phenocopies by Kruppel antisense RNA injection into *Drosophila* embryos, *Nature,* 313, 702, 1985.

77. **Cabrera, C. V., Alonso, M. C., Johnston, P., Phillips, R. G., and Lawrence, P. A.,** Phenocopies induced with antisense RNA identify the wingless gene, *Cell,* 50, 659, 1987.

78. **Ecker, J. R. and Davis, R. W.,** Inhibition of gene expression in plant cells by expression of antisense RNA, *Proc. Natl. Acad. Sci. U.S.A.,* 83, 5372, 1986.

79. **Rothstein, S. J., DiMaio, J., Strand, M., and Rice, D.,** Stable and heritable inhibition of the expression of nopaline synthase in tobacco expressing antisense RNA, *Proc. Natl. Acad. Sci. U.S.A.,* 84, 8439, 1987.

80. **Smith, C. J. S., Watson, C. F., Ray, J., Bird, C. R., Morris, P. C., Schuch, W., and Grierson, D.,** Antisense RNA inhibition of polygalacturonase gene expression in transgenic tomatoes, *Nature,* 334, 724, 1988.

81. **Rodermel, S. R., Abbott, M. S., and Bogorad, L.,** Nuclear-organelle interactions: nuclear antisense gene inhibits ribulose bisphosphate carboxylase enzyme levels in transformed tobacco plants, *Cell,* 55, 673, 1988.

82. **Green, P. J., Pines, O., and Inouye, M.,** The role of anti-sense RNA in gene regulation, *Annu. Rev. Biochem.,* 55, 569, 1986.

83. **Weintraub, H., Izant, J. G., and Harland, R. M.,** Anti-sense RNA as a molecular tool for genetic analysis, *Trends Genet.,* 1, 22, 1985.

84. **Zaitlin, M. and Hull, R.,** Plant virus-host interactions, *Annu. Rev. Plant Physiol.,* 38, 291, 1987.

85. **Coleman, J., Hirashima, A., Inokuchi, Y., Green, P. J., and Inouye, M.,** A novel immune system against bacteriophage infection using complementary RNA (micRNA), *Nature,* 315, 601, 1985.

86. **Hirashima, A., Sawaki, S., Inokuchi, Y., and Inouye, M.,** Engineering of the mRNA-interfering complementary RNA immune system against viral infection, *Proc. Natl. Acad. Sci. U.S.A.,* 83, 7726, 1986.

87. **To, R. Y.-L., Booth, S. C., and Neinam, P. E.,** Inhibition of retroviral replication by antisense RNA, *Mol. Cell Biol.,* 6, 4758, 1986.

88. **Rezaian, M. A., Skene, K. G. M., and Ellis, J. G.,** Anti-sense RNAs of cucumber mosaic virus in transgenic plants assessed for control of the virus, *Plant Mol. Biol.,* 11, 463, 1988.

88a. **Nelson, A., Roth, D. A., and Johnson, J. D.,** Tobacco mosaic virus infection of transgenic *N. tabacum* plants is inhibited by antisense RNA construction directed at the 5′ untranslated region of viral RNA, submitted

89. **Baulcombe, D. C., Saunders, G. R., Bevan, M. W., Mayo, M. A., and Harrison, B. D.,** Expression of biologically active viral satellite RNA from the nuclear genome of transformed plants, *Nature,* 321, 446, 1986.

90. **Francki, R. I. B.,** Plant virus satellites, *Annu. Rev. Microbiol.,* 39, 151, 1985.

91. **Murant, A F. and Mayo, M. A.,** Satellites of plant viruses, *Annu. Rev. Phytopathol.,* 20, 49, 1982.

91a. **Fritsch, C. and Mayo, M. A.,** Satellites of plant viruses, in *Plant Viruses,* Vol. 1, *Structure and Replication,* Mandahar, C. L., Ed., CRC Press, Boca Raton, FL, 1989, 289.

92. **Tien, P. and Chang, X. H.,** Control of two plant viruses by protection inoculation in China, *Seed Sci. Technol.,* 11, 969, 1983.

93. **Tien, P., Zhang, X., Qui, B., Qin, B., and Wu, G.,** Satellite RNA for the control of plant diseases caused by cucumber mosaic virus, *Ann. Appl. Biol.,* 111, 143, 1987.

94. **Collmer, C. W. and Kaper, J. M.,** Site-directed mutagenesis of potential protein-coding regions in expressible cloned cDNAs of cucumber mosaic viral satellites, *Virology,* 163, 293, 1988.

95. **Buzayan, J. M., Gerlach, W. L., and Bruening, G.,** Non-enzymatic cleavage and ligation of RNAs complementary to a plant virus satellite RNA, *Nature,* 323, 349, 1986.

96. **Forster, A. C. and Symons, R. H.,** Self-cleavage of virusoid RNA is performed by the proposed 55 nucleotide active site, *Cell,* 50, 9, 1987.

97. **Kaper, J. M., Tousignant, M. E., and Steger, G.,** Nucleotide sequence predicts circularity and self-cleavage of 300-ribonucleotide satellite of arabis mosaic virus, *Biochem. Biophys. Res. Commun.,* 154, 318, 1988.

98. **Uhlenbeck, O. C.,** A small catalytic oligoribonucleotide, *Nature,* 328, 596, 1987.

99. **Kaper, J. M. and Collmer, C. W.,** Modulation of viral plant diseases by secondary RNA agent, in *RNA Genetics,* Vol. 3, Domingo, E., Holland, J., and Ahlquist, P., Eds., CRC Press, Boca Raton, FL, 1988.

100. **Deom, C. M., Oliver, M. J., and Beachy, R. N.,** The 30-kilodalton gene product of tobacco mosaic virus potentiates virus movement, *Science,* 237, 389, 1987.

101. **Garcia, J. A., Hille, J., Pieter, V., and Goldbach, R.,** Transformation of cowpea, *Vigna unguiculata* with a full length DNA copy of cowpea mosaic virus M-RNA, *Plant Sci.,* 48, 89, 1987.

102. **van Dun, C. M. P., van Vloten-Doting, L., and Bol, J. F.,** Expression of alfalfa mosaic virus cDNA 1 and 2 in transgenic tobacco plants, *Virology,* 163, 572, 1988.

103. **Hanley-Bowdoin, L. K., Elmer, J. S., and Rogers, S. G.,** Expression of functional replication protein from tomato golden mosaic virus in transgenic tobacco plants, *Proc. Natl. Acad. Sci. U.S.A.,* 87, 1446, 1990.

# INDEX

## A

94360
M|L

X